Handbook of Composite Materials

Handbook of Composite Materials

Edited by
Nicholas Thomson

STATES
ACADEMIC PRESS
www.statesacademicpress.com

Published by States Academic Press,
109 South 5th Street,
Brooklyn, NY 11249, USA

ISBN: 978-1-63989-255-6

Cataloging-in-Publication Data

Handbook of composite materials / edited by Nicholas Thomson.
p. cm.
Includes bibliographical references and index.
ISBN 978-1-63989-255-6
1. Composite materials. 2. Materials. 3. Composite construction.
I. Thomson, Nicholas.
TA418.9.C6 H36 2022
620.118--dc23

For information on all States Academic Press publications
visit our website at www.statesacademicpress.com

STATES
ACADEMIC PRESS

Contents

Preface

This book was inspired by the evolution of our times; to answer the curiosity of inquisitive minds. Many developments have occurred across the globe in the recent past which has transformed the progress in the field.

Composite materials are those materials which are formed by two or more materials. The constituent materials individually have significantly different chemical and physical properties as compared to the properties of the composite material. The individual materials are divided into two components: matrix and reinforcement. To make a composite material a portion of both should be present. A wide variety of techniques are used for the production of composite materials such as open molding, close molding and cast polymer molding. These include advanced fibre placement, tailored fibre placement, filament winding, tufting and Z-pinning. Some of the characteristics of these materials are that they are lighter, stronger and less expensive as compared to traditional materials. Some of the most common composite materials are plywood, fiberglass and composite ceramics. These materials are usually used for the construction of bridges, buildings, racing car bodies, storage tanks and countertops. This book outlines the processes and techniques involved in the production of composite materials. From theories to research to practical applications, case studies related to all contemporary topics of relevance to this discipline have been included herein. The book is appropriate for the students as well as experts who are seeking detailed information in this field.

This book was developed from a mere concept to drafts to chapters and finally compiled together as a complete text to benefit the readers across all nations. To ensure the quality of the content we instilled two significant steps in our procedure. The first was to appoint an editorial team that would verify the data and statistics provided in the book and also select the most appropriate and valuable contributions from the plentiful contributions we received from authors worldwide. The next step was to appoint an expert of the topic as the Editor-in-Chief, who would head the project and finally make the necessary amendments and modifications to make the text reader-friendly. I was then commissioned to examine all the material to present the topics in the most comprehensible and productive format.

I would like to take this opportunity to thank all the contributing authors who were supportive enough to contribute their time and knowledge to this project. I also wish to convey my regards to my family who have been extremely supportive during the entire project.

Editor

1

A Study on the Fundamental Quality of Magnesia-Phosphate-Formed Mortar Composites Using Superabsorbent Polymer for Development of Concrete for Biological Panel

Yun-Wang Choi[1] **Sung-Rok Oh**[1] **Cheol-Gyu Kim**[1] **and Jae-Heun Lee**[2]

[1]*Department of Civil Engineering, Semyung University, Jecheon 27136, Republic of Korea*
[2]*Institute of Technology, Eugene Corporation Co. Ltd., Bucheon 14441, Republic of Korea*

Correspondence should be addressed to Sung-Rok Oh; cgdbs02@nate.com

Guest Editor: Rujie He

The purpose of this study was to investigate the base material of the biological panel as one of the researches for developing the biological panel using (micro)organisms. In order to select the base material for the biological panel, the applicability of the magnesia-potassium-phosphate-foamed mortar composites material using the superabsorbent polymer was evaluated through the evaluation of basic quality and characteristics. Experiments were conducted to evaluate the fluidity, compressive strength, pH, roughness (porosity distribution), and moisture retention (absorbency) of the magnesia-phosphate-foamed mortar composites according to the additional amount of the superabsorbent polymer. Experimental results showed that the addition of a suitable superabsorbent polymer can provide good results in terms of compressive strength, workability, and moisturizing properties (absorbency). As a result, it can be used as a basic quality study data of the biological panel design.

1. Introduction

Due to the rapid industrial revolutions around the world and indiscriminate urban development, there is a series of serious environmental problem such as increase of CO_2 emissions, fine dust generation, and increase of cooling energy use due to reduction of the green area [1, 2]. In order to solve this problem, various researches and efforts have been carried out both domestically and externally in order to develop eco-friendly construction materials. However, most of them have been focused on waste resource utilization, carbon abatement type, and environmentally friendly concrete technology [3–10]. These technologies can solve the problem of CO_2 reduction, but it is difficult to take alternative measures to heat island phenomenon and increase of cooling energy use due to increase of city's atmospheric temperature.

In order to compensate for this, there is a way to solve the problem by installing a vertical garden inside and outside the building, but it is difficult to integrate with the exterior wall of the structure. In addition, complex support structures and expensive maintenance and installation costs increase about three times as compared to the common garden of the same area, making it difficult to commercialize them economically [11]. Although there are sponge, mat type, block type, and lightweight soil, similar to the vertical garden, there are many durability problems [12–17].

To solve this problem, biological panels can be developed and integrated with the structure. Biological panels should be easy to put on (biology), etc., and environment should be created so that it can survive after it is rooted. However, when cement is used as a material for panel fabrication, it forms a high pH, which makes it unsuitable for survival and maintenance of living (micro)organisms [18]. In addition, smooth or dense surface makes it difficult to attach (micro) organisms, and similarly problems arise with nutrients and water supply for (micro)organisms [19].

As one of the basic studies for the development of biological panels, it was investigated whether the magnesia-phosphate-foamed mortar composites used the superabsorbent polymer. Unlike conventional cement, magnesium-phosphate composites can form neutral pH [19]. In addition, air bubbles can form pores on the surface and inside the matrix, and adsorption of nutrients in the air through the pores can be beneficial for the growth of living (micro)organisms [19].

In addition to these, porous surfaces are important, and their supply to moisture also plays a very important role in the growth of living (micro)organisms. In order to achieve such a purpose, the superabsorbent polymer can be applied. The superabsorbent polymer can absorb more than several times its own mass of moisture, and the absorbed moisture gradually dries. When the superabsorbent polymer is applied to the biological panel, the moisture retention of the panel can be improved.

Accordingly, even in a dry environment, the absorbed moisture of the superabsorbent polymer can be utilized for the purpose of moisture supply of (micro)organisms [19].

Therefore, in this study, a magnesium-phosphate-potassium-formed mortar composites using a superabsorbent polymer was prepared as a parent material of the biological panel, and basic quality characteristics and biological characteristics were examined. The results of this study are used to reflect the quality of the biological panel design.

2. Experiment Outline

2.1. Experimental Materials

2.1.1. Magnesia-Potassium-Phosphate Cement Mortar Composites. The magnesia cement used in this experiment was dead magnesia (MgO, hereinafter referred to as "M") with a density of 3.6 g/m^3 fired at a high temperature of 1,500°C or more. As a curing agent which reacts with dead magnesia, potassium monophosphate (KH_2PO_4, hereinafter referred to as "P") having a purity of 99% or more and a density of 2.34 g/m^3 was used. In addition, boric acid (H_3BO_3, hereinafter referred to as "R") was used as a retarding agent in consideration of its workability, and silicate no. 8 (hereinafter referred to as "S") of 95% or more of SiO_2 was used as an fine aggregate. Magnesia cement requires a retarder for workability due to fast curing time. And since it is made of a dense ceramic structure, it exhibits high strength of about 40 MPa in a short time.

Table 1 shows the chemical composition and physical properties of materials used in the experiment.

2.1.2. Foaming Agent. The foaming agent used in this experiment was sodium bicarbonate ($NaHCO_3$, hereinafter referred to as "F") which reacted with potassium ions and reacted with acid ions to generate bubbles. F is a suitable foaming agent that combines with ions generated by an acid-base neutralization reaction to generate a gas, and the chemical reaction formula can be expressed as follows:

$$HCO_3^- + H^+ \longrightarrow CO_2 + H_2O \quad (1)$$

The bubbles created through this bond create a rough surface through the voids inside and outside the mortar, which facilitates the absorption and storage of water and the adsorption of nutrients in the atmosphere [20].

Table 1: The physical and chemical composition of the materials used in the experiments.

Type	Purity (%)	Al_2O_3 (%)	SiO_2 (%)	MgO (%)	Density (g/m^3)
M	96	0.1	1.1	95.6	3.6
P	99	—	—	—	2.34
R	99.5	—	—	—	1.4
S	95	—	95	—	2.65

2.1.3. Superabsorbent Polymers (SAP). The superabsorbent polymer used in this study was an acrylic acid polymer, sodium salt (sodium polyacrylate). The superabsorbent polymer is a white powder which is capable of absorbing water several hundred times its own weight. When the water is absorbed, the volume expands and reaches gel state. The swelling characteristics of such a superabsorbent polymer vary depending on the interparticle density, the chemical structure of the superabsorbent polymer, the pH of the absorbed liquid, the temperature, and the ion concentration.

2.1.4. Experimental Plan. In this study, first, data were collected for magnesium-phosphate-formed mortar composites (SMPFC) using the superabsorbent polymer. Through preliminary experiments, the data of more than 10 MPa of the target strength were obtained by using the water/binder ratio (W/B) 30%, MgO : KH_2PO_4 ratio (P : M) 1 : 0.5, fine aggregate volume/total mortar volume ratio (Vs/Vm) 10%, retardation 3%, and foaming agent 0.5%. Therefore, 4 levels (0.25, 0.5, 0.75, and 1.0%) were prepared for data acquisition according to SAP addition. In addition, the flow characteristics, compressive strength, pH, roughness, and moisture retentivity were measured. The bubbles were applied in the postfoaming mode. The compressive strength was set to 10 MPa or more in terms of landscaping design at 6 hours, 1 day, 3 days, and 7 days of age in accordance with the characteristics of magnesia cement. Table 2 shows the mixing ratio of SMPFC used in the experiment.

2.2. Experimental Method

2.2.1. Mixing of SMPFC. The SMPFC was subjected to dry mixing with M, P, R, and SAP for about 2 minutes. Then, the blend was then mixed with the compounding water for 2 minutes at the low speed and then for 3 minutes at the high speed. Thereafter, additional water containing F was added, and the mixture was further mixed at the high speed for one minute.

2.2.2. Extra Water (eW) Decision. SAP rapidly absorbs moisture when exposed to moisture. Also, the amount of absorbed water increases with time. For this reason, the unit water quantity sharply decreases, the fluidity decreases, and the workability decreases. In addition, magnesia cement

Table 2: Mix design.

Type	W/B (%)	Binder		R	F	SAP	Extra water (eW)
		P	M				
SMPFC-P	35	1	0.5	3% on M	0.5% on binder	—	—
SMPFC-0.25						0.25% on binder	10% on W
SMPFC-0.5						0.5% on binder	16% on W
SMPFC-0.75						0.75% on binder	22% on W
SMPFC-1.0						1.0% on binder	27% on W

requires a large amount of unit water, unlike ordinary cement. This problem can be solved by adding additional water [21].

Previous studies have shown that SAP exhibits maximum absorption after 10 minutes of contact with water and is shaped like a gel [22]. Therefore, additional water should be added before reaching maximum absorption.

As a result of the preliminary test, it was found that, to achieve the target slump, extra water should add about 6% of the unit water when the SAP is increased by 0.25%.

2.2.3. Flow. The flow of SMPFC was measured in accordance with the "KS L 5105 Hydraulic Cement Mortar Compressive Strength Test Method".

2.2.4. Compressive Strength. The compressive strength of SMPFC was cubic-type specimens of 50 mm × 50 mm × 50 mm prepared according to "KS L 5105 Compressive Strength Test Method of Hydraulic Cement Mortar," followed by curing. Afterwards, the sample was demolded, and the compressive strength was measured by universal testing (UTM) at each age.

2.2.5. pH. The pH of SMPFC was measured in accordance with KS M 0011 "Method for measuring pH of aqueous solution." Also, the pH of SMPFC was measured according to age.

2.2.6. Surface Roughness. The roughness measurement of SMPFC was measured by relative comparison with the surface of the bubble-free reference specimen. Roughness is an important factor in the application of biological panels because it is one of the indicators to induce adsorption on the surface of rainwater or nutrients in the air.

2.2.7. Moisturizing Property. The moisture retention (moisture content) of SMPFC was measured by wetting curing of the specimen for one day according to KS F 2459 "Test method of apparent density, water content, water absorption, and compressive strength of foam concrete". Thereafter, the mass of the dried sample in the drying furnace was measured.

3. Experimental Results and Discussion

3.1. Flow. Figure 1 shows the flow measurement result according to the addition amount of SAP of SMPFC. As a result of the experiment, the flow value decreased by about 10% every time the SAP addition amount was increased by 0.25% without determining the additional extra water (eW). The cause of this phenomenon is considered to be the property of absorbing moisture of SAP. However, if eW is added as shown in Figure 2, the target flow can be satisfied.

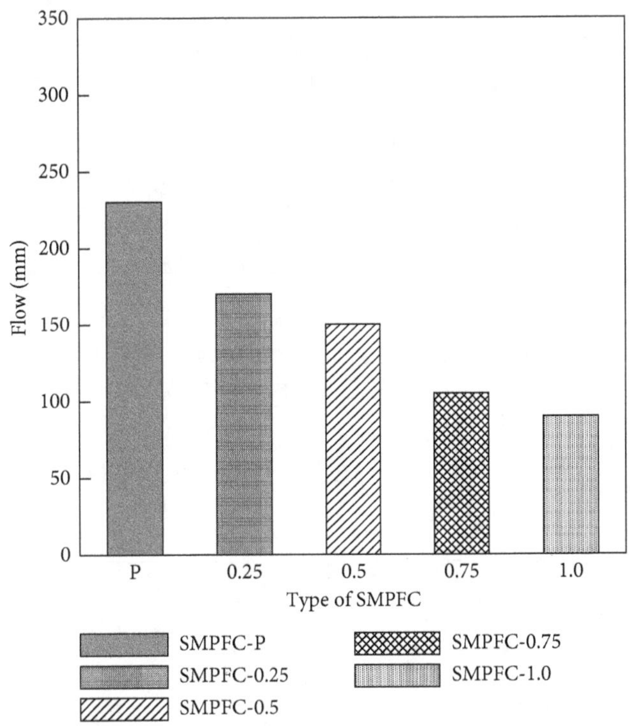

Figure 1: Test results of flow (extra water not added).

3.2. Compressive Strength. Figures 3–6 show the results of compressive strength measurement according to the additional amount of SAP of SMPFC, and the results of the compressive strength test at 6 hours, 1 day, 3 days, and 7 days of age are shown. The compressive strength at 6 hours did not achieve the target strength in all cases. Also, the compressive strength at 1 and 3 days was more than 10 MPa for plain. A mix with SAP that could achieve the target intensity was possible after 7 days. The compressive strength of SAP 0.25% at 7 days was above 12 MPa and decreased proportionally with increasing SAP. The compressive strength of SAP 0.5% was about 8 MPa, and the compressive strength of SAP 0.75 and 1.0% was less than 5 MPa lower than the target strength. Overall, as the amount of SAP increased, the compressive strength decreased, and the specimen with 1.0% SAP was easily destroyed. However, when the addition

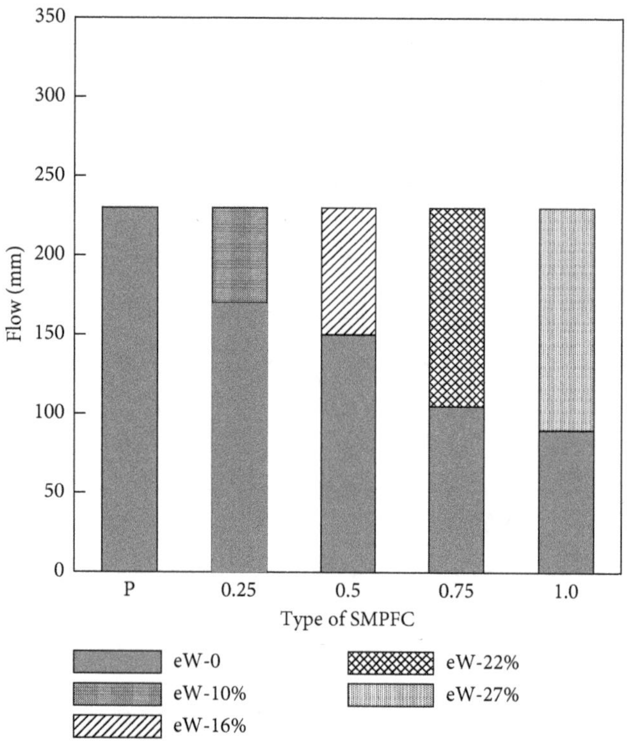

Figure 2: Test results of flow (extra water added).

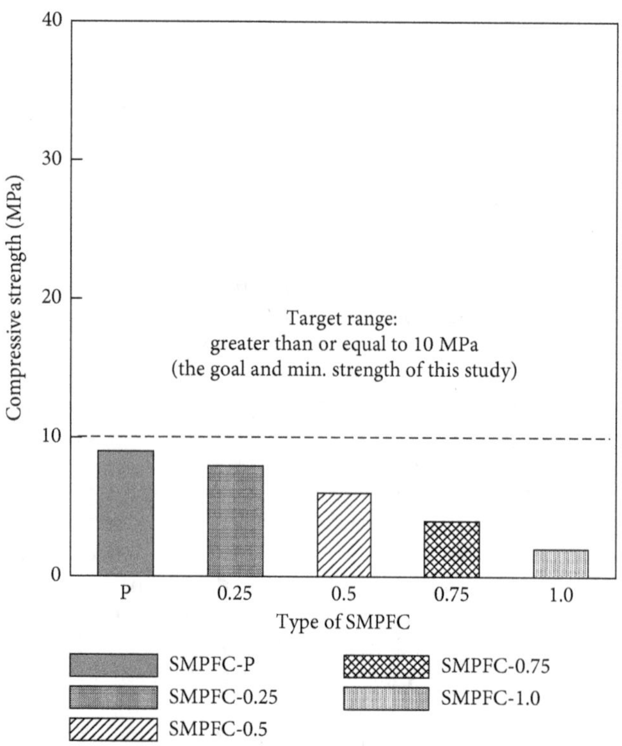

Figure 3: Test results of compressive strength (6H).

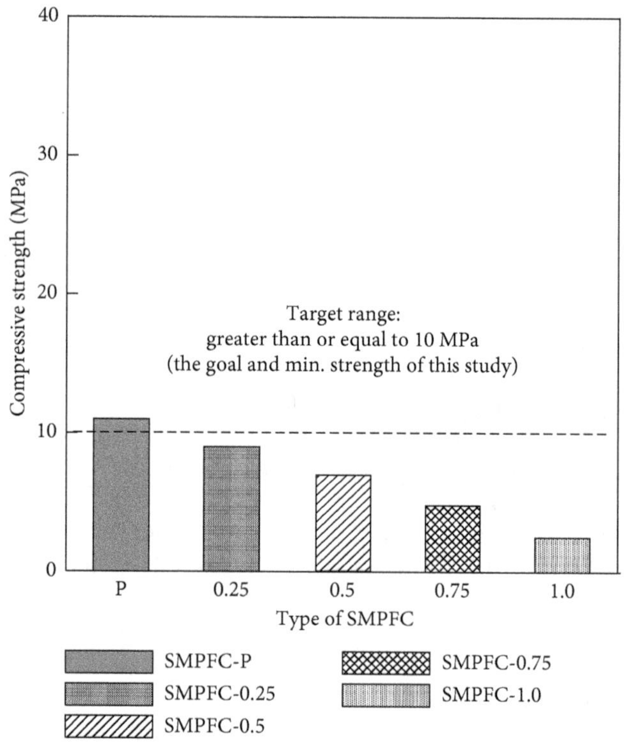

Figure 4: Test results of compressive strength (1D).

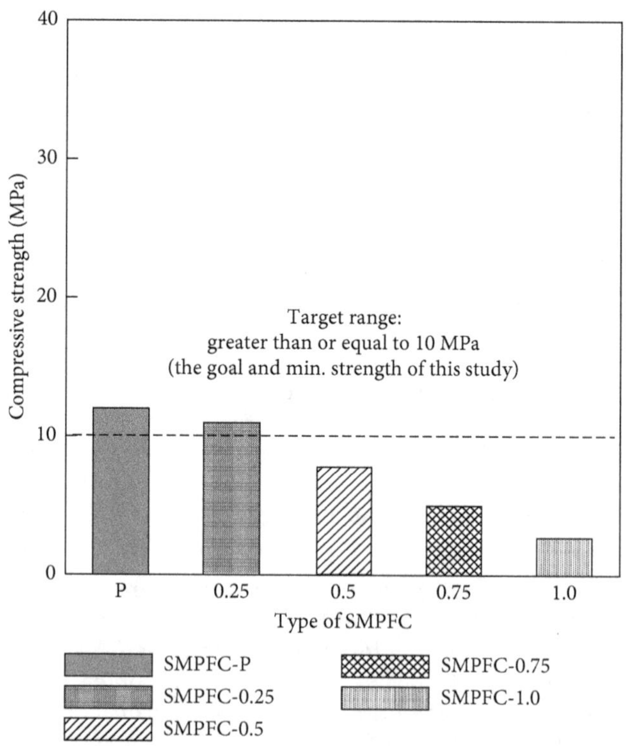

Figure 5: Test results of compressive strength (3D).

amount of SAP was 0.25%, similar results were obtained. Therefore, if the optimum amount of SAP is to be found, it will be possible to control the strength when applying the biological panel.

3.3. pH. Figure 7 shows the pH measurement results of SMPFC at 6 hours, 1 day, 3 days, and 7 days. As a result of the experiment, the pH of the specimen tended to increase with age. Magnesia cement was measured at pH 7–9

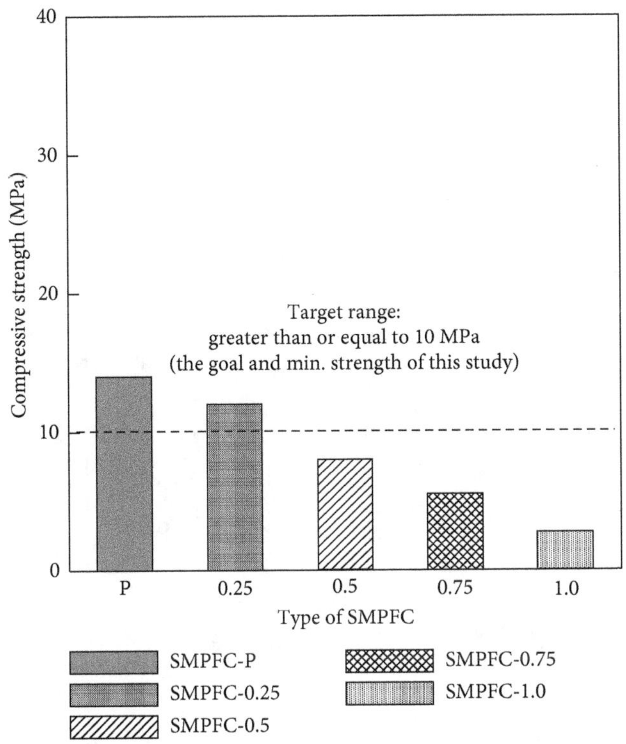

FIGURE 6: Test results of compressive strength (7D).

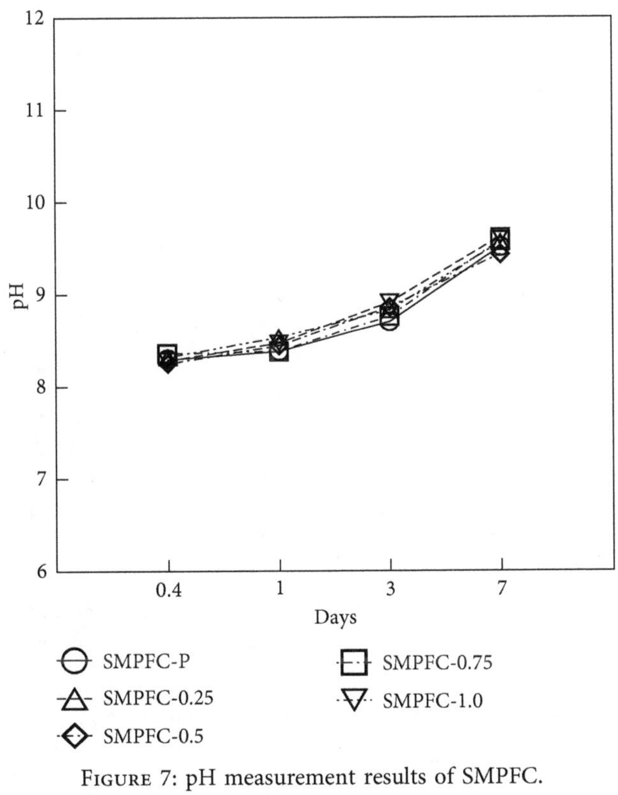

FIGURE 7: pH measurement results of SMPFC.

depending on the type of the phosphate used in the formulation. The SAP used in the experiment was a material with a pH of 6.5 and showed no significant effect on the pH change of SMPFC. The pH at 7 days was measured as 9. Since the matrix for biological panels is important for the survival of (micro)organisms, it is necessary to examine the pH. From these results, SMPFC, which is a neutral region, would be a favorable condition for the survival environment of (micro)organisms.

3.4. Surface Roughness. Figure 8 shows the result of measuring the surface roughness of the SMPFC. It found that the pore depths of the 0.25% are in the range of 1.0–0.005 mm and the pore depths of the 0.5% SAP additions are in 0.1 mm. And pores visible by the naked eye were observed from the micropores.

In the case of 0.75% and 1.0% of SAP additions, SAP was observed on the surface because the amount of SAP added was too large. In addition, it was confirmed that the effect of the foaming agent was hardly observed. The reason for this is that a considerable amount of SAP absorbs moisture quickly and suppresses bubble generation. Figure 9 shows the distribution of voids on the surface of the bubbled test piece. These results suggest that SMPFC with a rough surface due to bubbles will be an advantageous growth condition of (micro)organisms and will be advantageous for the adsorption of nutrients in the air.

3.5. Moisturizing Property. Figures 10–14 show the measurement results of moisture content (moisture content) according to the addition amount of SAP of SMPFC. The test roughly evaluated the amount of moisture absorbed by the test piece and SAP through the pores generated on the inside and the surface of the test piece. A qualitative evaluation was performed on the assumption that the larger the pores, the more the SAP could contain in the pores and absorb a large amount of water. The weight of the dried specimen was measured first, and then, the weight of the specimen immersed in water for one day was measured. Then, the period of reaching the dry weight at room temperature was measured.

As shown in Figures 10–14, the moisture retention of SMPFC-P was about 10% larger than that of dry weight. And SMPFC-0.25 was 5% higher than SMPFC-P. As the SAP increased by 0.25%, it increased continuously by about 5%. These results show that the amount of SAP added affects the moisture retention, and the sample without SAP showed a drying time about 3 days faster than the sample with SAP added. As the amount of SAP increased, the drying time at room temperature tended to increase by 2 to 3 days compared to that of SMPFC-P. In the case of the sample with 1.0% SAP content, the complete drying time was about 9 to 10 days. These results suggest that when applied as a matrix for biological panels, the addition of SAP improves the moisture retention and absorbs moisture in the air, which may have a positive effect on the growth of the organism. However, when the amount of SAP increases, it will decrease in compressive strength and durability. Therefore, the optimal amount of SAP should be reviewed. Figure 15 shows the SAP that absorbed moisture inside the test.

These results suggest that SMPEC with improved moisture retention by SAP may contribute to the water supply of (micro)organisms to some extent. In addition, it is

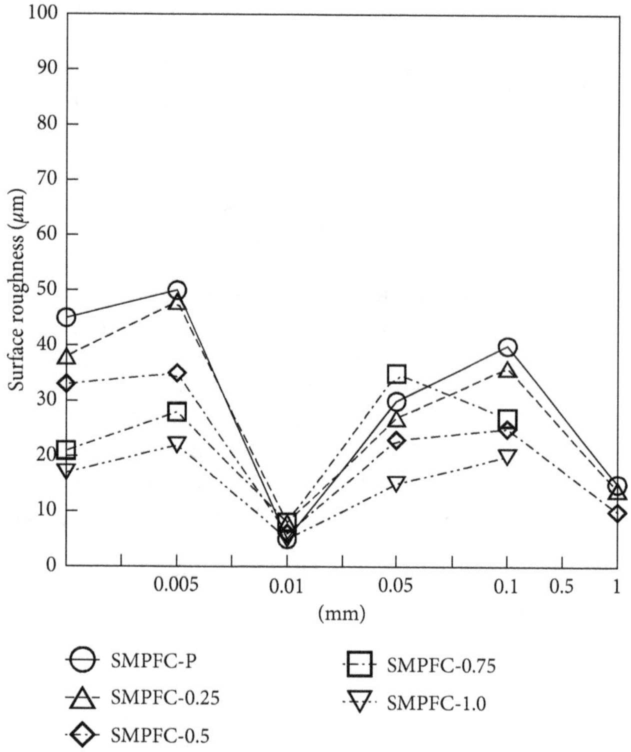

Figure 8: Surface roughness results of SMPFC.

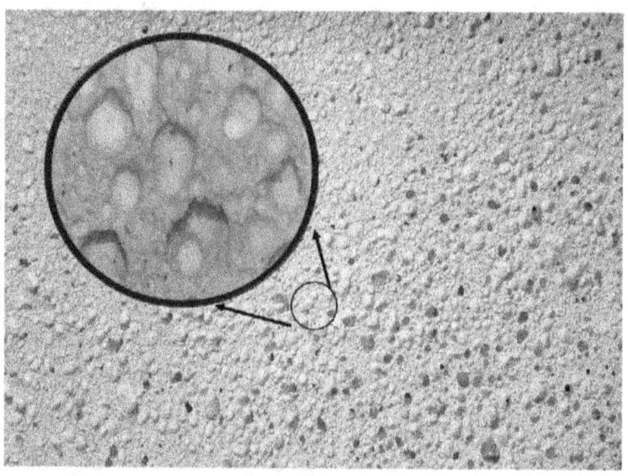

Figure 9: Surface of SMPFC.

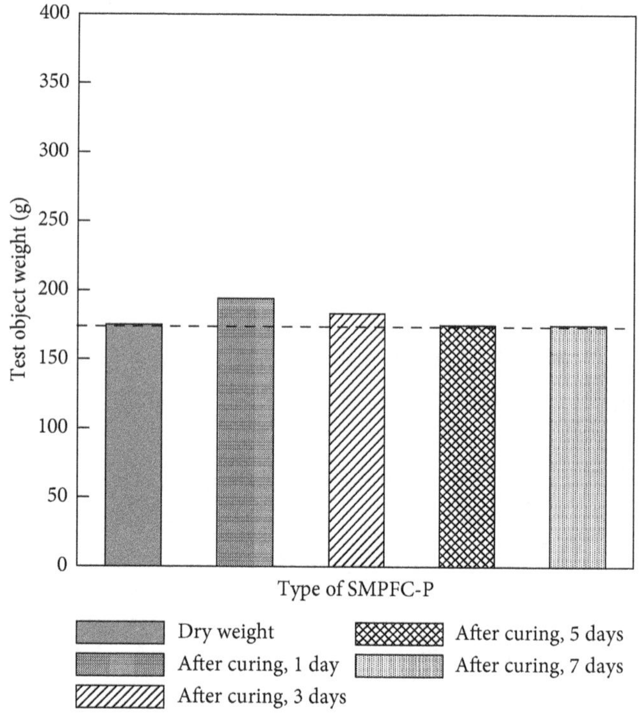

Figure 10: Test results of moisture content (plain).

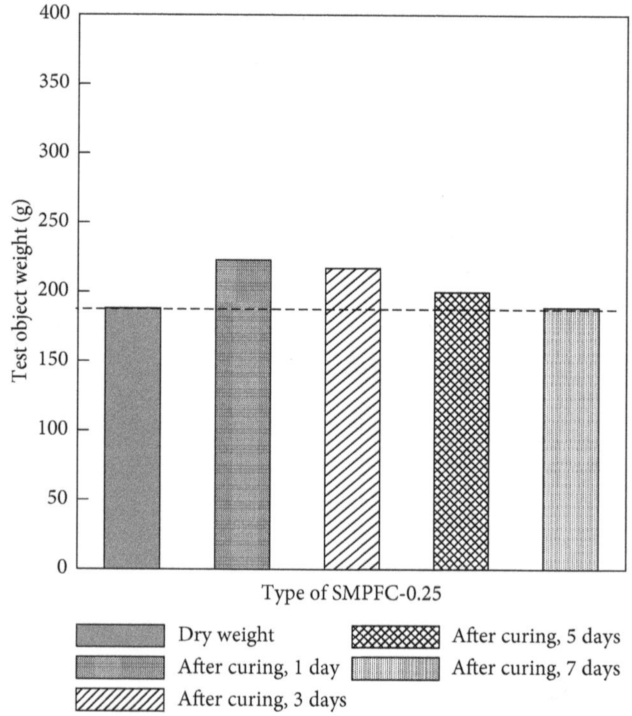

Figure 11: Test results of moisture content (SAP 0.25%).

considered that the moisturizing effect in the atmospheric moisture or rainy weather will be further increased.

4. Conclusion

As one of the basic studies for the development of biological panels, it is investigated that the magnesia-phosphate-foamed mortar composites using the superabsorbent polymer. The following conclusions were obtained:

(1) The magnesia-potassium-phosphate-foamed mortar composites using a superabsorbent polymer were able to satisfy the target flow when extra water was added according to the mixing ratio of the superabsorbent polymer.

(2) The magnesia-potassium-phosphate-foamed mortar composites using a superabsorbent polymer showed a tendency to decrease in compressive strength depending on the mixing ratio of the superabsorbent polymer. The mixing ratio of the superabsorbent polymer satisfying the target strength was found to be within 0.25%.

(3) The magnesia-potassium-phosphate-foamed mortar composites using a superabsorbent polymer

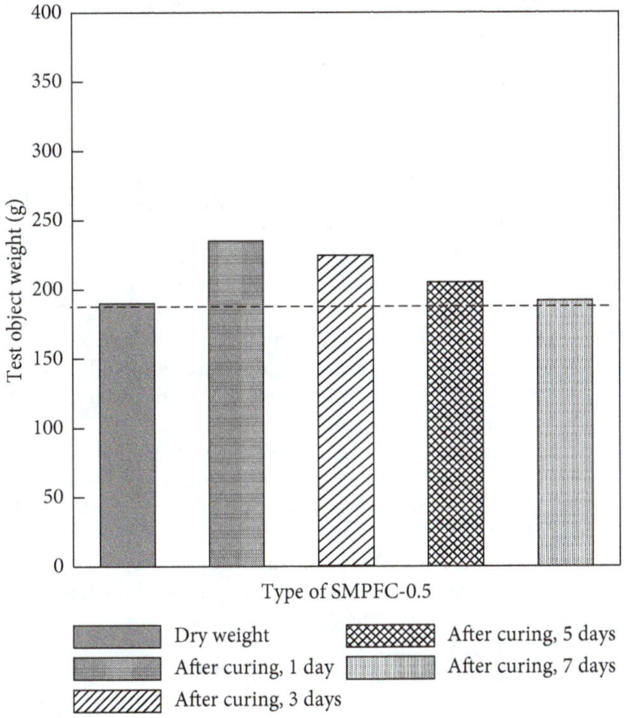

FIGURE 12: Test results of moisture content (SAP 0.50%).

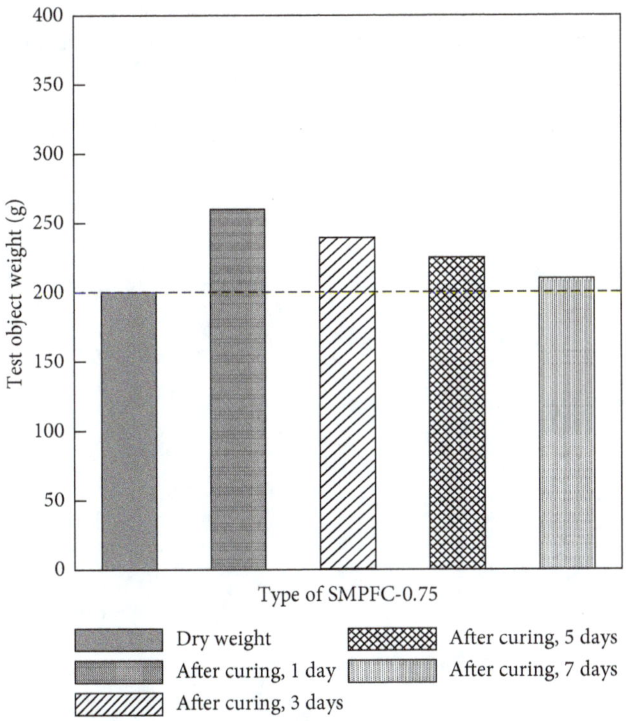

FIGURE 13: Test results of moisture content (SAP 0.75%).

showed a tendency to increase in water content according to the mixing ratio of the superabsorbent polymer. It also formed a roughened surface by bubbles.

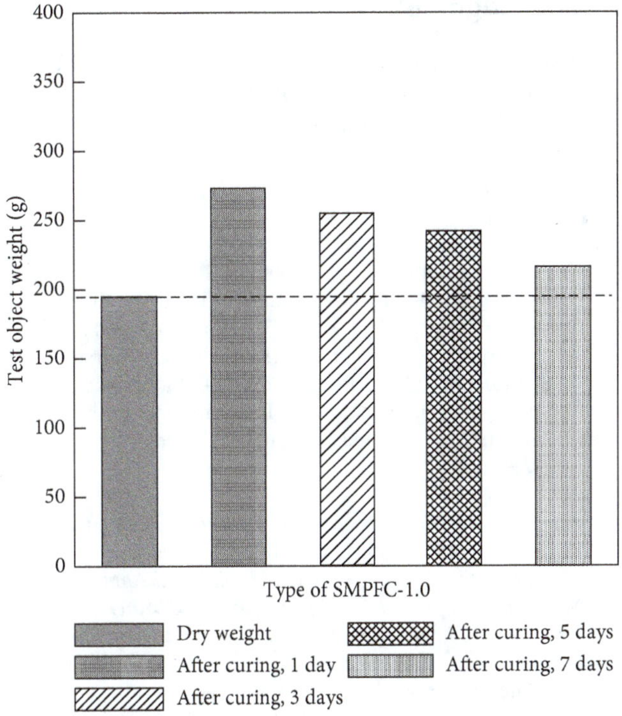

FIGURE 14: Test results of moisture content (SAP 1%).

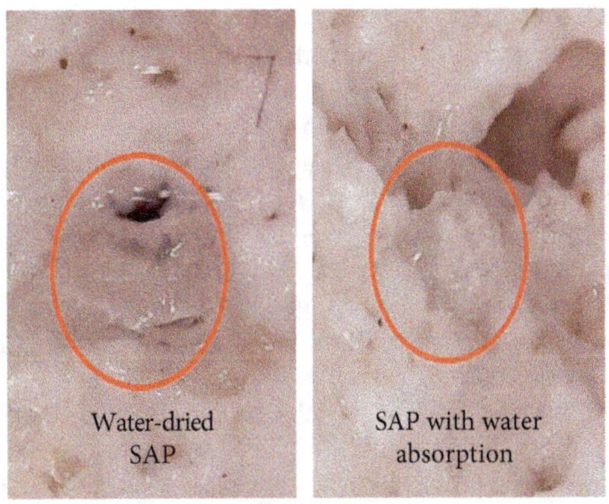

FIGURE 15: SAP absorbed inside the specimen.

From these results, it was confirmed that the optimum range of the superabsorbent polymer satisfying the target strength was 0.25%. In addition, the rough surface by the foaming agent and the moisture retention by a superabsorbent polymer will be beneficial to rooting and growth conditions of (micro)organisms. As a sequential study, studies on biological acceptability and the growth of (micro) organisms should be accompanied.

Conflicts of Interest

The authors declare that there are no conflicts of interest regarding the publication of this paper.

Acknowledgments

This research was supported by a grant (17CTAP-C114646-02) from Construction Technology Research Program funded by Ministry of Land, Infrastructure and Transport of Korean government.

References

[1] K. Perini, M. Ottelé, E. Haas, and R. Raiteri, "Greening the building envelope, façade greening and living wall systems," *Open Journal of Ecology*, vol. 1, no. 1, pp. 1–8, 2011.

[2] A. Eggert, N. Häubner, S. Klausch, U. Karsten, and R. Schumann, "Quantification of algal biofilms colonising building materials: chlorophyll a measured by PAM-fluorometry as a biomass parameter," *Biofouling*, vol. 22, no. 2, pp. 79–90, 2006.

[3] C. G. Han and S. G. Park, *Eco-Friendly Ready Mixed Concrete Using Waste Materials and by-Products*, Magazine of the Korea Concrete Institute, Republic of Korea, 2016.

[4] H.-G. Lee, M.-H. Gong, and Y.-S. Jung, *Field Application and Manufacturing Technology of Low Carbon Ultra High Strength Green PHC Pile using High Volume Ground Granulated Blast Furnace Slag*, Magazine of the Korea Concrete Institute, Republic of Korea, 2016.

[5] T. W. Kim and C. H. Kang, "The self-healing and ageing effect of OPC-GGBFS cement in sea-water and tap-water," *Journal of the Korea Concrete Institute*, vol. 29, no. 1, pp. 11–21, 2017.

[6] L. Ferrara, V. Krelani, and M. A. Carsana, "Fracture testing based approach to assess crack healing of concrete with and without crystalline admixtures," *Construction and Building Materials*, vol. 68, pp. 535–551, 2014.

[7] Z. Jiang, W. Li, and Z. Yuan, "Influence of mineral additives and environmental conditions on the self-healing capabilities of cementitious materials," *Journal of Cement and Concrete Composites*, vol. 57, pp. 116–127, 2015.

[8] I. You, C.-I. Jeong, C. Lyu, G. Lee, and G. Zi, *State of Research on High Durable Concrete with Waste Glass Powder for Road Structure*, Magazine of the Korea Concrete Institute, Republic of Korea, 2017.

[9] C.-W. Baek, H.-S. Kim, S.-W. Choi, H.-T. Jo, and D.-H. Ryu, "Durability assessment of high strength concrete with high volume mineral admixture," *Journal of the Korea Concrete Institute*, vol. 27, no. 6, pp. 641–649, 2015.

[10] Y. Halit, Y. Mert, Y. Husyinn, A. Serdar, and T. Selcuk, "Mechanical properties of reactive powder concrete containing high volumes of ground granulated blast furnace slag," *Journal of Cement and Concrete Composite*, vol. 32, no. 8, pp. 639–648, 2010.

[11] J.-K. Jeon, Y.-S. An, and E.-J. Kim, *Performance Evaluation Study on Green-wall System Using Moss*, Magazine of the Korea Concrete Institute, Republic of Korea, 2014.

[12] J. S. Lee, S. H. Kim, Y. S. Shin, and C. Hong, "A fundamental experiment of the artificial soil mixture using stone sludge burning pellet for applying to green roof soil," *Proceedings of the Korea Concrete Institute*, vol. 2012, no. 11, pp. 209-210, 2012.

[13] H.-R. Gkim and T.-M. Ahn, "Cafeteria users' preference for an indoor green-wall in a university dining Hall," *Journal of Korean Institute of Landscape Architecture*, vol. 43, no. 6, pp. 62–72, 2015.

[14] S. R. Choi, Y. S. Kim, and P. S. Yoon, "The management plan for the maintenance of the green-wall colors-focusing on the eco plaza," in *Proceedings of Korean Society of Color Studies*, pp. 38–44, 2013.

[15] M. Ottele, K. Perini, A. L. A. Fraaij, E. M. Haas, and R. Raiteri, "Comparative life cycle analysis for green facades and living wall systems," *Energy and Buildings*, vol. 43, no. 12, pp. 3419–3429, 2011.

[16] K. Perini, M. Ottele, E. M. Haas, and R. Raiteri, "Vertical greening systems, a process tree for green facades and living walls," *Urban Ecosystems*, vol. 16, no. 2, pp. 1–13, 2009.

[17] H. E Wright Wendel, R. K. Zarger, and J. R. Mihelcic, "Accessibility and usability: green space preferences, perceptions, and barriers in a rapidly urbanizing city in Latin America," *Landscape and Urban Planning*, vol. 107, no. 3, pp. 272–282, 2012.

[18] T. H. Kim, F. Q. Li, T. W. Ahn, I. S. Choi, and J. M. Oh, "Research on improvement of water purification efficiency by concrete using bio film," *Journal of Environmental Impact Assessment*, vol. 20, no. 6, pp. 815–821, 2011.

[19] S. Manso, W. D. Muynck, I. Segura et al., "Bioreceptivity evaluation of cementitious materials designed to stimulate biological growth," *Science of the Total Environment*, vol. 481, pp. 232–241, 2014.

[20] C. Ma and B. Chen, "Experimental study on the preparation and properties of a novel foamed concrete based on magnesium phosphate cement," *Construction and Building Materials*, vol. 137, pp. 160–168, 2017.

[21] S. H. Kang, J. Moon, and S. G. Hong, "Effect of internal curing by super-absorbent polymer (SAP) on hydration, autogenous shrinkage, durability and mechanical characteristics of ultra-high performance concrete (UHPC)," *Journal of the Korea Concrete Institute*, vol. 28, no. 3, pp. 317–328, 2016.

[22] O. M. Jensen and P. F. Hansen, "Water-entrained cement-based materials: II. Experimental observations," *Cement and Concrete Research*, vol. 32, no. 6, pp. 973–978, 2002.

2

The Impact Toughness and Hardness of Treated and Untreated Sisal Fibre-Epoxy Resin Composites

Wilson Webo,[1] **Leonard Masu**,[1] **and Maina Maringa**[2]

[1]*Department of Mechanical Engineering, Faculty of Engineering and Technology, Vaal University of Technology, Private Bag X021, Andries Potgieter Blvd., Vanderbijlpark 1911, South Africa*
[2]*Department of Mechanical and Mechatronics Engineering, Faculty of Engineering and Technology, Central University of Technology, Private Bag X20539, Bloemfontein 9300, South Africa*

Correspondence should be addressed to Wilson Webo; wilsonwebo@yahoo.com; Leonard Masu; leonard@vut.ac.za; and Maina Maringa; maina_maringa@yahoo.com

Academic Editor: Yuanshi Li

The effect of the combined chemical treatment of sisal fibres through the subsequent processes of mercerisation (alkali treatment), then silane treatment and eventually acid hydrolysis, on sisal fibre was investigated. The effect of the treated fibres on the impact toughness and hardness of their composites with epoxy resin was also studied. Scanning electron microscopy of the surfaces of the treated and untreated fibres showed that the chemical treatment processes enhanced the removal of surface impurities and therefore increased the roughness of the surfaces of the fibres. This avails an increased surface area for interlocking with matrix and is, therefore, expected to enhance adhesion of the two. The treated fibre reinforced composites were observed to have higher values of impact toughness and hardness than the untreated fibre reinforced composites. These higher values were attributed to better interfacial bonding due to better mechanical interlocking between the treated fibres and epoxy resin arising from the increased roughness of the treated fibres.

1. Introduction

A composite material is a physical mixture of two or more different materials that result in a component with superior properties to those of any individual component [1]. In contrast to metallic alloys, each material retains its separate chemical, physical, and mechanical properties [1, 2]. The main advantages of composite materials are their high specific strength and stiffness, when compared with other materials allowing for a reduction in weight of finished parts [3].

The constituents of a composite are categorized as the reinforcement and the matrix. The reinforcement is the discontinuous phase while the matrix is the continuous phase [1–3]. The reinforcement is usually embedded in the matrix. The reinforcement is usually in the form of fibres, sheets, or particles whereas the matrix is either a polymer, ceramic, or metal [2]. The reinforcement provides strength and stiffness to the composite material [4]. The matrix on the other hand transfers and distributes the applied loads to the fibres and in many cases contributes some needed properties such as ductility, toughness, and electrical insulation [1, 3]. Examples of polymer matrices include epoxy resins, polyester resins, and phenolic resins while metallic matrices include aluminium resin, magnesium resin, and titanium resin [3]. The most common methods used in the manufacture of composite materials are compression moulding, pultrusion, hand layup, resin transfer moulding (RTM), injection moulding, and filament moulding [1].

Composites may be classified as particulate reinforced composites or fibre reinforced composites [3]. The particulate reinforced composites consist of tiny particles having thickness in the micron [5, 6] and nanoscales [7] embedded into a matrix. The particles occur in the form of flakes or powder [8]. Fibre reinforced composites on the other hand are made of continuous or discontinuous fibres that are embedded in a matrix [3].

A reemerging area of fibre reinforced composites is that of natural fibre reinforced composites [9]. Natural fibres are increasingly being considered as an environmentally friendly substitute for synthetic fibres in the reinforcement of polymer-based composites [2, 9]. The use of these fibres instead of glass and carbon is, for example, under consideration in the automotive industry [10]. Although many different types of natural fibres are available, sisal (*Agave sisalana*) is a particularly attractive option due to its rapid growth over a wide range of climatic conditions [11] and its low cost [10]. Although the use of natural fibres partly satisfies the requirements of regulations that enforce the use of environmentally friendly and sustainable materials [10], the matrix system must also be considered in this regard. Clearly, completely biodegradable matrices such as polylactic acid (PLA) are preferable, but the current high cost of these matrices is a disadvantage in comparison to epoxy resin [10]. Sisal/epoxy composites are therefore of considerable commercial interest.

Even though natural fibres have the potential to supplement glass fibres in polymer composites [12], limitations arise with respect to mechanical performance and moisture absorption when natural fibres are used [9]. These limitations are present irrespective of whether thermoset or thermoplastic polymers are used as the matrix material. Previous studies [13–15] have reported that the main factor that limits the mechanical properties of natural fibre composites is the chemical incompatibility between the hydrophilic lignocellulosic molecules of the natural fibre and the hydrophobic molecule of the resin. This incompatibility leads to difficulties in ensuring effective fibre-matrix interface bonding which, in turn, causes ineffective load transfer between the reinforcing material and the matrix [10, 12, 13]. Various options, including chemical treatments of natural fibres and the use of compatibilisers, have been suggested in order to achieve the necessary compatibility of surface energies between the fibres and the matrix. Li et al. [16] presented a review article on the various chemical treatments used to improve compatibility between natural fibres and polymer matrices. Approaches such as mercerisation (alkali treatment), silane treatment, acetylation, benzoylation, use of maleated coupling agents, peroxide treatment, permanganate treatment, and isocyanate treatment were considered. Amongst the various methods presented, mercerisation and silane treatments have been widely reported on.

Mercerisation involves immersing the fibres in an alkaline solution for a period of time. It works by increasing the surface roughness of the fibre which improves mechanical bonding with matrices [16]. It also exposes more cellulose on the surface of the fibre to potential chemical bonding with matrices [17]. Silane treatment entails soaking the fibres in a weak solution of a silane diluted in water/alcohol or water/ketone mixture. In the presence of water, silane breaks down into silanol and alcohol. The silanol reacts with the OH groups of the cellulose in natural fibres, forming stable covalent bonds on the cell wall that are chemisorbed onto the fibre surface [16]. Use of silane improves the degree of cross-linking in the interface region and increases the fibre surface area, allowing for stronger bonding between the fibre and matrix [18]. Acid hydrolysis involves subjecting natural fibres to harsh acid treatment, which causes the amorphous lignin regions to break up thus releasing the individual crystallites of cellulose. Sulphuric acid and hydrochloric acid are extensively used in performing acid hydrolysis [19].

While most researchers have used either alkali treatment or silane treatment or acid hydrolysis individually, it is necessary to attempt using the three methods of treatment in combination, since the mechanisms by which they affect natural fibres are completely different. Herrera-Franco and Valadez-Gonzalez [20] considered high-density polyethylene [HDPE] reinforced with henequen fibres at a volume fraction of 20%. The effects of alkali and silane treatments on these fibres were studied individually in addition to the effect of the two treatments acting in combination. It was found that alkali treatment resulted into almost no improvement, whereas silane resulted into a 19% improvement. When the combination of the treatment was considered, there was an improvement of 30%.

Impact tests are usually carried out during the earliest stages of the design process in order to determine the integrity of the material under impact loading. Impact tests are usually done in order to determine the maximum impact load that a material can bear at a particular temperature [21]. Impact resistance of fibre-reinforced composites, like that of other materials, is measured by several test methods, namely, Charpy, Izod, drop weight, split Hopkinson bar (SHB), explosive, and ballistic impact [22]. The results of impact tests are presented in the form of fracture energy, damage accumulation, and/or measurement of the number of drops to achieve a determined damage or stress level. The results depend on many variables such as size of specimen, strain rate, type of instrumentation, and test setup [22].

Wang et al. [23] identified two different damage mechanisms for fibre reinforced composites that were loaded on a drop weight impact system. For fibre fractions lower than the critical fibre volume, fibre fracture dominated the failure mechanisms, while for higher values than the critical fibre volume, the fibre pull-out mechanism dominated the response.

Ramires and Frollini [24] used the Izod unnotched impact test to investigate the impact properties of sisal fibre reinforced tannin-phenolic composites. They carried out their investigations according to the ASTM D256 standard. The researchers found that the improvement in the interfacial bond was the main reason for the improvement of the impact strength. The authors reported that the increased acid sites at the surface of the sisal fibres improved interfacial adhesion. The researchers further reported that the Izod impact strength of the composites increased with an increase in the fibre content up to a maximum of 50% fibre content. However, beyond 50% fibre content, the Izod impact strength decreased due to ineffective impregnation of resin in the fibres. The researchers concluded that the use of sisal fibres as reinforcement material in tannin-phenolic resin was useful since it greatly improved the impact properties of the composite.

Han-Seung et al. [25] carried out a study on the effect of different compatibilising agents on the tensile strength and Izod impact strength of lignocellulosic materials. Low- and high-density polyethylene was used as the matrix, while rice husk flour and wood flour were used as reinforcement. Maleated propylene and maleated polyethylene were used as compatibilising agents to improve the fibre-matrix adhesion.

The researchers found that maleated polyethylene improved the Izod impact strength of notched specimens because of the enhancement in the fibre-matrix interfacial adhesion. On the other hand, the Izod impact strengths of composites produced from maleated polypropylene reinforced with rice husk dropped slightly. The researchers concluded that the Izod impact strength of the maleated polyethylene-incorporated composites improved slightly because of a better interfacial bonding between the matrix and the fibre.

Hardness refers to the resistance a solid shows to local deformation. A hard indenter is placed onto the surface of a material and is then pressed into the material. The size of the permanent indentation thus formed is then measured to determine the hardness of the material. The common methods used to measure hardness are the Brinell hardness test, Rockwell hardness test, Vickers hardness test, and the Barcol hardness test. While Brinell, Rockwell, and Vickers hardness test methods are commonly used, the Barcol hardness test is rarely performed.

Kumar et al. [26] studied the hardness of treated and untreated samples reinforced with sisal/glass epoxy-based hybrid composites using Rockwell hardness testing machine. In each case, five samples were tested and the average value was tabulated. Test specimens were made according to the ASTM D 785 test standard. The diameter of the ball indenter used was 0.25 inches, and all the readings were taken 10 seconds after the indenter made firm contact with the specimen. All the sample surfaces were rubbed with smooth emery paper, which facilitates accurate reading. It was observed that 2 cm fibre length composites had a higher hardness than 1 cm and 3 cm fibre length composites.

The effects of three combined chemical treatments, namely, mercerisation, silane treatment and acid hydrolysis, on sisal fibres are investigated in this paper. The effect of these combined treatments on the impact toughness and hardness of their composites with epoxy resin is also looked into. This is in contrast to other studies on impact and hardness that have used these treatments individually. The composites were manufactured using the vacuum infusion method of composite manufacture. This method prevents the entry of air into, and, formation of voids in the composites that are fabricated and is expected to facilitate the attainment of high strength and stiffness of the composites produced. Furthermore, this method differs from the other fabrication techniques by virtue of the fact that the creation of a vacuum in the mould cavity ensures that the resin is driven into the mould and through the laid out fibres [27] by atmospheric pressure and therefore ensures proper wetting of reinforcing fibres. Air bubbles can still be introduced into the system through the mixing process of the resin and hardener prior to charging into the mould, and through mould leaks. These, however, are weaknesses that are ameliorated in the infusion process.

2. Experimental Details

2.1. Materials. Epoxy Epolam 2015 resin and Epolam 2014 hardener were obtained from AMT composites of South Africa. The mix ratio of the resin and the hardener according to the data sheet obtained from the company was 100 : 32. Sisal fibres in a bundle of 10 kgs were obtained from Mogotio farm, Nakuru County, Republic of Kenya. The reagents which were used in this work were: sodium hydroxide, 3-glycidyloxypropyltrimethoxy-silane, methanol, and hydrochloric acid (HCL). Sodium hydroxide was provided by Minema chemicals (Pty) limited while 3-glycidyloxypropyltrimethoxy-silane, methanol, and hydrochloric acid were provided by Sigma Aldrich limited, all from South Africa.

FIGURE 1: The vacuum infusion process.

2.2. Preparation of Treated Sisal Fibres. The sisal fibres were mercerised by immersion in 5% sodium hydroxide solution for 20 hours. The fibres were then washed with distilled water in order to remove the sodium hydroxide from them and then further immersed in 1% acetic acid in order to neutralise any remaining sodium hydroxide. The fibres were, thereafter, immersed in a silane solution made up of 5% of 3 glycidyloxy-propyl-trimethoxy-silane diluted in a 95% aqueous solution of methanol, the later in order to hydrolyse the silane and make it active [16]. This treatment was then followed by immersion of the fibres in a 67.5% solution of hydrochloric acid for 1 hour [28]. The treated fibres were, thereafter, washed with deionized water and dried in an oven at 45°C for 24 hours.

2.3. Manufacture of Composites. The vacuum infusion method of manufacturing composites was used to fabricate the composites here as shown in Figure 1. This method supports proper wetting of the reinforcing fibres, which gives rise to better and stronger interfacial bonding of the fibre and matrix. Both treated and untreated sisal fibre-epoxy resin composites were manufactured using this method. The sisal fibres were cut using a pair of scissors and straightened using a comb in order to avoid bunching of fibres which would otherwise minimise wetting of the fibres with resin and, therefore, reduce the efficacy of reinforcement. The fibres were then weighed using an electronic balance with an accuracy of $\pm$ 0.5 gm and grouped into various masses corresponding to different fibre weight fractions. A thin layer of wax was smeared onto the base of a square glass mould of 50 cm by 50 cm. The wax ensures that the composite can be easily removed from the glass mould after curing. Predetermined weights of sisal fibres were spread longitudinally in the glass mould. The fibres were then covered with peel ply and infusion mesh simultaneously. The peel ply was made of polyester material and is used to wick away slight excesses of resin. The infusion mesh was made of a plastic material, and it aids the resin to flow efficiently throughout the fibres. A bleeder cloth was then laid near the tube exit to the vacuum so that excess resin could be sucked through the tube to the resin trap. Spiral tubings were then both connected, one to the tube from the resin beaker and one to the vacuum. Thereafter, a vacuum bag was used to cover the entire casting. A tacky tape was then used to secure the vacuum bag onto the mould. The vacuum pump motor was switched on, and the tube leading to the resin storage container was temporarily closed off using a G-clamp in order to avoid suction of air into the fibres before creating a vacuum. A break of one hour was allowed with the vacuum pump running as the resin was being prepared. The resin and the hardener were then measured in their appropriate ratios using an electronic balance of accuracy ±0.5 gm and subsequently mixed using a spatula. Air entrapment in the resin was eliminated through puncturing of air bubbles with a sharp needle. After the break, the resin suction pipe was placed into the resin container and the closing G-clamp on the pipe was removed in order to allow suction of the resin onto the fibres. The sisal fibre reinforced epoxy resin composites were subsequently cured in air for 24 hours. After this air curing, the composite was further cured in an oven at 80°C for four hours as recommended by the supplier in order to produce composites with excellent mechanical properties. The reinforcement was varied from 0 to 50 wt.%, inclusive in the different composites.

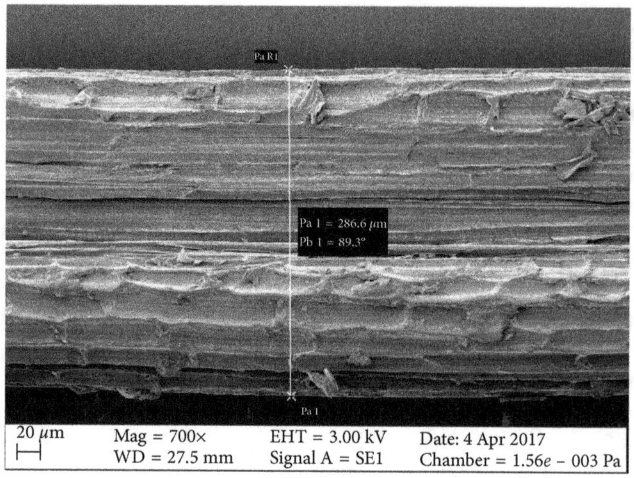

FIGURE 2: Untreated sisal fibre.

2.4. Impact Tests. In order to determine the fracture toughness of the fabricated composites, Charpy impact tests were performed at room temperature using a Hounsfield Balanced Impact Tester (Tensometer Ltd., Croydon, England). The Hounsfield Balanced Impact Tester imposes a three-point impact similar to that created by the Charpy apparatus specified in ASTM D6110-10. Test specimens with a geometry of $50 \times 12 \times 5$ mm in length, width, and thickness, respectively, were cut from the cast composites using a CNC machine. For each composition, a V-notch of $45^\circ \pm 1^\circ$, root radius of 0.25 ± 0.05 mm, and notch width of 2 mm was made. The specimens were, thereafter, sanded off using emery cloth of grade 220 in order to ensure that no out-of-plane notches were introduced onto the specimens. The Charpy impact test, also known as the Charpy V-notch test, is a standardised high-strain rate test which is used to determine the amount of energy absorbed by a material during fracture. The results obtained from Charpy impact test are a measure of a given material's toughness and can also be used to determine temperature-dependent brittle-ductile transition curve of a material [3]. When determining the fracture toughness using the Charpy impact test frame, the sample piece is loaded horizontally onto the apparatus, and the Charpy impact test pendulum hammer is released to strike the notched sample on the back side of the V-notch. The notch is introduced in the test samples to create a known point of stress concentration at which cracks are expected to initiate. After impact, the sample is removed and its failure surfaces examined. If the failure surface is flat and smooth, then failure is classified as brittle. If on the contrary, the failure surface is fibrous, then failure is classified as ductile. The scope of the current study was restricted to the determination of the impact toughness of the composites at room temperature.

2.5. Hardness. Hardness testing was done in accordance with ASTM D 2583 test standard specifications using the Barber Colman Barcol Impressor. Test specimens with a geometry of $12.7 \times 12.7 \times 3$ mm in length, width, and thickness, respectively, were cut from the cast composites using a CNC machine. The hardness test characterises the indentation hardness of materials by measuring the depth of penetration of the indenter point. The Barcol Impressor, model GYZJ-934-1 consists of a hardened steel truncated cone, with an angle of 26° and a flat tip of 0.157 mm diameter at the spring loading plunger, which is used to make indentation on materials.

2.6. Scanning Electron Microscopy. The surface morphologies of the composites of various fibre weight fractions for both treated and untreated composites were analysed using electron microscopy technique. The composite surfaces were analysed using the Zeiss Environmental SEM (ESEM: model EVO HD 15, operating at 20 kV), where the specimen was gold sputter coated using Quorum 150R ES model thin film coating equipment. The coating was applied in order to enable the specimens to become easily visible. The treated and untreated sisal fibres were scanned as well.

3. Results and Discussion

3.1. SEM Results of Untreated and Treated Sisal Fibres.

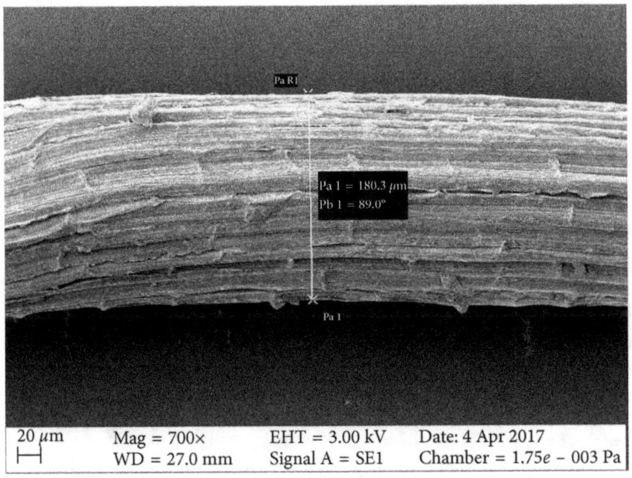

FIGURE 3: Treated sisal fibres.

The untreated and treated sisal fibres were subjected to scanning electron microscopy with the resulting images shown in Figures 2 and 3.

From the images, it is clear that the cross-sectional dimensions of the treated sisal fibres (180.3 μm) are smaller than that of the untreated sisal fibres (286.6 μm). A reduction in size of the treated sisal fibres leads to better mechanical properties of the resulting composites than those of the untreated sisal fibres due to an increase in number of reinforcing fibres (and therefore reinforcing fibre volume fraction) that can be packed in a cross section of a matrix.

3.2. Impact Test Results. The impact fracture surfaces were observed with a scanning electron microscope. Five scans were conducted for each sample, and representative micrographs are shown in Figures 4(a)–4(j) and 5.

When both the untreated and treated composites were compared at each weight fraction and the number of incidences of fibre pull-out counted in each case, it was found that there were higher incidences of fibre pull-out in the untreated sisal fibre-epoxy resin composites than in the treated sisal fibre-epoxy resin composites. Furthermore, the impact fracture surface of the epoxy resin exhibited a smooth fracture surface. Higher incidence of fibre pull-out is a clear indication of lower fibre-matrix adhesion. Lower interfacial adhesion results in the matrix absorbing higher fractions of the total energy imposed on a composite than would be the case where higher interfacial adhesion exists and, therefore, leads to lower values of impact energy, where the fracture toughness of the matrix is lower.

Average values of impact toughness for both the treated sisal fibre-epoxy resin composites and the untreated sisal fibre-epoxy resin composites which were determined using a Charpy impact test machine are presented in Table 1. The impact toughness of the pure epoxy resin was also determined and is recorded in the table as well.

The results in Table 1 clearly show that impact energy of reinforced composites was enhanced by the three combined treatment methods for the reinforcing fibres. Increase of impact energy due to an enhanced fibre-matrix adhesion was also reported by Ramires and Frollini [24]. The average impact energy of the treated sisal fibre-epoxy resin composites is seen in the table to have increased by 28.09%, 20.07%, 14.06%, 15.44%, and 22.34% for composites with 5 wt.%, 10 wt.%, 15 wt.%, 30 wt.%, and 50 wt.%, respectively, over the values obtained for the untreated sisal fibre-epoxy resin composites for the same weight fractions. The standard deviations of the impact energies for both the treated and the untreated composites are small, all falling within the range from 0.12 to 0.93, while the coefficient of variations for both the treated and untreated sisal fibre-epoxy resin composites are also low, ranging from 13.48% to 31.59%. The low values of the standard deviations and coefficient of variations implies that there was not much scatter in the experimental data obtained.

It is evident from the values in the above table that the average values of impact energy of the tested specimens increased from a value of 0.84 joules for pure epoxy resin to a maximum of 10.34 joules and 12.65 joules for a 50 wt.% untreated and treated sisal fibre-epoxy resin composites, respectively. These are 1130.95% and 1405.95% increases in the value of impact energy, respectively, over the values of pure epoxy resin. The increase of the average impact energy for the untreated and treated fibre reinforced composites over the values for the pure epoxy resin for the five percentage weights of 5 wt.%, 10 wt.%, 15 wt.%, 30 wt.%, and 50 wt.% were significant at 35.73%, 197.62%, 313.10%, 678.57%, and 1130.95% and 154.76%, 252.38%, 371.43%, 798.81% and 1405.95%, respectively, for the untreated and treated sisal fibre-epoxy resin composites. The higher values of impact energy for the treated sisal fibre-epoxy resin composites over the untreated sisal fibre-epoxy resin composites can be attributed to the improved mechanical interlocking due to rougher surfaces of the treated sisal fibres and improved interfacial bonding.

The values of impact toughness shown in Table 1 are plotted in Figure 6.

The curve plots in Figure 6, for both the treated and untreated sisal fibre-epoxy resin composites depict a continuous increase of the impact toughness with an increase in fibre loading. The magnitude of the impact energy of the plotted curve for the treated sisal fibre-epoxy resin composites is higher than that for the untreated sisal fibre-epoxy resin composites. This difference in magnitude of the two curves is because of the difference in reinforcing effects of the treated and untreated sisal fibre. In addition, both curves do not have minimum points. The absence of minimum points in both curves is because the range of percentage weights at which tests were done did not include those at or around the minimum point. The correlation coefficient in both graphs is very unity, implying a near perfect curve fit of the curves to the experimental data plotted in both case.

3.3. Hardness Test Results. The results arising from the measurements of hardness are presented in Table 2 for the treated and untreated sisal fibre-epoxy resin composites. The results at 0 wt.% represent pure epoxy resin.

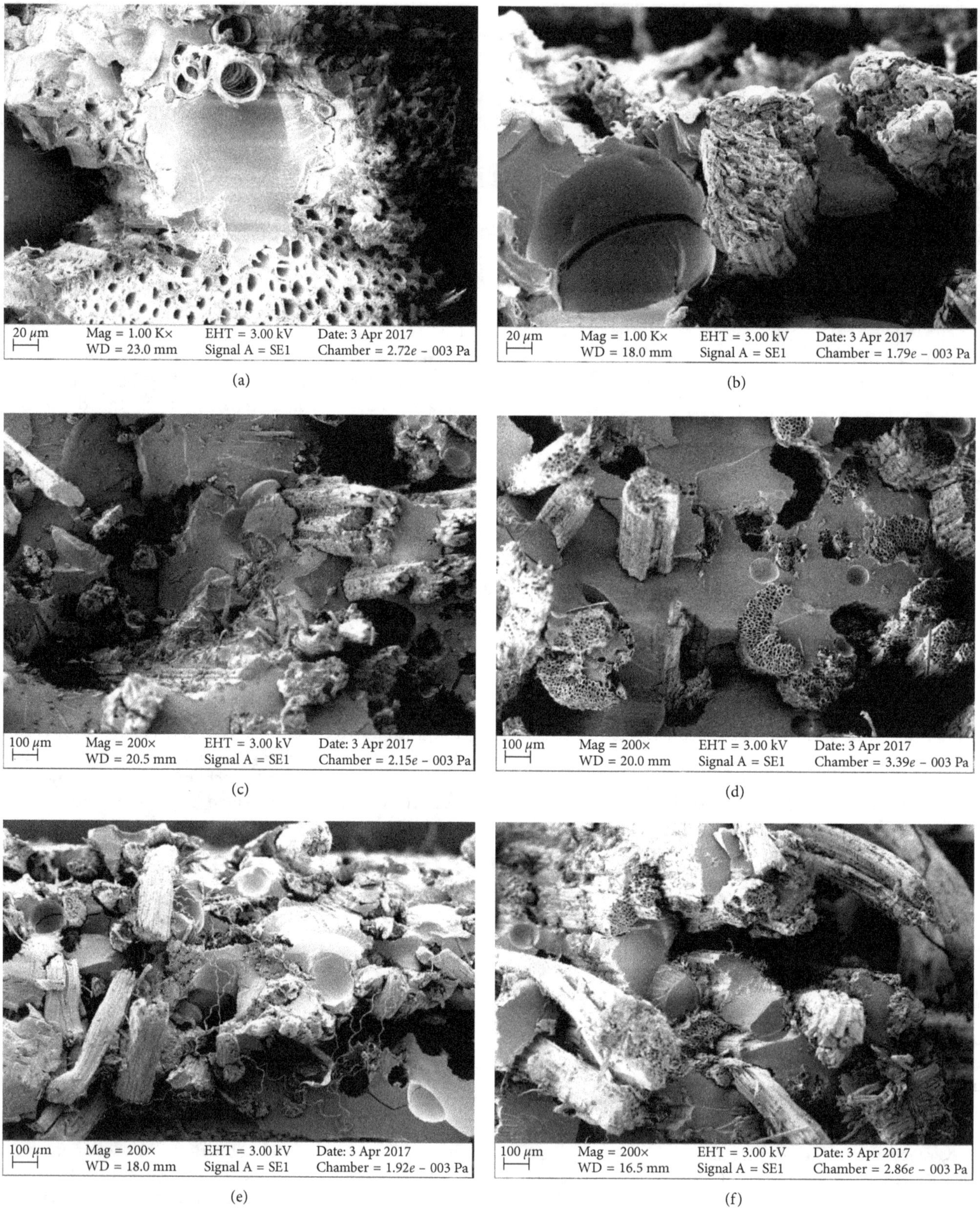

FIGURE 4: Continued.

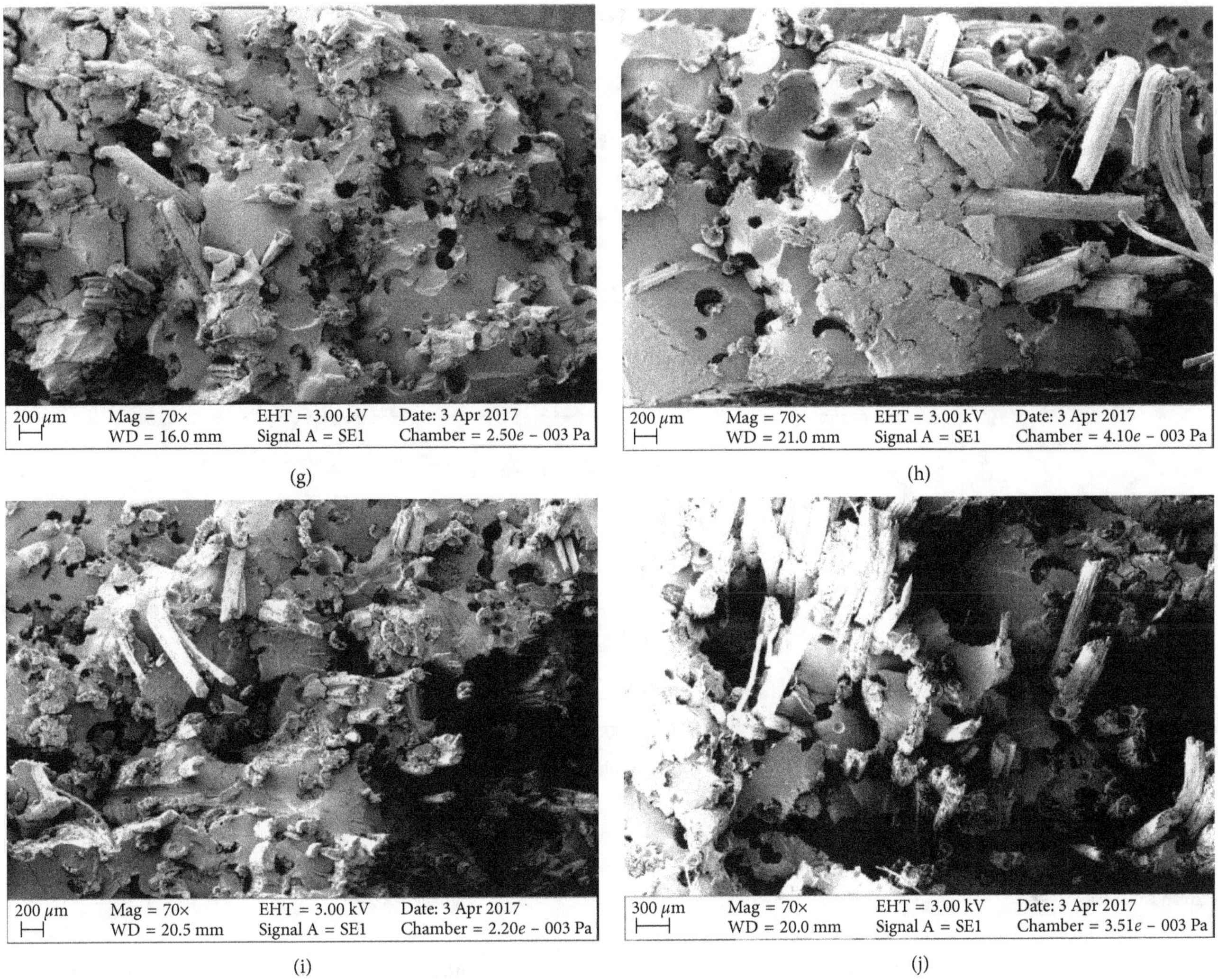

FIGURE 4: Impact scans of epoxy resin and untreated and treated sisal fibre-epoxy resin composites. (a) 5 wt.% untreated sisal fibre-epoxy resin reinforced composite. (b) 5 wt.% treated sisal fibre-epoxy resin reinforced composite. (c) 10 wt.% untreated sisal fibre-epoxy resin reinforced composite. (d) 10 wt.% treated sisal fibre-epoxy resin reinforced composite. (e) 15 wt.% untreated sisal fibre-epoxy resin reinforced composite. (f) 15 wt.% treated sisal fibre-epoxy resin reinforced composite. (g) 30 wt.% untreated sisal fibre-epoxy resin reinforced composite. (h) 30 wt.% treated sisal fibre-epoxy resin reinforced composite. (i) 50 wt.% untreated sisal fibre-epoxy resin reinforced composite. (j) 50 wt.% treated sisal fibre-epoxy resin reinforced composite.

Pure epoxy at 0 wt.% and treated and untreated sisal fibre-epoxy resin composites were tested at ten different points on their respective surfaces and their average values calculated. All the samples showed statistical similarity as indicated by the small values of coefficient of variation in the table. This similarity indicates the efficiency of the vacuum resin infusion manufacturing process of preparing composites with reference to the presence of pores and distribution of fibres.

The 50 wt.% sisal fibre-epoxy resin composite had the highest value of hardness of 45.00 BU for the untreated sisal fibre-epoxy resin composites and 49.00 BU for the treated sisal fibre-epoxy resin composites with a value of standard deviation of 0.82 for both sets of composites, and coefficients of variation of 2.72 and 2.15, respectively. These low values of standard deviation and coefficients of variation imply that there was no much scatter in the values obtained.

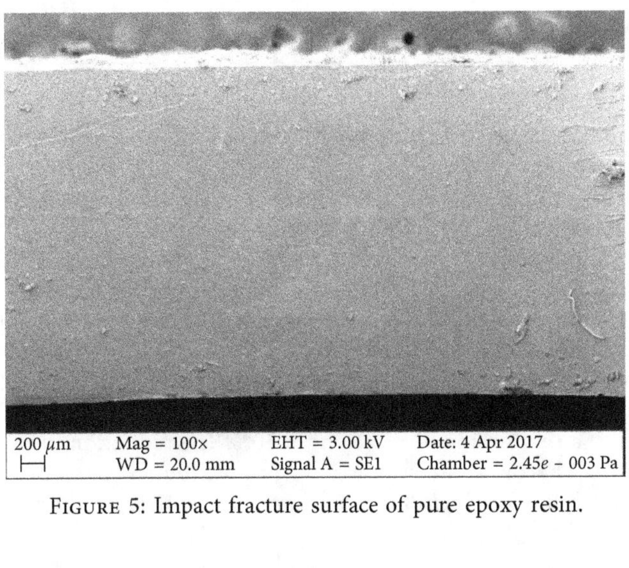

FIGURE 5: Impact fracture surface of pure epoxy resin.

Table 1: Values of impact toughness for pure epoxy as well as treated and untreated sisal fibre-epoxy resin composites.

Weight % of the reinforcing fibre	Impact energy of untreated sisal fibre-epoxy resin composites			Impact energy of treated sisal fibre-epoxy resin composites			% increase of impact energy
	Range (joules)	Average (joules)	Coefficient of variation (%)	Range (joules)	Average (joules)	Coefficient of variation	
Epoxy (0)	0.54–1.22	0.84 ± 0.24	28.33	0.54–1.22	0.84 ± 0.24	28.33	—
5	0.81–1.89	1.14 ± 0.29	25.47	1.63–2.44	2.14 ± 0.29	13.48	28.09
10	2.30–2.71	2.50 ± 0.12	14.86	1.08–4.75	2.96 ± 0.93	31.59	20.07
15	1.36–4.61	3.47 ± 0.86	24.81	2.03–5.56	3.96 ± 0.84	21.32	14.06
30	5.28–7.86	6.54 ± 0.67	22.45	6.34–8.15	7.55 ± 0.67	25.47	15.44
50	8.76–11.74	10.34 ± 0.74	18.44	10.77–13.15	12.65 ± 0.82	25.22	22.34

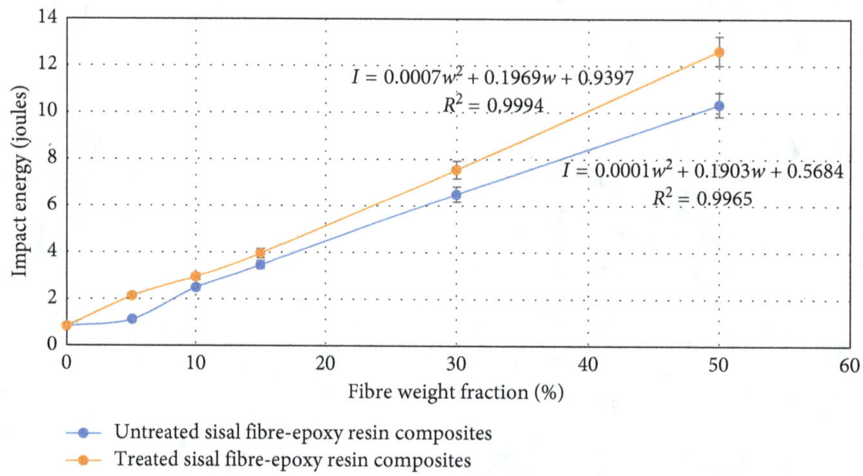

Figure 6: A plot of the average values of impact energy versus fibre weight fraction % for treated and untreated sisal fibre-epoxy resin reinforced composites.

Table 2: Barcol values of hardness for pure epoxy resin and treated and untreated sisal fibre-epoxy resin composites.

Weight % of the reinforcing fibre	Values of hardness for the untreated sisal fibre-epoxy resin composites (Barcol units-BU)			Values of hardness for the treated sisal fibre-epoxy resin composites (Barcol units-BU)		
	Range (BU)	Average (BU)	Coefficient of variation (%)	Range (BU)	Average (BU)	Coefficient of variation (%)
Epoxy (0)	25.00–27.00	26.00 ± 0.94	3.63	25.00–27.00	26.00 ± 0.94	3.63
5	27.00–31.00	29.00 ± 1.41	4.88	29.00–31.00	30.00 ± 0.94	3.14
10	29.00–31.00	30.00 ± 0.82	2.72	37.00–39.00	38.00 ± 0.82	2.15
15	27.00–31.00	29.00 ± 1.49	5.14	27.00–31.00	29.00 ± 1.56	5.39
30	38.00–41.00	39.00 ± 1.77	3.46	42.00–46.00	43.00 ± 1.43	3.54
50	44.00–47.00	45.00 ± 1.65	4.35	48.00–51.00	49.00 ± 1.32	3.26

The values of hardness presented in Table 2 are plotted in Figure 7.

From Figure 7, it is clear that there is an initial increase in the values of hardness of the treated and untreated sisal fibre-epoxy resin composites up to maximum values of 30 and 38 BU respectively, both at a reinforcing fibre weight of 10% fibre. After this, the values of hardness reduce down to a value of 29 BU at a reinforcing fibre weight of 15%, which is higher than the value of 26 BU for pure epoxy resin. Thereafter, the values of hardness increase continuously upto a reinforcing fibre weight of 50%. The reduction in the Barcol hardness after 10 wt.% can be attributed to the poor wettability between the fibres and the matrix leading to agglomeration as the fibre weight percentage increases upto 15 wt.%. From 15 wt.%, the values of hardness increases continuously upto 50 wt.% due to more even distribution of the reinforcing fibres and a corresponding increase in the stiffness of the composites. The correlation coefficient in both graphs is near unity, implying a near perfect curve fit to the experimental data plotted in both cases.

3.4. Comparison to Other Published Works. It is interesting to compare the current results with the previous works. Table 3 summarises the available impact and hardness properties of polymeric composites based on natural fibres.

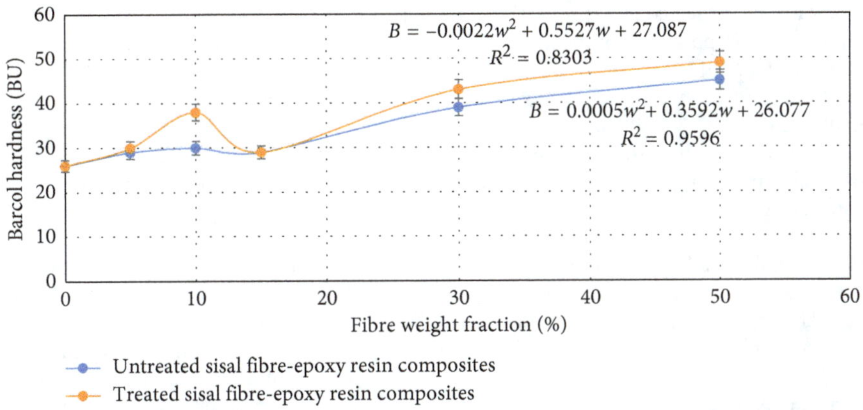

Figure 7: A plot of average values of the Barcol hardness (BU) versus fibre weight fraction % for treated and untreated sisal fibre-epoxy resin composites.

Table 3: A summary of impact strength and hardness of selected natural fibre composites.

Matrix	Fibre (type of orientation)	Fibre content (%)	Type of treatment	Impact strength (joules)	Hardness (Barcol units)	Reference
PLA	Sisal (random)	30.00	None	3.30	—	[29]
PLA/PEG-plasticized	Sisal	30.00	None	4.50	—	[29]
UFR	Sisal (random)	50.00	Alkali, MAPP	9.50	—	[30]
PP	Wood (flour)	30.00	Alkali, MAPP	5.00	—	[30]
PP/EPR/MAPP	Wood (flour)	30.00	Alkali, MAPP	7.00	—	[30]
PP/MAPP	Wood (flour)	30.00	Alkali, MAPP	5.00	—	[30]
Epoxy	Areca fibre	40.00	Alkali	—	45.00	[31]
Epoxy	Sisal (unidirectional)	0.00–50.00	Alkali, silane, acid hydrolysis	0.84–12.65	26.00–49.00	Present study

It is evident from Table 3 that the current results are an improvement of 7%–56% for impact and 9% for hardness on the values obtained from the previous results. Clearly, the combined chemical treatment of mercerisation, silane, and acid hydrolysis, leads to an improvement in the mechanical properties of impact and hardness of the order of 7%–56% for impact and 9% for hardness of sisal fibre reinforced epoxy resin composite.

4. Conclusions

The following conclusions were made:

(1) Chemical treatment of sisal fibre surfaces leads to continuous increases in the impact toughness of the treated sisal fibre-epoxy resin composites with increasing weight fraction of the reinforcing fibres.

(2) Chemical treatment of sisal fibre surfaces leads to an increase in the hardness of the sisal fibre-epoxy resin composites.

(3) The hardness of both the treated and untreated sisal fibre-epoxy resin composites increased gradually with an increase in fibre loading up to a maximum fibre weight fraction of 10% and then decreased beyond here up to fibre weight content of 15% as a result of fibre bunching, thereafter increasing gradually up to fibre weight fraction of 50% due to an increase in the spatial distribution of the reinforcing fibre and increase in the stiffness of the resulting composite.

(4) The combined chemical treatment of mercerisation, silane, and acid hydrolysis leads to an improvement in the mechanical properties of impact and hardness that is higher than in cases where only one treatment process is used on its own.

Conflicts of Interest

The authors have no conflicts of interest to disclose in relation to the current research.

Acknowledgments

The research was supported by Vaal University of Technology, South Africa.

References

[1] G. Jin and M. C. Jeffrey, "Polylactic acid composites incorporating casein functionalized cellulose nanowhiskers," *Journal of Biological Engineering*, vol. 7, no. 1, p. 31, 2013.

[2] R. A. V. Prasad and K. M. Rao, "Mechanical properties of natural fibre reinforced polyester composites: jowar, sisal and bamboo," *Materials and Design*, vol. 32, no. 8-9, pp. 4658–4663, 2011.

[3] M. Sakthivei and S. Ramesh, "Mechanical properties of natural fibre (banana, coir, sisal) polymer composites," *Science Park*, vol. 1, no. 1, pp. 1–6, 2013.

[4] F. C. Campbell, *Introduction to Composite Materials*, ASM International, Materials Park, OH, USA, 2010.

[5] M. Maringa and C. G. Injairu, "A study of the elastic behavior of some natural materials under applied bending loads," *Journal of Civil Engineering and Practice*, vol. 7, no. 1, pp. 13–46, 2010.

[6] S. Y. Fu, X. Q. Feng, B. Lauke, and Y. W. Mai, "Effect of particle size, particle-matrix adhesion and particle loading on mechanical properties of particulate-polymer composites," *Composites Part B: Engineering*, vol. 39, no. 6, pp. 933–961, 2012.

[7] R. Kaundal, P. Amar, and B. Sandhyarani, "Effect of red mud and copper slag particles on physical and mechanical properties of bamboo-fibre-reinforced epoxy composites," *Journal of Advances in Mechanical Engineering*, vol. 4, article 141248, 2012.

[8] N. Chawla and Y. Shen, "Mechanical behavior of particle reinforced metal matrix composites," *Journal of Advanced Engineering Materials*, vol. 3, no. 6, pp. 357–370, 2001.

[9] D. M. Panaitescu, N. F. Adriana, G. Marius, I. S. Catalin, R. Constantin, and D. I. Michaela, "Properties of polymer composites with cellulose microfibrils," in *Advances in Composite Materials*, pp. 103–122, National Institute of Research and Development in Chemistry and Petrochemistry, Bucharest, Romania, 2013.

[10] A. K. Mohanty, M. Misra, and L. T. Drzal, "Surface modification of natural fibres and performance of the resulting bio composites: an overview," *Composite Interfaces*, vol. 8, no. 5, pp. 313–343, 2001.

[11] K. Joseph, S. Vaeghese, G. Kalaprasad, S. Thomas, L. Prassanakumar, and P. Koshy, "Influence of interfacial adhesion on the mechanical properties and fracture behavior of short sisal fibre reinforced polymer composites," *European Polymer Journal*, vol. 32, no. 10, pp. 1243–1250, 1996.

[12] S. Raghavendra, P. Lingaraju, S. Balachandra, and P. G. Mukunda, "Mechanical properties of short banana fibre reinforced natural rubber composites," *International Journal of Innovative Research in Science, Engineering and Technology*, vol. 2, no. 5, pp. 1652–1655, 2013.

[13] C. Giuseppe, A. Latteri, R. Giuseppe, and C. Gianluca, "Composites based on natural fibre fabrics," in *Woven Fabric Engineering*, pp. 317–342, University of Catania, Catania, Italy, 2008.

[14] S. Rassman, "Effects of processing conditions on the mechanical and water absorption properties of resin transfer moulded kenaf fibre reinforced polyester composite laminates," *Composite Part A: Applied Science and Manufacturing*, vol. 41, no. 11, pp. 1612–1619, 2010.

[15] H. N. Dhakal, Z. Y. Zhang, and M. O. W. Richardson, "Effect of water absorption on the mechanical properties of hemp fibre reinforced unsaturated polyester," *Advanced Polymer and Composites*, vol. 67, no. 7-8, pp. 1674–1683, 2007.

[16] X. Li, G. Lope, and P. Satyanarayan, "Chemical treatments of natural fibre for use in natural fibre-reinforced composites: a review," *Composite Part A*, vol. 15, no. 1, pp. 25–33, 2007.

[17] N. Mokaloba and R. Batane, "The effects of mercerization and acetylation treatments on the properties of sisal fibre and its interfacial adhesion characteristics on polypropylene," *International Journal of Engineering, Science and Technology*, vol. 6, no. 4, pp. 83–97, 2014.

[18] J. Andersons and K. Tillman, "Strength and adhesion characteristics of elementary flax fibres with different surface treatment," *Applied Science Manufacturing*, vol. 34, no. 7, pp. 603–612, 2001.

[19] E. T. N. Bisanda and M. P. Ansell, "The effect of silane treatment on the mechanical and physical properties of composites," *Journal of Composite Science and Technology*, vol. 41, no. 2, pp. 165–178, 1991.

[20] P. J. Herrera-Franco and A. Valadez-Gonzalez, "A study of the mechanical properties of short natural-fiber reinforced composites," *Composites Part B: Engineering*, vol. 36, no. 8, pp. 597–608, 2005.

[21] M. D. Joshua, "Impact testing of advanced composites," *Journal of Advanced Topics in Characterization of Composites*, vol. 65, no. 5, pp. 97–112, 2008.

[22] F. A. Silva, N. Chawla, and R. D. Toledo Filho, "An experimental investigation of the fatigue behaviour of sisal fibres," *Journal of Material Science and Engineering A*, vol. 516, no. 1-2, pp. 90–95, 2009.

[23] B. Wang, S. Panigrahi, L. Tabil, and W. Crerar, "Pretreatment of flax fibres for use in rotationally moulded biocomposites," *Journal of Reinforced Plastics and Composites*, vol. 26, no. 5, pp. 447–463, 2007.

[24] E. C. Ramires and E. Frollini, "Tannin-phenolic resins: synthesis, characterization and application in bio based composites reinforced with sisal fibres," *Composites Part B: Engineering*, vol. 43, no. 7, pp. 2836–2842, 2012.

[25] Y. Han-Seung, M. P. Walcott, S. K. Hee-sookim, and K. Hyunjoong, "Effect of different compatibilizing agents on the mechanical properties of lignocellulosic material filled polyethylene bio composites," *Journal of Composite Structures*, vol. 79, no. 3, pp. 369–375, 2007.

[26] M. A. Kumar, G. R. Reddy, Y. S. B. Naidu, and V. N. P. Naidu, "Frictional co-efficient, hardness, impact strength and chemical resistance of reinforced sisal glass fibre epoxy hybrid composites," *Journal of Composite Materials*, vol. 44, no. 26, pp. 3195–3202, 2010.

[27] A. H. Saputra and G. Setyarso, "Vacuum infusion equipment design and the influence of reinforcement layers addition to the resin infusion time," *Journal of Polymers*, vol. 162, no. 1, pp. 1–6, 2017.

[28] S. Beck-candanedo, M. Roman, and D. G. Gray, "Effect of reaction conditions on the properties and behavior of wood cellulose nanocrystal suspensions," *Journal of Bio-Macromolecules*, vol. 6, no. 2, pp. 1048–1054, 2005.

[29] M. K. Gupta and R. K. Srivastava, "Properties of sisal fibre reinforced epoxy composites," *Indian Journal of Fibre and Textile Research*, vol. 41, pp. 235–241, 2016.

[30] G. Keledi, A. Sudar, C. Burgstaller, K. Renner, and B. Pukanszky, "Tensile and impact properties of three-component PP/wood/elastomer composites," *Express Polymer Letters*, vol. 6, no. 3, pp. 224–226, 2012.

[31] C. V. Srinivasa and K. N. Bharath, "Impact and hardness properties of areca fibre-epoxy reinforced composites," *Journal of Material Environmental Science*, vol. 2, no. 4, pp. 351–356, 2011.

3

Characteristics of Biodegradable Polylactide/ Thermoplastic Starch/Nanosilica Composites: Effects of Plasticizer and Nanosilica Functionality

Regina Jeziorska ,[1] Agnieszka Szadkowska,[1] Ewa Spasowka,[1] Aneta Lukomska,[2] and Michal Chmielarek[3]

[1]*Department of Polymer Technology and Processing, Industrial Chemistry Research Institute, Rydygiera 8 St., 01-793 Warsaw, Poland*
[2]*Department of Biomedical Technology, Cosmetic Chemicals and Electrochemistry, Industrial Chemistry Research Institute, Rydygiera 8 St., 01-793 Warsaw, Poland*
[3]*Faculty of Chemistry, Warsaw University of Technology, Koszykowa 75 St., 00-662 Warsaw, Poland*

Correspondence should be addressed to Regina Jeziorska; regina.jeziorska@ichp.pl

Academic Editor: Rajaram S. Mane

The effect of plasticizer (polydimethylsiloxanol) and neat (SiO_2) or modified (having amine functional groups) silica (A-SiO_2) on morphology, thermal, mechanical, and rheological properties of PLA/TPS blends compatibilized by maleated PLA (MPLA) was investigated. Toughened PLA/MPLA/TPS (60/10/30) blend containing 3 wt.% of plasticizer and various contents (1, 3, or 5 wt.%) of silica were prepared in a corotating twin-screw extruder. From SEM, it is clear that plasticized PLA/MPLA/TPS blend continuous porous structure is highly related to the silica content and its functionality. The results indicate that polydimethylsiloxanol enhances ductility and the initial thermal stability of the plasticized blend. DSC and DMTA analyses show that nucleation ability and reinforcing effect of A-SiO_2 on plasticized blend crystallization are much better than those of SiO_2. Silica practically had no effect on the thermo-oxidative degradation. However, the composites with A-SiO_2 had better thermal stability than those with SiO_2. Moreover, silica significantly improved the elongation at break.

1. Introduction

Polylactide (PLA), a biodegradable linear aliphatic polyester, is one of the most potential materials with great environmental benefits since it offers very low toxicity and high mechanical performance. Therefore, PLA can be broadly applied in the fields of drug delivery systems, sutures, orthopaedic implants, tissue engineering, and packaging materials [1–4]. However, a low glass transition temperature (about 60°C), relatively high brittleness, and higher cost, if compared to conventional polymers, limit its applications [5]. There are many ways of improving PLA toughness including copolymerization and melt blending with other more flexible commodity polymers such as linear low-density polyethylene or natural rubbers, as well as plasticizers [6–10]. Generally, an efficient plasticizer decreases the glass transition temperature (T_g) and melting point (T_m) of polymers. It was reported that the addition of plasticizers such as polyethylene glycol, citrate, oligomeric lactic acid, and triacetine could overcome PLA's brittleness [11–14]. Unluckily, plasticization decreases PLA stiffness, limiting its constructional applications. Therefore, an efficient impact modifier is required to improve toughness without extensive loosening of PLA stiffness.

Thermoplastic starch (TPS) is one of the well-known biopolymers and is often used in order to lower the cost of the final product and enhance the biodegradable characteristics of polymer composites [15–17]. Świerz-Motysia et al. reported that TPS increased the biodegradation rate of PLA in so-called soil test, which was enhanced as a function of TPS [18]. Thus, melt compounding TPS with PLA is one of the most promising methods to solve PLA limitations or to lower its high price [19]. Nevertheless, incompatibility between hydrophobic PLA and hydrophilic starch results in poor mechanical properties, due to inadequate interfacial interaction between these polymers. To improve the adhesion of PLA and TPS, chemical modification of the starch before blending is performed and compatibilizers such as maleic anhydride (MA) and epoxy resins have been used [4, 20–24].

It has been previously suggested that the intensive interfacial interactions among hydrogen bonds of the anhydride groups of the maleated polylactide (MPLA) and the hydroxyl groups of the starch are responsible for the increase in tensile, flexural, and impact properties of PLA and thermoplastic potato or corn starch blends [25].

Recent results show that the nanoscale distribution of montmorillonite [18, 26–28], fumed silica (SiO_2) [29–32], and calcium carbonate [33] within the PLA matrix significantly enhances their thermal, mechanical, and rheological properties, as well as reduces permeability and flammability, compared to conventional micro- and macrocomposites. In order to reach the nanoscale distribution, the naturally hydrophilic nanofiller has to be organically modified to be more compatible with the organic polymer matrix. However, only few studies have been conducted in the development of plasticized PLA/thermoplastic starch nanocomposites recently [34–36].

In a previous work [37], we have studied the influence of the core-shell polymeric nanofiller and maleated polylactide, used as a compatibilizer, on the structural, mechanical, and barrier properties of PLA and TPS blends. The addition of the core-shell polymeric nanofiller in the presence of MPLA improved adhesion between PLA and starch. Melting point and crystallinity temperature of PLA were found to decrease by adding TPS and the polymeric nanofiller. The nanocomposites showed significantly lower stiffness compared to neat PLA. However, the storage modulus increased as a function of nanofiller content. Moreover, the addition of the core-shell polymeric nanofiller strongly improved impact strength, the elongation at break and barrier properties of PLA/TPS blends.

To the best of our knowledge, no systematic studies have been done so far to investigate the properties of polydimethylsiloxanol toughened PLA/TPS/silica nanocomposites. Therefore, the objective of this paper is to investigate the thermal, mechanical, rheological, and morphological properties of polydimethylsiloxanol toughened PLA/TPS/silica nanocomposites in the presence of MPLA used as a compatibilizer. In particular, the effects of plasticizer, spherical silica content, and its functionality were also studied.

2. Materials and Methods

2.1. Materials. Polylactide polymer (PLA, 2003D) was provided by Nature Works, USA, with melt flow rate of 6.9 g/10 min (ISO 1133 at 210°C and 2.16 kg). The characteristics of PLA are described in Table 1. Maleated polylactide (MPLA) containing 0.68 wt.% of grafted maleic anhydride, with melt flow rate of 3.8 g/10 min (ISO 1133 at 210°C and 2.16 kg), was obtained according to the procedure published elsewhere [18, 38], and it was used as a compatibilizer for PLA/TPS blends at the concentration of 10 wt.%. Maleic anhydride (MA) as coupling agent and dicumyl peroxide (DCP) as initiator were purchased from Sigma-Aldrich, Italy.

Thermoplastic corn starch (TPS) was obtained by melt blending according to the procedure published elsewhere [18, 39] and used as a biodegradable additive. Standard corn starch with about 23% of amylase, 11–13% humidity, and pH 6.5 was purchased from Cargill, Germany. The glycerol with 99.5% purity was the product of Rafineria Trzebinia, Poland. The characteristics of TPS are shown in Table 1.

Plastosil M-2000, the commercial name of polydimetylsiloxanol, containing 0.04 wt.% of reactive silanol groups, $\eta = 2000 \pm 500$ cP, $d = 0.97$ g/cm^3, supplied by Chemical Plant Polish Silicones Ltd., was used as a plasticizer (P) for the PLA/MPLA/TPS (60/10/30) blend.

Neat (SiO_2) and modified (having amine functional groups) spherical silica (A-SiO_2) nanoparticles synthesized according to the previously reported sol-gel process [40–42] were used as nanofillers. Tetraethoxysilane (TEOS), technical grade, commercial product name TES 28, was supplied by Wacker Chemie (Germany). Ethyl alcohol (reagent grade) and aqueous ammonia (reagent grade, 25 wt.%, $d = 0.91$ g/cm^3) were supplied by POCh S.A. (Poland) and used as received. The characteristics of silica are described in Table 2.

2.2. PLA Functionalization. In order to improve compatibility between PLA and TPS, PLA was grafted with maleic anhydride via reactive extrusion to obtain a grafting yield of 0.68 wt.% according to the procedure published elsewhere [18, 38]. MPLA was prepared as follows: PLA, MA (2 wt.%, PLA basis), and DCP (18 wt.%, MA basis) were mixed in the internal mixer and, subsequently, the mixture was extruded using a corotating twin-screw extruder (Berstorff, Germany) with a screw diameter of 25 mm and length to diameter ratio of 33. To prevent the degradation of the polymer, during the processing time, a nitrogen purge flow was used. Different screw elements along the screw worked in order to induce polymer melting [43]. The three mixing sections enhanced the compounding and increased the residence time of the mixture in the barrel. The barrel pressure in these parts, as well as at the section before the die, could be increased. The extruder also had a vacuum degassing port to remove any moisture traces or other volatile products formed during compounding. The process was carried out at a barrel temperature 130–140°C. The output was 4 kg/h, and the screw speed was 100 rpm.

2.3. TPS Preparation. To obtain TPS, destructurization and plasticization of native corn starch were performed in one-step extrusion process in the earlier mentioned Berstorff twin-screw corotating extruder, using 30 wt.% of glycerol as a plasticizer [18, 39]. The process was carried out at a barrel temperature 130–170°C and screw speed of 80 rpm. The TPS

TABLE 1: Characteristics of PLA and TPS.

Sample	Tensile strength (MPa)	Elongation at break (%)	Tensile modulus (MPa)
PLA	67 ± 1.5	6.1 ± 0.3	3930 ± 64
TPS	19 ± 0.7	2.7 ± 0.6	1550 ± 60

TABLE 2: Characteristics of nanosilica fillers.

Nanosilica	Size (nm)	Polydispersity	Surface area (m^2/g)	Amine groups content (wt.%)
Neat silica	30	0.16	274.4	0
Modified silica	30	0.02	274.4	0.35

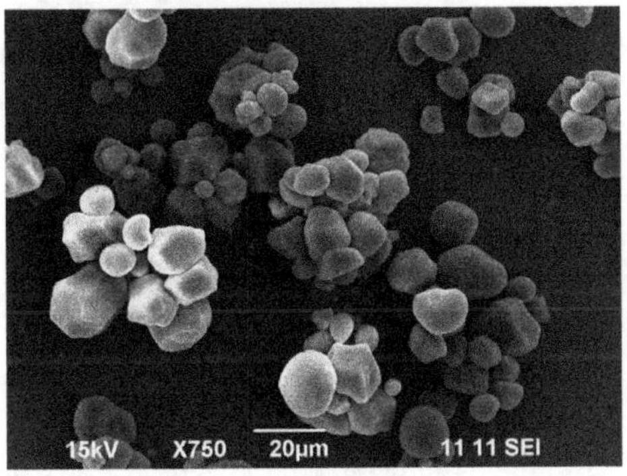

(a)

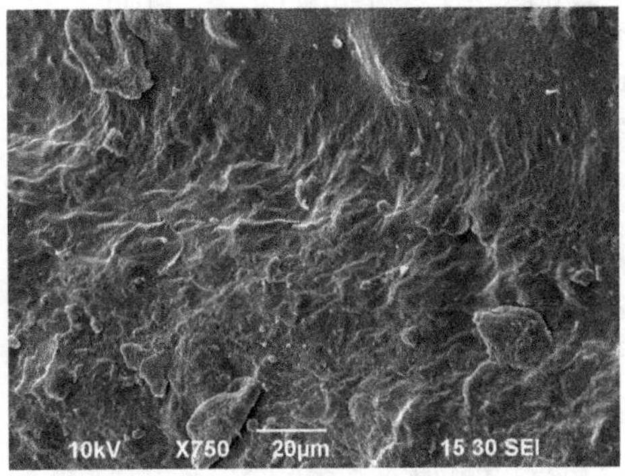

(b)

FIGURE 1: SEM images of (a) native corn starch and (b) thermoplastic corn starch.

was cooled in ambient air and then was pelletized. Figure 1 shows SEM micrographs of native and thermoplastic corn starch. It is obvious from Figure 1 that the starch particles grains disappeared after compounding with a glycerol, suggesting that thermoplastic starch was successfully plasticized and TPS presented a homogeneous morphology.

2.4. Synthesis of Neat and Modified Spherical Nanosilica. Neat (SiO_2) and modified (heavy amine functional groups) spherical silica (A-SiO_2) were synthesized according to the developed sol-gel process using tetraethoxysilane (TEOS) as alkoxysilane precursor at room temperature (23°C), with a stirring speed of 250 rpm [40–42]. The process was carried out in an aqueous ammonia-ethyl alcohol reaction mixture using molar ratio of TEOS/EtOH/H_2O as 0.023/0.500/0.477, and the initial pH ranged from 10.4 to 11.3. The final pH range was 7.5–10.8. The pH of the reaction mixture was measured using a pH meter (Schott Instruments LAB 850). Modified silica was synthesized by adding drop by drop γ-aminopropyltriethoxysilane (Momentive Performance Materials, USA) to the reaction mixture, according to the reaction (Figure 2).

The final product of neat or modified silica was dried in a spray dryer for 2 hours at 50–90°C. Particle size and particle size distribution in resulting sols were measured by photon correlation spectroscopy (PCS) using a Malvern apparatus (Zetasizer Nano ZS, UK). The monomodal particle size distribution and very low polydispersity of particle size were observed for homogeneous sols obtained by the sol-gel process [40]. It is clear from Figure 3 that spherical shape and uniform sized (about 30 nm) silica nanoparticles were obtained. The amine groups content was determined based on nitrogen content measurement by the Kjeldahl method. The characteristics of the neat and modified silica are presented in Table 2.

2.5. Blend Preparation. The PLA/MPLA/TPS 60/10/30 blends without or with 3 wt.% silicone rubber plasticizer were melt blended using the earlier mentioned twin-screw corotating extruder. PLA, MPLA, TPS, and plasticizer (P) were fed into the throat of the extruder using separate gravimetric feeders. The temperatures of the ten zones were 35°C/160°C/175°C/180°C/180°C/180°C/180°C/185°C/185°C, and 190°C for the sequential heating zones, from the hopper to the die. The screw speed was 100 rpm. The melt temperature and pressure were continuously recorded during compounding. The extrudate was immersed immediately in a cold-water bath (20°C) and pelletized with an adjustable rotating knife located behind the water bath, into 5 mm pellets.

2.6. Composite Preparation. The various composites containing 1, 3, or 5 wt.% of SiO_2 or A-SiO_2 silica nanoparticles were prepared through melt-direct dispersion by using a conventional polymer extrusion process and optimized parameters. Firstly, PLA pellets with 10 wt.% of silica were blended using the earlier mentioned twin-screw corotating extruder, based on the reported method [44]. Prior to the

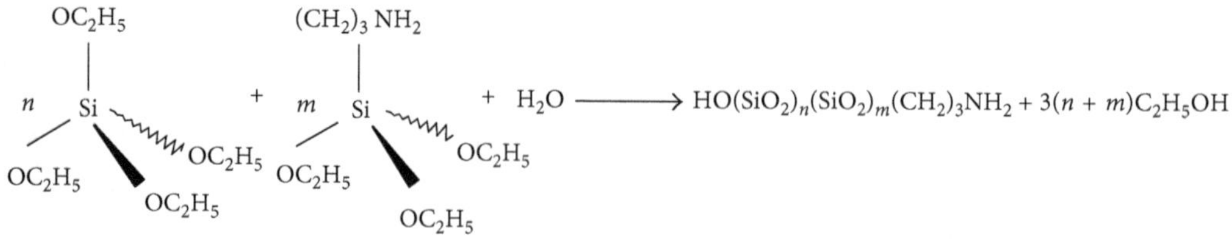

FIGURE 2: The reaction scheme for the preparation of A-SiO_2.

melt processing, this masterbatch was dried for 12 hours at 80°C and then melt blended with pure PLA, MPLA, TPS, and plasticizer. The process was carried out using a screw speed of 150 rpm and also a temperature profile of 35°C/165°C/175°C/180°C/180°C/180°C/175°C/185°C/185°C, and 200°C for the sequential heating zones, from the hopper to the die. Then, the material was cooled in water and pelletized.

2.7. Fourier Transform Infrared (FTIR) Spectra and the Grafting Degree. Fourier transform infrared spectroscopy (FTIR) spectra were recorded using a Thermo Scientific spectrometer, model Nicolet 6700 for a frequency range between 4000 cm^{-1} and 500 cm^{-1}. An average of 64 scans at a resolution of 2 cm^{-1} was conducted at room temperature. To collect the spectra of the polymers and composites, thin films were prepared in a hydraulic hot press. The nanosilica spectrum was taken using KBr pellets. The spectra presented were baseline corrected and converted to the absorbance mode.

The MA grafting degree was measured by Fourier transform infrared (FTIR) spectroscopy. FTIR was performed on a spectrometer PerkinElmer System 2000 on films 0.07 mm thick. The spectra were obtained by collecting 64 scans between 500 and 4000 cm^{-1} with a resolution of 2 cm^{-1}. The absorbance of the analytical band was determined, and the MA content was calculated from the calibration curve. The measured grafting yield was 0.68 wt.% [18].

2.8. Scanning Electron Microscopy. The microstructures of silica nanoparticles, TPS, blends, and composites were examined using a JEOL JSM-6490LV scanning electron microscope (SEM). Fractured surface of TPS, blends, and composites were gold-coated prior to observation to avoid electrical charging and to increase image contrast.

2.9. Differential Scanning Calorimetry. Thermal analysis was performed by differential scanning calorimetry (DSC) using a DSC-7A apparatus of PerkinElmer (Switzerland) under nitrogen. All measurements were carried out according to the following cycle: heating from 20 to 180°C at a rate of 10°C/min, 3 min isothermal step at 180°C, cooling down to 0°C at a rate of −10°C/min, 3 min isothermal step at 0°C, and final heating up from 0 to 180°C at a rate of 10°C/min. The amount of sample placed in the DSC aluminium pans was about 6 mg. An empty pan was used as a reference. The glass transition temperature (T_g), crystallization temperature (T_c), and melting temperature (T_m) were determined from the second heating scans. The degree of crystallinity (X_c) of PLA, blends, and composites was evaluated from the melting

FIGURE 3: SEM micrograph of A-SiO_2.

enthalpy results (ΔH_m) of each sample using (1), where ΔH_m is the experimental melting enthalpy and ΔH_m° is the melting enthalpy for 100% crystalline PLA, 93 J/g [45]:

$$X_c = \left(\frac{\Delta H_m}{\Delta H_m^{\circ} \left(1 - \left(\%_{\text{wt.filer}}/100\right)\right)} \right) 100. \quad (1)$$

2.10. Thermogravimetric Analysis. Thermogravimetric analysis (TGA) was performed using a thermogravimetric analyzer (TGA/SDTA 851e Mettler Toledo) at a rate of 10°C/min from 20 to 600°C under flowing air (50 ml/min). The precision on temperature measurements is ±0.5°C.

2.11. Tensile Properties. The test specimens were prepared by injection moulding using an Arburg 420 M single screw injection machine (Allrounder 1000-250, Germany) containing five different heating zones. The temperatures of these were 180/190/195/195/200°C, from the feeding zone to the die, when the mould was cooled with water at 25°C.

Tensile strength and elongation at break were determined using an extensometer clip-on incremental (Instron series 5500 R, UK) at a cross-head speed of 5 mm/min, whereas tensile modulus was measured at the speed of 2 mm/min. All tests were performed at standard atmosphere conditions (23°C and 50% HR). Prior to testing, the samples were stored at 23°C and 50% RH for 48 h, according to ISO 527 and ISO 179 standards. All the results represent an average value of a minimum 5 tests.

2.12. Dynamic-Mechanical Thermal Analysis (DMTA). The dynamic mechanical properties of samples were tested using a dynamic mechanical analyzer, model Rheometrics RDS 2. The torsion method was used with a frequency of 1 Hz, a strain level of 0.1% in the temperature range of −150°C to 100°C. The heating rate was 3°C/min. The testing was performed using rectangular bars measuring approximately $38 \times 10 \times 2$ mm, prepared by injection moulding.

3. Results and Discussion

3.1. FTIR Analysis. Figure 4 shows FTIR spectra of the pure polymers (PLA, MPLA, and TPS), plasticizer, A-SiO_2, and composite with 5 wt.% A-SiO_2. The strong absorption band, which appears at around 1750 cm^{-1} in the spectrum of both PLA and MPLA, is assigned to C = O carbonyl stretching vibration. The spectrum of MPLA shows the new relatively weak absorption band at around 1850 cm^{-1}, which is ascribed to the carbonyl group (C = O) stretching of the succinic anhydride ring (or saturated cyclic anhydride ring) [4, 18, 46]. The new absorption band indicated that MA was grafted onto the PLA backbone, during which MA was transformed to a saturated anhydride (succinic anhydride). The typical anhydride band at 1780 cm^{-1} could not be observed because of overlapping of the intense PLA band at 1750 cm^{-1} [47]. In the spectrum of TPS, the peaks at 1020 cm^{-1} and 1075–1150 cm^{-1} were attributed to C–O stretching of the C–O–C group and C–O stretching of the C–O–H group, respectively [48]. The peak at 1650 cm^{-1} was due to the bound water present in the starch. A broad band, due to the hydrogen bonded hydroxyl group (O–H), appeared at 3040–3640 cm^{-1} and is attributed to the complex vibrational stretching, associated with free, inter- and intramolecular bound hydroxyl groups [49]. The spectrum of the plasticizer shows a strong absorption band at 1260 cm^{-1} and a weak absorption band near 860 cm^{-1} attributed to the Si–CH_3 stretching. Moreover, the peaks at 2910 cm^{-1} and 2960 cm^{-1} were attributed to C–H stretching of the methyl group. However, the typical silanol groups (Si–OH) band at 3500 cm^{-1}could not be observed because they were in very small amount (0.04%). In the spectrum of A-SiO_2, there is a strong band at 1060 cm^{-1} attributed to the Si–O groups and a broad peak with a maximum at 3435 cm^{-1} corresponding to the surface hydroxyl groups [50]. However, the peak of amine functional groups was not recorded, due to their presence in very small amount (0.35%).

From Figure 4, it is clear that the characteristic peak assigned to the anhydride group (1850 cm^{-1}) in MPLA is diminished in the spectrum of composite, maybe due to the very small amount that the composite contains (0.068 wt.% MA groups of the whole composite). The new weak absorption band at 1260 cm^{-1} attributed to the Si–CH_3 stretching from the plasticizer was observed. The typical Si–O groups band at 1060 cm^{-1}could not be observed because of overlapping of the intense PLA band at 1180 cm^{-1}, TPS band at 1120 cm^{-1}, and plasticizer band at 996 cm^{-1}. The intensity of the peak centred at 3310 cm^{-1}, corresponding to OH groups of starch, is suppressed in the spectrum of the composite. This may be due to reaction between the OH groups of starch and anhydride groups of MPLA. The formation of hydrogen bonds between the hydroxyl groups of the carbonyl groups of TPS and PLA is one of the possible reactions [48]. The complex interactions between A-SiO_2, plasticizer, MPLA, and TPS also could have occurred.

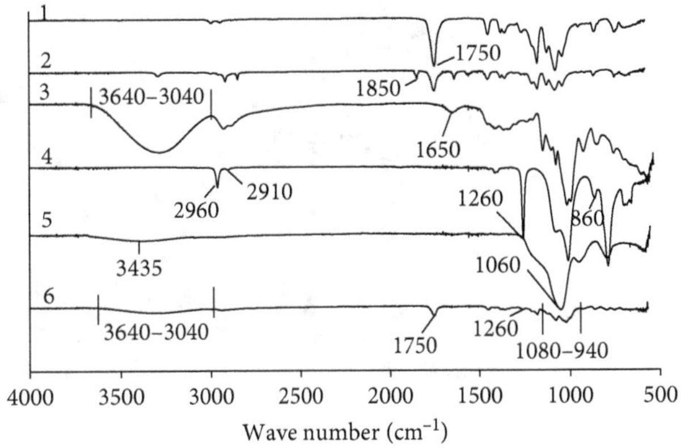

FIGURE 4: FTIR spectra of PLA (1), MPLA (2), TPS (3), P (4), A-SiO_2 (5), and PLA/MPLA/TPS/P/A-SiO_2 (6).

3.2. Morphological Analysis. It is well known for polymer blends that the morphology control of the respective phases is a key factor in achieving the desired material properties [26, 34, 37]. Figure 5 shows cross-sectional images of the nonplasticized and plasticized PLA/MPLA/TPS (60/10/30) blends. As expected, the PLA and thermoplastic starch blend in the presence of compatibilizer (MPLA) shows individual grains of starch forming dispersed phase in the PLA matrix. From those images, it is obvious that the phase morphology of the samples can be depicted as continuous and porous. Moreover, the addition of the plasticizer results in less porous structure, suggesting improved toughness. It is well known that filler dispersion and adhesion to the polymer matrix are of great importance for the mechanical properties of composites improvement. Good control of the interface morphology of the composite is one of the most critical parameters to achieve the desired mechanical properties of such materials [26, 34, 37]. Scanning electron microscopy was performed to explain the behaviour of the silica-filled PLA/MPLA/TPS/P composites.

Figure 6 shows the dispersion state of silica-filled PLA/MPLA/TPS/P composites.

In our previous work, we found that silica nanoparticles were agglomerated at higher content, thus reducing their available surface area for reinforcing effect [51]. Moreover, A-SiO_2 shows bigger tendency to form agglomerates than SiO_2.

From the SEM images (Figure 6), it is clear that the PLA/MPLA/TPS/P porous structure is highly related to the silica content as well as its functionality. However, pores size distribution could not be analyzed because of their irregular shape. The neat silica results in more porous structure than the modified one. Furthermore, the composites with of A-SiO_2 had more regular shape and smaller size of pores than those with SiO_2. This may be due to the result of complex chemical reactions that occur mainly among hydroxyl groups of neat silica, MPLA and TPS, which seems to be more favoured than reactions among amine functional groups of modified silica, MPLA and TPS. These behaviours are highly proportional to the trend observed in the stress-strain behaviour of the composites with neat or modified silica as depicted in Figure 7.

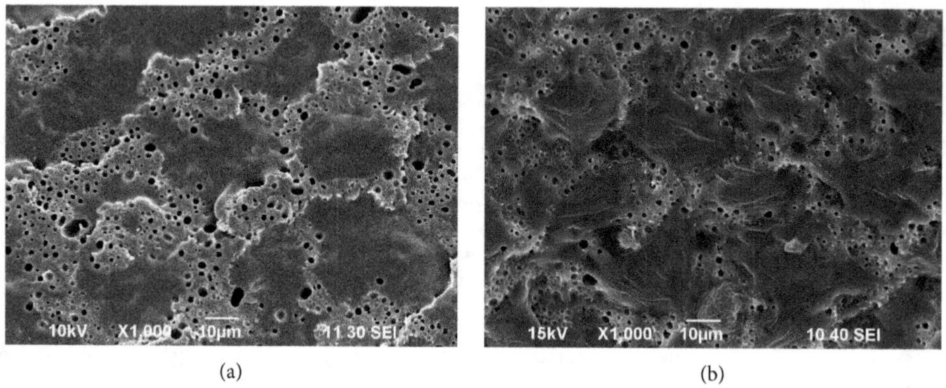

FIGURE 5: SEM images of PLA/MPLA/TPS (a) and PLA/MPLA/TPS/P (b) blends.

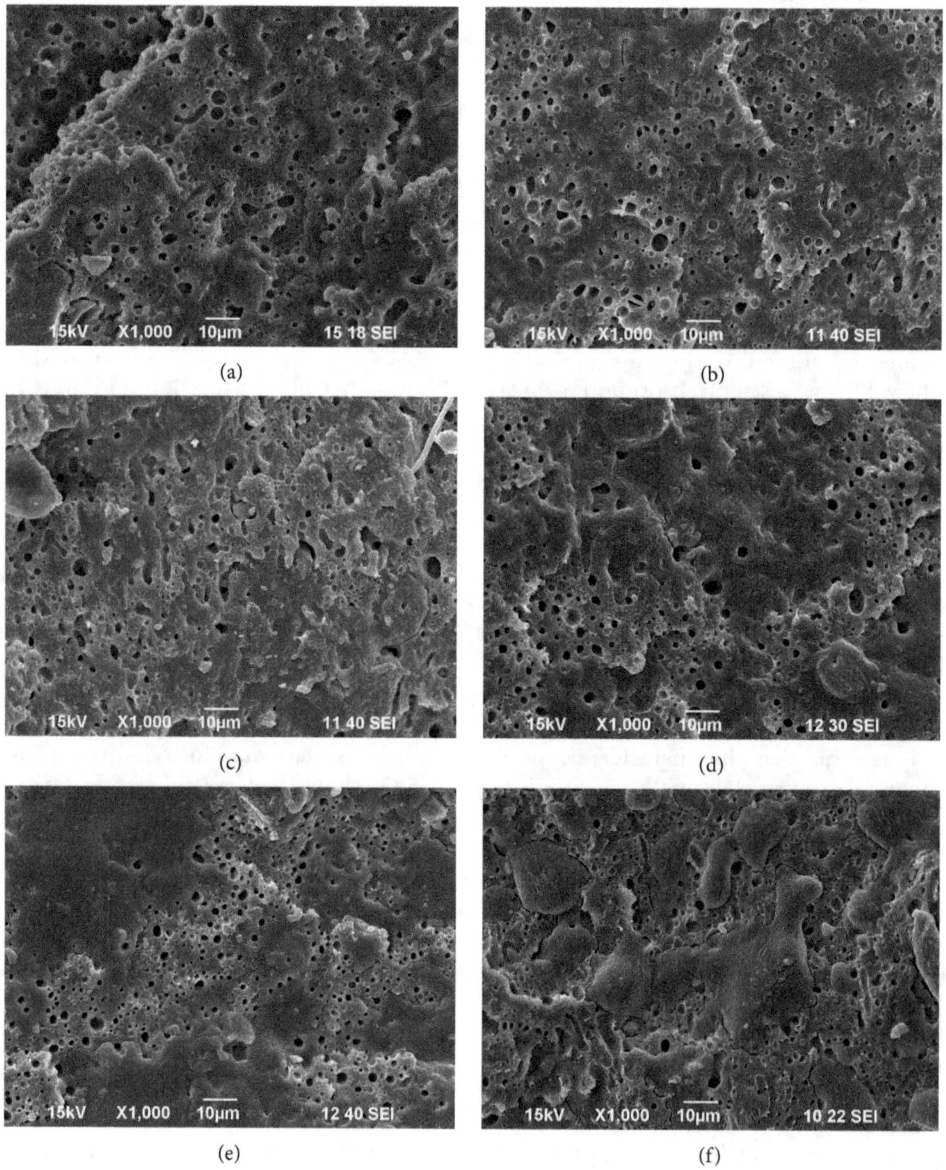

FIGURE 6: SEM images of the PLA/MPLA/TPS/P composite: (a) 1 wt.% of SiO_2, (b) 3 wt.% of SiO_2, (c) 5 wt.% of SiO_2, (d) 1 wt.% of A-SiO_2, (e) 3 wt.% of A-SiO_2, and (f) 5 wt.% of A-SiO_2.

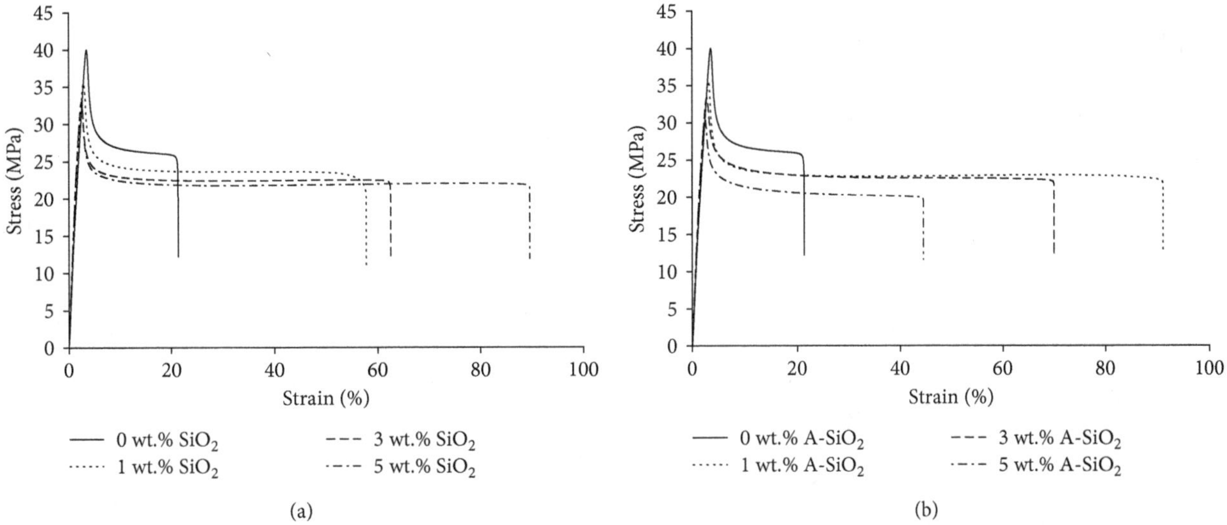

FIGURE 7: Stress-strain curves of PLA/MPLA/TPS/P composites with (a) SiO_2 and (b) A-SiO_2.

3.3. Thermal Properties. The second heating of differential scanning calorimetry (DSC) thermograms of neat PLA, TPS, nonplasticized, and polydimethylsiloxanol plasticized PLA/MPLA/TPS (60/10/30) blends are displayed in Figure 8. It is clear from the above figure that the temperature according to the endothermic peak for each sample is considered to be the glass transition temperature (T_g) of PLA. Moreover, all the samples show an exothermic peak that can be correlated to the crystallization of PLA; the corresponding temperature is known as crystallization temperature (T_c). The neat PLA showed a clear glass transition temperature at 62.1°C, crystallization temperature at 112.5°C, and melting point (T_m) at 153.6°C, corresponding to residual crystallinity, as also discussed by Martin and Averous [14]. It is interesting to know that the T_c peak of PLA did not appear during cooling of PLA and its blends. It is believed that the main reason for this occurrence was due to a very slow crystallization rate of PLA during cooling [52]. The crystallinity (X_c) of pure PLA is only 3.3% after melt blending, which indicates that the material is almost amorphous. It can be also observed in Figure 8 that there are no discernible changes in the DSC thermograms of TPS, suggesting that the thermoplastic starch is in the amorphous phase.

The DSC data for various blends and composites are summarized in Table 3. It is well known that the value of T_g depends primarily on chain flexibility, molecular weight, branching/crosslinking, intermolecular attraction, and steric effects. The glass transition temperature of PLA (62.1°C) was reduced to 60.3°C with the introduction of TPS and MPLA. This emphasized that the chain mobility of PLA has been increased, owing it to the plasticizing effect brought by gelatinized starch with glycerol [53]. Moreover, a small reduction of melting temperature of PLA was observed (from 154°C to 153.2°C), and there was a significant increase in crystallization temperature (from 112.5°C to 128°C) as well as in the degree of crystallinity (from 3.3 to 7.2%). The addition of plasticizer to the PLA/MPLA/TPS blend further decreased the T_g value from 60.3°C to 59.6°C. The reduction of T_g affected other two temperatures, that is, T_m and T_c. The addition of plasticizer decreased the T_m value from 153.2°C to 148.9°C, while the T_c value was reduced from 128°C to 126.7°C resulting in significantly lower crystallinity of the PLA phase. This may have occurred due to the preferable interaction between plasticizer silanol groups and TPS hydroxyl groups and carboxyl groups of the PLA chains [54]. Hence, the thermal characteristics of plasticized PLA/MPLA/TPS blend compared to that unplasticized demonstrated that polydimethylsiloxanol could be regarded as the efficient plasticizer for the PLA/MPLA/TPS blend. These results are in agreement with their tensile properties. Figure 8(b) shows DSC traces for the PLA/MPLA/TPS/P blend and composites differing in A-SiO_2 content. It can be observed from Table 3 that the addition of neat as well as modified silica had only little influence on the glass transition temperature of the composites. However, the T_g of PLA increased from 59.6 to 61.5°C at 5 wt.% of A-SiO_2. It should be noted that T_c strongly depends upon the content as well as functionality of silica. Table 3 also shows that the T_c decreases with increasing loading of silica and is lower for the composites with modified silica. Moreover, the degree of crystallinity increases with silica content and is significantly higher for the composites with modified silica. This behaviour indicates that the large surface of the dispersed silica acts as a nucleating agent for the PLA phase crystallization. It should be noted that modified silica is a more efficient nucleating agent for the toughened PLA/MPLA/TPS blend than neat silica. Wu et al. [55] also reported similar results, where the T_c of PLA decreased with increasing MMT loadings. They also suggested that the difference in the dispersion state of MMT might also be an important factor influencing the crystallization behaviour of PLA nanocomposites.

Table 3 also shows that the melting temperature of the composites with neat silica is higher (2.8–3.3°C) when compared to the PLA/MPLA/TPS/P blend. However, T_m decreases in SiO_2 function. In the contrary, T_m increases

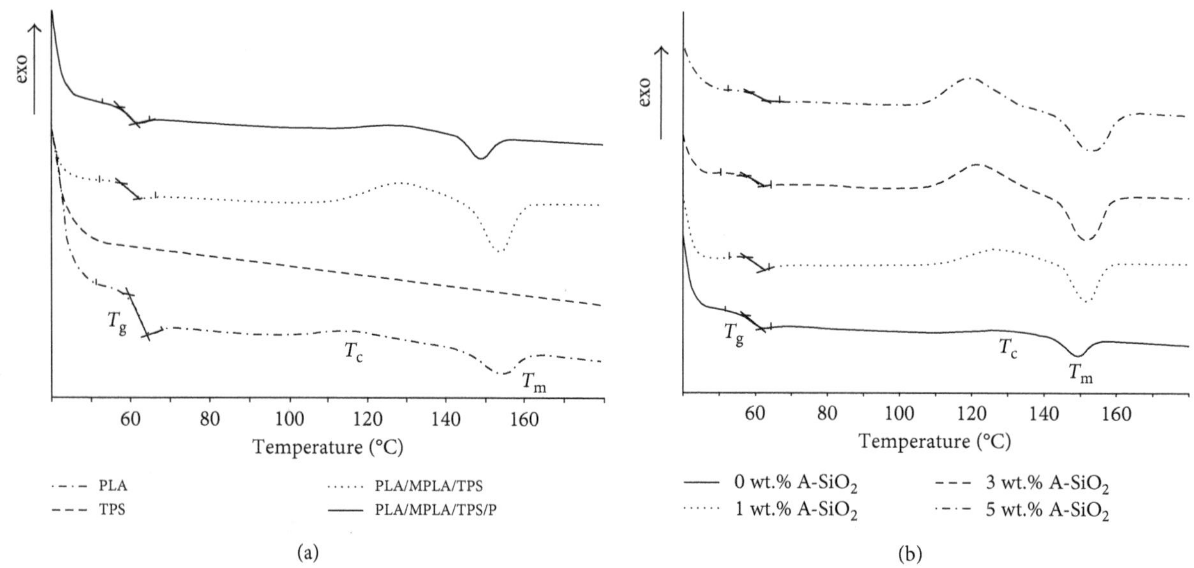

FIGURE 8: DSC curves of PLA, TPS, and various blends (a) and composites (b); second heating scan.

TABLE 3: DSC data of PLA, TPS, and various blends and composites.

Sample	T_g (°C)	T_m (°C)	T_c (°C)	X_c (%)
PLA	62.1	154.0	112.5	3.3
TPS	—	—	—	—
PLA/MPLA/TPS 60/10/30	60.3	153.2	128.0	7.2
PLA/MPLA/TPS/P 60/10/30/3	59.6	148.9	126.7	4.1
PLA/MPLA/TPS/P/1 wt.% SiO_2	60.3	152.3	129.4	12.2
PLA/MPLA/TPS/P/3 wt.% SiO_2	60.5	151.9	129.2	13.7
PLA/MPLA/TPS/P/5 wt.% SiO_2	60.7	151.7	127.8	19.5
PLA/MPLA/TPS/P/1 wt.% A-SiO_2	60.5	151.2	128.4	16.7
PLA/MPLA/TPS/P/3 wt.% A-SiO_2	60.8	151.7	122.3	23.0
PLA/MPLA/TPS/P/5 wt.% A-SiO_2	61.5	153.0	119.4	22.2

with A-SiO_2 content. The melting temperature of the composites with modified silica is 2.3–4.1°C higher than that for the blend without silica. However, T_m of PLA/MPLA/TPS/P/A-SiO_2 composites is 1–3°C lower as compared to that of pure PLA.

Figure 9 shows the TGA curves of neat PLA, TPS, nonplasticized, and plasticized PLA/MPLA/TPS (60/10/30) blends without or with 5 wt.% A-SiO_2. The initial thermal stability is characterized by the temperature that occurred at 10% weight losses, referred to as T_{10} (see Table 4). As shown in Figure 8, TPS dramatically reduces thermal stability of PLA. Petinakis et al. [56] reported that small molecules such as CO, CO_2, H_2O, CH_4, C_2H_4, and CH_2O were produced when starch was decomposed. These molecules could break down the PLA chain resulting in lower thermal decomposition temperatures of PLA. During thermal degradation, the TGA curves display triple-step degradation processes for all the blends. Addition of polydimethylsiloxanol resulted in significant improvement in the initial thermal stability of the PLA/MPLA/TPS blend. As shown in Table 4, the T_{10} increases dramatically from 236°C to 296°C with the incorporation of 3 wt.% of plasticizer. However, there is practically no effect of the plasticizer on the other degradation temperatures (at 50% weight loss, T_{max1} and T_{max2}). Moreover, the addition of silica practically had no effect on the thermal degradation of the plasticized PLA/MPLA/TPS blend. However, the composites with A-SiO_2 had better thermal stability than those with SiO_2. The best thermal stability showed the composite with 5 wt.% of A-SiO_2 (Figure 8), where T_{10} reached 296°C (an increment of 3°C). Moreover, the T_{max2} was 5°C higher when compared to neat PLA.

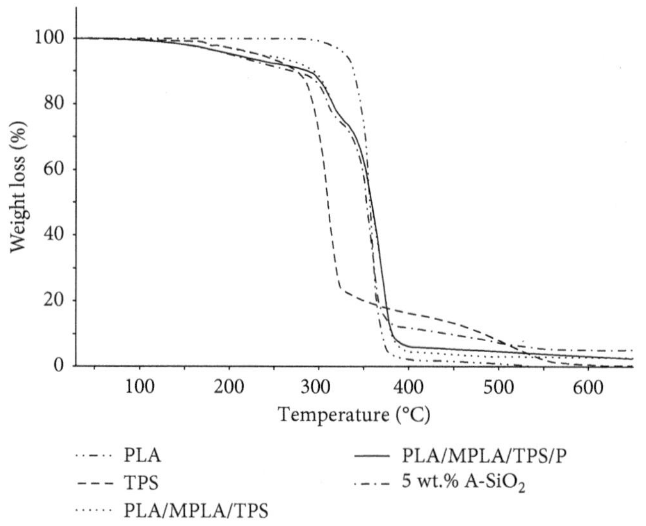

FIGURE 9: TGA curves of PLA, TPS, and various blends and composites.

3.4. Mechanical Properties. Neat PLA has high tensile modulus (3930 MPa) and tensile strength (67 MPa). However, it is a brittle material with the elongation at the break of 6%. Thus, it is needed to be improved by some

TABLE 4: TGA data of PLA, TPS, and various blends and composites in air.

Sample	T_{10} (°C)	T_{50} (°C)	T_{max1} (°C)	T_{max2} (°C)	Weight loss (%)
PLA	330	362	—	364	100.0
TPS	226	320	318	—	99.7
PLA/MPLA/TPS	236	356	311	367	97.5
PLA/MPLA/TPS/P	296	357	312	366	97.7
1 wt.% SiO_2	296	357	310	365	98.0
3 wt.% SiO_2	295	356	310	365	96.5
5 wt.% SiO_2	295	357	311	365	95.0
1 wt.% A-SiO_2	296	357	311	365	97.8
3 wt.% A-SiO_2	296	358	311	366	96.6
5 wt.% A-SiO_2	299	358	312	369	95.0

additives. To investigate the effect of interfacial modification on the mechanical performance of the polydimethylsiloxanol toughened PLA/MPLA/TPS blends, the tensile stress-strain behaviour was characterized. It is clear from Table 5 that the addition of plasticizer caused a significant increase in an elongation at break indicating that the polydimethylsiloxanol is an efficient plasticizer for PLA/MPLA/TPS blends. Based on the abovementioned results, it can be supposed that using plasticizer can improve intermolecular interactions among the blend components through the reactive =SiOH groups.

Moreover, tensile strength and modulus increased compared to the blend without plasticizer.

The similar results and the dependency of poly(ethylene glycol) (PEG) on mechanical properties of PLA/TPS blends were reported by other researchers [57]. The optimized mechanical properties were obtained for the blend with 3 wt. % PEG. Szadkowska et al. [54] used two types of reactive plasticizers with silanol groups to obtain blends of PLA, maleinated PLA (MPLA), and thermoplastic corn starch (TPS). The incorporation of plasticizer with silanol groups into PLA/MPLA/TPS blends resulted in enhanced mechanical properties (especially elongation at break and impact strength) compared to the nonplasticized blend. This is a consequence of the chemical reactions that occurred between silanol groups of the plasticizer and functional groups of TPS and MPLA, which improved compatibility between PLA and TPS.

Silica further increased the elongation at break of the material. However, the improvement strongly depends on silica content as well as its functionality. For neat silica, elongation increased whereas for the modified one, it decreased as silica content increased. Significantly higher value was observed for 1–3 wt.% of the chemically modified silica, indicating, most probably, intensive interfacial interactions among the hydrogen bonds of the anhydride groups of the MPLA, the hydroxyl groups of the starch, and amine groups of the silica. The improvement in elongation was also observed for plasticized PLA and organically modified montmorillonite nanocomposites [27]. However, Arroyo et al. reported an opposite behaviour for PLA/thermoplastic starch/montmorillonite nanocomposites [28]. Moreover, composites show lower tensile strength and modulus than that of the plasticized PLA/MPLA/TPS blend, which increased with increasing silica content. This anomalous behaviour may be resulting from preferential or virtually unpredictable distribution characteristics of silica around the micropores within the toughened PLA/MPLA/TPS structure with respect to weight content and functionality [18]. Recalling DSC findings, we should also take into account the PLA crystallization behaviour as well. In polymers, surfaces are known to act as catalysts for the nucleation of crystals. In polymers patterned with pores, as in our case, it is possible that the shape of the pores, sizes of which shows dependence on silica content and functionality, can control the kinetics of surface-induced crystal nucleation. Moreover, it is well known that the degree of crystallinity may significantly influence on the mechanical properties, since it affects the extent of the intermolecular secondary bonding. Theoretically, higher crystallinity could reduce the elongation at break. But the kind of physical crosslinking formed through physical hypobonds made the composite elastic and have a high elongation at break [58]. For crystalline regions wherein molecular chains are packed in an ordered arrangement, wide-ranging secondary bonding occurs between adjacent chain segments. These bonds lead to significant increase of polymer tensile modulus with the growing degree of crystallinity [26, 51].

3.5. Dynamic Mechanical Thermal Analysis (DMTA). DMTA measures the response of a given material to an oscillatory deformation (here in torsion mode) as a function of temperature. DMTA results are expressed by two main parameters: the storage modulus (G'), corresponding to the elastic response to the deformation, and $\tan\delta$, that is the G''/G' ratio, useful for determining the occurrence of molecular mobility transitions such as the glass transition temperature. Figure 10 shows the temperature dependence of G' and $\tan\delta$ of pure PLA, TPS, and unplasticized and plasticized PLA/MPLA/TPS blends. As can be seen, the storage modulus of the toughened blend was lower than that of the blend without a plasticizer. It is known that the storage modulus detected by DMTA relates to composite stiffness. The stiffness of the PLA/MPLA/TPS blend decreased with the addition of the plasticizer (Table 6). This is a typical behaviour for plasticized thermoplastics.

In general, the storage modulus decreased as the temperature increased. However, in the region corresponding to the maximum of $\tan\delta$ plot, the decrease in storage modulus was usually rapid. Figures 10(b) and 10(c) show the curves of

TABLE 5: Tensile properties of PLA/MPLA/TPS and various blends and composites.

Sample	Tensile strength (MPa)	Elongation at break (%)	Tensile modulus (MPa)
PLA/MPLA/TPS	33 ± 1.5	4 ± 0.3	3264 ± 64
PLA/MPLA/TPS/P	41 ± 0.7	22 ± 0.6	3450 ± 60
1 wt.% SiO_2	36 ± 0.6	60 ± 1.8	2660 ± 29
3 wt.% SiO_2	33 ± 0.4	63 ± 7.3	3060 ± 51
5 wt.% SiO_2	34 ± 0.4	88 ± 5.9	3210 ± 33
1 wt.% A-SiO_2	34 ± 1.4	91 ± 0.2	2990 ± 44
3 wt.% A-SiO_2	33 ± 1.0	72 ± 2.0	3110 ± 35
5 wt.% A-SiO_2	30 ± 0.7	45 ± 0.2	3130 ± 28

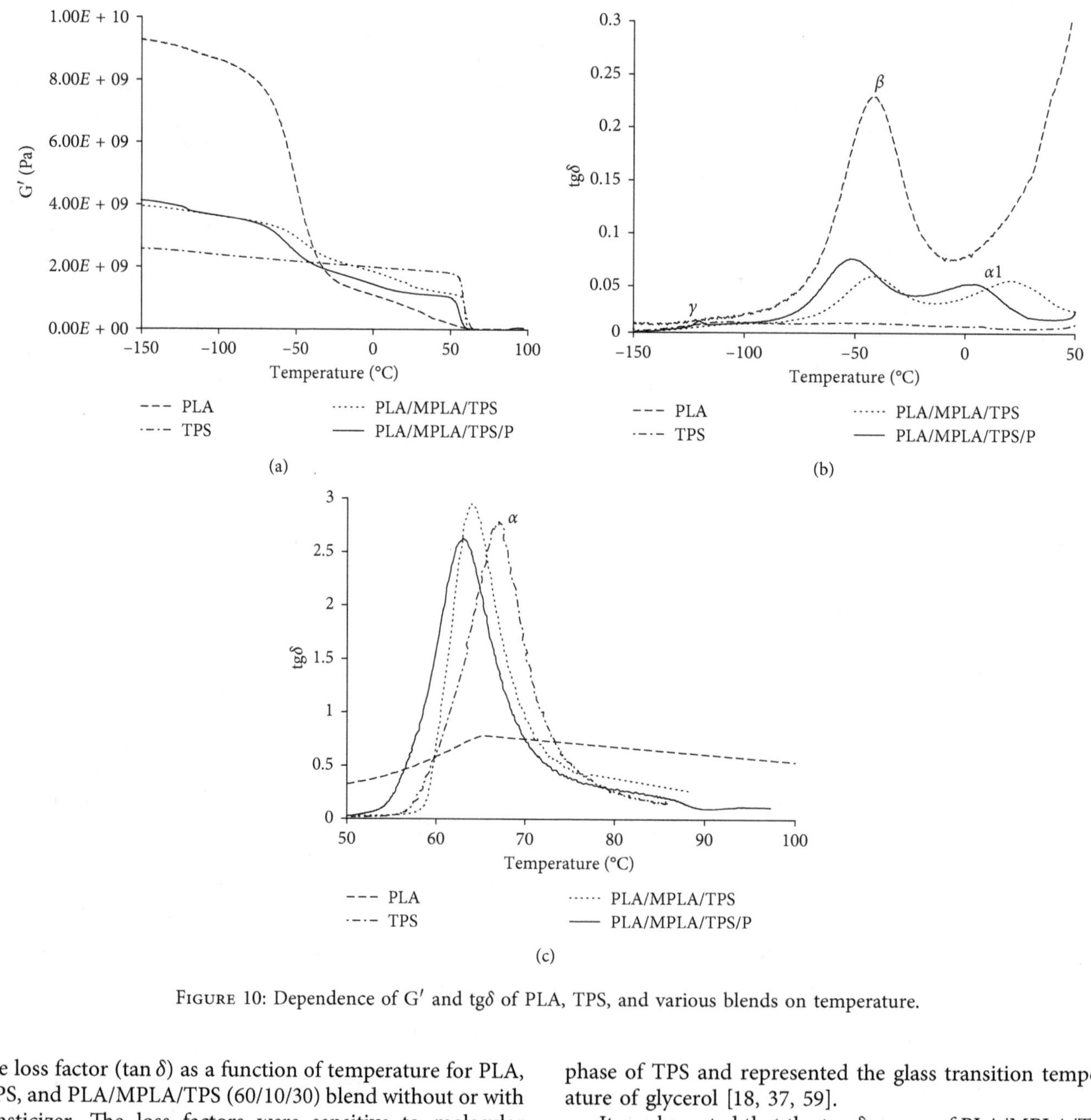

FIGURE 10: Dependence of G′ and tgδ of PLA, TPS, and various blends on temperature.

the loss factor (tan δ) as a function of temperature for PLA, TPS, and PLA/MPLA/TPS (60/10/30) blend without or with plasticizer. The loss factors were sensitive to molecular motion, and their peak was related to the glass transition temperature. The curve of TPS revealed one thermal transition located at −42°C corresponding to a glycerol-rich phase of TPS and represented the glass transition temperature of glycerol [18, 37, 59].

It can be noted that the tan δ curves of PLA/MPLA/TPS blends revealed three thermal transitions (α, β, and γ). In α transition, for the blend without the plasticizer, a tan δ peak located at about 22°C could be ascribed to the glass transition

Table 6: DMTA results of PLA, TPS, and various PLA/MPLA/TPS blends and composites.

Sample	Storage modulus (MPa)			tgδ peak position (°C)			
	At −120°C	At −50°C	At 23°C	α	α_1	β	γ
PLA	2480	2170	1920	67	—	—	−109
TPS	8990	4690	711	—	—	−42	—
PLA/MPLA/TPS	3780	2960	1490	64	22	−42	−109
PLA/MPLA/TPS/P	3830	2470	1190	63	5	−52	−121
1 SiO_2	3670	2570	1310	64	14	−47	−120
3SiO_2	3650	2630	1340	64	19	−47	−121
5 SiO_2	3910	2730	1390	63	20	−48	−121
1 A-SiO_2	3810	2640	1320	64	19	−48	−121
3 A-SiO_2	3780	2580	1380	63	28	−49	−121
5 A-SiO_2	4000	2690	1480	63	30	−49	−120

temperature of the starch-rich phase [13, 18, 60], whereas that of the blend containing plasticizer occurred at 5°C. This indicates that the T_g value of the TPS phase has shifted to significantly lower temperatures after the plasticization due to the enhanced chain mobility of the starch molecules. Another tan δ peak, in α transition, located at 64°C, represented the glass transition temperature of PLA. The value is slightly lower than that of the pure PLA (67°C). This is a typical behaviour of plasticized thermoplastics, where plasticizers can reduce the T_g by increasing the free volume, and thus chain mobility of the polymeric molecules may also promote their crystallinity due to the enhanced chain mobility. These results are in agreement with thermal and mechanical properties (Tables 3–5).

As shown in Figures 11(a) and 11(b), the storage modulus for the various composites increased with the addition of silica nanoparticles. However, the modified silica is more efficient reinforcement for the investigated blend comparing with neat silica, due to the higher storage modulus (Table 6). Figures 11(c)–11(f) show the curves of the loss factor (tan δ) as a function of temperature for the composites containing neat or modified silica, respectively. The addition of silica significantly increases α and β-relaxation temperatures of the TPS phase, due to the improved intermolecular interaction of TPS in both the starch-rich and starch-poor phases. However, the values of the above temperatures strongly depend on the silica functionality and content. It can be observed from Table 6 that the T_{a1} of TPS increases dramatically from 5 to 20°C and to 30°C with 3 wt. % of neat or modified silica, respectively. Since modified silica particles were more reactive than neat silica, A-SiO_2 increased the glass transition temperature higher than that of SiO_2.

Moreover, Table 6 shows that the T_a of the PLA phase of the composites with neat silica is slightly higher as compared to that of the PLA/MPLA/TPS/P blend as well as to the composites with A-SiO_2. However, there is practically no effect of the silica content and functionality on the α relaxation temperature.

Simultaneously, the β-relaxation of a glycerol-rich phase of TPS can be observed as a clear maximum in tan δ (Figure 11(c)) for the blends and represents the glass transition temperature of glycerol [59]. The β-relaxation appears as a maximum at −42°C in tan δ for the blend without plasticizer, whereas β-relaxation of the plasticized blend T_β is about 10°C lower. This is related to increased interfacial interactions between the TPS phase and plasticizer.

Moreover, for the blend without the plasticizer, a tan δ peak located at about −107°C could be ascribed to the β transition temperature of the PLA, whereas that of the blend containing the plasticizer occurred at about −121°C. This indicates that the T_β value of the PLA has shifted to lower temperatures after the plasticization. Functioning like a physical joint, the plasticizer enhanced the chain mobility of the PLA molecules and hence increased the free volume and reduced glass transitions of the blends. These results indicate that the polydimethylsiloxanol played a significant role as a plasticizer in the PLA/TPS blends.

4. Conclusions

Toughened PLA/MPLA/TPS/silica composites were successfully prepared by melt blending in the twin-screw corotating extruder. The morphological analysis of the plasticized blends showed that the addition of plasticizer resulted in less porous structure with increase in its content, suggesting improved toughness. More porous structure was observed for neat silica than that for the modified one. Moreover, at the same contents of silica, the shape of pores was more regular and the size was smaller in the composites with modified silica than with the neat one. Thermal and mechanical properties demonstrated that polydimethylsiloxanol can be used as the efficient plasticizer for the PLA/MPLA/TPS blend. DSC analysis showed that glass transition temperature and melt temperature were not significantly affected by the addition of silica. However, the strong effect of silica content and functionality on crystallization temperature and crystallinity of PLA/MPLA/TPS/P blend was observed. From TGA analysis, it is clear that the plasticizer greatly improved the initial thermal stability of the PLA/MPLA/TPS blend. However, the addition of silica had no effect on the thermal degradation of the plasticized blend. The tensile strength and tensile modulus of plasticized PLA/MPLA/TPS blends decreased as the content of plasticizer increased. However, the plasticized blend had higher tensile properties as compared to the unplasticized one. Meanwhile, the elongation at break reached up to 26%. The addition of silica significantly increased the elongation at break of the material (up to 91%). However, the improvement strongly depends on the silica content and functionality. For neat silica, elongation increased, whereas for the modified one, it decreased as silica content increased. The highest value was observed for 1 wt.% of the A-SiO_2. This was due to the intensive interfacial interactions among the hydrogen bonds of the anhydride groups of the MPLA, the hydroxyl groups of the starch, and amine groups of the silica. Moreover, the composites showed lower tensile strength and tensile modulus than that of the plasticized PLA/MPLA/TPS blend, which increased with the increasing silica content. This was resulting from preferential or virtually unpredictable

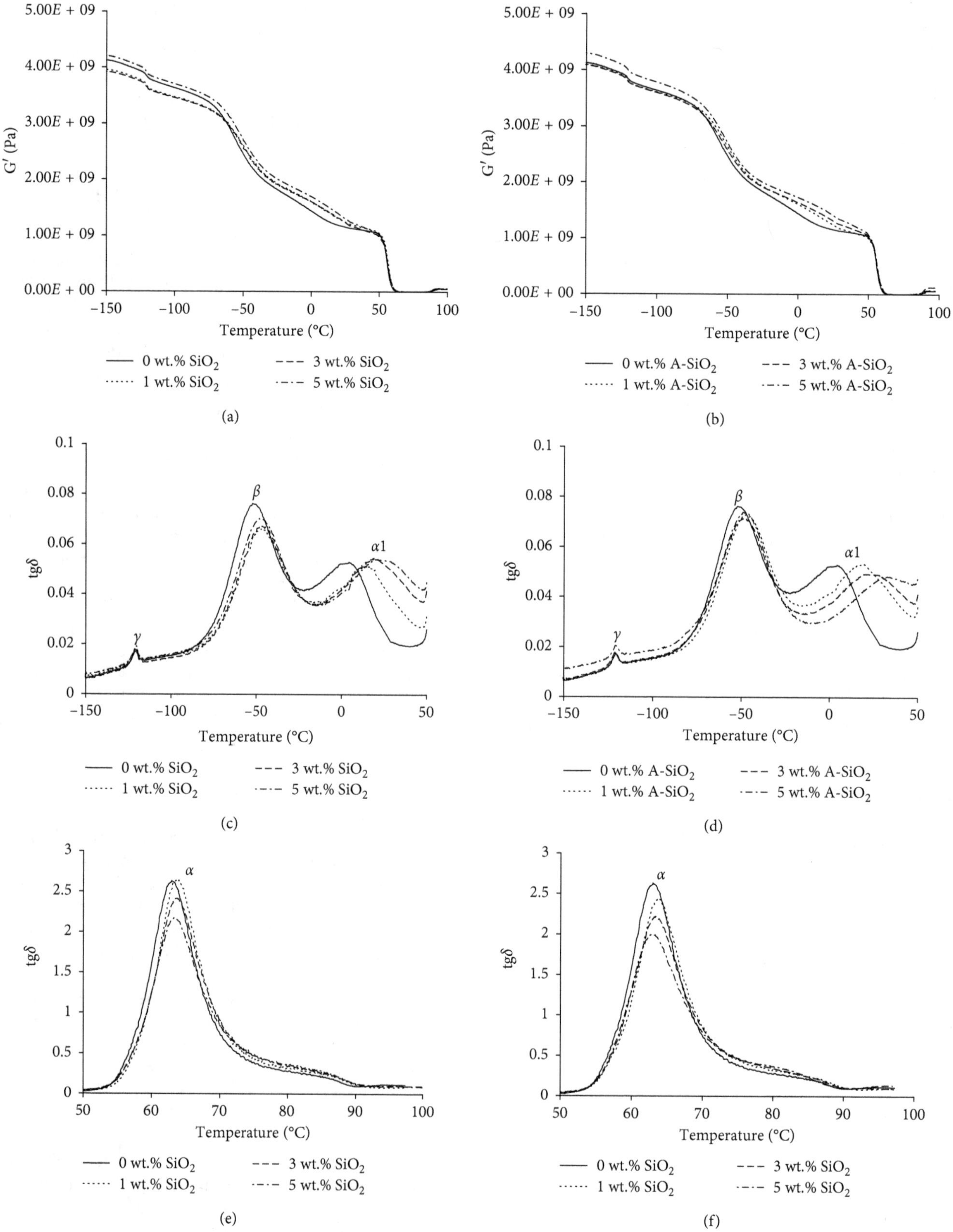

FIGURE 11: Dependence of G′ and tgδ of various composites on temperature.

distribution characteristics of silica around the micropores within the toughened PLA/MPLA/TPS structure with respect to concentration and functionality. From DMTA, it was found that the modified silica is more efficient reinforcement than the neat one, due to the higher storage modulus of the composites.

The plasticized PLA/MPLA/TPS/silica composites can be one of the good candidates in the potential application, such as packaging materials, disposable goods, electronics materials, tissue engineering materials, surgical sutures, and drug delivery systems.

Conflicts of Interest

The authors declare that there are no conflicts of interest regarding the publication of this paper.

Acknowledgments

The authors are very grateful to Professor Maria Zielecka for silica modification and the evaluation of the results and to Maciej Studziński, MSc, for providing DMTA analyses. The work was fonded by the statutory resources of the Industrial Chemistry Research Institute.

References

[1] H. Tsuji and Y. Ikada, "Blends of aliphatic polyesters. II. Hydrolysis of solution-cast blends from poly(L-lactide) and poly(E-caprolactone) in phosphate-buffered solution," *Journal of Applied Polymer & Science*, vol. 67, no. 3, pp. 405–415, 1998.

[2] C. Zhang, L. Wang, T. Zhai, X. Wang, Y. Dan, and L.-S. Turng, "The surface grafting of grapheme oxide with poly (ethylene glycol) as a reinforcement for poly(lactic acid) nanocomposite scaffolds for potential issue engineering applications," *Journal of Mechanical Behaviour of Biomedical Materials*, vol. 53, pp. 403–413, 2016.

[3] Z. Liu, Y. Lei, Z. Hu, W. Kong, C. Zhou, and J. Lei, "Preparation, characterization and properties of poly(lactic acid)/poly(1,4-butylene adipate) blends for biodegradable packaging materials," *Macromolecular Research*, vol. 25, no. 5, pp. 439–445, 2017.

[4] M. R. A. Moghaddam, S. M. A. Razavi, and Y. Jahani, "Effect of compatibilizer and thermoplastic starch (TPS) concentration on morphological, rheological, tensile, thermal and moisture sorption properties of plasticized polylactic acid/tPS blends," *Journal of Polymers and the Environment*, vol. 26, no. 8, pp. 3202–3212, 2018.

[5] S. Jacobsen and H. G. Fritz, "Plasticizing polylactide—the effect of different plasticizers on the mechanical properties," *Polymer Engineering and Science*, vol. 39, no. 7, pp. 1303–1310, 1999.

[6] G. Theryo, F. Jing, L. M. Pitet, and M. A. Hillmayer, "Though polylactide graft copolymers," *Macromolecules*, vol. 43, no. 18, pp. 7394–7397, 2010.

[7] S. Ishida, R. Nagasaki, K. Chino et al., "Toughening of poly(l-lactide) by melt blending with rubbers," *Journal of Applied Polymer & Science*, vol. 113, no. 1, pp. 558–566, 2009.

[8] C. Zhang, W. Wang, Y. Huang et al., "Thermal, mechanical and rheological properties of polylactide toughened by epoxidized natural rubber," *Materials and Design*, vol. 45, pp. 198–205, 2013.

[9] S. Pichaiyut, C. Nakason, and S. Wisunthorn, "Biodegradability and thermal properties of novel natural rubber/linear low density polyethylene/thermoplastic starch ternary blends," *Journal of Polymers and the Environment*, vol. 26, no. 7, pp. 2855–2866, 2018.

[10] H. Younes and D. Cohn, "Phase separation in poly(ethylene glycol)/poly(lactic acid) blends," *European Polymer Journal*, vol. 24, no. 8, pp. 765–773, 1988.

[11] M. Bariado, G. Frisoni, M. Scandola et al., "Thermal and mechanical properties of plasticized poly(L-lactic acid)," *Journal of Applied Polymer & Science*, vol. 90, no. 7, pp. 1731–1738, 2003.

[12] Z. Kulinski and E. Piotrowska, "Crystallization, structure and properties of plasticized poly(L-lactide)," *Polymer*, vol. 46, no. 23, pp. 10290–10300, 2005.

[13] N. Ljungberg and B. Wesslen, "The effects of plasticizers on the dynamic mechanical and thermal properties of poly(lactic acid)," *Journal of Applied Polymer & Science*, vol. 86, no. 5, pp. 1227–1234, 2002.

[14] O. Martin and L. Averous, "Poly(lactic acid): plasticization and properties of biodegradable multiphase systems," *Polymer*, vol. 42, no. 14, pp. 6209–6219, 2001.

[15] J. Korol, J. Lenża, and K. Formela, "Manufacture and research of TPS/PE biocomposites properties," *Composites Part B: Engineering*, vol. 68, pp. 310–316, 2015.

[16] H. Pan, D. Ju, Y. Zhao et al., "Mechanical properties, hydrophobic properties and thermal stability of the biodegradable poly(butylenes adipate-co-terephthalate)/maleated thermoplastic starch blow films," *Fibers and Polymers*, vol. 17, no. 10, pp. 1540–1549, 2016.

[17] M. Sabetzadech, R. Bagheri, and M. Masoomi, "Effect of oxidized starch on morphology, rheological and tensile properties of low-density polyethylene/linear low-density polyethylene/thermoplastic oxidized starch blends," *Journal of Polymers and the Environment*, vol. 26, no. 6, pp. 2219–2226, 2018.

[18] B. Świerz-Motysia, R. Jeziórska, A. Szadkowska et al., "Synthesis and properties of biodegradable polylactide and thermoplastic starch blends," *Polimery*, vol. 56, no. 4, pp. 271–280, 2011.

[19] L. Chen, X. Y. Qiu, Z. G. Xie et al., "Poly(L-lactide)/starch blends compatibilized with poly(L-lactide)-g- starch copolymer," *Carbohydrate Polymers*, vol. 65, no. 1, pp. 75–80, 2006.

[20] C. W. Obiro, M. E. Naushad, and S. R. Suprakas, "Including PLA/starch compatibility through butyl-etherification of waxy and high impact amylose starch," *Carbohydrate Polymers*, vol. 112, pp. 216–224, 2014.

[21] P. Dubois and R. Narayan, "Biodegradable compositions by reactive processing of aliphatic polyester/polysaccharide blends," *Macromolecular Symposia*, vol. 198, no. 1, pp. 233–244, 2003.

[22] J. F. Zhang and X. Sun, "Mechanical properties of poly(lactic acid)/starch composites compatibilized by maleic anhydride," *Biomacromolecules*, vol. 5, no. 4, pp. 1446–1451, 2004.

[23] J. Wootthikanokkhan, N. Wongta, N. Sombatsompop et al., "Effect of blending conditions on mechanical, thermal and rheological properties of plasticized poly(lactic acid)/maleated

thermoplastic starch blends," *Journal of Applied Polymer & Science*, vol. 124, no. 2, pp. 1012–1019, 2012.
[24] S. W. Hwang, J. K. Shim, S. Selke, H. Soto-Valdez, M. Rubino, and R. Auras, "Effect of maleic-anhydride grafting on the physical and mechanical properties of poly(L-lactic acid)/starch blends," *Macromolecular Materials and Engineering*, vol. 298, no. 6, pp. 624–633, 2013.
[25] Z. Xiong, S. Ma, L. Fan et al., "Surface hydrophobic modification of starch with bio-based epoxy resins to fabricate high-performance poly(lactide) composite materials," *Composite Science and Technology*, vol. 94, no. 1, pp. 16–22, 2014.
[26] H. Balakrishnan, A. Hassan, M. U. Wahit et al., "Novel toughened polylactic acid nanocomposite: mechanical, thermal and morphological properties," *Materials and Design*, vol. 31, no. 7, pp. 3289–3298, 2010.
[27] C. Thellen, C. Orroth, D. Froio et al., "Influence of montmorillonite layered silicate on plasticized poly(L-lactide) blown films," *Polymer*, vol. 46, no. 25, pp. 11716–11727, 2005.
[28] O. H. Arroyo, M. A. Huneault, B. D. Favis, and M. N. Bureau, "Processing and properties of PLA/thermoplastic starch/montmorillonite nanocomposites," *Polymer Composites*, vol. 31, pp. 114–127, 2010.
[29] M. R. Aghjeh, V. Asadi, P. Mehdijabbar, H. A. Khonakdar, and S. H. Jafari, "Application of linear rheology in determination of nanoclay localization in PLA/EVA/Clay nanocomposites: correlation with microstructure and thermal properties," *Composites Part B: Engineering*, vol. 86, pp. 273–284, 2016.
[30] J. Zhang, J. Lou, S. Ilias, P. Krishnamachari, and J. Yan, "Thermal properties of poly(lactic acid) fumed silica nanocomposites: experiments and molecular dynamics simulations," *Polymer*, vol. 49, no. 9, pp. 2381–2386, 2008.
[31] F. Wu, X. Lan, D. Ji, Z. Liu, W. Yang, and M. Yang, "Grafting polymerization of polylactic acid on the surface of nano-SiO_2 and properties of PLA/PLA-grafted-SiO_2 nanocomposites," *Journal of Applied Polymer & Science*, vol. 129, no. 5, pp. 3019–3027, 2013.
[32] J. Ch. H. Lai, M. R. Rahman, and S. Hamdan, "Comparative studies of thermo-mechanical and morphological properties of polylactic acid/fumed silica/clay (1.28E) and polylactic acid/fumed silica/clay (1.34 TCN) nanocomposites," *Polymer Bulletin*, vol. 75, no. 1, pp. 135–147, 2018.
[33] V. Kumar, A. Dev, and A. P. Gupta, "Studies of poly(lactic acid) based calcium carbonate nanocomposites," *Composites Part B: Engineering*, vol. 56, pp. 184–188, 2014.
[34] E. Jalalvandi, R. A. Majid, and T. Ghanbari, "Processing, morphological, thermal and absorption behavior of PLA/thermoplastic starch/montmorillonite nanocomposites," *World Academy of Science, Engineering and Technology*, vol. 6, no. 12, pp. 715–719, 2012.
[35] B. Ayana, S. Suin, and B. B. Khatua, "Highly exfoliated eco-friendly thermoplastic starch (TPS)/poly(lactic acid)(PLA)/clay nanocomposites using unmodified nanoclay," *Carbohydrate Polymers*, vol. 110, pp. 430–439, 2014.
[36] I. Petersson and K. Oksman, "Biopolymer based nanocomposites: comparing layered silicates and microcrystalline cellulose as nanoreinforcement," *Composites Science and Technology*, vol. 66, no. 13, pp. 2187–2196, 2006.
[37] R. Jeziórska, A. Szadkowska, B. Świerz-Motysia, and J. Kozakiewicz, "Effect of "core-shell" polymeric nanofiller structure and properties of polylactide and thermoplastic corn starch blend," *Polimery*, vol. 57, no. 5, pp. 354–363, 2012.
[38] B. Świerz-Motysia, R. Jeziórska, A. Szadkowska et al., "Biodegradable composition with thermoplastic starch," Polish Patent 214 329, 2013.
[39] B. Świerz-Motysia, R. Jeziórska, A. Szadkowska et al., "Method for obtaining thermoplastic starch," Polish Patent 216 930, 2014.
[40] M. Zielecka, K. Bajdor, A. Szulc et al., "Method for obtaining silica nanopowder also functionalized," Polish Patent 198 188, 2007.
[41] M. Zielecka, E. Bujnowska, B. Kepska et al., "Antimicrobial additives for architectural paints and impregnates," *Progress in Organic Coatings*, vol. 72, no. 1-2, pp. 193–201, 2011.
[42] M. Zielecka, E. Bujnowska, K. Suwała, and M. Wenda, "Sol-gel-derived silicon-containing hybrids," in *Recent Applications in Sol-Gel Synthesis*, INTECH, London, UK, 2017.
[43] R Jeziórska, "Functionalization of low density polyethylene with ricinol-2-oxazoline methyl maleate in a twin-screw extruder," *International Polymer Processing*, vol. 22, no. 2, pp. 122–131, 2007.
[44] B Świerz-Motysia, R. Jeziórska, A. Szadkowska et al., "Biodegradable polymer composites containing powder nanofiller," Polish Patent 216 295, 2014.
[45] J. F. Turner, A. Riga, A. O'Connor, J. Zhang, and J. Collis, "Characterization of drawn and undrawn poly-L-lactide films by differential scanning calorimetry," *Journal of Thermal Analysis and Calorimetry*, vol. 75, no. 1, pp. 257–268, 2004.
[46] R. Mani, M. Bhattacharya, and J. Tang, "Functionalization of polyesters with maleic anhydride by reactive extrusion," *Journal of Polymer Science Part A: Polymer Chemistry*, vol. 37, no. 11, pp. 1693–1702, 1999.
[47] J. Prachayawarakorn, P. Ruttanabus, and P. Boonsom, "Effect of cotton fiber contents and lengths on properties of thermoplastic starch composites prepared from rice and waxy rice starches," *Journal of Polymers and the Environment*, vol. 19, no. 1, pp. 274–282, 2010.
[48] M. P. Guaras, V. A. Alvarez, and L. N. Luduena, "Processing and characterization of thermoplastic starch/polycarbonate/compatibilizer ternary blends for packaging applications," *Journal of Polymer Research*, vol. 22, no. 9, pp. 165–176, 2015.
[49] J. Prachayawarakorn, S. Chaiwatyothin, S. Mueangta, and A. Hanchama, "Effect of jute and kapok fibers on properties of thermoplastic cassava starch composites," *Materials and Design*, vol. 47, pp. 309–315, 2013.
[50] R. Jeziórska, B Swierz-Motysia, M. Zielecka, A. Szadkowska, and M. Studzinski, "Structure and mechanical properties of low-density polyethylene/spherical silica nanocomposites prepared by melt mixing: the joint action of silica's size, functionality, and compatibilizer," *Journal of Applied Polymer Science*, vol. 125, no. 6, pp. 4326–4337, 2012.
[51] R. Jeziórska, B. Świerz-Motysia, M. Zielecka et al., "Polyamide/spherical nanosilica nanocomposites," *Polimery*, vol. 54, no. 10, pp. 647–656, 2009.
[52] W. S. Chow and S. K. Lok, "Flexural, morphological and thermal properties of poly(lactic acid)/organo-montmorillonite nanocomposites," *Polymers and Polymer Composites*, vol. 16, no. 4, pp. 263–270, 2008.
[53] H. J. Lee, T. G. Park, H. S. Park et al., "Thermal and mechanical characteristics of poly(l-lactic acid) nanocomposite scaffold," *Biomaterials*, vol. 24, no. 16, pp. 2773–2778, 2003.
[54] A. Szadkowska, R. Jeziórska, M. Żubrowska, E. Spasówka, and P. Rościszewski, "Effect of plasticizer with silanol groups on the structure and properties of polylactide and thermoplastic corn starch blend," *Polimery*, vol. 61, no. 10, pp. 683–692, 2016.

[55] T. M. Wu and C. Y. Wu, "Biodegradable poly(lactic acid)/chitosan-modified montmorillonite nanocomposites: preparation and characterization," *Polymer Degradation and Stability*, vol. 91, no. 9, pp. 2198–2204, 2006.

[56] E. Petinakis, X. Liu, L. Yu et al., "Biodegradation and thermal decomposition of poly(lactic acid)-based materials reinforced by hydrophilic fillers," *Polymer Degradation and Stability*, vol. 95, no. 9, pp. 1704–1707, 2010.

[57] X. Ping, W. Kejian, J. Mingryin, and Y. Meijuan, "Biodegradation and mechanical property of polylactic acid/thermoplastic starch blends with poly(ethylene glycol)," *Journal of Wuhan University of Technology-Materials Science Edition*, vol. 28, no. 1, pp. 157–162, 2013.

[58] Z. Ren, L. Dong, and Y. Yuming, "Dynamic mechanical and thermal properties of plasticized poly(lactic acid)," *Journal of Applied Polymer Science*, vol. 101, no. 3, pp. 1583–1590, 2006.

[59] D. Lourdin, H. Bizot, and P. Colonna, "Antiplasticization in starch-glycerol films," *Journal of Applied Polymer Science*, vol. 63, no. 8, pp. 1047–1053, 1997.

[60] P. Sarazin, G. Li, W. J. Orts, and B. D. Favis, "Binary and ternary blends of polylactide, polycaprolactone and thermoplastic starch," *Polymer*, vol. 49, no. 2, pp. 599–609, 2008.

4

Flexural Behaviors of Concrete/EPS-Foam/Glass-Fiber Composite Sandwich Panel

Qi Liu,[1] Wenfeng Du,[1] Nasim Uddin,[2] and Zhiyong Zhou[1]

[1]*College of Civil Engineering and Architecture, Henan University, Kaifeng, China*
[2]*Department of Civil, Construction and Environmental Engineering, University of Alabama at Birmingham, Birmingham, Alabama, USA*

Correspondence should be addressed to Wenfeng Du; dwf0925@hotmail.com

Academic Editor: Alexander Kromka

Composite structural insulated panels (CSIPs) have been developed for structural floor applications instead of traditional structural insulated panels (SIPs). However, the load bearing capacity of CSIPs is low due to the debonding between the top face sheet and the core when they are used for floors. To overcome this drawback, an improved composite structural insulated panel (ICSIP) was proposed and analyzed in this paper. In ICSIPs, a thick layer of concrete is used as the top face sheet instead of glass-fiber-reinforced polymer (GFRP) in CSIPs to increase the stiffness of the top compression face sheet. However, the bottom GFRP face sheet and EPS cores in CSIPs are preserved to reduce the weight of the structure and act as a template for the top concrete panels. Full-scale experimental testing and finite-element analysis were conducted to predict the flexural strength and deflection of the ICSIP floor member. Good agreement has been observed between the numerical results and experimental response up to the failure. The cause of failure of ICSIPs is the crushing of concrete face sheet rather than debonding. Moreover, the calculation formula for the ultimate bearing load and deflection was also developed based on the classical sandwich theory. The theoretical predictions reflect well the linear flexural response of the ICSIPs, while deviate as the load increases up to failure due to the theory limitations.

1. Introduction

The sandwich panels named structural insulated panels (SIPs) were proposed in 1935 by Forest Products Laboratory (FPL) in the United States [1]. And the first large-scale SIPs manufacturing effort appeared in 1959 when Koopers Company converted an automotive production plant in Detroit into a SIP production facility [2]. Then in the mid-1980s, a significant number of manufactures began producing SIPs on a large scale with the improvement of production technology. To date, traditional SIPs have been used for many different applications, such as exteriors wall, roof, floor, and foundation system throughout the world [3, 4]. SIPs have brought many benefits for the modular panelized construction industry because of characteristics of high quality and precast. However, SIPs still have a lot of drawbacks such as termite attack, low permeability, and weak resistance against wind loading because the oriented strand board (OSB) face sheet is a wood-based composite laminate [5].

To overcome the shortcomings of traditional SIPs, a new composite structural insulated panel (CSIP) was proposed in 2010 [6]. CSIPs utilized low-cost orthotropic thermoplastic glass-fiber-reinforced polymer (GFRP) as face sheets instead of OSB. They can provide much higher strength, stiffness, and creep resistance than traditional ones [7–9]. A comprehensive explanation of CSIP concept, materials characteristics, and manufacturing techniques is proved by Vaidya and Uddin [10]. Mousa and Uddin carried out a series of study in order to determine the global buckling, debonding interfacial tensile stress, critical wrinkling stress, equivalent stiffness, and deflection for CSIPs [10, 11]. Theoretical and experimental studies have demonstrated that CSIPs are characterized by high strength-to-weight ratio, excellent thermomechanical properties, and superior corrosion resistance.

However, two drawbacks of CSIPs are still noticeable when they are used as floors or large span roofs. One is the deflection of CSIPs is large relatively in despite of the high strength [10–13]. The elastic modulus of GFRP is comparatively small, which leads to a large deflection of CSIPs. Several investigations show that it is difficult for CSIPs to satisfy the requirements of deflection in building codes if they have the same thickness with concrete floors panel [14, 15]. The other is the debonding between the compression face sheet and the core. This debonding phenomenon always appears under a small load due to the weak bonding strength of adhesive and low compressive performance of GFRP face sheet, which makes only 17% of carrying capacity of CSIPs be used [10, 15].

An improved composite structural insulated panel (ICSIP) was proposed to expand the application of CSIPs in this study. ICSIPs retains the bottom GFRP face sheet and EPS core in the CSIPs, but recycled aggregate concrete was used as the top face sheet instead of the top GFRP face sheet. The recycled aggregate concrete face sheet has higher compression performance than the GFRP face sheet. Therefore, ICSIPs effectively overcome the problem of large structural displacement and low bearing capacity due to the low compressive performance of the top face sheet in CSIPs. Meanwhile, the recycled aggregate concrete has lower cost than the GFRP, which makes the ICSIPs have more advantageous in terms of production cost than the CSIPs. Similarly, the cost of ICSIPs is still attractive compared to traditional SIP while maintaining the same thickness of EPS core.

The characteristic and manufacturing of ICSIPs were studied in this paper and flexural behaviors of full-scale ICSIP floor members were investigated. In order to understand the load bearing capacity and deflection of the ICSIP floor member, the experimental testing, finite-element analysis, and theoretical evaluation were performed.

2. Characteristic and Manufacturing of ICSIPs

2.1. Characteristic of ICSIPs. The concept of ICSIPs is still based on a sandwich structure, in which a soft light-weight thick core is sandwiched between two strong face sheets. The ICSIPs that are developed and evaluated in this research are composed of low-cost orthotropic thermoplastic GFRP laminate as the bottom face sheet, expanded polystyrene (EPS) foam as the core, and recycled aggregate concrete as the top face sheet (Figure 1). The dimension of the ICSIPs is consistent with the traditional CSIPs. The thickness of GFRP face sheet is 3.04 mm which is the same thickness as the bottom face sheet in the CSIPs. The thickness of the recycled aggregate concrete face sheet is consistent with the EPS core, which can effectively ensure that most of the concrete material is under pressure. Therefore, the thickness of both the recycled aggregate concrete face sheet and the EPS core is 70 mm.

The top face sheet in ICSIPs carries the compressive stress, and the bottom face sheet in ICSIPs carries the tensile stress. However, the core stabilizes the face sheets against buckling and increases the stiffness by holding the face sheets apart. Therefore, structural layout of ICSIPs is

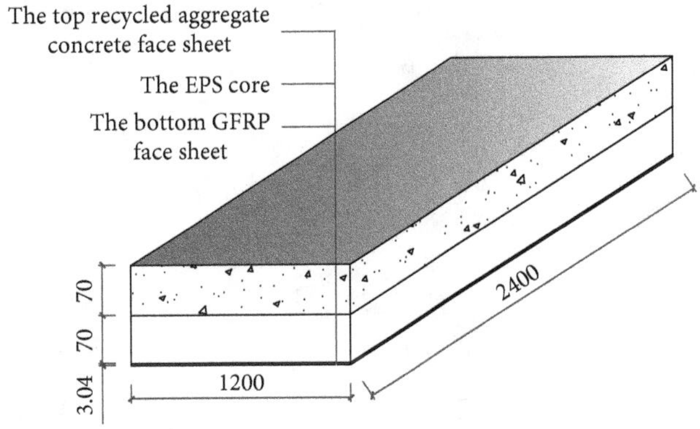

FIGURE 1: Schematic of the ICSIPs (unit: mm).

reasonable for GFRP as the bottom face sheet provides high tensile strength, high durability, and fire resistance; EPS as the core are characterized by light weight, thermal insulation, and excellent impact properties, and recycled aggregate concrete as the top face sheet provides high compressive strength and low cost. The assembly of the bottom GFRP face sheet and EPS core can serve as a template for the top concrete face sheet. The application of recycled aggregate concrete reduces the construction waste materials effectively and achieves the purpose of environmental protection. Moreover, ICSIPs have another advantage related to corrosion, for the bottom GFRP face sheet, and the EPS core can prevent moisture entering the cracks in concrete.

Figure 2 illustrates the fabrication process of ICSIPs. The GFRP face sheet is bonded to the EPS core using a hot-melt thermoplastic spray adhesive. Then, the assembly is used as the template supporting the setup of reinforced steel mesh and the cast of concrete. The recycled aggregate concrete can bond with the EPS core naturally during the hardening process of concrete. High-quality and attractive panels can be produced in a short duration by large-scale assembly lines.

2.2. Materials of ICSIPs. The thermoplastic composite bottom face sheet used for ICSIPs consists of 70% bidirectional glass-fibers impregnated with polypropylene resin. The production of GFRP is directly obtained from the manufacturer with mechanical properties listed in Table 1. The EPS foam core has a high *R* value, which represents better resistance to the flow of heat. Good thermal resistance property is valuable when CSIPs are used as roof covers. The properties of EPS foam bought from the manufacturer and used in this study are provided in Table 1.

Recycled aggregate concrete is used as the top face sheet. The use of recycled aggregate concrete can not only reduce the construction waste but also decrease the cost. Ordinary Portland cement with a strength of 32.5 MPa and natural river sand fine aggregate with a fineness modulus of 2.67 are used in the recycled aggregate concrete. The recycled coarse aggregate is a commercial concrete with a strength of C20~C40 discarded by a testing station. The concrete was crushed using a small jaw crusher, which produced

(a) (b) (c) (d)

FIGURE 2: Manufacturing of ICSIPs: (a) the EPS core was bonded on GFRP; (b) steel mesh was setup; (c) the recycled aggregate concrete was cast; (d) a piece of ICSIP was made.

TABLE 1: Mechanical properties of materials.

Items	Glass-PP face sheet	EPS foam core	Recycled aggregate concrete face sheet	Reinforced steel
Nominal thickness, t	3.04 mm	70 mm	70 mm	/
Density, ρ_f	980 kg/m^3	36 kg/m^3	2500 kg/m^3	7850 kg/m^3
Weight percentage of glass-fiber	70%	/	/	/
Longitudinal modulus, Ex	15,169 MPa	75 MPa	26 GPa	206 GPa
Transverse modulus, Ey	15,169 MPa	75 MPa	26 GPa	206 GPa
Thickness direction modulus, Ez	1050 MPa	75 MPa	26 GPa	206 GPa
Poisson's ratio (ν_{xy}, ν_{yz}, ν_{xz})	0.11, 0.22, 0.22	0.25, 0.25, 0.25	0.2, 0.2, 0.2	0.3, 0.3, 0.3
Shear modulus (G_{xy}, G_{yz}, G_{xz})	1800 MPa, 1800 MPa, 750 MPa	50 MPa, 50 MPa, 50 MPa	10.4 GPa, 10.4 GPa, 10.4 GPa,	79 GPa, 79 GPa, 79 GPa,
Tensile strength (MPa)	690 MPa	0.6 MPa	1.20 MPa	235 MPa
Compression strength (MPa)	317 MPa	0.8 MPa	20.4 MPa	235 MPa

aggregates with a maximum nominal size of 30 mm. After that, the aggregates were separated according to their dimension, by mechanical sieving, and only the diameter between 5 mm and 20 mm was used. After determining the basic performance indexes such as void ratio, water absorption rate, and apparent density of recycled coarse aggregate, the mix design was determined (Table 2). Standard concrete test blocks are produced in the laboratory

TABLE 2: Recycled aggregate concrete composition.

Items	Composition
P.O 32.5 cement (kg/m^3)	476
River sand (kg/m^3)	742.5
Coarse aggregate (kg/m^3)	1090.5
Water (l/m^3)	191
W/C ratio	0.40

according to the Chinese building code [16]. After a 28-day period of wet curing, the material properties of recycled aggregate concrete were tested in the laboratory and the test results are presented in Table 1. According to the construction requirements, a steel-mesh layer is used for reinforcing concrete. The diameter of steel bars is 8 mm and the space is 200 mm. The mechanical properties of reinforced steel bars are also listed in Table 1.

It is well known that debonding between the core and face sheet is the main failure mode of sandwich composite structural. According to the pull-off testing of the three kinds of adhesives for EPS core and GFRP face sheet, epoxy spray adhesive was the most cost-effective candidate adhesive [2]. Thus, the epoxy spray adhesive was chosen for bonding the bottom GFRP face sheet and EPS core in this paper. The adhesive used in ICSIPs is a two-component high-temperature adhesive based on epoxy resin. This adhesive was purchased from USA Osborne (China) Co., Ltd., and the performance specifications of the material are shown in Table 3.

3. Experimental Investigations

3.1. Specimen Preparations and Test Setup. Five specimens were produced in the laboratory and the production process is shown in Figure 2. At first, the EPS core was bonded to the bottom GFRP face sheet using the adhesive. Then, a layer of steel mesh was setup on the top of the EPS core. Finally, the recycled aggregate concrete was cast. After curing for 28 days, the ICSIPs were finished. The standard concrete blocks were made for measuring the material properties which are shown in Table 1. The dimension of the specimen is shown in Figure 1 which is the same with traditional SIPs used in the modular buildings nowadays [10].

The experimental testing was performed according to the ASTM E-72-5 standard [17]. And the load was applied at the one-third and two-thirds span of the specimen through a 100 kN SHIMDZU universal testing machine with a loading rate of 2 mm/min. Figure 3 shows the actual test setup for the flexural test of ISCIPs. Electric resistance strain gauges were used to monitor longitudinal strains in the tension and compression skins at midspan along the two center lines of the specimen. A displacement transducer with sensitivity 0.1 mm was used to record the midspan deflections, which were recorded by the testing machine automatically.

3.2. Experimental Results. The load-deflection relationship is meaningful to reflect the flexural bearing capacity of ICISPs.

TABLE 3: Performance specifications of the epoxy adhesive.

Items	Parameters
Viscosity (cps)	103000
Curing time (h)	24
Shore hardness	75 ± 2
Glass transition temperature (°C)	170
Shear strength (MPa)	≥250
Compressive strength (MPa)	≥5
Tensile strength (MPa)	≥2.5

FIGURE 3: Experimental setup for the test.

The experimental results of the five test pieces are basically the same. And the average of the results of the five test pieces was calculated to evaluate the mechanical properties of the ICISPs. As seen from Figure 4, the load-deflection curve for the experimental panels can be divided into three stages. In stage I, the curve is linear. The load is 15.8 kN in the end of stage I and the corresponding displacement is 3.9 mm. Then in stage II, the characteristic of nonlinear can be observed. The load arrives the peak value at 27.7 kN with a midspan deflection of 8.8 mm. Before the load reaches the peak value, cracks can be observed at the recycled aggregate concrete top skin at the end of stage II. The debonding between recycled aggregate concrete face sheet and EPS core appears at the end of the stage II. In stage III, the deflection keeps increasing but the load decreases. It can be observed that the concrete is crushed in this stage. And the ICISPs reach the ultimate bearing capacity.

Figure 5 shows the typical failure states of the ICSIPs. The results of the experiments show that the specimen fails in a brittle manner due to the crush of concrete. The appearance of the initial cracks and the crack propagation in the concrete layer can be observed during this stage with the expansion of the crack debonding between the recycled aggregate concrete faces and EPS core happens. There is no debonding between EPS core and GFRP face sheet, and there is no visible failure of the bottom face sheet. In the end of stage III, the recycled aggregate concrete top skin is crushed fully and the deflection arrives to 164 mm. But the bottom skin showed good capacity of large deformation.

The tensile and compressive strains are recorded from the strain gauges attached to the ICSIPs specimen, as plotted in Section 6.2. The results show that the longitudinal strains of GFRP skin and recycled aggregate concrete skin increase linearly with load before any failure is observed. This shows

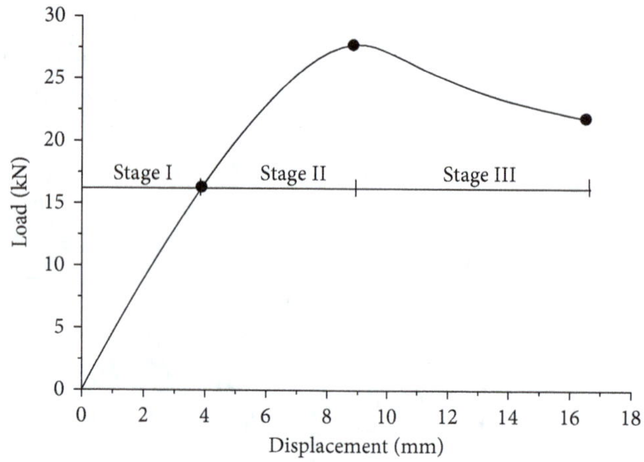

FIGURE 4: Load-deflection in the midspan position.

FIGURE 5: Typical failure mode of the ICSIPs.

the elastic behavior of ICSIP floor panels. When the specimen fails, the tensile strain in the bottom GFRP skin is 0.000972 and the compressive strain in the top concrete skin is 0.000442.

4. Investigation by Numerical Simulation

4.1. Finite-Element Models. The ICSIPs in this paper were analyzed using the universal finite-element software ANSYS [18]. The structural scheme of the panel is 2.4 m in span and 140.04 mm in thickness referring to the CSIP [10]. The materials of the core and the bottom GFRP skin of ICSIP are the same to those of CSIP. The panel is constrained on simple supports. The loads are applied on the nodes at one-third and two-thirds span of the top layer to simulate the testing situation.

The properties of all materials are listed in Table 1. Standard Solid45 element of the ANSYS element library is used for the models of the EPS core and the GFRP skin, and Solid65 element is used to simulate the recycled aggregate concrete [15, 19, 20]. This element is capable of simulating plastic deformation, calculating cracking in three orthogonal directions, and judging crushing. In ICSIPs models, there is no relative movement between face sheets and EPS foam cores as they are glued together. Ideal linear elastic relationships are assumed for the GFRP and EPS foam. Multilinear stress-strain relationship for recycled aggregate concrete is adopted (Figure 6) because both cracking and crushing failure modes are accounted for. The shear transfer coefficient introduced into the crack model is 0.3 [21–24]. Reinforcement steels are discredited into elements, and the effects of reinforcing are averaged within the pertaining element. Cracks are idealized into the smeared type. 19800 elements are required for the ICSIPs model because the face sheets are very thin and the materials are different between face sheets and core.

4.2. Static Analysis. Based on the previously described finite-element model, numerical calculations with ANSYS were conducted. The experimental limit load 27.7 kN is averaged and applied at the loading points in addition to structural dead weight. The results of linear static analysis are summarized in Figure 7.

The maximum vertical displacement of the ICSIP is 7.76 mm. It is about 1/309 the span and can satisfy the requirement of codes with surplus. If only gravity is considered, the displacement is 0.09 mm. This proportion in the total displacement is about 1.2%. It shows that the structural self-weight is light relatively.

From the side-by-side comparison of stresses in three components of ICSIPs (Figure 7), it can be observed that the maximum stress of GFRP face sheet is 11.5 MPa, which occurs at the middle span. It is much less than the allowable stress of GFRP. The maximum stress of the EPS foam at the middle span is much smaller, only 0.01 MPa. But the maximum stress reaches 0.62 MPa at the region besides supports. The compression stress in recycled aggregate concrete is 6.77 MPa at the middle span, much less than the allowable stress. However, the maximum stress of the recycled aggregate concrete is tensile 2.02 MPa at the middle span. It shows the tensile stress is dramatic even if the panel bends downwards. The value of the tensile stress is beyond the allowable stress and cracks occur, so it is necessary to carry on the geometrical nonlinear analysis considering the development of cracks in recycled aggregate concrete.

4.3. Nonlinear Analysis. Nonlinear buckling analysis can satisfy the real-world situation such as large deflection, initial imperfection, and cracks better [15]. A nonlinear constitutive relationship of recycled aggregate concrete was adopted (Figure 6), while the GFRP face sheet and EPS core still use linear elastic constitutive relationship because their maximum stress is much less than their allowable values. The Newton–Raphson equilibrium iteration method was activated, and stress stiffness was considered. Convergence criteria were based on force and displacement, and the convergence tolerance limits were initially selected by the ANSYS program.

The results obtained from finite-element models are summarized in Figure 8. It can be observed that the initial cracking of the panel in the FE model corresponds to a load of 16.4 kN. At this time the, stress is just beyond 1.20 MPa, the modulus of rupture of the recycled aggregate concrete.

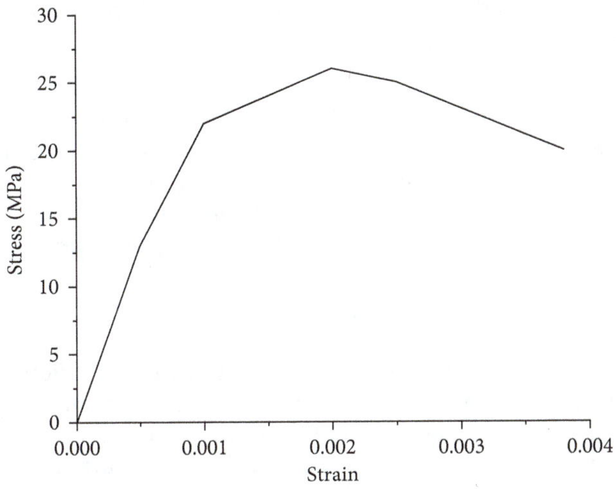

FIGURE 6: Stress-strain relationship of recycled aggregate concrete.

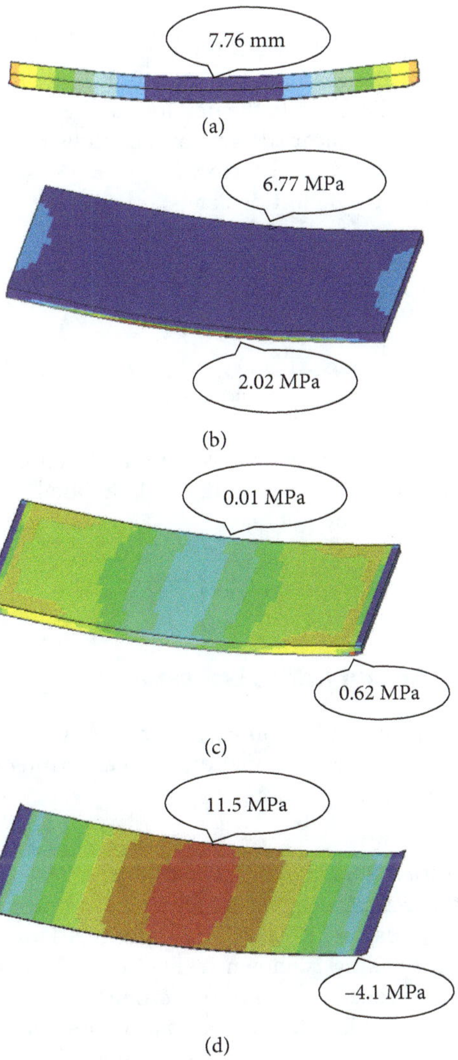

FIGURE 7: Results of static analysis: (a) structural displacement; (b) stress in the recycled aggregate concrete layer; (c) stress in the EPS core; (d) stress in the GFRP skin.

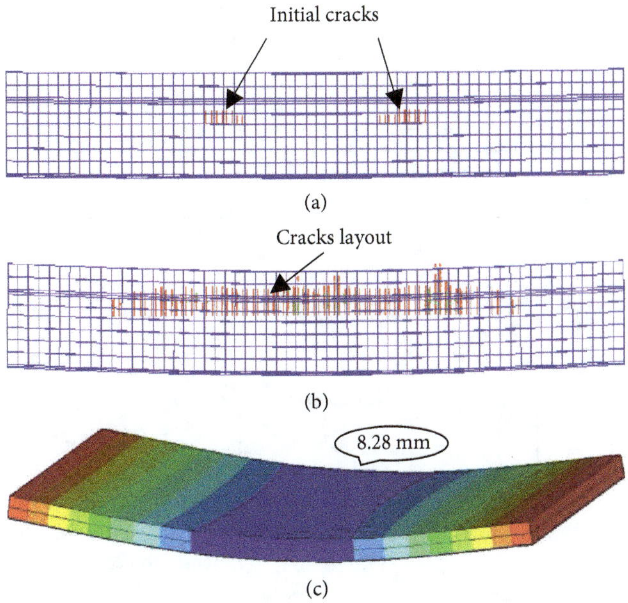

FIGURE 8: Results of nonlinear analysis: (a) crack appears; (b) cracks layout at limit state; (c) structural deformation.

This agrees well with the load of the experiment. The initial crack can be seen to occur in the constant moment region in the middle one-third span, and it is a flexural crack.

Subsequently, cracks expand as the load increases, and these cracks extend out from the initial one. Displacements of the panel begin to increase at a higher rate with the increase of the load. The characteristic of nonlinear is obvious from the load-displacement curve. At 29.4 kN, one crack has reached the top of the panel and failure is soon to follow. At last, a few severe cracks throughout the entire thickness arise, and the compressive stress in concrete has been beyond the ultimate strength. The panel can no longer bear the additional load as indicated by an insurmountable convergence failure. In the last stage, it shows the crushing of recycled aggregate concrete. The maximum displacement at midspan is 8.28 mm.

5. Theoretical Evaluation of the Flexural Response

5.1. Analysis of Load Bearing Capacity. Figures 1 and 9 show the geometrical characteristic and details of the test specimen. The width is $B = 1200$ mm, the depth is $H = 143.04$ mm, and the test span is $L = 2400$ mm. The depth of the top recycled aggregate concrete layer is $t_1 = 70$ mm, the depth of the EPS core is $t_2 = 70$ mm, and the depth of the GFRP layer is $t_3 = 3.04$ mm. The distance between the recycled aggregate concrete of the top layer and the core is $d_1 = 70$ mm, and the distance between the center of the core and the bottom layer is $d_2 = 36.52$ mm.

The failure mode of ICSIPs is the crack of the recycled aggregate concrete top skin when the applied load reaches the ultimate strength. The classical theory of sandwich panels provides the calculation formula corresponding to this failure mode [3, 25–29].

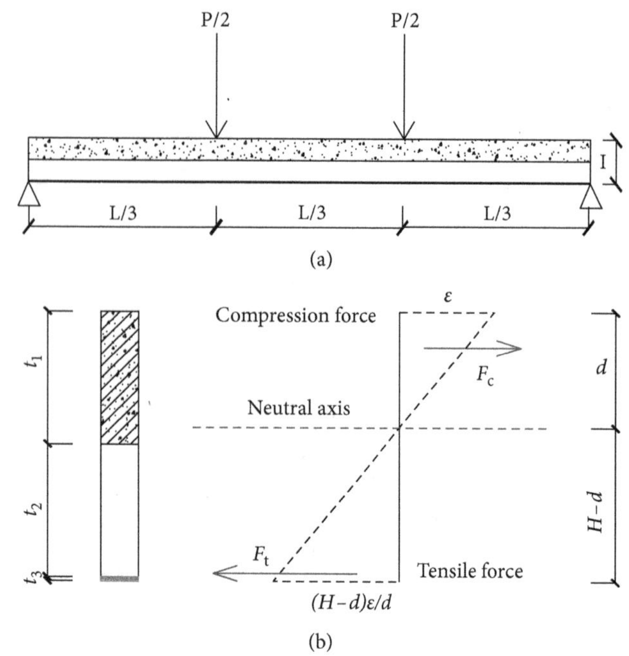

FIGURE 9: Force diagram: (a) schematic illustration the loading; (b) forces for cross section.

$$\sigma_1 = \frac{M}{W} = \frac{MH}{2\mathrm{EI}}E_1, \quad (1)$$

where σ_1 is the bending strength of the concrete compressive skin, M is the moment about the neutral axis, W is the flexural section factor, E_1 is the elasticity modulus of the recycled aggregate concrete compressive skin, and I is the second moment of the neutral axis of the composite sandwich section. EI is the flexural stiffness of the entire section. Then the crest value of the load P can be expressed by the following equation:

$$P = \frac{12\mathrm{EI}\sigma_1}{LE_1H} = \frac{\mathrm{EB}\sigma_1}{LE_1}. \quad (2)$$

And the flexural stiffness of ICSIPs was calculated using the sum of the flexural stiffness of the constituent parts about the center axis of the entire section. It can be obtained by the following equation:

$$\mathrm{EI} = \frac{Bt_1^3}{12}E_1 + Bt_1\left(d - \frac{1}{2}t_1\right)^2 E_1 + \frac{Bt_2^3}{12}E_2 + Bt_2\left(\frac{3}{2}t_2 - d\right)^2 E_2 + \frac{Bt_3^3}{12}E_3 + Bt_3\left(H - d - \frac{1}{2}t_3\right)^2 E_3, \quad (3)$$

where E_2 and E_3 are the modulus of elasticity of the core and the bottom skin, respectively, and d is the distance between compression force line and the neutral axis (Figure 9), and it can be obtained by the following equation:

$$\frac{d}{2}\varepsilon E_1 = F_c = F_t = \frac{\varepsilon E}{2d\eta}(H-d)^2, \quad (4)$$

where E is the equivalent modulus of elasticity of the entire section and η is the reduction factor of the equivalent elastic modulus. And considering the stress state of the ICSIPs, η takes 0.57.

5.2. Analysis of Deflection. In ICSIPs, the shear stiffness of EPS core is very low compared with the face sheets, so the shear deformation should be accounted in the total deflection. And the total deformation of the ICSIPs can be predicted as the sum of bending and the shear deflections:

$$\Delta = \Delta_b + \Delta_s, \quad (5)$$

where Δ, Δ_b, and Δ_s indicate the total deflection, deflection owing to bending, and deflection owing to shear, respectively.

As we know, the deflection at the midspan of the beam is the maximum for a simply supported beam under four-point static bending. So, the total deflection can be calculated by the following equation, which is based on the principle of virtual work from the classical structural analysis:

$$\Delta = \int \frac{M_p\overline{M}}{\mathrm{EI}}dx + \int \frac{kQ_p\overline{Q}}{\mathrm{GA}}dx, \quad (6)$$

where M_p and Q_p indicate the bending moment and the shear due to the load and $\overline{M}$ and $\overline{Q}$ indicate the bending moment and shear due to the unit load at the midspan. GA is referred to as the shear stiffness of the sandwich beams, and k is the shear correction factor.

Therefore, the deflection can be calculated by integrating the deflection along the beam as follows:

$$\Delta_b = 2 \times \frac{PL}{6\mathrm{EI}} \times \left(\int_0^{L/3} \frac{3x^2}{2L}dx + \int_{L/3}^{L/2} \frac{x}{2}dx\right) = \frac{23PL^3}{1296\mathrm{EI}}, \quad (7)$$

$$\Delta_s = 2 \times \frac{K}{\mathrm{GA}} \int_0^{L/3} \left(\frac{P}{2} \times \frac{1}{2}\right)dx = \frac{KPL}{6\mathrm{GA}}. \quad (8)$$

For the specimen with a rectangular section, the shear correction factor K is 1.2. In the end, the total deflection at the midspan can be obtained:

$$\Delta = \frac{23PL^3}{1296\mathrm{EI}} + \frac{PL}{5\mathrm{GA}}. \quad (9)$$

6. Comparison and Discussion

The analysis of results comparison between the theoretical prediction, numerical simulations, and experimental testing is discussed in this section.

6.1. Discussions on Load-Deflection Curves. The comparison of the theoretical analytical, numerical and experimental loads, and midspan deflection curves for ICSIPs tested under four-point bending is shown in Figure 10. To describe the average difference between the experimental data points and the mean, we added standard deviation information in the form of error bars. Overall, the theoretical results are in good agreement with the experimental and numerical results. The theoretical predictions reflect well the linear flexural response of the ICSIPs, while deviate as the load

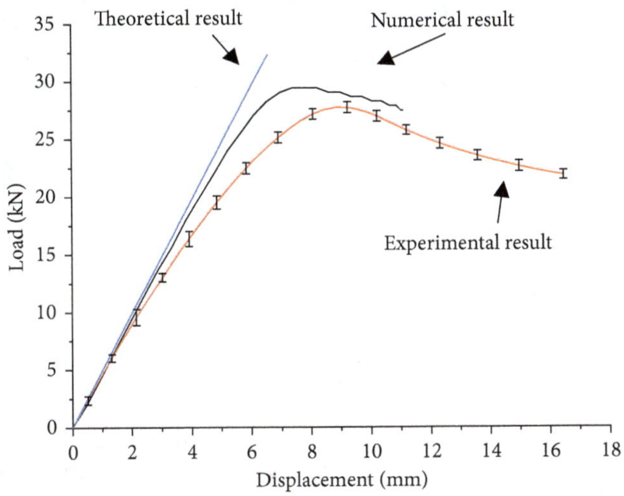

FIGURE 10: The load-deflection relation curves of the experimental (mean ± SD, $n = 5$), numerical, and theoretical results.

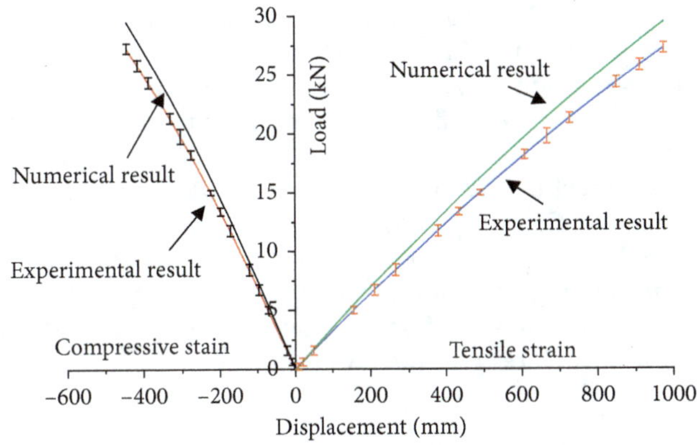

FIGURE 11: Load-strain relationships. The bar graphs are presented as mean ± SD ($n = 5$).

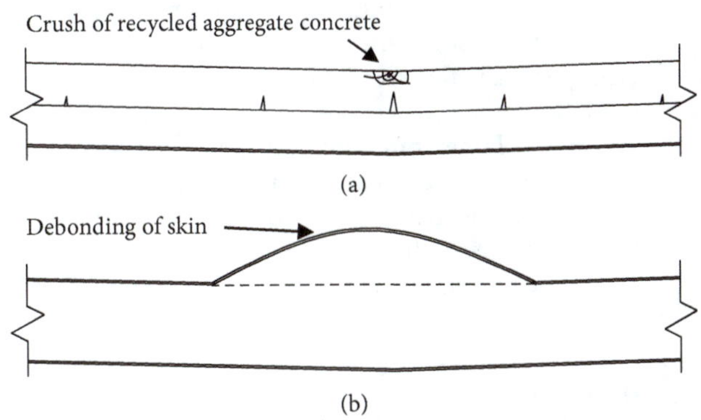

FIGURE 12: Failure modes of ICSIP and CSIP: (a) crush of recycled aggregate concrete for ICSIP; (b) debonding of skin for CSIP.

increases up to failure due to the theory limitations. The load increases linearly with deflection until the load reaches the maximum values. Then, the obvious decrease of stiffness is seen due to the appearance of cracking of the top recycled aggregate concrete skin. The cracks extend along the thickness of the top skin, which leads to dramatic increase of deflection.

The maximum load of experiment is 27.7 kN. However, the theoretical peak value of the load is 32.3 kN which is 16.6% larger than the experimental results. Correspondingly, the maximum deflection of theoretical results is 6.72 mm, which is 13.4% smaller than the experimental results. The reason is owing to the assumption of linear elastic behavior of materials.

The FE analysis considering the double nonlinear behavior of material and geometry provides the process of failure in good agreement with the experiments. The numerical value of the maximum load is 29.4 kN, which is a little larger than that of the experimental result. And the corresponding deflection of midspan is 8.28 mm, which is a little smaller than that of the experimental result. The small divergence between the experimental and numerical results is attributed to two reasons. One is the effect of the debonding phenomenon was not considered in the FE analysis, and the other is the errors during the manufacture and test of specimen.

6.2. Discussions on Load-Strain Relationship. Figure 11 shows the comparison of the load-strain relationship between the numerical and experimental results (mean ± SD, $n = 5$). The linear relationship can be observed in both the tension and compression skins. The numerical and experimental results have good agreement. The good match between the experimental and the simulated response of the ICSIPs indicated the good interaction between the core and double skins.

6.3. Failure Mode. The experimental results showed that the failure mode of the specimen under four-point bending test is the compressive failure of the top recycled aggregate concrete skin. The failure mechanism analysis from the FE analysis is shown in Figure 12. The maximum stress in the top recycled aggregate concrete skin reaches the critical value, while the stress in the bottom GFRP skin is far less than their allowable value. The cracks appear at the bottom of the concrete layer at the midspan and extended upwards quickly until the crush of the recycled aggregate concrete layer. When the specimen reached the ultimate load, the debonding occurs between the top recycled aggregate concrete face sheet and the EPS core due to crushing of the recycled aggregate concrete face sheet. However, there is no debonding between EPS core and GFRP face sheet. In all specimens, the failure mechanisms predicted from the FE simulations are similar to the failure mechanisms observed in the experimental investigation.

6.4. Comparison with the CSIPs. The research results of the improved CSIPs in this paper show three obvious differences from CSIPs [10]. The first is the load bearing capacity of ICSIP 3.46 times of that of CSIP. The frontier is 27.7 kN while the latter is 8 kN. It shows that ICSIP is strong enough to bear the load 7.6 kN/m^2 except of self-weight when they are used in floors with the span 2.4 m.

The second is that the maximum deflection of ISCIP (8.8 mm) is about 6 times smaller than that of CSIP (52 mm). The deformation of 1/273 of the span shows that it can satisfy the requirement of stiffness when they are used in floors or roofs.

The third is that the failure mode of ICSIP is different from that of CSIP (Figure 12). The frontier is the crush of the top recycled aggregate concrete layer. However, the latter is the debonding between the compression face sheet and the core which causes the face sheet wrinkling failure and the core shear failure immediately.

6.5. Comparison with the Recycled Aggregate Concrete Panel. If only the recycled aggregate concrete top layer is kept while the EPS core and the bottom GFRP skin are moved away, the limit load is 2.8 kN. Even if enough reinforced steel bars are set inside to abide by the building codes for panels, its ultimate load is no more than 12.5 kN. It shows the ICSIP is much stronger than the recycled aggregate concrete panel with nearly the same self-weight. The sandwich structure of ICSIP has much larger moment resistance stiffness by enhancing thickness of the panel. And advantages of compressive concrete and tensile GFRP materials are made full use. The sandwich structure of ICSIP can also be used for reference to strengthen the damaged concrete floors or roofs in service.

7. Conclusions

New improved composite structural insulated panels (ICSIPs) were presented for the structural floor of roof applications in this paper. A full-scale experimental test of four-point bending was investigated. The FE analysis was conducted based on the ANSYS program. And theoretical formulas were also developed to calculate the ultimate bearing load and deflection. The main conclusions from this study can be summarized as follows:

(1) The ICSIPs are applicable for building floors or roofs with enough structural strength and stiffness in comparison with CSIPs. The ICSIPs can overcome the two shortcomings of CSIPs; one is the large deflection, and the other is the low bearing capacity. It is especially suitable for large-span roofs owing to excellent thermomechanical properties and superior durability except of high strength.

(2) The general mode of failure of ICSIPs was the crush of the top recycled aggregate concrete layer at the midspan.

(3) The nonlinear FE analysis is effective in the engineering design of ICSIPs in engineering practice. And the theoretical predictions reflect well the linear flexural response of the ICSIPs, while deviates as the load increases up to failure due to the theory limitations.

(4) ICSIPs are strong enough to bear heavy loads. From the analysis of stiffness, it is still the requirement of deflection that controls the design of ICSIPs. Therefore, the strength of the recycled aggregate concrete is feasible as the top face sheet in ICSIPs, and it is conducive to save resources and to protect the environment.

Conflicts of Interest

The authors declare that they have no conflicts of interest.

Authors' Contributions

QL designed and performed the experiments, as well as drafted the manuscript. WD have carried the literature study and participated in experiments design and result discussions. NU provided some advices on the manuscript. ZZ performed the experimental works. All authors read and approved the final manuscript.

Acknowledgments

The authors would like to express sincere thanks to Yun Sun., Zhihao Du., and Shuailiang Zhang., for supporting the fabrication of test specimens. This research was supported by the National Science Foundation (NSF) (CMMI-0825938) and National Science Foundation of China (NSFC) (U1704141).

References

[1] C. G. Weber, "Fiber building boards," *Journal of Industrial and Engineering Chemistry*, vol. 27, no. 8, pp. 896–898, 1935.

[2] N. Uddin, A. Vaidya, U. Vaidya, and S. Pillay, "Thermoplastic composite structural insulated panels (CSIPs) for modular panelized construction," *Developments in Fiber-Reinforced Polymer (FRP) Composites for Civil Engineering*, vol. 12, pp. 302–316, 2013.

[3] H. G. Allen and B. G. Neal, *Analysis and Design of Structural Sandwich Panels*, Elsevier, Amsterdam, Netherlands, 1969.

[4] M. Morley, *Building with Structural Sandwich Pannel*, The Taunton Press, Newtown, CT, USA, 2000.

[5] J. M. Davies, *Lightweight Sandwich Construction*, University of Manchester press, Manchester, UK, 2001.

[6] N. Uddin, A. Vaidya, N. Uddin, and U. Vaidya, "Design and analysis of composite structural insulated panels (CSIPs) for exterior wall applications," *Journal of Composites for Construction*, vol. 14, no. 4, pp. 1313–1322, 2010.

[7] N. M. Long, D. Vo-Le, D. Tran-Thanh, T. M. Pham, C. Ho-Huu, and M. Rovňák, "Shear capacity of unbonded post-tensioned concrete T-beams strengthened with CFRP and GFRP U-wraps," *Composite Structures*, vol. 184, pp. 1011–1029, 2018.

[8] G. Boscato and S. Ientile, "Experimental and numerical investigation on dynamic properties of thin-walled GFRP buckled columns," *Composite Structures*, vol. 189, pp. 273–285, 2018.

[9] D. Y. Yoo, N. Banthia, and Y. S. Yoon, "Flexural behavior of ultra-high-performance fiber-reinforced concrete beams

reinforced with GFRP and steel rebars," *Engineering Structures*, vol. 111, pp. 246–262, 2016.

[10] M. A. Mousa and N. Uddin, "Flexural behavior of full-scale composite structural insulated floor panels," *Advanced Composite Materials*, vol. 20, no. 6, pp. 547–567, 2011.

[11] M. A. Mousa and N. Uddin, "Structural behavior and modeling of full-scale composite structural insulated wall panels," *Engineering Structures*, vol. 41, pp. 320–334, 2012.

[12] N. Uddin and W. Du, "New thin shells made of composite structural insulated panels," *Journal of Reinforced Plastics and Composites*, vol. 33, no. 21, pp. 1954–1965, 2014.

[13] M. A. Mousa and N. Uddin, "Debonding of composites structural insulated sandwich panels," *Journal of Reinforced Plastics and Composites*, vol. 29, no. 12, pp. 3380–3391, 2010.

[14] N. Uddin, M. Mousa, and N. Uddin, "Performance of composite structural insulated panels (CSIPs) after exposure to floodwater," *Journal of Performance of Constructed Facilities*, vol. 27, no. 4, pp. 424–436, 2010.

[15] W. Du and N. Uddin, "Innovative composite structural insulated panels (CSIPs) folded shell structures for large-span roofs," *Materials and Structures*, vol. 50, no. 1, p. 51, 2017.

[16] Ministry of Housing and Urban-Rural Construction of the People's Republic of China, *Standard for Evaluation of Concrete Compressive Strength*, AQSIQ, China, 2010.

[17] Annual Book of ASTM Standards E-72-05, *Standard Test Method of Conducting Strength Tests of Panels for Building Construction, 100 Barr Harbor Drive*, West Conshohocken, PA, USA, 2005.

[18] ANSYS, *ANSYS 11.0 Manual Set*, ANSYS Inc., Canonsburg, PA, USA, 2008.

[19] J. Tejchman and J. Bobiński, *Continuous and Discontinuous Modelling of Fracture in Concrete Using FEM*, Springer, Berlin, Germany, 2013.

[20] C. Borsellino, L. Calabrese, and A. Valenza, "Experimental and numerical evaluation of sandwich composite structures," *Composites Science and Technology*, vol. 64, no. 10-11, pp. 1709–1715, 2004.

[21] F. Y. Li, L. Y. Li, Y. Dang, and P. F. Wu, "Study of the effect of fibre orientation on artificially directed steel fibre-reinforced concrete," *Advances in Materials Science and Engineering*, vol. 2018, Article ID 8657083, 11 pages, 2018.

[22] N. Grover, D. K. Maiti, and B. N. Singh, "A new inverse hyperbolic shear deformation theory for static and buckling analysis of laminated composite and sandwich plates," *Composite Structures*, vol. 95, no. 1, pp. 667–675, 2013.

[23] R. Sun, B. Xie, R. Perera et al., "Modeling of reinforced concrete beams exposed to fire by using a spectral approach," *Advances in Materials Science and Engineering*, vol. 2018, Article ID 6936371, 12 pages, 2018.

[24] S. Mangalathu and J. S. Jeon, "Classification of failure mode and prediction of shear strength for reinforced concrete beam-column joints using machine learning techniques," *Engineering Structures*, vol. 160, pp. 85–94, 2018.

[25] J. R. Vinson, *The Behavior of Sandwich Structures of Isotropic and Composite Materials*, Technomic, Lancaster, PA, USA, 1999.

[26] R. A. Hawileh, H. A. Rasheed, J. A. Abdalla, and A. K. Al-Tamimi, "Behavior of reinforced concrete beams strengthened with externally bonded hybrid fiber reinforced polymer systems," *Materials and Design*, vol. 53, no. 1, pp. 972–982, 2014.

[27] H. T. Thai and D. H. Choi, "A simple first-order shear deformation theory for laminated composite plates," *Composite Structures*, vol. 106, no. 12, pp. 754–763, 2013.

[28] A. J. M. Ferreira, "A formulation of the multiquadric radial basis function method for the analysis of laminated composite plates," *Composite Structures*, vol. 59, no. 3, pp. 385–392, 2003.

[29] M. M. Kheirikhah, S. M. R. Khalili, and K. M. Fard, "Biaxial buckling analysis of soft-core composite sandwich plates using improved high-order theory," *European Journal of Mechanics-A/Solids*, vol. 31, no. 1, pp. 54–66, 2012.

5

Al6061 Hybrid Metal Matrix Composite Reinforced with Alumina and Molybdenum Disulphide

G. Pitchayyapillai,[1] P. Seenikannan,[1] K. Raja,[2] and K. Chandrasekaran[3]

[1]*Department of Mechanical Engineering, Sethu Institute of Technology, Kariapatti, Tamil Nadu, India*
[2]*Department of Mechanical Engineering, University College of Engineering, Anna University, Dindigul, Tamil Nadu, India*
[3]*Department of Mechanical Engineering, MAM School of Engineering, Trichy, Tamil Nadu, India*

Correspondence should be addressed to G. Pitchayyapillai; gp_pillai05@yahoo.com

Academic Editor: Luigi Nicolais

Aluminum Hybrid Reinforcement Technology is a response to the dynamic ever-increasing service requirement of industries such as transportation, aerospace, automobile, and marine, due to its attractive properties like high ductility, highly conductivity, light weight, and high strength to weight ratio. In this evolution, an attempt has been made to investigate the wear rate of Al6061 hybrid metal matrix composite reinforced with the hard ceramic alumina (4, 8, and 12 wt.% of Al_2O_3) and soft solid lubricant of molybdenum disulphide (2, 4, and 6 wt.% of MoS_2) is fabricated by using stir casting method. The unlubricated pins on disc wear tests were conducted to examine the wear behaviour of Al6061/12 wt.% of Al_2O_3/MoS_2 composites. The sliding wear tests were carried out at various loads of 15, 30, and 45 N, sliding velocity (1.25, 2.50, and 3.25 m/sec), and different MoS_2 wt.% (2, 4, and 6 wt.%). In addition, the CNC turning experiments were conducted on Al6061/12 wt.% Al_2O_3/6 wt.% MoS_2 using CNMG 120408 uncoated carbide cutting tool under cutting of 100, 150, and 200 m/min, feed of 0.1, 0.2, and 0.3 mm/rev, and depth of cut of 1, 1.5, and 2 mm.

1. Introduction

Aluminum hybrid matrix composites have become better substitutes for the conventional aluminum alloys because of their characters like improved strength to weight ratio, energy saving, and better wear resistance [1]. Reinforced aluminum hybrid matrix, a composite for the use of the automotive industry and other structural applications, has been increased because of the physical and mechanical properties it possesses [2]. Aluminum based hybrid matrix composites with a variety of particulate reinforcements such as Al_2O_3, TiC, Ti_2B, B_4C, TiN, Si_3N_4, graphite, and industrial waste byproduct have been conceived and developed for various potential applications [3]. Researchers have developed aluminum hybrid matrix composites and they are used in many commercial and industrial applications [4]. Newer compositions have been continuously explored and the related valuable studies presented by the past researchers were discussed. Veeresh Kumar et al. [5] conducted the experiment on Al6061/SiC and Al7075/Al_2O_3 metal matrix composites and presented the experimental results of the studies regarding hardness, tensile strength, and wear resistance properties of Al6061/SiC and Al7075/Al_2O_3 composites. The SiC and Al_2O_3 resulted in improving the hardness and density of their respective composites.

Anilkumar et al. [6] have conducted the mechanical properties of fly ash reinforced Al6061 composites and found that the tensile strength, compressive strength, and hardness of the Al6061 composites decrease with the increase in particle size of reinforced fly ash. Increase in the weight fractions of the fly ash particles increases the ultimate ensile strength, compressive strength, and hardness and decreases the ductility of the composite. Asif et al. [7] have investigated the development of aluminum based hybrid metal matrix composites for heavy duty applications and investigated the dry sliding wear behaviour of aluminum alloy based composites, reinforced with silicon carbide particles and solid lubricants such as graphite/antimony trisulphide. The results revealed that wear rate of hybrid composites is lower than the binary composite. The wear rate decreases with the increasing load and increases with increasing speed.

Singh et al. [8] have conducted an experimental investigation on mechanical behaviour of aluminum by adding SiC and alumina. This work was focused on studying the change in behaviour of aluminum by adding a different percentage age amount of SiC and Al_2O_3 composite and it was concluded that the mechanical properties such as hardness, yield strength, and ultimate strength increasing with a weight percentage of reinforcement increase. But at the same time elongation is decreased and the behaviour of material is changed from ductile to brittle. Gaitondel et al. [9] have done an experiment on wear and corrosion properties of Al/Al_2O_3/graphite hybrid composites. The effects of reinforcement, time duration, and particle size on prepared samples of composites have been studied on slurry erosive wear. The static and accelerated corrosion tests have been performed and the microhardness of the developed composites was also investigated. The experimental results on Al5083/Al_2O_3/graphite hybrid composites revealed that the addition of reinforcement improves the hardness and reduces corrosion and wear rates.

Sharanabasappa and Motgi [10] were conducting experiments on mechanical properties of fly ash and alumina reinforced aluminum alloy composites. The mechanical properties of fly ash and alumina reinforced aluminum alloy (LM25) composite samples, processed by stir casting route, were reported. It was found that the tensile strength and hardness of the aluminum alloy composites increase with the increase in %wt. of Al_2O_3. Kuma and Singh [11] investigated a comparative investigation of mechanical properties of aluminum based hybrid metal matrix composites. The result indicated that there is an increase in the value of tensile strength, ultimate tensile strength, hardness value, and flexural strength of newly developed composite having SiC and B_4C particulates in comparison to the SiC, graphite reinforced composite.

Wang and Yan [12] investigated the feasibility of machining Al6061/Al_2O_3 composite materials by electro discharge machining for blind hole drilling and evaluated the material removal rate, tool wear rate, and surface roughness with various input parameters. Yan and Wang [13] observed that material removal rate is increased with peak current and was erratic with pulse duration. The material removal rate, tool wear rate, and surface roughness are increased with flushing pressure during machining of Al6061/Al_2O_3 composites using rotary electro discharge machining with a tube electrode. Yan et al. [14] found that, in case of rotary electro discharge machining process, the main challenge is using a disk like electrode for machining Al6061/Al_2O_3 composites. The influence of reinforcement of SiC and Al_2O_3 in different matrix of Al6061 and Al7075 on mechanical properties was investigated by Veeresh kumar et al. [15] and they found that microhardness of the composites was increased linearly with the increase of filler content. The surface roughness increases with increase in volume fraction of Al_2O_3 particles reinforced with aluminium metal matrix composites when machined by carbide coated tools [16]. The cutting velocity has 20.8% influences on the surface roughness during machining of A356/20/SiC/T6 composite [17]. Better surface finish was achieved while machining aluminum metal matrix composites with 5% SiCp at higher cutting speed produces [18]. Metin Kok [19] concluded that while machining Al2024/Al_2O_3 composites using coated and uncoated carbide cutting tools the surface roughness of the work piece was mostly affected by cutting speed. The machined surfaces of Al-Si alloy-graphite composites tend to be rougher than the surfaces machined at identical conditions on matrix material without reinforcement because of the presence of deeper holes or valleys [20]. The surface roughness value decreases with increasing cutting speed.

However many researchers have carried out the mechanical and wear properties and machining characteristics of aluminum metal matrix composites with SiC and Al_2O_3 as reinforcing materials. In the case of hybrid Al6061/Al_2O_3/MoS_2 composites, limited literature is available, encompassing various aspects such as mechanical properties and wear behaviour and conducting the machining study of the composites. Aluminum based Al_2O_3 composite material has properties such as low weight, heat resistant, wear resistant, and low cost. These are found in various engineering applications such as cylinder block liners, vehicle drive shafts, automotive pistons, and bicycle frames. These materials are known as difficult to machine materials, because of the hardness and abrasive nature of reinforcement element like alumina particle [21]. The addition of MoS_2 reinforcement in composites as a hybrid reinforcement further increases the wear resistance of the composite. This is attributed to the stable and MoS_2 rich mechanically mixed layer, which prevents metal to metal contact and reduces the wear of the composite. The unstable mechanically mixed layer in the absence of the lubricant phase leads to lower wear resistance in hybrid composite [22]. So an attempt was made to study the tribological, mechanical, and machining characteristics on Al6061/Al_2O_3/MoS_2 hybrid composite.

2. Experimental Details

The stir casting method is simple and the most economical way of fabricating particulate reinforced composites. In this technique, to accomplish the optimum properties of the hybrid composites, the distribution of the reinforced particles in the base material should be homogeneous and the wettability among the olden materials and particulates ought to be optimized. The moisture level in the cast composite must be reduced and the element reactions among the particles and the base material have to be avoided. The whirlpool method is one of the enhanced recognized tactics used to build a high quality allocation of the reinforced material in the base matrix. In this method, the base material is melted followed by a forceful stirring by automatic agitator to form a whirlpool at the face of dissolve, and the particle is subsequently introduced at the region of the vortex.

2.1. Material Fabrication. The chemical composition of Al6061 alloy is given in Table 1 and Al6061 alloy was

Table 1: Chemical composition of Al6061 alloy.

Si	Fe	Cu	Mn	Mg	Cr	Zn	Ti	Al
0.65	0.7	0.25	0.15	0.8	0.07	0.25	0.15	Remaining

Table 2: Affecting factors and levels selected for tensile strength Al6061/Al_2O_3/MoS_2.

Factors/levels	1	2	3
X: wt.% Al_2O_3 (wt.%)	4	8	12
Y: wt.% MoS_2 (wt.%)	2	4	6

melted in a crucible furnace at 725°C. Subsequent to melting and degassing by nitrogen, the molten metal was stirred by an alumina coated stainless steel stirrer at 600 rpm for 20 min time duration. During stirring, it is preheated at 600°C; particles alumina and MoS_2 powder were added. After the addition of MoS_2, the composite was poured into the preheated (250°C) permanent mould. The Al6061 alloy with different wt.% of hybrid composites is produced and test specimens were machined according to ASTM standards.

2.2. Mechanical Behaviour. The hardness tests were carried out according to ASTM E10-07 standards using Brinell hardness testing machine with a 10 mm ball indenter and 500 kg load for 30 sec. The test was carried out at atmospheric temperature (30°C) and the measurement of hardness was taken at three different locations on each sample to obtain an average value of hardness. As per the ASTM E08-8 standard, the tensile strength was evaluated on the cylindrical rod of casted composites. The 1200 grit grindings silicon carbide paper was used to polish the test specimens in order to decrease the machining scratches and the effects of surface defects on the sample. The universal testing machine was loaded with 10 KN; load cell was used to conduct the tensile test. The affecting factors and levels selected for mechanical behaviour Al6061/Al_2O_3/MoS_2 are given in Table 2.

2.3. Tribological Behaviour. Dry sliding wear behaviour of Al6061/12 wt.% Al_2O_3/MoS_2 hybrid composites is studied in pin on disc test apparatus. Pin specimens of 6 mm diameter and 15 mm height, for wear test, were prepared from the above composites and the composites were machined and polished. The test was conducted with various loads of 15 N, 30 N, and 45 N at a sliding speed of 125, 2.50, and 3.25 m/s and 2, 4, and 6 MoS_2 wt.%. It was conducted at room temperature (30°C) and relative humidity of 60–65%. The affecting factors and levels selected for Tribology Al6061/12 wt.% Al_2O_3/MoS_2 are given in Table 3.

2.4. CNC Turning Test. The work material used for the present investigation is Al6061/12 wt.% Al_2O_3/6 wt.% MoS_2 and diameter of the material is 20 mm and machined length is 60 mm for all trials. The experiments were conducted on Fanuc CNC lathe and CNMG 120408; uncoated carbide cutting tool is used as the insert for all machining operations. The affecting factors and levels selected for turning properties Al6061/12 wt.% Al_2O_3/6 wt.% MoS_2 are given in Table 4. SR is measured by the Mitutoyo surface roughness tester. The three cutting parameters selected for the present investigation are cutting speed, feed, and depth of cut. Minitab14 statistical software has been used for the analysis of the experimental work.

3. Result and Discussion

3.1. Optimum Setting for TS and BHN of Al6061/Al_2O_3/MoS_2. The tensile strength and hardness of the Al6061/4, 8, and 12 wt.% Al_2O_3/2, 4 wt.% MoS_2 hybrid composite are shown in Figures 1(a) and 1(b). It is observed that an addition of alumina particles improves the tensile strength and hardness of hybrid composites. It is evident that tensile strength and hardness clearly increase with the addition of alumina particles. The tensile strength and hardness of the Al6061/4, 8, and 12 wt.% Al_2O_3/6 wt.% MoS_2 hybrid composites are shown in Figure 1(c). The tensile strength and hardness of hybrid composites were decreased with increasing content of molybdenum disulphide.

The measured values of tensile strength and BHN for Al6061/Al_2O_3/MoS_2 under different wt.% and corresponding signal to noise ratio for all experimental runs are given in Table 5. The Taguchi analyses of tensile strength and BHN for Al6061/Al_2O_3/MoS_2 under different wt.% are given in Table 6. It clearly shows the maximum tensile strength is obtained at 12 wt.% of Al_2O_3 and 2 wt.% of MoS_2 and the maximum BHN is obtained at 12 wt.% of Al_2O_3 and 4 wt.% of MoS_2. Table 7 shows the results of ANOVA for tensile strength and BHN on Al6061/Al_2O_3/MoS_2 hybrid composite. It is observed that Al_2O_3 is the most significant parameter affecting the tensile strength of Al6061/Al_2O_3/MoS_2 with $F : P$ value of 1092.26 : 0.000 and Al_2O_3 is significantly affecting the BHN of Al6061/Al_2O_3/MoS_2 with $F : P$ value of 480.69 : 0.000.

3.2. Tribological Behaviour of Al6061/Al_2O_3/MoS_2. The variation of wear rate for Al6061/12 wt.% Al_2O_3/MoS_2 hybrid composite with respect to different sliding velocity and applied load is shown in Figures 2(a), 2(b), and 2(c). The applied load affects the wear rate of hybrid composites significantly. The wear rate varying with applied load is an indicative of Archard's law and significantly lower in the case of hybrid composites. The wear rate increases with increasing applied load and it is less at 6 wt.% of MoS_2 hybrid composites as compared to 2 wt.% of MoS_2 [23]. This is mainly due to the presence of Al_2O_3 in Al6061. However, at all load conditions

TABLE 3: Affecting factors and levels selected for Tribology Al6061/12 wt.% Al_2O_3/MoS_2.

Factors/levels	1	2	3
A: load (N)	15	30	45
B: sliding velocity (m/s)	1.25	2.50	3.25
C: wt.% of MoS_2 (wt.%)	2	4	6

TABLE 4: Affecting factors and levels selected for turning properties Al6061/12 wt.% Al_2O_3/6 wt.% MoS_2.

Factors/levels	1	2	3
V (cutting speed (m/min))	100	150	200
F (feed (mm/rev))	0.1	0.2	0.3
D (depth of cut (mm))	1	1.5	2

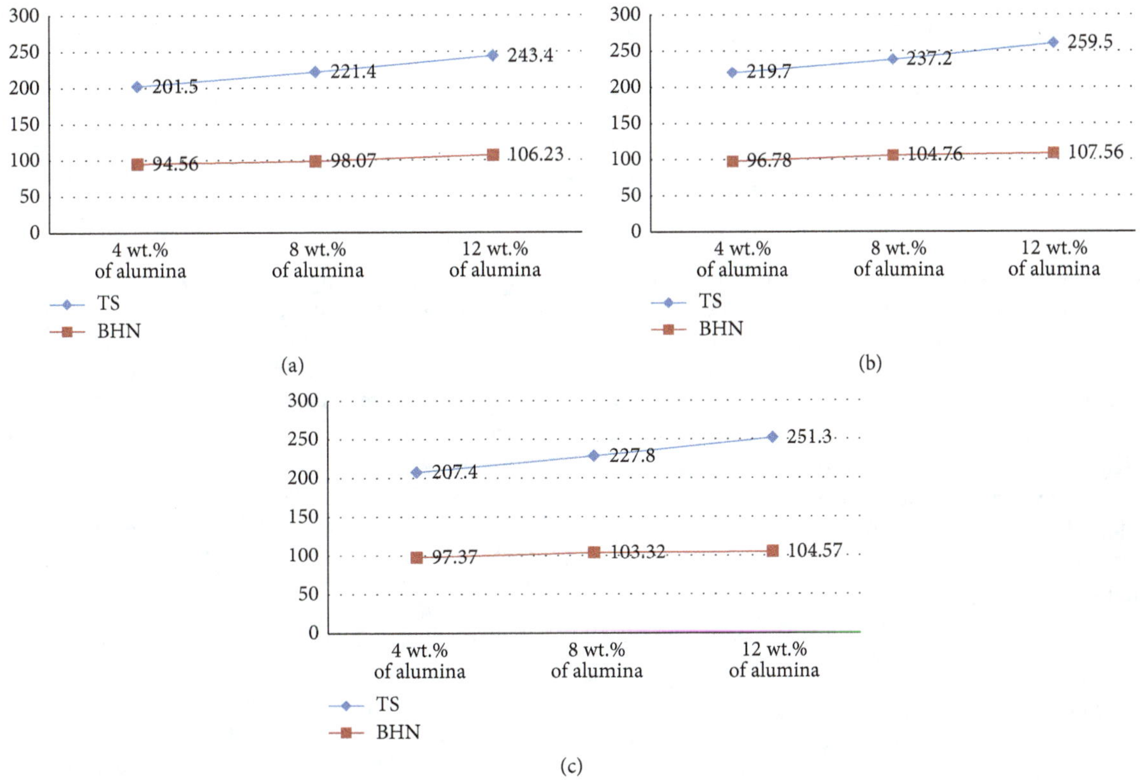

FIGURE 1: (a) Mechanical behaviour of Al6061/2 wt.% of MoS_2/4 : 8 : 12 wt.% of Al_2O_3. (b) Mechanical behaviour of Al6061/4 wt.% of MoS_2/4 : 8 : 12 wt.% of Al_2O_3. (c) Mechanical behaviour of Al6061/6 wt.% of MoS_2/4 : 8 : 12 wt.% of Al_2O_3.

TABLE 5: Experimental result for tensile strength of Al6061/Al_2O_3/MoS_2.

Trial	*X*	*Y*	Al_2O_3 (wt.%)	MoS_2 (wt.%)	TS (N/mm^2)	S/N ratio	BHN	S/N ratio
1	1	1	4	2	201.5	46.0855	94.56	33.2713
2	1	2	4	4	219.7	46.8366	96.78	33.4117
3	1	3	4	6	207.4	46.3362	97.37	33.3184
4	2	1	8	2	221.4	46.9036	98.67	33.4241
5	2	2	8	4	237.2	47.5023	104.76	33.5343
6	2	3	8	6	227.8	47.1511	103.32	33.4698
7	3	1	12	2	243.4	47.7264	106.23	33.5752
8	3	2	12	4	259.5	48.2827	107.56	33.6758
9	3	3	12	6	251.3	48.0038	104.57	33.6255

Table 6: Taguchi analysis: TS & BHN versus X, Y of Al6061/Al_2O_3/MoS_2.

Level	TS		BHN	
	X	Y	X	Y
1	46.42	**46.91**	33.33	33.42
2	47.19	47.54	33.48	**33.54**
3	**48.00**	47.16	**33.63**	33.47
Delta	1.58	0.64	0.29	0.12
Rank	1	2	1	2

Table 7: Analysis of variance for TS & BHN of Al6061/Al_2O_3/MoS_2.

Source	DF	TS				BHN			
		Seq SS	Adj MS	F	P	Seq SS	Adj MS	F	P
X	2	2634.78	1317.39	1092.26	0.000	3.7691	1.8845	480.69	0.000
Y	2	423.56	211.78	175.59	0.000	0.6125	0.3063	78.12	0.001
Error	4	4.82	1.21			0.0157	0.0039		
Total	8	3063.17				4.3973			
		$R^2 = 99.84\%$; R^2 (adj) = 99.69%				$R^2 = 99.64\%$; R^2 (adj) = 99.29%			

considered, the wear resistance of the hybrid composites is superior to the matrix alloy. The wear rate is increased with increase in sliding speed Al6061/12 wt.% Al_2O_3/MoS_2 hybrid composite. Increasing sliding speed temperature is raised to the sliding surfaces; this leads to softening of the matrix and composite pin surfaces and it is due to the enhancement in the hardness. The increase in hybrid composite hardness results in improvement of wear resistance. The presence of MoS_2 and Al_2O_3 particles in the hybrid composite will lead to further reduction of wear rate [24].

The coefficient of friction for Al061/12 wt.% Al_2O_3/MoS_2 hybrid composites under varying load and sliding velocity is shown in Figures 3(a), 3(b), and 3(c). The coefficient of friction is decreased with increasing wt.% of MoS_2 and increasing the due to load. The Al_2O_3 particles act as load bearing elements in the hybrid composites and also this results in formation of more stable lubricating film on the tribosurface of the hybrid composites. The decreased coefficient of friction of hybrid composite with MoS_2 content can be credited to the collective effects of MoS_2 and Al_2O_3 particles in formation of a more resistant tribolayer on the contact surface. The graphite tribofilm minimizes the degree of shear stress transferred to the sliding material underneath the sliding contact area which results in less plastic deformation in the subsurface region and reduces the wear rate in the hybrid composites [25].

3.3. Effect of Alumina and Molybdenum Disulfide Content in Al6061. The wear surface of the Al6061/12 wt.% Al_2O_3/6 wt.% MoS_2 under sliding velocity of 2.50 m/sec is given in Figures 4(a) and 4(b). It clearly reveals the formation of tribolayer on the contact surfaces of aluminum composites tested at 15 N and 45 N conditions. Figure 4(a) reveals that the tribolayer is formed only to a limited extent in Al6061/12 wt.% Al_2O_3/4 wt.% MoS_2 hybrid composite at 15 N load, 2.50 m/s sliding speed. At a higher load of 45 N and 2.50 m/s sliding speed the tribolayer that is formed increases, due to squeezing out of solid lubricant from the hybrid composites. The extent of tribolayer formation also increases with increasing percentage of molybdenum disulphide located immediately below the contact surface.

The experimental values of wear rate and coefficient of friction for Al6061/12 wt.% Al_2O_3/MoS_2 under different parameters and corresponding signal to noise ratio for all experimental runs are given in Table 8. The Taguchi analysis for Al6061/12 wt.% Al_2O_3/MoS_2 under different load, sliding velocity, and wt.% of MoS_2 is given in Table 9. It clearly shows the minimum wear rate and coefficient of friction is obtained at 15 N of load, 3.25 m/sec of sliding velocity, and 6 wt.% of MoS_2. Table 10 shows the results of ANOVA for wear rate and coefficient of friction on Al6061/12 wt.% Al_2O_3/MoS_2 hybrid composite. It is observed that load is the most significant parameter affecting the wear rate and coefficient of friction of Al6061/12 wt.% Al_2O_3/MoS_2 with $F : P$ value of 29.06 : 0.033 and 756.27 : 0.001.

3.4. Optimum Setting for CNC Tuning Al6061/12 wt.% Al_2O_3/6 wt.% MoS_2. The experimental results of arithmetic mean value for turning Al6061/12 wt.% Al_2O_3/6 wt.% MoS_2 under different cutting conditions and corresponding signal to noise ratio for all experimental runs are given in Table 11. The Taguchi analysis for Al6061/12 wt.% Al_2O_3/6 wt.% MoS_2 using uncoated carbide cutting tool is given in Table 12. It clearly shows the minimum surface roughness is obtained at 100 m/min of cutting speed, 0.1 mm/rev of feed, and 1.5 mm of depth of cut for tuning Al6061/12 wt.% Al_2O_3/6 wt.% MoS_2. Table 13 shows the results of ANOVA for surface roughness on Al6061/12 wt.% Al_2O_3/6 wt.% MoS_2 hybrid

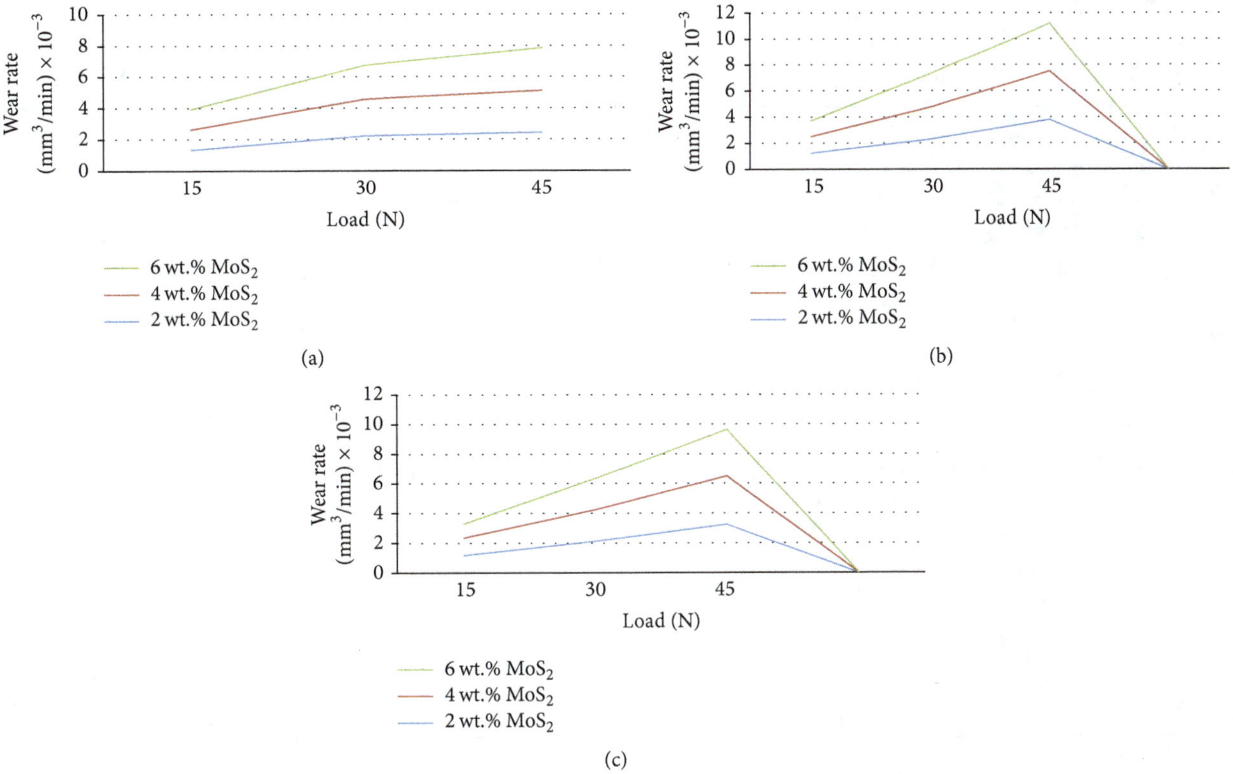

FIGURE 2: (a) Wear rate of Al6061/12 wt.% Al_2O_3/MoS_2 during sliding velocity of 1.25 m/sec. (b) Wear rate of Al6061/12 wt.% Al_2O_3/MoS_2 during sliding velocity of 2.50 m/sec. (c) Wear rate of Al6061/12 wt.% Al_2O_3/MoS_2 during sliding velocity of 3.25 m/sec.

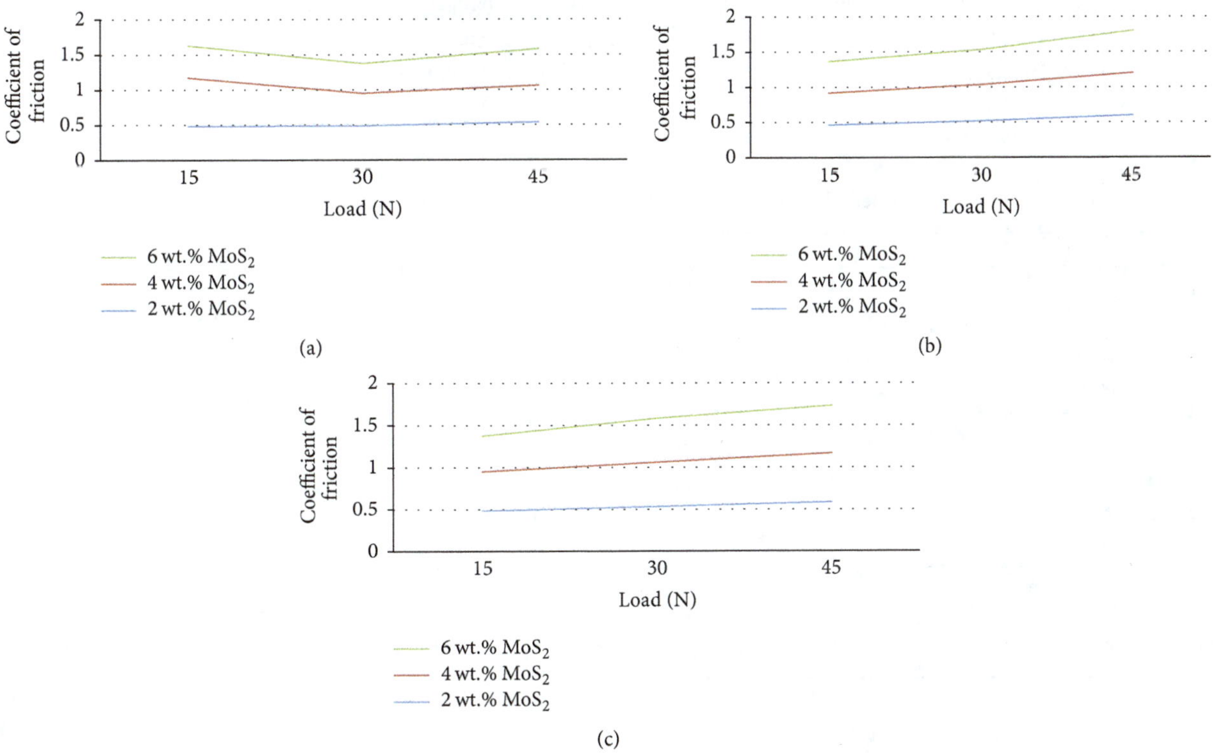

FIGURE 3: (a) Coefficient of friction of Al6061/12 wt.% Al_2O_3/MoS_2 during sliding velocity of 1.25 m/sec. (b) Coefficient of friction of Al6061/12 wt.% Al_2O_3/MoS_2 during sliding velocity of 2.50 m/sec. (c) Coefficient of friction of Al6061/12 wt.% Al_2O_3/MoS_2 during sliding velocity of 3.25 m/sec.

TABLE 8: Experimental results for Al6061/12 wt.% Al_2O_3/MoS_2 of wear study.

Trial	*A*	*B*	*C*	Load (N)	Sliding velocity (m/s)	wt.% of MoS_2 (wt.%)	Wear rate $(mm^3/min) \times 10^{-3}$	S/N	Coefficient of friction	S/N
1	1	1	1	15	1.25	2	1.346	−2.580	0.487	6.249
2	1	2	2	15	2.50	4	1.250	−1.938	0.455	6.839
3	1	3	3	15	3.25	6	0.966	0.300	0.423	7.473
4	2	1	2	30	1.25	4	2.344	−7.399	0.523	5.629
5	2	2	3	30	2.50	6	2.576	−8.218	0.496	6.090
6	2	3	1	30	3.25	2	2.133	−6.579	0.537	5.400
7	3	1	3	45	1.25	6	2.756	−8.805	0.568	4.913
8	3	2	1	45	2.50	2	3.813	−11.625	0.603	4.393
9	3	3	2	45	3.25	4	3.245	−10.224	0.579	4.746

TABLE 9: Taguchi analysis: wear rate and coefficient of friction versus *A*, *B*, and *C* of Al6061/12 wt.% Al_2O_3/MoS_2.

Level	Wear rate			Coefficient of friction		
	A	*B*	*C*	*A*	*B*	*C*
1	**−1.406**	−6.262	−6.929	**6.854**	5.597	5.348
2	−7.399	−7.261	−6.521	5.707	5.775	5.739
3	−10.218	**−5.501**	**−5.575**	4.684	**5.873**	**6.159**
Delta	8.812	1.760	1.354	2.170	0.276	0.811
Rank	1	2	3	1	3	2

TABLE 10: Analysis of variance for wear rate of Al6061/12 wt.% Al_2O_3/MoS_2.

Source	*DF*	Wear rate				Coefficient of friction			
		Seq SS	Adj MS	*F*	*P*	Seq SS	Adj MS	*F*	*P*
A	2	6.5442	3.2721	**29.06**	**0.033**	0.0247047	0.0123523	**756.27**	**0.001**
B	2	0.3456	0.1728	1.53	0.395	0.0002580	0.0001290	7.90	0.112
C	2	0.1651	0.0826	0.73	0.577	0.0032667	0.0016333	100.00	0.010
Error	2	0.2252	0.1126			0.0000327	0.0000327		
Total	8	7.2801				0.0282620			
		$R^2 = 96.91\%$; R^2 (adj) = 87.63%				$R^2 = 99.88\%$; R^2 (adj) = 99.54%			

TABLE 11: Experimental result of surface roughness for Al6061/12 wt.% Al_2O_3/6 wt.% MoS_2.

Trial	*V*	*F*	*D*	*V*	*F*	*D*	Ra (μm)	S/N ratio
1	1	1	1	100	0.1	1	2.21	−6.887
2	1	2	2	100	0.2	1.5	2.35	−7.421
3	1	3	3	100	0.3	2	3.56	−11.029
4	2	1	2	150	0.1	1.5	2.48	−7.889
5	2	2	3	150	0.2	2	2.65	−8.464
6	2	3	1	150	0.3	1	3.62	−11.174
7	3	1	3	200	0.1	2	2.88	−9.187
8	3	2	1	200	0.2	1	3.02	−9.600
9	3	3	2	200	0.3	1.5	3.87	−11.754

TABLE 12: Taguchi analysis: Ra versus *V*, *F*, and *D* for Al6061/12 wt.% Al_2O_3/6 wt.% MoS_2.

Level	*V*	*F*	*D*
1	**−8.446**	**−7.988**	−9.221
2	−9.176	−8.495	**−9.022**
3	−10.181	−11.319	−9.561
Delta	1.735	3.331	0.539
Rank	2	1	3

(a) (b)

FIGURE 4: (a) Wear surface of Al6061/12 wt.% Al_2O_3/6 wt.% MoS_2 tested at 15 N load. (b) Wear surface of Al6061/12 wt.% Al_2O_3/6 wt.% MoS_2 tested at 45 N load.

TABLE 13: Analysis of variance for Ra of Al6061/12 wt.% Al_2O_3/6 wt.% MoS_2.

Source	DF	Seq SS	Adj SS	Adj MS	*F*	*P*
V	2	0.46220	0.46220	0.23110	24.85	0.039
F	2	2.38820	2.38820	1.19410	**128.40**	**0.008**
D	2	0.02580	0.02580	0.01290	1.39	0.419
Error	2	0.01860	0.01860	0.00930		
Total	8	2.89480				
$S = 0.0964365$; $R^2 = 99.36\%$; R^2 (adj) = 97.43%						

composite. It is observed that feed is the most significant parameter affecting the surface roughness of Al6061/12 wt.% Al_2O_3/6 wt.% MoS_2 with $F : P$ value of 128.40 : 008.

4. Conclusion

In the present investigation, the Al6061/Al_2O_3/MoS_2 hybrid composite is successfully fabricated using stir casting process. The mechanical behaviour, tribological behaviour, and machinability behaviour are evaluated. The obtained results can be summarized as follows:

(i) Mechanical properties of hybrid composites increase with an increase in weight fraction of alumina particles. An increase in weight fraction of molybdenum disulphide reinforcement decreases the mechanical properties like tensile strength and BHN. The optimum parameter for maximization of tensile strength is obtained at 12 wt.% of Al_2O_3 and 2 wt.% of MoS_2 and the maximum BHN is obtained at 12 wt.% of Al_2O_3 and 4 wt.% of MoS_2.

(ii) The incorporation of Al_2O_3 reinforcement to Al6061 increases the wear resistance of the composites. The addition of MoS_2 reinforcement in Al6061/Al_2O_3 composites as a hybrid reinforcement further increases the wear and friction resistance of the composite. This is attributed to the stable and MoS_2 rich mechanically mixed layer, which prevents metal to metal contact and reduces the wear of the composite. The unstable mechanically mixed layer in the absence of the lubricant phase leads to lower wear resistance in Al6061/Al_2O_3/MoS_2 hybrid composite. The optimum parameter for minimization wear rate and coefficient of friction are obtained at 15 N of load, 3.25 m/sec of sliding velocity, and 6 wt.% of MoS_2.

(iii) The optimum parameter for minimization surface roughness is obtained at 100 m/min of cutting speed, 0.1 mm/rev of feed, and 1.5 mm of depth of cut for CNC turning Al6061/12 wt.% Al_2O_3/6 wt.% MoS_2 hybrid composite.

Competing Interests

The authors declare that they have no competing interests.

References

[1] L. Poovazhagan, K. Kalaichelvan, A. Rajadurai, and V. Senthilvelan, "Characterization of hybrid silicon carbide and boron carbide nano particles reinforced aluminum alloy composites," *Procedia Engineering*, vol. 64, pp. 681–689, 2013.

[2] A. M. Al-Qutub, A. Khalil, N. Saheb, and A. S. Hakeem, "Wear and friction behavior of Al6061 alloy reinforced with carbon nanotubes," *Wear*, vol. 297, no. 1-2, pp. 752–761, 2013.

[3] K. Umanath, K. Palanikumar, and S. T. Selvamani, "Analysis of dry sliding wear behaviour of Al6061/SiC/Al_2O_3 hybrid metal matrix composites," *Composites Part B: Engineering*, vol. 53, pp. 159–168, 2013.

[4] K. Umanath, ST. Selvamani, and K. Palanikumar, "Friction and wear behaviour of Al6061alloy (SiCp/Alp) hybrid composite," *International Journal of Engineering Science and Technology*, vol. 3, pp. 5441–5551, 2011.

[5] G. B. Veeresh Kumar, C. S. P. Rao, N. Selvaraj, and M. S. Bhagyashekar, "Studies on Al6061/SiC and Al7075/Al_2O_3 metal

matrix composites," *Journal of Minerals Materials Characterization Engineering*, vol. 9, pp. 454–461, 2010.

[6] H. C. Anilkumar, H. S. Hebbar, and K. S. Ravishankar, "Mechanical properties of fly ash reinforced aluminum alloy composites," *International Journal of Mechanical and Materials Engineering*, vol. 6, pp. 225–233, 2011.

[7] M. Asif, K. Chandra, and P. S. Misra, "Development of aluminum based hybrid metal matrix composites for heavy duty applications," *Journal of Minerals and Materials Characterization and Engineering*, vol. 10, pp. 115–134, 2011.

[8] D. Singh, H. Singh, S. Kumar, and G. Singh, "An Experimental investigation of Mechanical behaviour of aluminum by adding SiC and Alumina," *International Journal on Emerging Technologies*, vol. 3, pp. 178–184, 2012.

[9] V. N. Gaitondel, S. R. Karnik, and M. S. Jayaprakash, "Some studies on wear and corrosion properties of Al5083/Al_2O_3/graphite hybrid composites," *Journal of Minerals and Materials Characterization and Engineering*, vol. 11, pp. 695–703, 2012.

[10] R. Sharanabasappa and B. S. Motgi, "A study on mechanical properties of fly ash and alumina reinforced aluminum alloy (LM25) composites," *IOSR Journal of Mechanical and Civil Engineering*, vol. 7, pp. 14–21, 2013.

[11] D. Kuma and J. Singh, "Comparative Investigation of Mechanical Properties of aluminum based hybrid metal matrix composites," *International Journal of Engineering Research and Applications*, vol. 1, pp. 31–36, 2014.

[12] C. C. Wang and B. H. Yan, "Blind-hole drilling of Al_2O_3/6061Al composite using rotary electro-discharge machining," *Journal of Materials Processing Technology*, vol. 102, no. 1, pp. 90–102, 2000.

[13] B. H. Yan and C. C. Wang, "Machining characteristics of Al_2O_3/6061 Al composite using rotary electro-discharge machining with a tube electrode," *Journal of Materials Processing Technology*, vol. 95, no. 1–3, pp. 222–231, 1999.

[14] B. H. Yan, C. C. Wang, W. D. Liu, and F. Y. Huang, "Machining characteristics of Al_2O_3/6061Al composite using rotary EDM with a disklike electrode," *International Journal of Advanced Manufacturing Technology*, vol. 16, no. 5, pp. 322–333, 2000.

[15] G. B. Veeresh Kumar, C. S. P. Rao, N. Several, and M. S. Bhagyshekar, "Studies on A16061-SiC and Al7075-Al_2O_3 metal matrix composites," *Journal of Minerals and Materials Characterization & Engineering*, vol. 9, pp. 43–55, 2010.

[16] Y. Sahin, M. Kok, and H. Celik, "Tool wear and surface roughness of Al_2O_3 particle-reinforced aluminium alloy composites," *Journal of Materials Processing Technology*, vol. 128, no. 1–3, pp. 280–291, 2002.

[17] J. P. Davim, "Design of optimisation of cutting parameters for turning metal matrix composites based on the orthogonal arrays," *Journal of Materials Processing Technology*, vol. 132, no. 1–3, pp. 340–344, 2003.

[18] E. Kiliçkap, O. Çakir, M. Aksoy, and A. Inan, "Study of tool wear and surface roughness in machining of homogenised SiC-p reinforced aluminium metal matrix composite," *Journal of Materials Processing Technology*, vol. 164-165, pp. 862–867, 2005.

[19] A. Metin Kok, "A study on the machinability of Al_2O_3 particle reinforced aluminum alloy composite," in *Proceedings of the 11th Inorganic Bonded Fiber Composites Conference*, vol. 1, pp. 272–281, Madrid, Spain, November 2008.

[20] C. A. Brown and M. K. Surappa, "The machinability of a cast aluminium alloy-graphite particle composite," *Materials Science and Engineering*, vol. 102, no. 1, pp. 31–37, 1988.

[21] M. Kök and K. Özdin, "Wear resistance of aluminium alloy and its composites reinforced by Al_2O_3 particles," *Journal of Materials Processing Technology*, vol. 183, no. 2-3, pp. 301–309, 2007.

[22] H. Kato, M. Takama, Y. Iwai, K. Washida, and Y. Sasaki, "Wear and mechanical properties of sintered copper-in composites containing graphite or molybdenum disulfide," *Wear*, vol. 255, no. 1–6, pp. 573–578, 2003.

[23] A. Baradeswaran, S. C. Vettivel, A. Elaya Perumal, N. Selvakumar, and R. Franklin Issac, "Experimental investigation on mechanical behaviour, modelling and optimization of wear parameters of B_4C and graphite reinforced aluminium hybrid composites," *Materials and Design*, vol. 63, pp. 620–632, 2014.

[24] A. Baradeswaran and A. Elaya Perumal, "Study on mechanical and wear properties of Al 7075/Al_2O_3/graphite hybrid composites," *Composites Part B: Engineering*, vol. 56, pp. 464–471, 2014.

[25] S. Mahdavi and F. Akhlaghi, "Effect of the graphite content on the tribological behavior of Al/Gr and Al/30SiC/Gr composites processed by in situ powder metallurgy (IPM) method," *Tribology Letters*, vol. 44, no. 1, pp. 1–12, 2011.

6

Experimental Analysis of Tensile Mechanical Properties of Sprayed FRP

Zhao Yang and Kun Wu

Institute of Urban Construction, Wuhan University of Science and Technology, Wuhan 430065, China

Correspondence should be addressed to Zhao Yang; yzwh77@163.com

Academic Editor: Kaveh Edalati

To study the tensile mechanical properties of sprayed FRP, 13 groups of specimens were tested through uniaxial tensile experiments, being analyzed about stress-strain curve, tensile strength, elastic modulus, breaking elongation, and other mechanical properties. Influencing factors on tensile mechanical properties of sprayed FRP such as fiber type, resin type, fiber volume ratio, fiber length, and composite thickness were studied in the paper too. The results show that both fiber type and resin type have an obvious influence on tensile mechanical properties of sprayed FRP. There will be a specific fiber volume ratio for sprayed FRP to obtain the best tensile mechanical property. The increase of fiber length can lead to better tensile performance, while that of composite thickness results in property degradation. The study can provide reference to popularization and application of sprayed FRP material used in structure reinforcement.

1. Introduction

Fiber reinforced polymers (FRP) composite materials have been largely used as strengthening and rehabilitation materials for decades in civil engineering. The main forms of these materials are sheet and plate, which are usually externally bonded on the surfaces of structure members. These materials have been proven to be very efficient and convenient in structure strengthening due to their high strength-to-weight ratio with excellent corrosion resistance [1–5]. But there are still some disadvantages in the use of external bonded FRP. The first is that the interfacial debonding between FRP sheet or plate and structure member will occur for the degradation of the bonding quality, which will lead to a premature failure of FRP [6–13]. Although some anchorage devices or mechanical strategies have been studied to delay such kind of debonding failure, this method will make the construction process more difficult. The second is that the ultimate strain of FRP sheets or plates used in structure strengthening cannot reach the ultimate strain in the material tensile test, which means the high strength of FRP sheets or plates cannot be fully used [14–16]. The third is that the external bonding method of FRP sheets or plates will be hardly conducted in some structure members such as beam-column joints, which cannot provide enough flat surfaces for FRP to be pasted continuously and the core areas of which are difficult for such forms of FRP to wrap efficiently [17–22].

To solve the abovementioned problems, the advanced strengthening method using sprayed FRP consisting of randomly oriented reinforcing fibers in a matrix has been suggested. Spray operation was conducted through special equipment (Figure 1). Resins were transferred to a spray gun and sprayed onto the structure surface. Simultaneously, fiber filaments were chopped to short ones in the spray gun and sprayed onto the structure surface, mixed with the resins. Then, the sprayed FRP material can be formed continuously on the structure surface. This kind of FRP has some advantages compared with the external bonded ones. Firstly, when being used on an uneven concrete surface, external bonded FRP needs some material to level the surface, and then the resin is used to bond the FRP and the leveled substrate. Thus, there are resins, leveling materials, and concrete substrate bonded together. This will increase the debonding possibility and make the application process complicated. While using sprayed FRP, the sprayed FRP material can level the surface as well, and thus there are only sprayed materials and concrete substrates that are bonded together and decrease the debonding possibility.

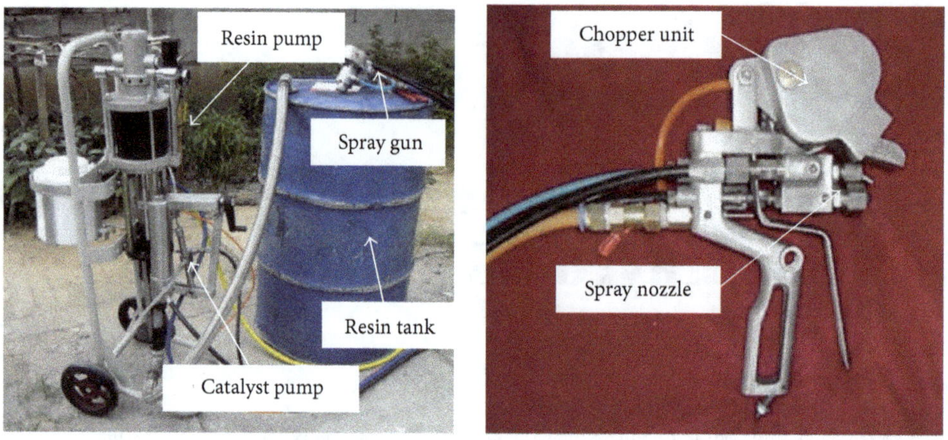

(a) Complete spray equipment (b) Spray gun

FIGURE 1: Special equipment of sprayed FRP.

The application process is simpler as well without a special leveling process [23]. Secondly, when being used to reinforce a beam-column joint, multilayers of external bonded FRP are required to be bonded on the joint along a different loading direction, which makes the application process much complicated and difficult to assure the construction quality. While using the sprayed FRP, the construction process can be much easier than external bonded method due to the spraying method.

In 2000, sprayed FRP was used for the first time to strengthen the reinforced concrete beams by Boyd. The study showed that the flexural reinforcing effect of sprayed glass FRP was similar to that of FRP fabric wrap or plate bonding onto the beams, while the shear strengthening effect of sprayed FRP was better than other techniques [24]. In 2004, Lee and Hausmann investigated the load capacity, ductility, and energy absorption aspect of RC beams retrofitted with sprayed FRP. The study showed that sprayed FRP was capable of substantially increasing the performance of the beams and was effective in repairing damaged RC beams [25]. Then, Soleimani SM studied sprayed FRP in the shear strengthening and enhancement of impact resistance of reinforced concrete beams [26]. Boyd et al. studied sprayed FRP used in prestressed concrete girders [27]. Ha et al. studied the bond characteristics of sprayed FRP composites bonded to concrete substrate considering various concrete surface conditions [28]. Though the studies above got satisfying results, the use of sprayed FRP for structure strengthening has never been fully investigated.

This paper is written to study the tensile properties and influencing factors of sprayed FRP. A series of specimens of sprayed FRP were tested through tensile experiments. Tensile strength, elastic modulus, breaking elongation, and Poisson ratio of the specimens were studied through the tests. And fiber type, resin type, fiber length, fiber volume ratio, composite thickness, and specimen cutting orientation were compared as influencing factors in the paper too. The results can provide theoretical basis for tensile property calculation and actual application of sprayed FRP.

2. Tensile Tests

2.1. Specimens and Experiment Method. In this paper, 3 types of fiber, which are glass fiber, carbon fiber, and basalt fiber, and 2 kinds of resin, which are vinyl ester resin (901) and unsaturated polyester resin (196), were used to make sprayed FRP specimens. The fiber yarns and resins used in the study are shown in Figures 2 and 3. The properties of the fibers and resins are shown in Tables 1 and 2, where the values were obtained from material suppliers. The special equipment used to conduct the spraying process is shown in Figure 1.

The specimens of sprayed FRP were designed in accordance with the shape recommended by Boyd [24] because no specific test method has been published by ASTM so far. However, the dimensions have been changed a little considering the length of the chopped fibers and the fit for the experiment equipment. The detailed dimensions of the actual specimens are shown in Figure 4, in which all dimensions are in millimeters (mm).

The fabrication process of the tensile specimens was divided into several steps. Firstly, a 1000 mm × 800 mm rectangle glass plate was put on the flat ground with a thin sheet of polyethylene terephthalate paved on it to prevent the adhesion of resin and glass. Then, the chopped fibers and resins were sprayed onto the glass plate to form a sprayed FRP plate. After spraying, a ribbed aluminum compaction roller was used to eliminate the entrapped air voids and make the composite thickness consistent. After hardening, the sprayed FRP plate can be removed from the glass plate and the polyethylene terephthalate sheet can be peeled off. Then, the sprayed FRP plate was cured for seven days in open air at 25°C. After that, the plate was polished and cut to the designed dimensions. In order to study the isotropic behavior of this sprayed composite, the specimens were cut from the plate in three different orientations, shown in Figure 5.

There are 13 groups of specimens fabricated in the tensile test to study the tensile properties and influencing factors of sprayed FRP material. Each group has 6 specimens to compare the test results and the influencing factors are

Table 1: Properties of fibers.

Type	Tensile strength, σ (MPa)	Elastic modulus, E (GPa)	Breaking elongation, ε (%)	Poisson ratio, μ	Density, ρ (kg/m^3)
Glass fiber[a]	1970	79	4.8	0.20	2550
Carbon fiber[b]	3500	205	1.5	0.31	1750
Basalt fiber[c]	2200	93	2.6	0.25	2630

[a]Data from Jilin Carbon Co., Ltd., in China.
[b]Data from Zhejiang GBF Basalt Fiber Co., Ltd., in China.
[c]Data from Jushi Group Co., Ltd., in China.

Table 2: Properties of resins.

Type	Tensile strength, σ (MPa)	Elastic modulus, E (GPa)	Breaking elongation, ε (%)	Poisson ratio, μ	Density, ρ (kg/m^3)
901[a]	85	3.4	5.5	0.3	1050
196[b]	60	3.0	3.0	0.35	1230

[a]Data from Swancor Renewable Energy Co., Ltd., in China.
[b]Data from Shenzhen Tonglai Material Technology Co., Ltd., in China.

(a) Glass fiber (b) Carbon fiber (c) Basalt fiber

Figure 2: Fiber yarns used in the experiments.

(a) Vinyl ester resin 901 (b) Unsaturated polyester resin 196

Figure 3: Resins used in the experiments.

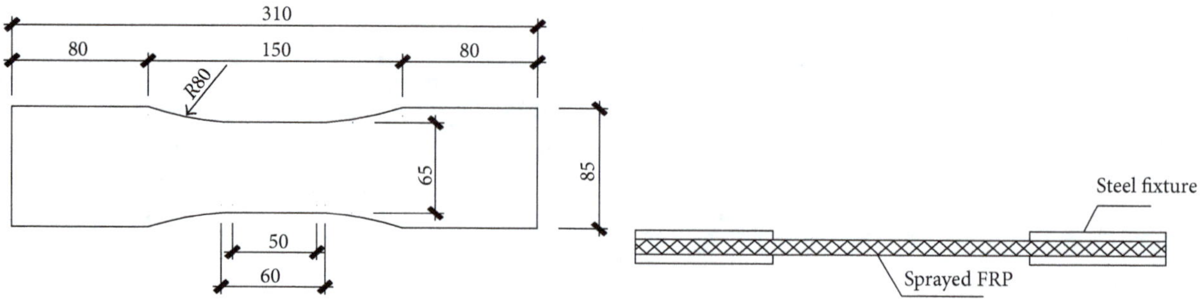

(a) Top view (b) Front view

FIGURE 4: Dimensions of the actual specimens.

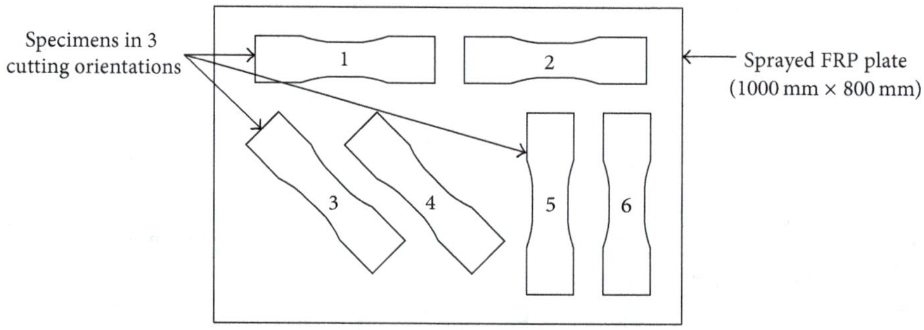

FIGURE 5: Specimen cutting orientation.

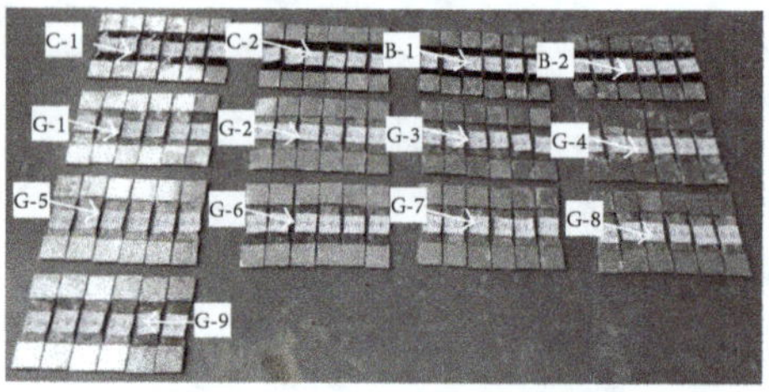

FIGURE 6: 13 groups of specimens.

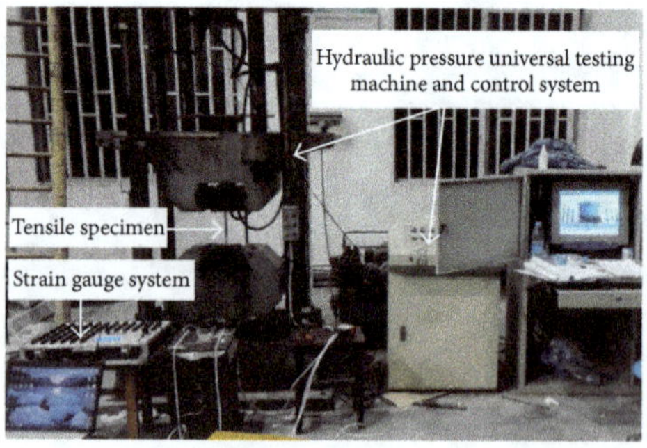

FIGURE 7: Loading and measurement apparatus.

considered as fiber type, resin type, fiber volume ratio, fiber length, and composite thickness in the paper. The specimen plan is shown in Table 3 and the pictures of the 13 specimen groups are shown in Figure 6.

The composite thickness was controlled by using some plastic plates surrounding the rectangle glass plate on the ground, and the required thickness was marked on the plastic plates, and then the mark could be used to control the composite thickness when spraying. Some concrete blocks were used to fix the plastic plates and prevent deformation. In addition, after cutting the specimens, the thickness of each specimen was measured at both sides of the middle section. The measurement errors were observed to be below 3% for all specimens, which indicated that the actual thickness met well the designed value.

The fiber length can be easily controlled by adjusting the chopped unit. The fiber volume ratio can be well controlled by adjusting the spraying speed of chopped fibers and resins, respectively, as well. It was observed that error percentages were both lower than 3%, which could meet the design requirements well.

The tensile tests were conducted using a hydraulic pressure universal testing machine, which is shown in Figure 7. In the middle of every tensile specimen, 2 strain gauges in horizontal and vertical direction, respectively, were bonded on the specimen surface to measure the strain of the composite during loading process through a strain gauge system. The loading speed was set to be 2 mm/min in accordance with Chinese code GB/T1447 (2005) [29].

3. Test Results

3.1. Failure Mode. Through the observation of failure processes of the specimens, all the 13 groups of specimens are similar in failure mode. The failure process can be divided

Table 3: Specimen plan.

Group number	Group name	Fiber type	Resin type	Fiber volume ratio	Fiber length (mm)	Composite thickness (mm)
1	C-1	Carbon	901	20%	30	4
2	C-2	Carbon	196	20%	30	4
3	B-1	Basalt	901	20%	30	4
4	B-2	Basalt	196	20%	30	4
5	G-1	Glass	901	20%	30	4
6	G-2	Glass	196	20%	30	4
7	G-3	Glass	901	15%	30	4
8	G-4	Glass	901	25%	30	4
9	G-5	Glass	901	30%	30	4
10	G-6	Glass	901	20%	15	4
11	G-7	Glass	901	20%	20	4
12	G-8	Glass	901	20%	40	4
13	G-9	Glass	901	20%	30	7

(a) Glass fiber (b) Basalt fiber (c) Carbon fiber

Figure 8: Through cracks in sprayed FRP specimens of different fiber types.

into 3 stages. In the first stage, the stress in the composite increased quickly with the increase of tensile load, while strain increased slowly. It was observed that there are no signs of cracking or debonding inside the specimens, which always appeared like some small white lines. In the second stage, initial crack could be observed. The strain in composite increased fast, while stress increased slowly. Cracks usually started at the edge of the composite and progressed rapidly to the center. Fracture always began in resin and then debonding occurred between fiber and resin, which made the fiber pull out of the resin and led to crack development. In the third stage, the specimen reached the ultimate load. The stress began to decrease rapidly and strain increased a little. Through cracks were formed (Figure 8) and resulted in a total failure of the specimens. Some fibers along the through cracks were pulled out and some still remained in the resin (Figure 9). There were 3 main modes for the through cracks, which are shown in Figure 10.

3.2. Stress-Strain Relationship. The tensile stress of sprayed FRP can be calculated by dividing the load value by the cross-sectional area of the specimen. The strain of sprayed FRP can be collected automatically by the strain gauge system. Thus, the experimental stress-strain relationship can be achieved. The relationship curves of the 13 groups are very similar and 3 groups of them are shown as examples in Figure 11, from which we can see the 3 stages of the failure process clearly. From Figure 11, we can see that, in the first stage, fibers and resins were working together to bear the tensile load, no cracking or debonding appeared, the strain increased slower than stress, and the first stage of the curve was formed. While the initial cracks appeared, some resins ruptured and debonding between fibers and resins appeared. The deformation of specimens increased more quickly than the loads, and thus the second stage of the curve was formed. When the cracks developed, more and more resins ruptured, the loads were mainly borne by fibers until reaching the bearing capacity, and the ultimate stress of specimens was reached. Then, the stress started to decrease and strain continued to increase. The third stage of the curve was formed. So, the sprayed FRP composites showed a trilinear constitutive behavior.

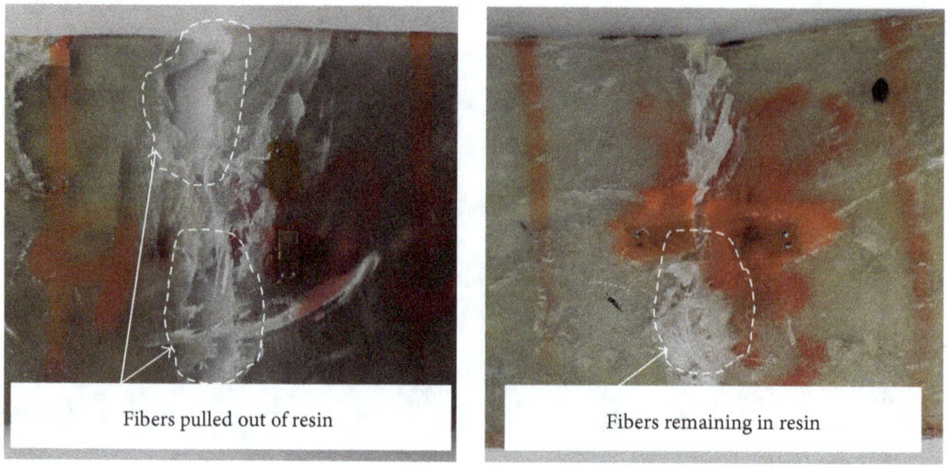

(a) (b)

FIGURE 9: Failure modes of fibers along the through crack.

(a) (b) (c)

FIGURE 10: Failure modes of through cracks.

3.3. Tensile Mechanical Properties of Sprayed FRP. Through the data gained from the tensile experiment, several tensile mechanical properties can be achieved using the calculation method recommended in Chinese code GB/T1447 (2005) [29], as shown in Table 4. Due to the difficulty in obtaining the actual initial modulus, the values of elastic modulus reported in Table 4 referred to a secant modulus. The secant endpoints were the endpoints of 50 $\mu\varepsilon$ and 40% of ultimate stress [24]. Note that the values reported in the table represent the averages of all the six specimens in each group and all the error percentages for each group were lower than 3%.

4. Influencing Factors on Tensile Mechanical Properties of Sprayed FRP

4.1. Fiber Type. Carbon fiber, glass fiber, and basalt fiber were used, respectively, to make sprayed FRP in the study, and their tensile performances were different. As is shown in Table 3, C-1, B-1, and G-1 are made of different fibers but the same resin 901 and other influencing factors are all the same. So, these 3 groups can be used to compare the effect of different fiber types using resin 901. In the same way, C-2, B-2, and G-2 are made of different fibers but the same resin 169 and other influencing factors are all the same too. So, these 3 groups can be used to compare the effects of different fiber types using resin 169.

As is shown in Table 4, C-1 using carbon fiber has the highest strength value of 105.4 MPa, and the strength of G-1 using glass fiber is 91.3 MPa, while B-1 using basalt fiber has the lowest strength value of 87.2 MPa. Among the 3 groups, C-1 also has the highest tensile elastic modulus value of 49.1 GPa, which is 54.4% higher than G-1 and 74.1% higher than B-1. As to breaking elongation, the case is just the opposite. B-1 has the highest value of 2.6%, which is 8.3% higher than G-1 and 23.8% higher than C-1.

For C-2, B-2, and G-2, the comparative results are similar to the above 3 groups. The strength value of C-2 is the highest

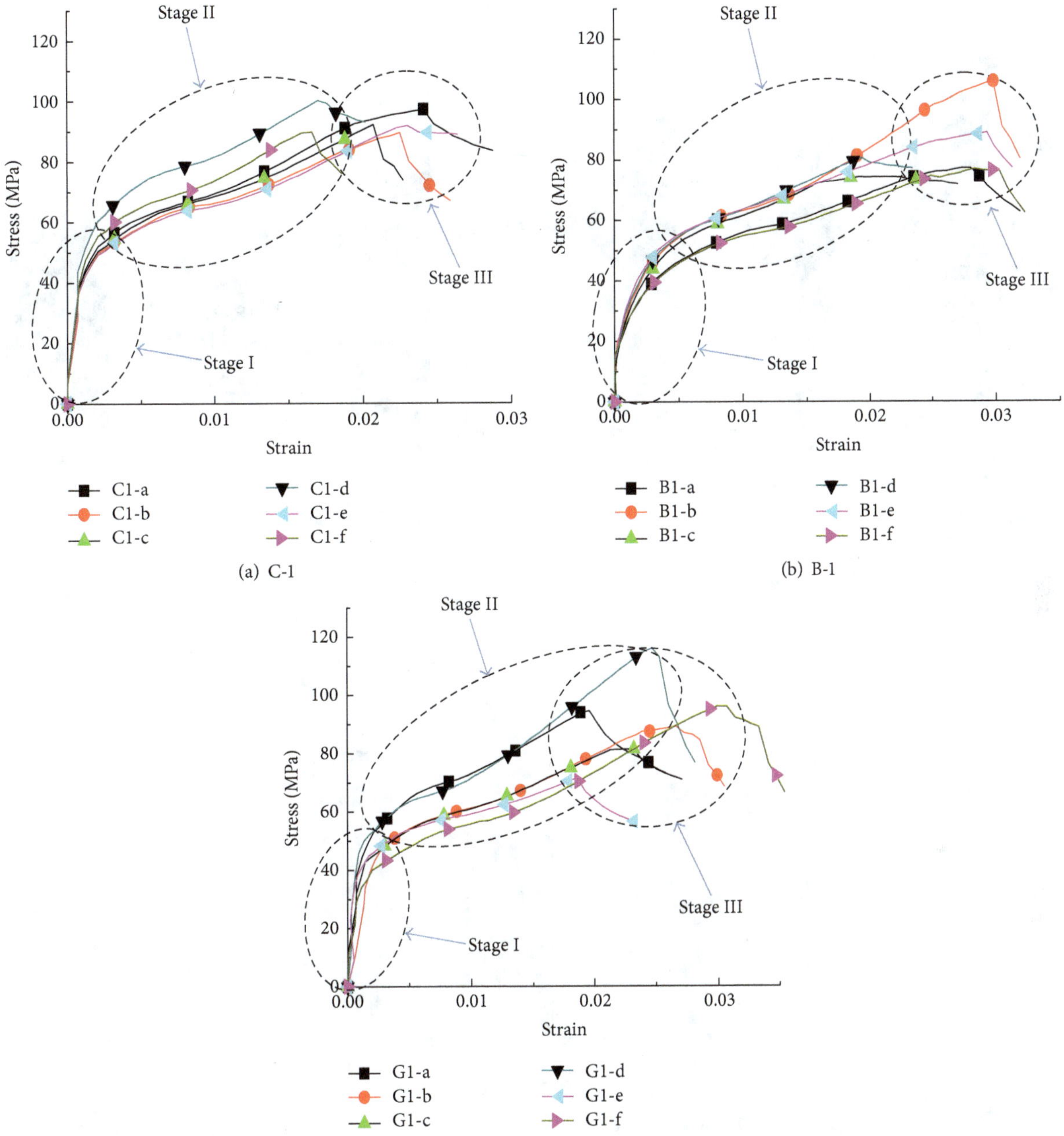

(a) C-1

(b) B-1

(c) G-1

FIGURE 11: Stress-strain relationship curves.

at 89.3 MPa and G-2 is the lowest at 75.2 MPa. The differences in tensile elastic modulus are more significant than the above 3 groups. The value of C-2 is up to 50.7 GPa, G-2 follows by 28.8 GPa, and B-2 is the minimum at 26.1 GPa. While the case of breaking elongation is also opposite, B-2 has the highest value of 3.1%, G-2 follows by 2.4%, and C-2 has the lowest value at 1.9%.

According to the data in Table 4, the effect of different fiber types on tensile properties of sprayed FRP is compared in Figure 12. It can be seen that tensile strength and tensile elastic modulus of sprayed FRP using carbon fiber are higher than those sprayed FRP using glass fiber and basalt fiber, no matter what kind of resin is used. But the breaking elongation of sprayed FRP using carbon fiber is the lowest among the three. These indicate that the tensile performance of sprayed FRP is affected greatly by fiber type and the performance feature of sprayed FRP is always similar to the fiber used, which is in accordance with the conclusion of Lee [30]. It should be noted that the differences in the strength of sprayed FRP using glass and basalt fiber are very small with the maximum value of only 4.9%.

4.2. Resin Type. The effect of resin type on tensile properties of sprayed FRP was studied by comparing C-1 and C-2, B-1

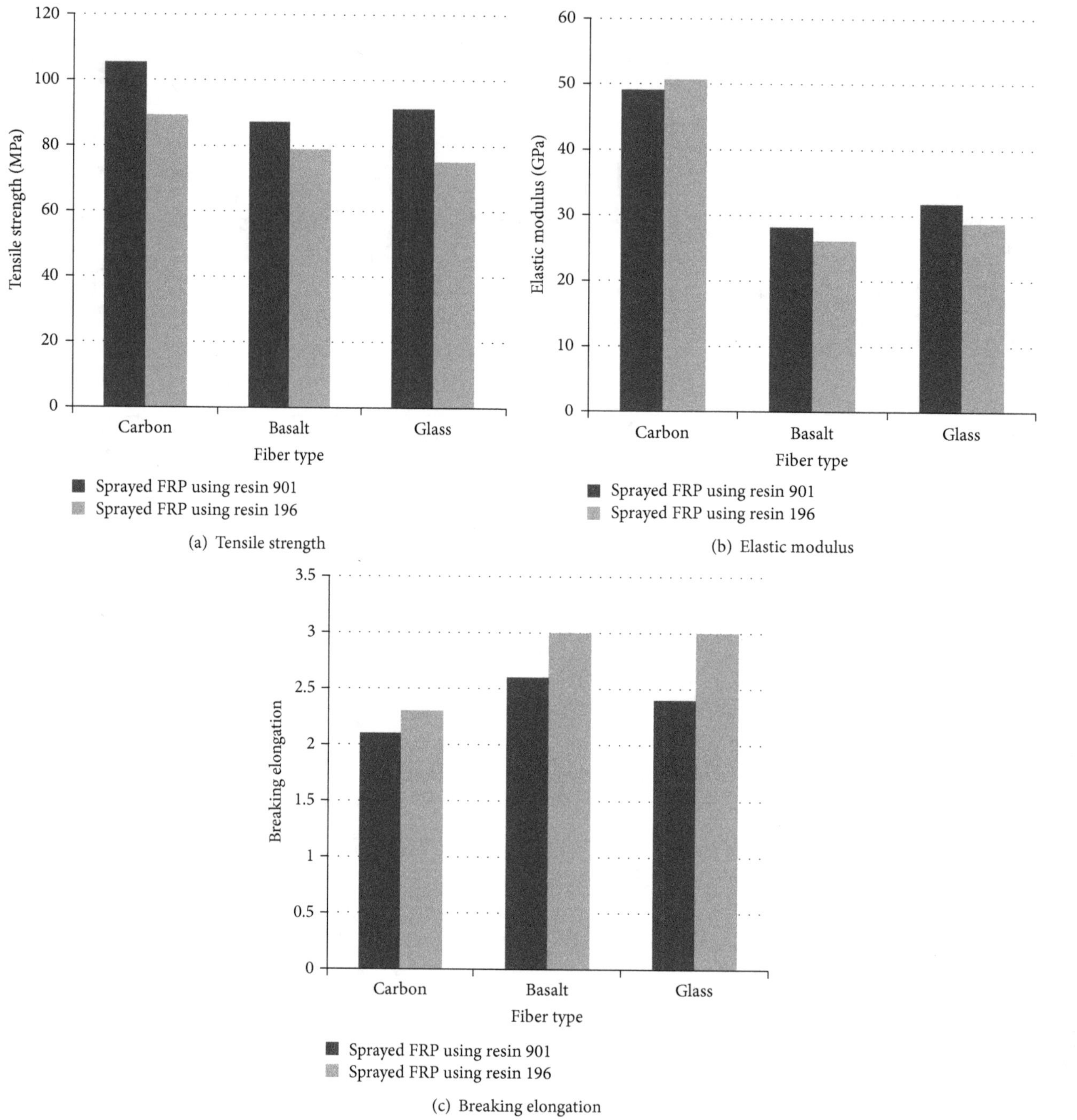

FIGURE 12: Comparison of tensile properties of sprayed FRP using different fiber types.

and B-2, and G-1 and G-2, separately, which are different in resin type but the same in all other factors. C-1, B-1, and G-1 used vinyl ester resins. C-2, B-2, and G-2 used unsaturated polyester resins. As is shown in Table 4, tensile strength of C-1 is 18.2% higher than C-2, B-1 is 6.3% higher than B-2, and G-1 is 21.4% higher than G-2. Regarding tensile elastic modulus, the values of C-1, G-1, and B-1 are also higher than those of C-2, B-2, and G-2. But, in the comparison of breaking elongation, the cases are different. The value of C-1 is 10.5% higher than C-2, the value of B-1 is 8.3% higher than B-2, and the value of G-1 is 20% lower than G-2.

The above results show that using vinyl ester resin as matrix and adhesive for sprayed FRP can get better tensile mechanical properties generally, which is similar to the result of Lee [30]. But it is worth noting that the breaking elongation of the sprayed FRP using glass fiber will be reduced when using this kind of resin.

4.3. Fiber Volume Ratio. From Table 3, there are 4 groups of specimens used to study the effect of fiber volume ratio on tensile properties of sprayed FRP, which are G-3, G-1, G-4,

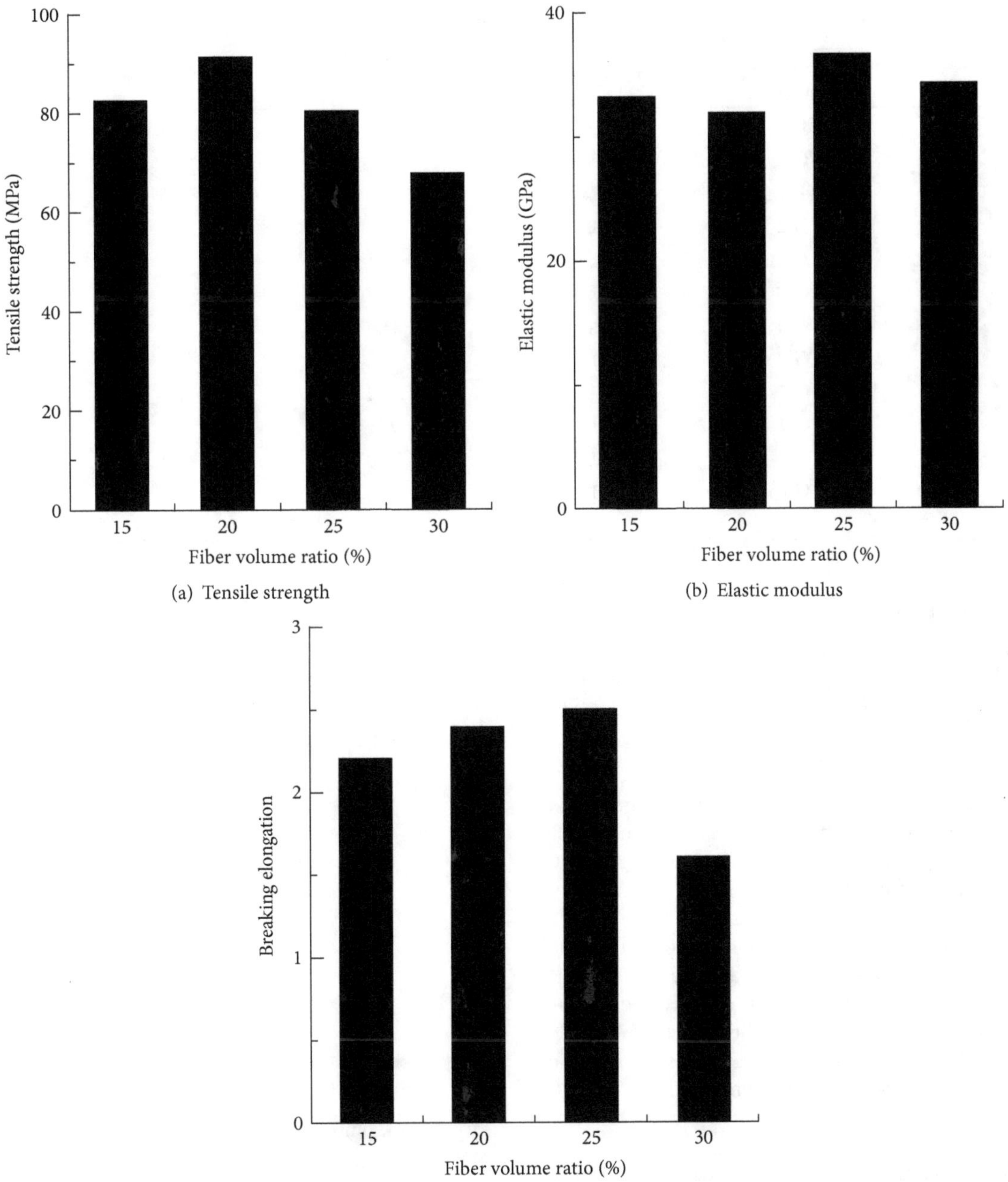

(a) Tensile strength

(b) Elastic modulus

(c) Breaking elongation

FIGURE 13: Comparison of tensile properties of sprayed FRP in different fiber volume ratios.

and G-5, with a fiber volume ratio of 15%, 20%, 25%, and 30%, respectively.

It can be seen from Table 4 that the tensile strengths of the 4 groups from high to low are G-1, G-3, G-4, and G-5, with a value of 91.3 MPa, 82.6 MPa, 80.3 MPa, and 67.7 MPa, respectively. The highest value of G-1 is 34.9% higher than the lowest one G-5. As to tensile elastic modulus, the differences of the 4 groups became very small. The differences between G-4 with the highest value and the other 3 groups ranged from 6.7% to 15.4%. In the case of breaking elongation, the highest value comes from G-4 too, which is 56.3% higher than that of G-5 with the lowest value.

From the data in Table 4, Figure 13 was drawn to compare the effects of different fiber volume ratios on tensile properties of sprayed FRP. It shows that there will be a specific value of fiber volume ratio that can lead to a maximum value of a tensile property. And when the ratio is lower than the specific value, the tensile property value will increase with the increase of volume ratio. While the fiber volume ratio is higher than the specific value, the case is just the opposite. This is because when the fiber volume ratio is low, there will be more fibers participating in bearing tensile stress with the increase of ratio, which results in an increase of tensile capacity of sprayed FRP. While the ratio is higher than the

TABLE 4: Summary of tensile properties of sprayed FRP.

Group number	Group name	Tensile strength, σ (MPa)	Elastic modulus, E (GPa)	Breaking elongation (%)	Poisson ratio, μ
1	C-1	105.4	49.1	2.1	0.30
2	C-2	89.3	50.7	2.3	0.31
3	B-1	87.2	28.2	2.6	0.22
4	B-2	78.9	26.1	3.0	0.25
5	G-1	91.3	31.8	2.4	0.24
6	G-2	75.2	28.8	3.0	0.26
7	G-3	82.6	33.3	2.2	0.22
8	G-4	80.3	36.7	2.5	0.25
9	G-5	67.7	34.4	1.6	0.24
10	G-6	78.1	27.1	2.1	0.21
11	G-7	87.1	30.8	2.2	0.26
12	G-8	101.2	38.3	2.4	0.24
13	G-9	66.1	23.7	1.9	0.24

specific value, fibers will be too much, and thus the resin around the fibers will not be enough to transfer and balance the stress of fibers effectively, which will lead to partial failure in some fibers or resins earlier and make the tensile property reduced.

4.4. Fiber Length. From Table 3, there are 4 groups of specimens used to study the effect of fiber length on tensile properties of sprayed FRP, which are G-6, G-7, G-1, and G-8, with a fiber length of 15 mm, 20 mm, 30 mm, and 40 mm, respectively.

From Table 4, tensile strengths of the 4 groups from high to low are G-8, G-1, G-7, and G-6, with a value of 101.2 MPa, 91.3 MPa, 87.1 MPa, and 78.1 MPa, respectively. For tensile elastic modulus, the highest value also belongs to G-8, which is 41.3% higher than the lowest one G-6. As to breaking elongation, the highest value is 2.4% which comes from G-8 and G-1, followed by 2.2% from G-7, and the lowest one is 2.1% from G-6.

The relationship curves of tensile stress and strain of the 4 groups are compared in Figure 14. It shows that the longer fiber can bear more stress than the shorter ones when tensile strain is the same, which is in accordance with Boyd's result [24]. This can be explained as the longer length fibers can obtain a larger bonding area with resins, and thus the bonding effect will be better than the shorter ones, which can lead to better tensile properties.

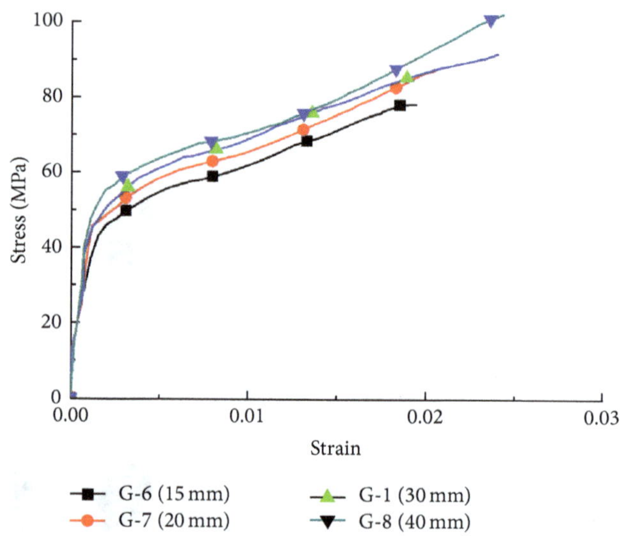

FIGURE 14: Comparison of stress-strain curves of sprayed FRP with different fiber lengths.

4.5. Composite Thickness. From Table 3, there are 2 groups of specimens used to study the effect of composite thickness on tensile properties of sprayed FRP, which are G-1 with a 4 mm thickness and G-9 with a 7 mm thickness.

From Table 4, the tensile strength of G-1 is 91.3 MPa and that of G-9 is 66.1 MPa. The tensile elastic modulus of G-1 is 31.8 GPa and that of G-9 is 23.7 GPa. As to breaking elongation, G-1 is 2.4% and G-9 is 1.9%. It is shown that tensile strength, elastic modulus, and breaking elongation will all be reduced with the increase of composite thickness. This is because the initial defects in sprayed FRP materials will become more with the increase of composite thickness, which will lead to degradation of tensile properties.

4.6. Specimen Cutting Orientation. All the specimens were cut from the sprayed FRP plate in three different orientations, which is shown in Figure 5. As mentioned before, all the error percentages for each group in Table 4 were lower than 3%, so there was not any obvious difference in tensile properties of the specimens in different cutting orientations in each group. Comparing the Poisson ratios of the 13 groups of specimens in Table 4, most values are close to the average value of 0.25. The above results indicate that the sprayed FRP composite can be seen as an isotropic material.

5. Summary and Conclusions

This paper has studied the tensile properties of sprayed FRP through tensile experiments of 13 groups of specimens, which include tensile strength, elastic modulus, breaking elongation, and Poisson ratio. The effects of several influencing factors are compared too. The following conclusions are drawn:

(1) Fiber type has an important influence on tensile properties of sprayed FRP material. The composites made of carbon fiber have higher values of tensile strength and elastic modulus than those made of glass fiber or basalt fiber, while the breaking elongation of carbon fiber composites is lower. The results equate the tensile properties of the three types of fibers.

(2) The test results of the 2 resins used in this paper show that using vinyl ester resins to make sprayed FRP can get better tensile properties than using unsaturated polyester resins generally. But the breaking elongation of glass fiber specimens using vinyl ester resins is lower compared with using unsaturated polyester resin.

(3) There will be a specific value of fiber volume ratio for every kind of sprayed FRP material, which can lead to the highest value of a tensile property. Less or more than this volume ratio can make the tensile property of this sprayed FRP degrade.

(4) The increase of fiber length can make the tensile properties of sprayed FRP better, while the increase of composite thickness will lead to property degradation.

(5) Cutting orientation of sprayed FRP specimens has no obvious influence on tensile properties and the Poisson ratios are almost the same for all the specimens, which shows that sprayed FRP can be looked at as a kind of isotropic material.

Competing Interests

The authors declare that there are no competing interests regarding the publication of this paper.

Acknowledgments

This research received specific grant from the Research Foundation of Wuhan Construction Committee and was supported by China Scholarship Council as well.

References

[1] W. Li, "The study and application of FRP material in concrete structures," *Research of Materials Science*, vol. 3, no. 1, pp. 17–21, 2014.

[2] H. Karpate, H. Wheat, J. Jirsa, D. Fowler, and D. Whitney, "Repair and rehabilitation of corrosiondamaged concrete elements using FRP composite wraps," *World Journal of Engineering*, vol. 8, no. 2, pp. 147–150, 2011.

[3] G. Portnov, C. E. Bakis, E. Lackey, and V. Kulakov, "FRP Reinforcing bars—designs and methods of manufacture (Review of Patents)," *Mechanics of Composite Materials*, vol. 49, no. 4, pp. 381–400, 2013.

[4] A. Prota, G. Manfredi, and F. P. Nardone, "Assessment of design formulas for in-plane FRP strengthening of masonry walls," *Journal of Composites for Construction*, vol. 12, no. 6, pp. 643–649, 2008.

[5] S. S. Pendhari, T. Kant, and Y. M. Desai, "Application of polymer composites in civil construction: a general review," *Composite Structures*, vol. 84, no. 2, pp. 114–124, 2008.

[6] J. G. Teng, H. Yuan, and J. F. Chen, "FRP-to-concrete interfaces between two adjacent cracks: theoretical model for debonding failure," *International Journal of Solids and Structures*, vol. 43, no. 18-19, pp. 5750–5778, 2006.

[7] B. Ferracuti, M. Savoia, and C. Mazzotti, "Interface law for FRP-concrete delamination," *Composite Structures*, vol. 80, no. 4, pp. 523–531, 2007.

[8] C. Mazzotti and M. Savoia, "FRP-Concrete bond behaviour under cyclic debonding force," *Advances in Structural Engineering*, vol. 12, no. 6, pp. 771–780, 2009.

[9] H. C. Biscaia, I. S. Borba, C. Silva, and C. Chastre, "A nonlinear analytical model to predict the full-range debonding process of FRP-to-parent material interfaces free of any mechanical anchorage devices," *Composite Structures*, vol. 138, pp. 52–63, 2016.

[10] H. W. Zhang, S. T. Smith, and R. J. Gravina, "Analysis of FRP-to-concrete interfaces using a displacement driven partial interaction model," *International Journal of Mechanical Sciences*, vol. 117, pp. 210–217, 2016.

[11] T. Xu, Z. J. He, C. A. Tang, W. C. Zhu, and P. G. Ranjith, "Finite element analysis of width effect in interface debonding of FRP plate bonded to concrete," *Finite Elements in Analysis and Design*, vol. 93, pp. 30–41, 2015.

[12] M. Coelho, A. Caggiano, J. Sena-Cruz, and L. Neves, "Fracture-based interface model for NSM FRP systems in concrete," *Composite Structures*, vol. 152, pp. 816–828, 2016.

[13] H. C. Biscaia, D. Cruz, and C. Chastre, "Analysis of the debonding process of CFRP-to-timber interfaces," *Construction and Building Materials*, vol. 113, pp. 96–112, 2016.

[14] S. F. Breña and G. N. McGuirk, "Advances on the behavior characterization of FRP-Anchored Carbon Fiber-Reinforced Polymer (CFRP) sheets used to strengthen concrete elements," *International Journal of Concrete Structures and Materials*, vol. 7, no. 1, pp. 3–16, 2013.

[15] S. C. Chin, N. Shafiq, and M. F. Nuruddin, "FRP as strengthening material for Reinforced Concrete beams with openings—a review," *KSCE Journal of Civil Engineering*, vol. 19, no. 1, pp. 213–219, 2014.

[16] W.-W. Wang, J.-G. Dai, and K. A. Harries, "Performance evaluation of RC beams strengthened with an externally bonded FRP system under simulated vehicle loads," *Journal of Bridge Engineering*, vol. 18, no. 1, pp. 76–82, 2013.

[17] S. S. Mahini and H. R. Ronagh, "Strength and ductility of FRP web-bonded RC beams for the assessment of retrofitted beam-column joints," *Composite Structures*, vol. 92, no. 6, pp. 1325–1332, 2010.

[18] K. Amin, "Investigation on the effects of L-shaped FRPs for strengthening of exterior RC joints to relocate the plastic hinge place away from the column face," *International Journal of Advances in Engineering Sciences*, vol. 3, no. 2, pp. 13–18, 2013.

[19] Z. Yang and C. Wu, "Analysis on the seismic performance of RC frame beam-column joints strengthened by sprayed FRP," *Applied Mechanics and Materials*, vol. 578-579, pp. 835–838, 2014.

[20] P. Agarwal, A. Gupta, and R. G. Angadi, "Effect of FRP wrapping on axial behavior of concrete and cyclic behavior of external RC beam column joints," *KSCE Journal of Civil Engineering*, vol. 18, no. 2, pp. 566–573, 2014.

[21] A. Bossio, F. Fabbrocino, G. P. Lignola, A. Prota, and G. Manfredi, "Simplified model for strengthening design of beam-column internal joints in reinforced concrete frames," *Polymers*, vol. 7, no. 9, pp. 1732–1754, 2015.

[22] E. Z. Beydokhty and H. Shariatmadar, "Behavior of damaged exterior RC beam-column joints strengthened by CFRP composites," *Latin American Journal of Solids and Structures*, vol. 13, no. 5, pp. 880–896, 2016.

[23] H. K. Lee, "Effectiveness of anchorage in concrete beams retrofitted with sprayed fiber-reinforced polymers," *Journal of Reinforced Plastics and Composites*, vol. 23, no. 12, pp. 1285–1300, 2004.

[24] A. J. Boyd, *Rehabilitation of reinforced concrete beams with sprayed glass giber reinforced polymers [Ph.D. thesis]*, University of British Columbia, British Columbia, Canada, 2000.

[25] H. K. Lee and L. R. Hausmann, "Structural repair and strengthening of damaged RC beams with sprayed FRP," *Composite Structures*, vol. 63, no. 2, pp. 201–209, 2004.

[26] S. M. Soleimani, *Sprayed glass fibre reinforced polymers in shear strengthening and enhancement of impact resistance of reinforced concrete beams [Ph.D. thesis]*, University of British Columbia, Vancouver, Canada, 2006.

[27] A. J. Boyd, N. F. Liang, P. S. Green, and K. Lammert, "Sprayed FRP repair of simulated impact in prestressed concrete girders," *Construction and Building Materials*, vol. 22, no. 3, pp. 411–416, 2008.

[28] S. K. Ha, S. Na, and H. K. Lee, "Bond characteristics of sprayed FRP composites bonded to concrete substrate considering various concrete surface conditions," *Composite Structures*, vol. 100, pp. 270–279, 2013.

[29] GB/T 1447, *Fiber-Reinforced Plastics Composites- Determination of Tensile Properties*, 2005.

[30] K. S. Lee, "A seismic strengthening technique for reinforced concrete columns using Sprayed FRP," *Journal of Advanced Concrete Technology*, vol. 10, pp. 219–230, 2012.

7

Correlation between the Compressive, Tensile Strength of Old Concrete under Marine Environment and Prediction of Long-Term Strength

Weilai Yao,[1] Shiyong Jiang,[2] Wei Fei,[1] and Tao Cai[1]

[1] *Civil Engineering Department, Logistical Engineering University, Chongqing 401311, China*
[2] *Training Department, Logistical Engineering University, Chongqing 401311, China*

Correspondence should be addressed to Shiyong Jiang; jiangshiy@163.com

Academic Editor: Andrey E. Miroshnichenko

Compressive strength and tensile strength are important mechanical properties of concrete. The long-term strength of concrete under real service environment is an important parameter when evaluating existing buildings, which should also be properly considered in structural design. In this study, the relationship between compressive and splitting tensile strength of old concrete existing for long period under marine environment was investigated. At a deserted harbour, concrete cores samples were drilled by pairs in site. For each pair of samples, the two cores were drilled from the adjacent location and conducted to compressive, splitting tensile test, respectively. 48 compressive and splitting tensile strengths were finally obtained. From the test results, tensile strength presents general uptrend with compressive strength, and the two parameters are well positively correlated. Exponential model generally recommended by building codes or literatures is still capable of describing the relationship between compressive and tensile strength of old deteriorated concrete, when function parameters are properly determined. Based on statistical theory and the experimental result of this study, a method for predicting long-term tensile strength of concrete is developed and an example is given, which may provide a potential way to estimate long-term concrete strength under real marine environment.

1. Introduction

The value of tensile strength of concrete affects the performance of structures, such as the shear force resistance and the load-bearing capacity of anchors with concrete cone failure as decisive failure mode. Hence, as concrete compressive strength, tensile strength is another critical mechanical property, which should be appropriately considered in structure design. The real value of the in situ tensile and compressive strengths of concrete deteriorated under authentic service environments for long time provide a necessary basis for the evaluation of structures, especially for those, which have existed for long period already. Considering the complexity, cost, and time-consuming nature of performing tensile tests, many researchers and building guidelines are interested to predict the tensile strength from compressive strength and their relationship in a simplified method with satisfactory accuracy [1–5]. Most correlations proposed by building codes or literatures are based on normal curing conditions and testing age of 28 days. In reality, there are many different time-, load-, and environment-dependent impacts on concrete existing in real service conditions for long term. Therefore, the compressive-tensile strength relationship in building codes cannot be assumed without further observations and investigations [5, 6].

On the other hand, so far, the concrete strength is considered as time invariant, when conducting structural design and calculation. In this way, constructions should be completely reliable and the service life should be infinite, as long as $R > S$ (R and S represent resistance and stress in broad sense) is satisfied in designing stage. But in fact, resistance of material, such as concrete strength, may degrade with time in outdoor environments, which leads to degradation of reliability and overall structure performance. In this respect, it is worthwhile to underline that the time-varying characteristics of concrete exposed to outdoor environment are necessary to be further researched. Moreover, in construction durability assessment or structure life prediction, the long-term concrete strength, along with its time-varying features, is essential information. However, it is not simple to carry out studies to research the changing law of concrete strength in long term, since it may takes several years or decades to run the tests.

Accelerated aging tests and simulative environmental tests are mainly adopted to study the environmental impacts on concrete strength. Li et al. [7] researched acidification corrosion depth and compressive strength of concrete subjected to sulphuric acid environments by accelerated tests. The results show that the remaining compressive strength values for the concrete exposed to pH = 2 and 3 sulphuric acid water environments and wet-dry cyclic sulphuric acid environment with pH = 2 are 74%, 72%, and 80% of initial strength, respectively. An accelerated corrosion by a sulphate solution in a dry-wet cycle was introduced to simulate the external sulphate corrosion environment by Zhou et al. [8]. The accelerated corrosion test results indicate that the strength degradation of concrete by sulphate attack had a significant relation with the corrosion depth. Cai et al. [9] assessed the concrete strength and deformation property under artificial sea water erosion environment. Concrete stress and strain curve was researched with the number of wet and dry cycles of 0 times, 10 times, 20 times, 30 times, 40 times, 50 times, and 60 times. A biaxial concrete strength criterion was established taking into account the stress ratio and the number of dry-wet cycles. Ramli et al. [10] studied the environmental resistance of concrete incorporating coconut fibres. Three types of simulated aggressive environments were adopted: air environment in a tropical climate, alternate air, and seawater environments in a 14-day cycle and continuous immersion in seawater. Shang et al. [11] investigated mechanical properties of plain concrete and air-entrained concrete subjected to freeze-thaw cycles (F-T-C). The influences of F-T cycles on ultrasonic velocity and mechanical properties of air-entrained concrete and plain concrete were analysed. Also, accelerated aging test or simulative environmental test is adopted in [12–18], which shows that it is convenient and time saving to adopt such experimental method to study the environmental effects on concrete properties. However, there are apparent differences between real service condition and artificial environment, since the latter can merely reproduce several environmental factors, such as sulphate, chloride ion, acid, high temperature, and humidity. In fact, the real environment is far more intricate and cannot be simply represented by simulative environmental test. In addition, in the previous literatures above, the longest aging time was 365 days [17], which cannot be regarded as a long enough period when comparing to the working life in practical engineering. Moreover, tensile strength of concrete was scarcely discussed in those previous studies.

There are only a few existing studies that investigated long-term environmental effects on concrete strength under real service conditions. However, only compressive strength was researched. Ismail et al. [19] investigated compressive strength loss of concrete exposed to outdoor environment by 6 year's monitoring. Studies by Mehta [20] Shekarchi et al. [21] and Washa and Wendt [22] discussed the long-term concrete compressive strength under actual marine environment. None of them addressed tensile strength and the correlation between compressive and tensile strength of concrete.

The major thrust of this study is to obtain both compressive and tensile strengths of concrete under real marine environment for long period (up to 61 years) and identify the correlation between the two mechanical properties. Concrete cores were drilled out by pairs at a deserted harbour in Zhoushan city, Zhejiang province, China. Every two cores of each pair were conducted in compressive and tensile splitting test, respectively; then, the relationship between long-term compressive and tensile strength was concluded. The applicability of compressive-tensile strength relationship recommended by literatures or building codes for concrete without deterioration (fresh concrete) was examined. A better fitted relationship function was presented. After that, based on statistical theory and a compressive strength predicting model from literature, a tensile strength predicting method was established and a calculating example was provided for interpreting, which may offer a potential way to estimate long-term tensile strength of concrete under real marine environment.

TABLE 1: The atmosphere chloride content of Zhoushan city.

Distance to sea (m)	2	100	140	200	260	340	500
Mean precipitation value of chloride per day (μg/cm^{2}d)	11.130	10.856	3.998	3.638	2.941	2.427	2.951

2. Experimental Program

2.1. Concrete Cores Drilling. Zhoushan is a coastal city in China. Constructions and buildings there are deteriorated by typical marine environment, like seawater, marine atmosphere, tidal waves, and so on. The yearly average temperature is 15 to 20 degrees Celsius. The content of chloride ion in atmosphere of Zhoushan city is given in Table 1.

Concrete cores samples were drilled from 4 different sites at a deserted harbour of Zhoushan city. Site A is the concrete slab of a deserted dock built in 1955. When tide rises, the distance between the dock slab and sea level is 4 m approximately. Site B is a large piece of plain concrete on seabeach, which was left as construction waste when nearby new harbour was established in 2011. Site C is a concrete pedestal of streetlight casted in 2000. Site D is the reinforced concrete roof of a deserted building constructed in 2005. The above four sites (A, B, C, and D) which had existed for 61, 5, 16, and 11 years till our test began are in adjacent areas. All of them are not immersed by sea water due to their high enough places and in atmosphere all time. Concrete cores

FIGURE 1: The cores drilling site.

drilled from each site were marked as group A, group B, group C, and group D, respectively.

In this study, the operations of drilling and testing refer to the technical specification for testing concrete strength with drilled core [23]. Location of steel bars was detected by magnetic spectroscopy steel detector and proper drilling places (away from bars) were selected, in order to avoid damage to reinforcement. Cores with diameter as 75 mm were mostly drilled, which are smaller than standards core samples in the formerly mentioned technical specification [23]. Another 4 standard cores with diameter as 100 mm were drilled, since the 75 mm-diameter drilling barrel was severely worn out. The concrete cores were obtained by pairs. For each pair of samples, the two cores were conducted to compressive and tensile test respectively, aiming to investigate the relationship between compressive and tensile strengths. The 2 cores of each pair were drilled from close enough places, which ensures that they consist of same batch of concrete with identical deterioration. The place (site A) for drilling cores is shown in Figure 1.

2.2. Cores Processing. The drilled cores were carefully processed including incising and end faces dealing before mechanical tests.

2.2.1. Cores Incising. Based on the mentioned test specification [23], after processing, the height-diameter ratio of core samples should be 1 theoretically, which can value from 0.95~1.05 in actual operation. Incision machine with double parallel blades was used and the distance between two blades will determine the height of each core. Before incising, the distance of double blades was carefully adjusted and then fixed, in order to ensure that the height-diameter ratio of each core is in the allowed range as 0.95~1.05. The incising procession is shown in Figure 2(a).

After incision, cores were paired up. Two cores of each pair were confirmed to be obtained from close enough locations. The matched cores were placed in pairs shown in Figure 2(b).

2.2.2. Cores Selecting and Loading Test Preparing. According to the test specification [23], there should be no more than 1 steel bar in qualified core and the diameter of steel bar should be less than 10 mm. The possibly existing steel bar should be vertical to the shaft of core. Samples with obvious cracks or flaws were excluded. According to the above prerequisites, 96 eligible cores (48 pairs) were finally selected. The end faces of all cores were levelled. Belt grinder machine or polymer cement mortar was adopted to make end faces smooth and level, as shown in Figure 3(a).

The scales of each qualified core sample were measured. Vernier caliper was applied to measure the diameters. Diameter was measured twice for each core at two vertical places and the average value was adopted as the final diameter, which would be applied in the further strength calculation (Figure 3(b)). Steel tap was applied to measure the core height (Figure 3(c)). Lines were drawn on those samples that would be tensile split to indicate the failure areas, as shown in Figure 3(d).

2.3. Loading Test. All of the 96 core samples were numbered before loading test. Every two cores of each pair were the same batch of concrete and deteriorated by environment equally: one sample was conducted to compressive test, while the other was conducted to tensile splitting test. Two universal testing machines with measuring range as 500 KN and 100 KN are applied to compressive and tensile splitting test, respectively. Finally, 48 compressive strengths were obtained, as well as 48 tensile splitting strengths. Compressive and tensile splitting tests are shown in Figures 4(a) and 4(b).

Test information as serial number, sample scale, and the maximum applied force of each core sample is given in Table 2. Data number consists of a three-part code, and the aleph (A, B, C, and D) represents the group of samples, which also indicates the places where concrete cores were drilled. The second figure in data number presents the sequence of sample pair in each group. The third figure in data number indicates the type of mechanical experiment: 1 represents tensile splitting test and 2 represents compressing test. d and h represent the diameter and height of core sample, respectively; d/h is the height-diameter ratio, which are all in the allowable range as 0.95~1.05; F is the maximum applied force of each sample in loading test.

2.4. Strength Calculation. According to the technical specification for testing concrete strength with drilled core [23], compressive strength of core sample was calculated as follows:

$$f_{cu,cor} = \frac{F_c}{A}, \tag{1}$$

where $f_{cu,cor}$ is compressive strength; F_c is the measured maximum compressive force in the test; A is the area of compressive section.

The tensile split strength was calculated as [23]

$$f_{cts} = 0.637 \cdot \frac{F_{spl,cor}}{A_{ts}}, \tag{2}$$

where f_{cts} is tensile splitting strength; $F_{spl,cor}$ is the measured maximum splitting force of each sample; A_{ts} is the area of split section; 0.637 is a coefficient derived from elastic theory.

(a) (b)

FIGURE 2: Incision progress and the incised cores.

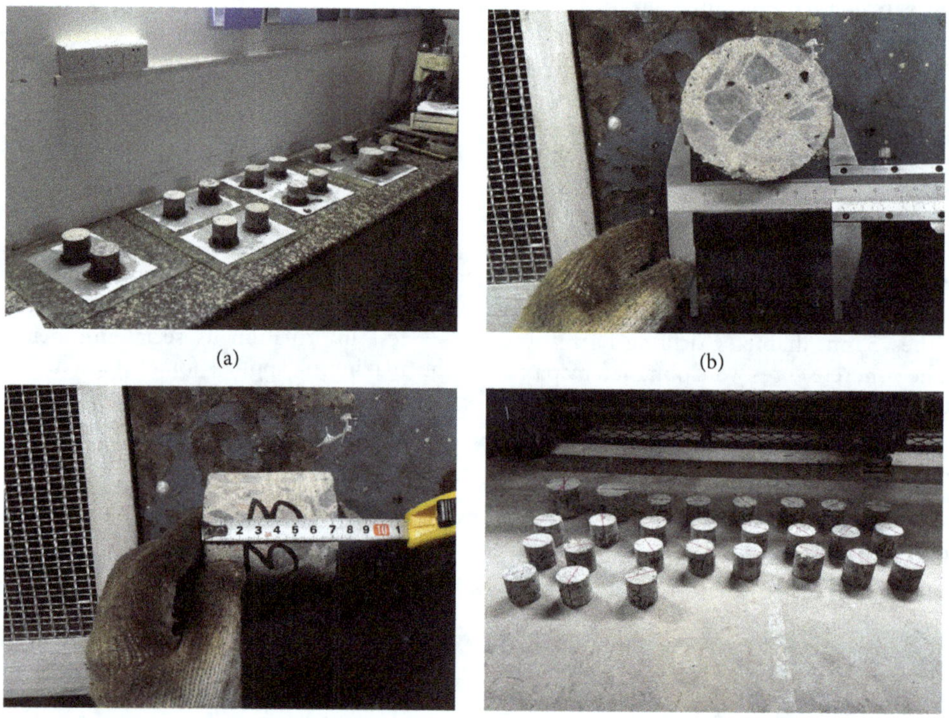

(a) (b) (c) (d)

FIGURE 3: End face levelling, measuring scale, and preparing for loading tests.

(a) (b)

FIGURE 4: Loading test.

TABLE 2: The primary data of experiment.

Data number	Test type	d/mm	h/mm	h/d	F/KN	Data number	Test type	d/mm	h/mm	h/d	F/KN
A-1-1	Tensile splitting	99.5	98	0.98	44.16	A-1-2	Compressive	99.0	98	0.99	242.12
A-2-1	Tensile splitting	73.0	73	1.00	41.08	A-2-2	Compressive	73.0	73	1.00	183.45
A-3-1	Tensile splitting	73.0	73	1.00	49.48	A-3-2	Compressive	73.0	74	1.01	217.26
A-4-1	Tensile splitting	73.5	73	0.99	28.06	A-4-2	Compressive	73.5	74	1.01	73.46
A-5-1	Tensile splitting	72.5	73	1.01	25.36	A-5-2	Compressive	73.0	73	1.00	120.43
A-6-1	Tensile splitting	73.0	73	1.00	40.26	A-6-2	Compressive	73.0	74	1.01	195.54
A-7-1	Tensile splitting	73.5	73	0.99	26.33	A-7-2	Compressive	73.5	73	0.99	144.40
A-8-1	Tensile splitting	73.0	72	0.99	41.62	A-8-2	Compressive	73.5	74	1.01	161.85
A-9-1	Tensile splitting	73.5	75	1.02	34.80	A-9-2	Compressive	72.5	73	1.01	161.14
A-10-1	Tensile splitting	72.5	73	1.01	22.77	A-10-2	Compressive	73.0	73	1.00	125.92
A-11-1	Tensile splitting	73.5	73	0.99	20.44	A-11-2	Compressive	73.5	73	0.99	120.44
A-12-1	Tensile splitting	73.0	73	1.00	23.69	A-12-2	Compressive	72.5	73	0.99	111.41
B-1-1	Tensile splitting	73.0	73	1.00	36.26	B-1-2	Compressive	73.5	72	0.98	160.48
B-2-1	Tensile splitting	73.5	73	0.99	34.26	B-2-2	Compressive	73.0	73	1.00	130.01
B-3-1	Tensile splitting	73.5	70	0.95	30.72	B-3-2	Compressive	73.5	70	0.95	131.43
B-4-1	Tensile splitting	73.5	70	0.95	29.20	B-4-2	Compressive	73.5	70	0.95	196.10
B-5-1	Tensile splitting	73.0	73	1.00	17.35	B-5-2	Compressive	73.5	72	0.98	186.28
B-6-1	Tensile splitting	73.5	74	1.01	38.18	B-6-2	Compressive	73.5	73	0.99	132.79
B-7-1	Tensile splitting	73.0	74	1.01	37.64	B-7-2	Compressive	73.0	73	1.00	129.80
B-8-1	Tensile splitting	73.5	73	0.99	34.42	B-8-2	Compressive	73.5	73	0.99	111.68
B-9-1	Tensile splitting	73.5	73	0.99	42.64	B-9-2	Compressive	73.5	73	0.99	153.36
B-10-1	Tensile splitting	73.5	73	0.99	40.48	B-10-2	Compressive	73.0	72	0.99	129.65
B-11-1	Tensile splitting	73.5	73	0.99	48.48	B-11-2	Compressive	73.5	72	0.98	135.82
B-12-1	Tensile splitting	73.5	72	0.98	42.62	B-12-2	Compressive	73.5	73	0.99	275.46
B-13-1	Tensile splitting	73.5	73	0.99	47.34	B-13-2	Compressive	73.5	74	1.01	192.14
B-14-1	Tensile splitting	73.5	72	0.98	39.76	B-14-2	Compressive	73.5	74	1.01	136.07
B-15-1	Tensile splitting	73.5	70	0.95	20.53	B-15-2	Compressive	73.5	70	0.95	120.86
B-16-1	Tensile splitting	73.0	73	1.00	20.32	B-16-2	Compressive	73.5	72	0.98	108.14
C-1-1	Tensile splitting	73.5	73	0.99	33.78	C-1-2	Compressive	73.0	73	1.00	100.45
C-2-1	Tensile splitting	73.0	73	1.00	33.70	C-2-2	Compressive	73.0	73	1.00	89.58
C-3-1	Tensile splitting	73.5	73	0.99	37.94	C-3-2	Compressive	73.5	72	0.98	76.64
C-4-1	Tensile splitting	73.5	72	0.98	34.64	C-4-2	Compressive	73.5	73	0.99	80.27
C-5-1	Tensile splitting	73.5	73	0.99	40.94	C-5-2	Compressive	73.5	73	0.99	110.00
C-6-1	Tensile splitting	73.5	73	0.99	26.33	C-6-2	Compressive	73.5	74	1.01	138.09
C-7-1	Tensile splitting	73.5	73	0.99	36.10	C-7-2	Compressive	73.5	73	0.99	68.36
C-8-1	Tensile splitting	73.0	73	1.00	34.38	C-8-2	Compressive	73.0	72	0.99	90.91
C-9-1	Tensile splitting	73.5	73	0.99	28.38	C-9-2	Compressive	73.5	73	0.99	107.07
C-10-1	Tensile splitting	73.5	73	0.99	30.80	C-10-2	Compressive	73.0	72	0.99	61.20
C-11-1	Tensile splitting	73.0	73	1.00	32.08	C-11-2	Compressive	73.5	73	0.99	86.87
C-12-1	Tensile splitting	73.5	73	0.99	34.54	C-12-2	Compressive	73.0	73	1.00	87.90
D-1-1	Tensile splitting	73.0	70	0.96	32.70	D-1-2	Compressive	73.5	70	0.95	82.07
D-2-1	Tensile splitting	73.5	71	0.97	25.61	D-2-2	Compressive	74.0	73	0.99	138.60
D-3-1	Tensile splitting	73.5	72	0.98	29.28	D-3-2	Compressive	73.0	73	1.00	68.38
D-4-1	Tensile splitting	73.0	72	0.99	23.20	D-4-2	Compressive	73.0	72	0.99	65.78
D-5-1	Tensile splitting	73.5	73	0.99	25.18	D-5-2	Compressive	73.5	74	1.01	110.32
D-6-1	Tensile splitting	73.5	73	0.99	31.02	D-6-2	Compressive	73.0	72	0.99	65.35
D-7-1	Tensile splitting	73.0	72	0.99	21.62	D-7-2	Compressive	73.5	73	0.99	44.14
D-8-1	Tensile splitting	99.5	98	0.98	37.06	D-8-2	Compressive	99.5	95	0.95	139.54

2.5. Size Effect. In the above technical specification [23], diameter of standard concrete core is 100 mm. However, there are usually large amount of steel bars in the field testing structures; as a result, drilling standard cores may cause damage to the reinforcement. Therefore, nonstandard cores with smaller scale are widely adopted in actual engineering, as well as in this study. Besides, different guidelines or codes may also adopt specimens with different shapes or sizes. In order to make the test results of this study more generally applicable, the size effect of nonstandard core samples was discussed, and then, strength of core was converted to that of 150 × 150 × 150 mm concrete cube specimen, which is defined as standard specimen in common construction standards, such as Europe codes [24] and China standards [25].

2.5.1. Size Effect of Compressive Strength. Great deal of experimental studies conducted by China Academy of Building Research demonstrated that when the height-diameter ratio of core equals 1, compressive strength of core samples with diameters ranging from 70 to 75 mm is identical to that of standard core samples, as well as 150 × 150 × 150 mm concrete cube specimens, which manifests that the size effect on compressive strength may not be considered [23]. Aitcin et al. [26] also showed that when the height equals diameter, no significant difference is observed between compressive strength of cores and cube specimens. Hence, this paper does not take account of the size effect on compressive strength. Furthermore, the compressive strengths of core samples in this study were regarded as the strengths of 150 × 150 × 150 mm concrete cube specimens [23].

2.5.2. Size Effect of Tensile Split Strength. Through experimental research and theoretical analysis, Bazant et al. [27], Duan et al. [28], and Rocco et al. [29, 30] show that the tensile splitting strength of concrete is strongly determined by the size of test specimen. Wide range of experimental studies were carried out on cubes, cylinder, and prisms to investigate the size effect of tensile splitting strength by Kadleček et al. [31]. It was demonstrated that the size effect highly depends on the areas of splitting fracture section rather than the shape of test specimens, which creates a unifying measurement for size effect. A generalized size effect formula was presented, which can be applied to cores or cubes obtained from existing structures. According to Kadleček et al. [31], the tensile splitting strength of concrete cores was converted to the strength of standard test specimen (150 × 150 × 150 mm). The size effect formula is given as [31]

$$k_p = \frac{f_{\mathrm{ts,cor}}}{f_{\mathrm{ts}}} = 2 \cdot A^{-0.128}, \tag{3}$$

where k_p is coefficient of size effect; $f_{\mathrm{ts,cor}}$ is the tensile splitting strength of drilled core samples; f_{ts} is the tensile splitting strength of standard test specimen (150 × 150 × 150 mm); A is the area of splitting fracture section of concrete core (cm^2).

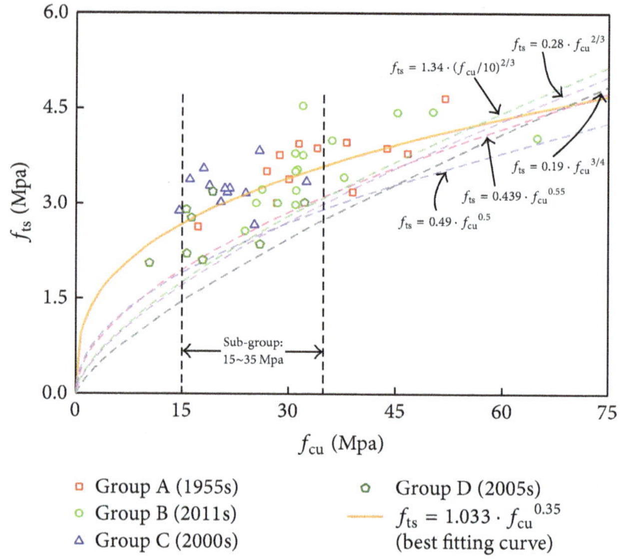

FIGURE 5: Compressive and splitting tensile strength.

3. Results and Discussion

3.1. The Relationship between Long-Term Compressive and Tensile Strength. According to Section 2.5, strengths of core samples were converted to strengths of standard test specimens, in order to improve practicability of the test result in this research. All of the 48 data points are presented in a coordinate system (Figure 5), with compressive strength as x-axis and tensile splitting strength as y-axis.

The experimental result signifies that compressive and tensile strength of concrete deteriorated by the marine environment of Zhoushan city for long term still present a certain regularity. The data points in Figure 5 distribute to the right oblique upward, forming a zonal shape. Although data points present discreteness, it is clear that as the raising of compressive strength, the tensile strength of the same concrete tends to increase as well. The correlation coefficient between compressive and splitting tensile strength is 0.72, demonstrating that the two types of strength are still well positively correlated under environmental effects for long period of time.

Generally, exponential law is recommended by literatures or building guidelines to express the relationship between compressive and tensile strength of concrete cured under normal conditions and without environmental deterioration (fresh concrete) [25, 32–34] (Table 3, left column). In Table 3, f_{cu} and f_{c} are the compressive strength of 150 × 150 × 150 mm standard concrete cube and 150/300 mm cylinder, respectively. f_{ts} is the splitting tensile strength of 150 × 150 × 150 mm standard concrete cube. f_{t} represents the axial tensile strength of concrete. f_{c} and f_{t} are converted into f_{cu} and f_{ts} based on the conversion formula borrowed from other pieces of literature: $f_{\mathrm{cu}} = 1.25 f_{\mathrm{c}}$ and $f_{\mathrm{t}} = 0.9 f_{\mathrm{ts}}$ [32]. Then, for fresh concrete, the relationships between compressive and tensile strength in left column, Table 3, are rewritten as the right column, where the strengths are f_{cu} and f_{ts}, corresponding to the mechanical indexes of the test results

Table 3: Compressive-tensile relationship.

$f_t = 0.26 \cdot f_{cu}{}^{2/3}$ [32]		$f_{ts} = 0.28 \cdot f_{cu}{}^{2/3}$
$f_{ts} = 0.55 \cdot \sqrt{f_c}$ [34]		$f_{ts} = 0.49 \cdot \sqrt{f_{cu}}$
$f_t = 1.4 \cdot \left(\frac{f_c}{10}\right)^{2/3}$ [33]	$\Longrightarrow$	$f_{ts} = 1.34 \cdot \left(\frac{f_{cu}}{10}\right)^{2/3}$
$f_t = 0.395 \cdot f_{cu}{}^{0.55}$ [25]		$f_{ts} = 0.439 \cdot f_{cu}{}^{0.55}$
—		$f_{ts} = 0.19 \cdot f_{cu}{}^{3/4}$ [32]

Table 4: The SSE of recommended functions.

Proposed functions	SSE
$f_{ts} = 0.19 f_{cu}{}^{3/4}$	58.0081
$f_{ts} = 0.439 f_{cu}{}^{0.55}$	25.4530
$f_{ts} = 1.34(f_{cu}/10)^{2/3}$	31.1486
$f_{ts} = 0.28 f_{cu}{}^{2/3}$	36.1975
$f_{ts} = 0.49 f_{cu}{}^{0.5}$	34.7893

in this study. Whether these existing compressive-tensile strength relationships (Table 3) for fresh concrete are still suitable for the deteriorated concrete in this paper or not was further explored.

As observed from Figure 5, all the correlations between compressive and tensile strength for fresh concrete recommended by building codes or literatures are increasing functions that basically present the tendency of the experimental results in this research. However, from the deviation distance between test results and these proposed function curves in Figure 5, none of them expresses the compressive-tensile strength relationship of the deteriorated concrete accurately enough. The SSE (sum of squares due to error, calculated as (4)), which is one of important indexes presenting goodness of curve fitting, is shown in Table 4.

$$\mathrm{SSE} = \sum_{i=1}^{n} (y_i - \hat{y}_i)^2, \quad (4)$$

where y_i and $\hat{y}_i$ represent the experimental and predicted (through the proposed functions) tensile strength of concrete and n is the number of data analysed.

The SSE statistic is the least-squares error of the fit, with a value closer to zero indicating a better fit. Nevertheless, the SSE of all proposed functions listed in this paper are large, together with quite small R-square statistics, indicating poor fitting effects to the experimental results. In addition, the index as (theoretical value − experimental value)/experimental value is used to reflect the errors between experimental values and model predicted values (Figures 6(a)~6(e)). It is clear that all the relations for fresh concrete $f_{ts} = 0.19 f_{cu}{}^{3/4}$, $f_{ts} = 0.439 f_{cu}{}^{0.55}$, $f_{ts} = 1.34(f_{cu}/10)^{2/3}$, $f_{ts} = 0.28 f_{cu}{}^{2/3}$, and $f_{ts} = 0.49 f_{cu}{}^{0.5}$ are inclined to underestimate the tensile strength. One possible reason may be that, after environmental deterioration for long time, compressive strength exhibits more degradation than tensile strength, which increases the ratio of tensile strength to compressive strength.

From the above discussion, it is shown that all the compressive-tensile strength relationships for fresh concrete listed in this study are not applicable to the environmentally deteriorated concrete with satisfactory accuracy, which may imply that appropriate revisions are needed. Hence, adjustment for parameters in exponential function was made aiming to promote the goodness of curve fitting. The compressive-tensile strength relationship in exponential form can be written as

$$f_{ts} = a \cdot f_{cu}{}^{b}, \quad (5)$$

where a and b are fitting parameters.

According to the test results, the function which best presents the relationship between compressive strength and tensile strength in this study was obtained through nonlinear curve fitting (Figure 5):

$$f_{ts} = 1.02 \cdot f_{cu}{}^{0.36}. \quad (6)$$

The SSE and R-square of such fitting are 8.1256 and 0.5344. Also, the predicting error shown in Figure 6(f) is reasonably moderate, which confirms that, even for the concrete deteriorated by marine environment for long period of time, the compressive-tensile strength relationship can also be rationally described by exponential function with acceptable accuracy, if the parameters of function were properly determined. This best fitting function (see (6)) can be regarded as a modified version of the compressive-tensile strength relationship recommended by guidelines or literatures for fresh concrete.

Furthermore, although concrete cores from groups A, B, C, and D are deteriorated by marine environment for different periods (61, 5, 16, and 11 years) and present various ranges of strengths, they commonly share the same subgroup (compressive strength ranging from 15 Mpa to 35 Mpa, Figure 7), which may imply that the compressive-tensile strength correlation $f_{ts} = 1.02 f_{cu}{}^{0.36}$ is adequately suitable for concrete with compressive strength from 15 Mpa to 35 Mpa, regardless of the exact deteriorated period within several decades.

In concrete technology, the aggregate strength is usually not a factor considered in normal strength concrete because, with the exception of lightweight aggregates, the aggregate partial is several times stronger than the matrix and the interfacial transition zone in concrete. In other words, with most natural aggregates the strength of the aggregates is hardly utilized because the failure is determined by the other two phases (matrix and interfacial transition zone). However, aggregate characteristics other than strength, such as the size, shape, surface texture, grading, and mineralogy, are known to affect concrete strength in varying degrees. It may be anticipated that, independent of the water-cement ratio, the size, shape, and surface texture of aggregate particles would influence the characteristics of the interfacial transition zone and therefore affect concrete strength. Furthermore, since the interfacial transition zone characteristics have more effect on the tensile strength of concrete compared to the compressive strength, it is to be expected that with a given concrete mixture any changes

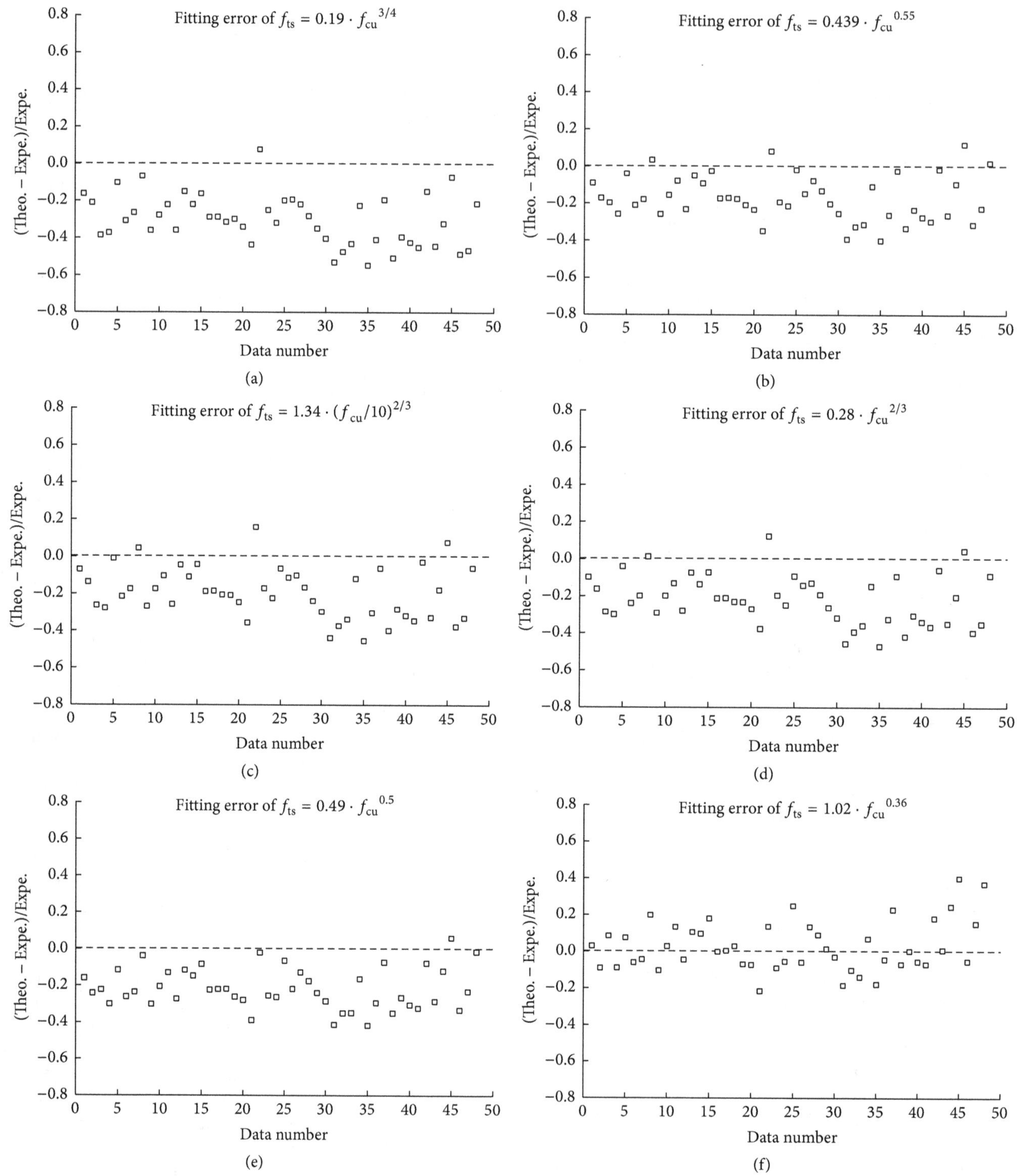

FIGURE 6: Prediction errors.

in the coarse aggregates properties would influence the tensile-compressive strength relationship of the material. Wang et al. [35] investigated the effects of the maximum aggregate size d_{max} on concrete tensile strength and fracture toughness. Considering that concrete is highly heterogonous, inclusion of d_{max} in a final relation of concrete fracture modelling is necessary. A simple methodology for analysing quasibrittle fracture of small notched three-point-bend concrete specimen was presented to determine tensile strength and fracture toughness. Fictitious crack growth before the maximum load is considered by the maximum aggregate size d_{max}. Also, the tensile strength controlled (f_t-controlled)

	Group A	Group B	Group C	Group D
The number of data points in subgroup (15~35 Mpa)	7	11	11	7
Total number of data points	12	16	12	8

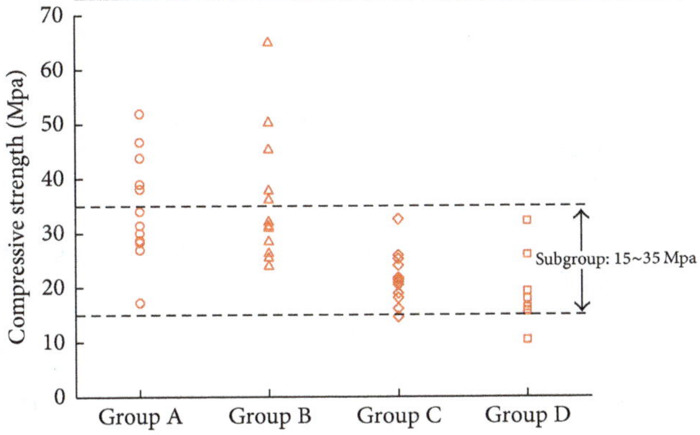

FIGURE 7: The subgroup shared by groups A, B, C, and D.

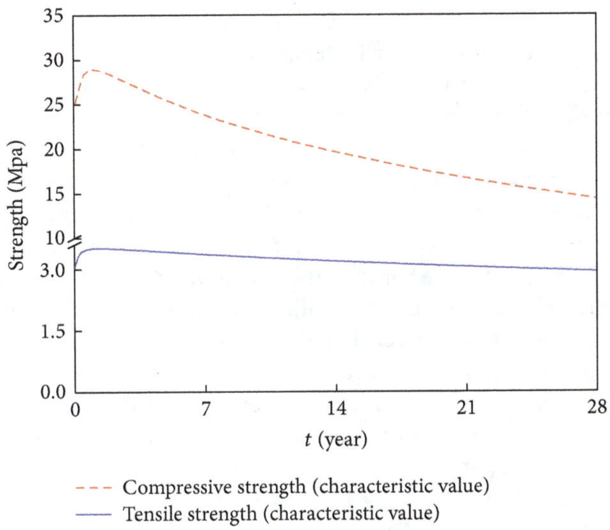

FIGURE 8: The predicted compressive and tensile strength.

failure and fracture toughness controlled (K_{IC}-controlled) failure link d_{max} to the fictitious crack growth at the peak load [36]. However, in the experimental research of this study, since the places for drilling core samples had existed for years, the authors of this paper were not capable of obtaining enough original engineering data in detail, such as mix proportion, size of aggregate, and aggregate grading. Although, aggregates near surface are roughly shown in the photographs of core sample (e.g., Figure 3), the information of aggregate such as the maximum size, shape, and distribution cannot be accurately obtained since each sample has been incised and processed. Hence the characteristics of aggregate are not particularly considered in Section 2.4. In future research, taking aggregate into analysis may be the further target, which might add more profound explanations to the test result of this study.

3.2. Prediction of Long-Term Strength under Marine Environment

3.2.1. Prediction of Long-Term Compressive Strength. Based on sufficient investigations on existing costal buildings in China and Japan, Niu [37] empirically concluded a possible model for estimating long-term compressive strength of concrete under marine environment. A nonstationary stochastic process is applied to describe the changing law of concrete compressive strength with time. At a certain time t, the compressive strength is regarded as a random variable obeying normal distribution. The probability density function at any time is given as follows:

$$p_f(x,t) = \frac{1}{\sqrt{2\pi}\sigma_f(t)} \exp\left[-\frac{\left(x-\mu_f(t)\right)^2}{2\sigma_f^2(t)}\right], \quad (7)$$

where $\mu_f(t)$, $\sigma_f(t)$ is mean value and standard derivation of concrete compressive strength after time t (year), which can be calculated as

$$\mu_f(t) = \xi(t)\,\mu_0, \quad (8)$$

$$\sigma_f(t) = \eta(t)\,\sigma_0, \quad (9)$$

where μ_0, σ_0 are mean value and standard derivation of concrete compressive strength at 28 days ($t = 0$); $\xi(t)$ and $\eta(t)$ are two functions of time, which are given as

$$\xi(t) = 1.2488e^{-0.0347(\ln t - 0.3468)^2},$$
$$\eta(t) = 0.0143t + 1.6024. \quad (10)$$

Therefore, when μ_0, σ_0 are given, the probability density function (PDF) of concrete compressive strength at any time is determined by (7)~(10). Then, compressive strength with specific probability assurance rate at given time can be determined through the PDF. For example, as stipulated in the Europe code [24] and China standard [25], the probability assurance of the concrete characteristic compressive strength is 95%. Based on the PDF, characteristic compressive strength can be obtained by numerical calculation or simply by the following formula (see (11)), since the probability is normally distributed.

$$f_{cu,k}(t) = \mu_f(t) - 1.645 \cdot \sigma_f(t). \quad (11)$$

For instance, the mean value and standard derivation of concrete compressive strength at 28 days are 31.15 Mpa and 3.738 Mpa, respectively. According to the above Niu's model [37], the characteristic compressive strength ($f_{cu,k}$) at any time is worked out, as shown in Figure 8. Strength with other probability assurance rates can be calculated similarly. Niu's model provides a possible method that is able to predict long-term concrete compressive strength under marine conditions including marine atmosphere, tidal areas, and immersed regions by seawater [37, 38]. According to the predicted results by Niu's model, the compressive strength of concrete under marine environment shows increment in first 1 or 2 years after casting. This could be due to the continuing internal chemical reactions within concrete, which gives rise to strength growth. For instance, formation of salt crystals contributes towards enhancing cohesion among aggregates which induces strength increase. Under long-term environmental effects, corrosive substance attacks and the expansions or shrinkages caused by environments would induce cracks within concrete, especially for those with inadequate mix design. Once the cracks are initiated, it allows the further penetration of more harmful and corrosive substances, which ultimately reduces the durability and strength of concrete in a substantial manner. For that reason, concrete strength tends to degrade with time in later period. Such possible characteristic of strength change was concluded by Niu [37] through numerous investigations on coastal constructions in China and Japan.

3.2.2. Prediction of Long-Term Tensile Strength. In Section 3.1, for concrete with compressive strength ranging from 15 Mpa to 35 Mpa, the recommended exponential function by this study $f_{\text{ts}} = 1.02 f_{\text{cu}}^{0.36}$ is applied to describe the compressive-tensile strength correlation within acceptable error, regardless of the specific deteriorated period among several decades.

Similar to Niu's model, both compressive and tensile strength are regarded as stochastic processes in this study, which are two random variables at certain time. The experimental research in this paper essentially presents the relationship (see (6)) between those two random variables. Then, at given time t, the PDF of tensile strength can be derived through the theory of probability. An example is provided herein to elucidate the method for calculating the statistic distribution of tensile strength, and then, the long-term tensile strength is predicted.

For tensile strength, the Cumulate Distribution Function (CDF) is derived as

$$\begin{aligned} F_{f_{\text{ts}}(t)}(x) &= P\{f_{\text{ts}}(t) < x\} = P\{1.02 \cdot f_{\text{cu}}(t)^{0.36} < x\} \\ &= P\left\{f_{\text{cu}}(t) < \frac{1}{1.02} \cdot x^{1/0.36}\right\} \\ &= F_{f_{\text{cu}}(t)}\left(\frac{1}{1.02} \cdot x^{1/0.36}\right), \end{aligned} \quad (12)$$

where $f_{\text{ts}}(t)$ and $f_{\text{cu}}(t)$ are the tensile strength and compressive strength at given time t, which are both random variables. $F_{f_{\text{ts}}(t)}(x)$ and $F_{f_{\text{cu}}(t)}(x)$ are the Cumulate Distribution Function (CDF) of random variable $f_{\text{ts}}(t)$, $f_{\text{cu}}(t)$.

Then, the probability density function of tensile strength can be obtained through differential operation as

$$\begin{aligned} p_{f_{\text{ts}}(t)} &= \frac{dF_{f_{\text{ts}}(t)}(x)}{dx} = \frac{dF_{f_{\text{cu}}(t)}\left((1/1.02) \cdot x^{1/0.36}\right)}{dx} \\ &= \frac{1}{1.02} \cdot \frac{1}{0.36} \cdot x^{(1/0.36-1)} p_{f_{\text{cu}}(t)}\left(\frac{1}{1.02} \cdot x^{1/0.36}\right). \end{aligned} \quad (13)$$

According to Niu's research [35, 36], at given time t, concrete compressive strength obeys normal distribution as (8); hence

$$\begin{aligned} & p_{f_{\text{cu}}(t)}\left(\frac{1}{1.02} x^{1/0.36}\right) \\ & = \frac{1}{\sqrt{2\pi}\sigma_f(t)} \exp\left[-\frac{\left((1/1.02)\, x^{1/0.36} - \mu_f(t)\right)^2}{2\sigma_f^2(t)}\right]. \end{aligned} \quad (14)$$

Bringing (14) into (13), the PDF of tensile strength at certain time t is obtained:

$$\begin{aligned} & p_{f_{\text{ts}}(t)} \\ & = \frac{2.72 \cdot x^{1.78}}{\sqrt{2\pi}\sigma_f(t)} \exp\left[-\frac{\left((1/1.02)\, x^{1/0.36} - \mu_f(t)\right)^2}{2\sigma_f^2(t)}\right]. \end{aligned} \quad (15)$$

Based on the statistic distribution of $f_{\text{ts}}(t)$ shown as (15), tensile strength with specific probability assurance rate such as 95% (characteristic value) can be calculated through numerical calculation, as well as tensile strength with some other probability assurance rates. The predicted tensile strength is shown in Figure 8. Resembling to the long-term compressive strength, the long-term tensile strength presents slight increase in the initial period and then decreases with time; however, it appears as a relatively moderate tendency. This section provides a possible method for estimating long-term tensile strength of concrete under marine environment. Using the method proposed herein, the concrete tensile strength is predicted in several years or decades, rather than several days or weeks only.

4. Conclusions and Prospects

An experimental research was carried out in this study to investigate the relationship between the long-term compressive and tensile strength of concrete under realistic marine environment. Exponential function obtained through non-liner fitting is recommended to quantitatively describe such correlation. Based on the tests results, a possible method for predicting long-term tensile strength is developed. According to the finds in this study, the following conclusions can be drawn:

(1) For concrete exposing to marine environments of Zhoushan city for long period as several years or decades, compressive and tensile strength present certain regularity: as compressive strength increasing, the tensile strength tends to increase as well. The correlation coefficient between the two types of strength is 0.72, indicating that they are well positively correlated after environmental deterioration.

(2) The exponential functions for fresh concrete ($f_{\text{ts}} = 0.19 f_{\text{cu}}^{3/4}$, $f_{\text{ts}} = 0.439 f_{\text{cu}}^{0.55}$, $f_{\text{ts}} = 1.34(f_{\text{cu}}/10)^{2/3}$, $f_{\text{ts}} = 0.28 f_{\text{cu}}^{2/3}$, and $f_{\text{ts}} = 0.49 f_{\text{cu}}^{0.5}$) proposed by building guidelines or literatures only roughly describe the overall tendency of compressive-tensile

strength relationship for deteriorated concrete. Large SSE along with quite small R-square statistics indicate their poor fitting effects. By adjusting the parameters of exponential function, a best fitting curve ($f_{ts} = 1.02 f_{cu}^{0.36}$) is recommended by this paper, which demonstrates that the exponential model can be extensively used to describe the compressive-tensile strength relationships for both fresh and deteriorated concrete. Moreover, the best fitting function ($f_{ts} = 1.02 f_{cu}^{0.36}$) may be suitable for concrete with compressive strength ranging from 15 Mpa to 35 Mpa, regardless of the specific deteriorating period within several decades.

(3) A potential tensile strength predicting method is established within the frame-work of probability analysis. The compressive and tensile strengths of concrete are regarded as two stochastic processes. The test in this study essentially investigated the relationship of them. Based on formerly known probability distributions of compressive strength from literature, the probability distributions of tensile strength at given time t is determined through statistical analysis. Then, tensile strength with specific probability assurance rate (such as 95%) can be worked out and prediction of long-term tensile strength is accomplished, which may provide meaningful information in life-through estimations, durability assessment, or structural health monitoring.

In future studies, aiming to experimentally characterize a more representative relationship between long-term compressive and tensile strength of concrete under marine environment, the authors of this paper will obtain added concrete samples from various coastal areas, not limited to Zhoushan city. The characteristics of aggregate may be taken into analysis. Also, the effectiveness of the method in this paper for predicting long-term tensile strength will be further verified.

Conflicts of Interest

The authors declare that they have no conflicts of interest.

Acknowledgments

The study was financially supported by the project named Service Life Prediction of Strengthened RC Beam (KJ-2015030) and provided by Chongqing Administration of Land, Resources and Housing. The test conducted in this paper was under the help of Yong Cheng Construction Detection Research Co. Ltd., Ningbo.

References

[1] C. Gaedicke, A. Torres, K. C. T. Huynh, and A. Marines, "A method to correlate splitting tensile strength and compressive strength of pervious concrete cylinders and cores," *Construction and Building Materials*, vol. 125, pp. 271–278, 2016.

[2] K. Yan, H. Xu, G. Shen, and P. Liu, "Prediction of splitting tensile strength from cylinder compressive strength of concrete by support vector machine," *Advances in Materials Science and Engineering*, vol. 2013, Article ID 597257, 2013.

[3] M. H. Severcan, "Prediction of splitting tensile strength from the compressive strength of concrete using GEP," *Neural Computing and Applications*, vol. 21, no. 8, pp. 1937–1945, 2012.

[4] M. Saridemir, "Empirical modeling of splitting tensile strength from cylinder compressive strength of concrete by genetic programming," *Expert Systems with Applications*, vol. 38, no. 11, pp. 14257–14268, 2011.

[5] F. A. Oluokun, "Prediction of concrete tensile strength from its compressive strength. Evaluation of existing relations for normal weight concrete," *ACI Materials Journal*, vol. 88, no. 3, pp. 302–309, 1991.

[6] M. Weber and C. Thiele, "Correlation between the compressive and tensile strength of old concretes: applicability of the relationship described by DIN EN 1992-1-1," *Beton- Und Stahlbetonbau*, vol. 111, no. 10, pp. 635–644, 2016.

[7] B. Li, L. Cai, K. Wang, and Y. Zhang, "Prediction of the residual strength for durability failure of concrete structure in acidic environments," *Journal Wuhan University of Technology, Materials Science Edition*, vol. 31, no. 2, pp. 340–344, 2016.

[8] Y. Zhou, H. Tian, L. Sui, F. Xing, and N. Han, "Strength deterioration of concrete in sulfate environment: an experimental study and theoretical modeling," *Advances in Materials Science and Engineering*, vol. 2015, Article ID 951209, 13 pages, 2015.

[9] J. Cai, F. Zhang, W. Cui, S. Chen, and P. Liu, "Concrete strength and deformation property under sea water erosion environment," *Advanced Materials Research*, vol. 446-449, pp. 2554–2559, 2012.

[10] M. Ramli, W. H. Kwan, and N. F. Abas, "Strength and durability of coconut-fiber-reinforced concrete in aggressive environments," *Construction and Building Materials*, vol. 38, pp. 554–566, 2013.

[11] H.-S. Shang, T.-H. Yi, and X.-X. Guo, "Study on strength and ultrasonic velocity of air-entrained concrete and plain concrete in cold environment," *Advances in Materials Science and Engineering*, vol. 2014, Article ID 706986, 7 pages, 2014.

[12] B. S. Thomas, R. C. Gupta, P. Mehra, and S. Kumar, "Performance of high strength rubberized concrete in aggressive environment," *Construction and Building Materials*, vol. 83, article no. 6441, pp. 320–326, 2015.

[13] J. O. Okeniyi, I. O. Oladele, O. M. Omoniyi, C. A. Loto, and A. P. Popoola, "Inhibition and compressive-strength performance of Na2Cr2O7 and C10H14N2Na2O8·2H2O in steel-reinforced concrete in corrosive environments," *Canadian Journal of Civil Engineering*, vol. 42, no. 6, pp. 408–416, 2015.

[14] W. H. Kwan, M. Ramli, and C. B. Cheah, "Flexural strength and impact resistance study of fibre reinforced concrete in simulated aggressive environment," *Construction and Building Materials*, vol. 63, pp. 62–71, 2014.

[15] M. Ramli, W. H. Kwan, and N. F. Abas, "Application of non-corrosive barchip fibres for high strength concrete enhancements in aggressive environments," *Composites Part B: Engineering*, vol. 53, pp. 134–144, 2013.

[16] X. Yuan, B. Li, G. Cui, S. Zhao, and M. Zhou, "Effect of fly ash and early strength agent on durability of concrete exposed to the cyclic sulfate environment," *Journal Wuhan University of Technology, Materials Science Edition*, vol. 25, no. 6, pp. 1065–1069, 2010.

[17] E. Güneyisi, M. Gesoğlu, and K. Mermerdaş, "Strength deterioration of plain and metakaolin concretes in aggressive sulfate

environments," *Journal of Materials in Civil Engineering*, vol. 22, no. 4, pp. 403–407, 2010.

[18] J. P. Gorninski, D. C. Dal Molin, and C. S. Kazmierczak, "Strength degradation of polymer concrete in acidic environments," *Cement and Concrete Composites*, vol. 29, no. 8, pp. 637–645, 2007.

[19] M. Ismail, B. Muhammad, and M. E. Ismail, "Compressive strength loss and reinforcement degradations of reinforced concrete structure due to long-term exposure," *Construction and Building Materials*, vol. 24, no. 6, pp. 898–902, 2010.

[20] P. K. Mehta, "Durability of concrete exposed to marine environment - A fresh look," *ACI Structure Journal*, vol. 109, pp. 1–30, 1988.

[21] M. Shekarchi, A. Rafiee, and H. Layssi, "Long-term chloride diffusion in silica fume concrete in harsh marine climates," *Cement and Concrete Composites*, vol. 31, no. 10, pp. 769–775, 2009.

[22] G. W. Washa and K. F. Wendt, "Fifty year properties of concrete," *American Concrete Institute*, vol. 72, no. 1, pp. 20–28, 1975.

[23] China Engineering Construction Standardization Association, *CECS 03: 2007 Technical Specification for Testing Concrete Strength with Drilled Core*, China Building Industry Press, Beijing, China, 2007.

[24] European Committee for Standardization, *Concrete-Part 1: Specification, Performance, Production and Conformity*, European Committee, 2006.

[25] Ministry of Housing and Urban-Rural Development of the People's Republic of China, *GB 50010-2010 Code for Design of Concrete Structures*, China Building Industry Press, Beijing, China, 2010.

[26] P. C. Aitcin, S. L. Sarkar, and P. Laplante, "Long-term characteristics of a very high strength concrete," *Concrete International*, vol. 12, no. 1, pp. 40–44, 1990.

[27] Z. P. Bazant, M. T. Kazemi, T. Hasegawa, and J. Mazars, "Size effect in Brazilian split-cylinder tests. Measurements and fracture analysis," *ACI Materials Journal*, vol. 88, no. 3, pp. 325–332, 1991.

[28] K. Duan, X. Hu, and F. H. Wittmann, "Scaling of quasi-brittle fracture: Boundary and size effect," *Mechanics of Materials*, vol. 38, no. 1-2, pp. 128–141, 2006.

[29] C. Rocco, G. V. Guinea, J. Planas, and M. Elices, "Size effect and boundary conditions in the Brazilian test: experimental verification," *Materials and Structures*, vol. 32, no. 217, pp. 210–217, 1999.

[30] C. Rocco, G. V. Guinea, J. Planas, and M. Elices, "Size effect and boundary conditions in the brazilian test: theoretical analysis," *Materials Structures*, vol. 32, pp. 437–444, 1999.

[31] V. Kadleček, S. Modrý, and V. Kadleček, "Size effect of test specimens on tensile splitting strength of concrete: general relation," *Materials and Structures*, vol. 35, no. 1, pp. 28–34, 2002.

[32] Z. Guo, *Theory of Reinforced Concrete*, Tsinghua University Press, Beijing, China, 2013.

[33] International Federation for Structural Concrete (FIB), *FIB056-Model Code 2010, Concrete Structures*, DCC Document Competence Center Siegmar Kästl e.K., Germany, Lausanne, Switzerland, 2010.

[34] ACI committee 318-14, *Building Code Requirements for Structural Concrete*, American Concrete Institute, Farmington Hills, Mich, USA, 2014.

[35] Y. Wang, X. Hu, L. Liang, and W. Zhu, "Determination of tensile strength and fracture toughness of concrete using notched 3-p-b specimens," *Engineering Fracture Mechanics*, vol. 160, pp. 67–77, 2016.

[36] J. Guan, X. Hu, and Q. Li, "In-depth analysis of notched 3-p-b concrete fracture," *Engineering Fracture Mechanics*, vol. 165, pp. 57–71, 2016.

[37] D. Niu, "Changing models of concrete strength along with time in marine environment," *Journal of Xi'an University of Architecture and Technology*, vol. 27, no. 03, pp. 49–52, 1995.

[38] Q. Zhao, Y. Yong, and G. Sen, "Reliability estimation of in-service tunnel shield capacity," in *Proceedings of the International Conference on Pipelines and Trenchless Technology 2009, ICPTT 2009: Advances and Experiences with Pipelines and Trenchless Technology for Water, Sewer, Gas, and Oil Applications*, pp. 1783–1792, Shanghai, China, 2009.

8

Effects of Nano-$CaCO_3$/Limestone Composite Particles on the Hydration Products and Pore Structure of Cementitious Materials

Huashan Yang and Yujun Che

[1]*School of Materials and Architecture Engineering, Guizhou Normal University, Guiyang 550025, China*

Correspondence should be addressed to Huashan Yang; 13368600935@163.com

Academic Editor: Robert Cerný

The agglomeration of nano-$CaCO_3$ (NC) is the largest bottleneck in applications in cementitious materials. If nano-$CaCO_3$ modifies the surface of micron-scale limestone powder (LS), then it will form nano-$CaCO_3$/limestone composite particles (NC/LS). It is known that micron-scale limestone is easily dispersed, and the "dispersion" of NC is governed by that of LS. Therefore, the dispersion of nano-$CaCO_3$ can be improved by the NC/LS in cementitious materials. In this work, the preparation of NC/LS was carried out in a three-necked flask using the $Ca(OH)_2$-H_2O-CO_2 reaction system. The morphology of NC/LS was observed by a field emission scanning electron microscope (FE-SEM). The effects of NC/LS on the hydration products and pore structure of cementitious materials are proposed. 5% NC/LS was added into cement paste and mortar, and the mechanical properties of the specimens were measured at a certain age. Differential scanning calorimetry (DSC), thermal gravimetric analysis (TG), and backscattered electron imaging (BSE) were conducted on the specimens to investigate the hydration products and pore structure. The properties of specimens with NC/LS were compared to that of control specimens (without NC/LS). The results revealed that NC/LS reduced the porosity and improved the mechanical properties of the cementitious materials.

1. Introduction

The development and increasing applications of green high-performance concrete (GHPC) have led to higher requirements for concrete additions, resulting in a variety of mineral additions, such as fly ash (FA), and ground granulated blast-furnace slag (GGBFS) [1–3]. The most successful of these mineral additions, FA, has seen its cost steadily increasing and its availability decreasing. One of the major drawbacks of concrete containing high volumes of FA is known to have negative impact on the early age properties [4]. It significantly slows setting and early-age compressive strength gain. An alternative solution is to use limestone powder (LS).

Considerable research has been carried out in recent years on the use of LS in Portland cement concrete [5–10]. Studies have revealed that limestone powders exhibit a filling effect on concrete [5]. Meanwhile, these powders can react with the aluminium phase in cement paste to create calcium carboaluminate hydrates [11–13] and accelerate C_3S hydration [14, 15]. In this way, LS can improve the mechanical properties of cement, thus widely applied through various industries. It has been reported that the surface structure is an important property of LS [16]. However, there have been only few published studies concerning the effects of surface structure of LS on the fluidity and strength of Portland cement concrete [17].

In recent years, nano-$CaCO_3$ has been introduced in concrete [18], and some studies have suggested a potential benefit of physical properties of nano-$CaCO_3$ on the development of cementitious system [19–24]. However, it is believed that the most important challenge in the nanocomposite research is to disperse properly the nanomaterials into the matrix [25, 26]. Due to the high surface energy of nanoparticles, nano-$CaCO_3$ is easy to agglomerate to form secondary particles, which will reduce the properties of cementitious materials. It is believed that nanomicron composite particle is a solution for this question [16]. If nano-$CaCO_3$ grows on the surface of LS, then the nano-$CaCO_3$/limestone composite particles (NC/LS) are prepared.

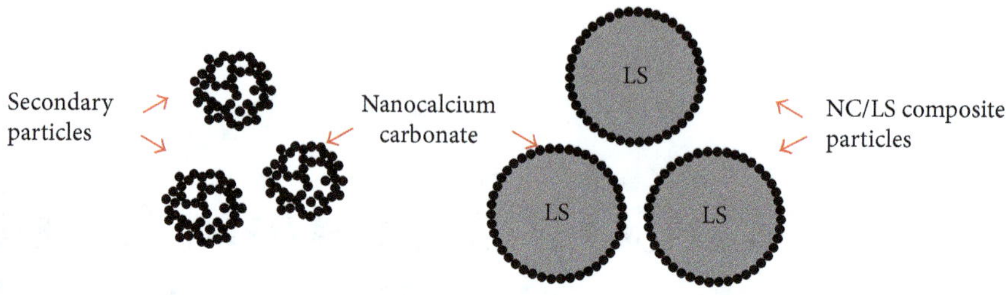

Figure 1: Secondary particles and nanomicron composite particles.

As it can be seen from Figure 1, this approach can readily achieve surface modification of LS. NC on the surfaces of LS means that NC is a part of LS. Thus, the "dispersion" of NC is governed by that of LS in cementitious materials. Compared to the nanoparticles, micron-scale LS is easy to obtain better dispersity due to lower surface energy. Therefore, the dispersity of nano-$CaCO_3$ will significantly be improved by NC/LS in cementitious materials. Furthermore, due to the sophisticated surface structures of nanomicron composite particles, it can have a significant effect on the hydration and microstructure of cementitious materials. Camiletti et al. [20] found that nano-$CaCO_3$ accelerates the setting and hardening process of ultra-high-performance concrete through providing nucleation sites, increasing contact points, and increasing the effective water-to-cement ratio. Yang et al. [16] also believes that nano-$CaCO_3$ particles on the surface of LS can act as heterogeneous nuclei for cement paste, making the size of calcium hydroxide crystals smaller, leading to a denser microstructure. Nevertheless, the effects of NC/LS on the hydration products and pore structure of cement pastes have received little attention.

This article explores the effects of NC/LS on the mechanical properties of cement paste and mortar. Differential scanning calorimetry, thermal gravimetric analysis, and backscattered electron imaging were employed to investigate the effects of NC/LS on the hydration products of cementitious materials. Cementitious materials are considered as porous composite materials that consist essentially of a binding medium within which are embedded particles or fragments of aggregate [27]. The complex pore structure of cementitious materials has great significance on their strength, permeability, and durability. The pore structure of cementitious materials was also investigated by BSE images analysis.

2. Materials and Methods

2.1. Materials. The materials involved in this study included nano-$CaCO_3$/limestone composite particles, limestone powder, calcium hydroxide, carbon dioxide, ordinary Portland cement type P.O. 42.5, and ISO standard sand. Carbon dioxide gas with a purity of 99.9% was purchased from standard commercial sources. Calcium hydroxide, P.O. 42.5 ordinary Portland cement, and ISO standard sand are also commercial products.

Nano-$CaCO_3$/limestone composite particles were prepared by heterogeneous nucleation. Via changing the reaction parameters such as mixing rate, the morphology of micron-scale limestone particles after coating has been altered. The preparation of NC/LS was carried out in a three-necked flask using the $Ca(OH)_2$-H_2O-CO_2 reaction system (Figure 2). The mixtures comprising limestone powder and saturated solution of calcium hydroxide were stirred at 2400 rpm in a three-necked flask for 10 minutes. Then, carbon dioxide gas was blown into the three-necked flask with the same stirring. The pH value of the reaction solution was inspected using a pH meter. When the pH value reached 6, the reaction was completed, and then the carbon dioxide flow was stopped. The NC/LS was obtained after the slurry was washed by deionized water, filtered, and dried at 120°C in drying oven for at least 24 h. It is reported that some nanoparticles are associated with a high level of toxicity. As shown in Figure 2, the nano-$CaCO_3$/limestone composite particles were prepared by heterogeneous nucleation in a three-necked flask using the $Ca(OH)_2$-H_2O-CO_2 reaction system. Nanosized calcium carbonate particles are attached to the surface of micron-scale limestone particles. Also, acute and subchronic toxicity tests were performed to establish the safety of these products after oral administration [28]. Results showed that nano-$CaCO_3$ did not produce any obvious symptoms of toxicity, even at the highest dose administered. Thus, as a raw material of concrete, nano-$CaCO_3$ is safety.

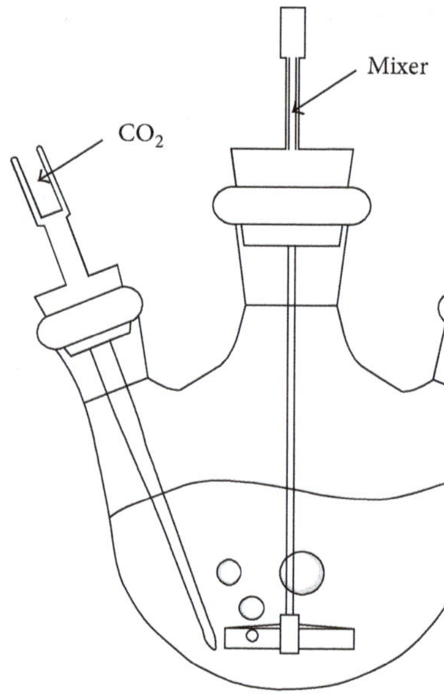

Figure 2: $Ca(OH)_2$-H_2O-CO_2 reaction system.

The particle size distribution (PSD) of the LS are presented in Figure 3, which was determined by laser diffraction

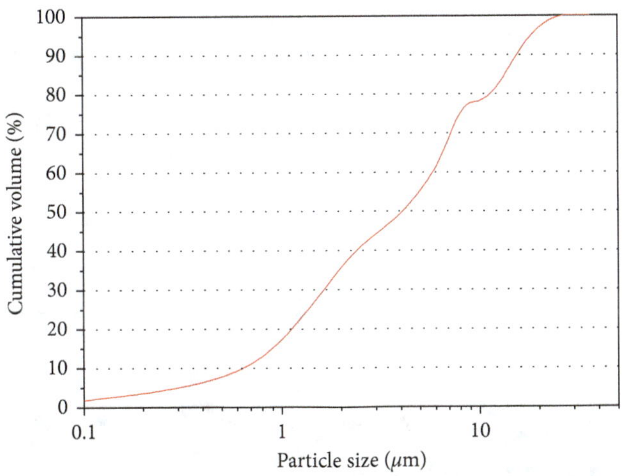

FIGURE 3: PSD of limestone powder.

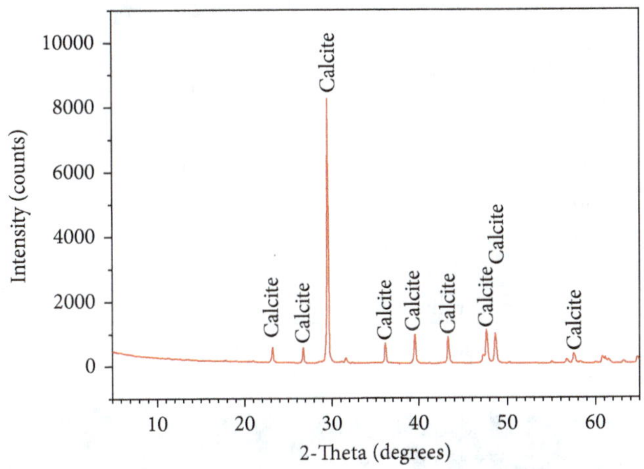

FIGURE 4: XRD pattern of limestone powder.

TABLE 1: Chemical composition of cement and limestone (%).

Oxide	Portland cement	Limestone
SiO_2	21.06	4.38
Al_2O_3	6.13	1.56
Fe_2O_3	2.61	0.76
CaO	61.18	48.00
MgO	1.66	0.49
K_2O	1.06	0.14
Na_2O	0.21	0.03
SO_3	3.18	—
Loss	2.12	44.17
Others	0.80	0.13

using a laser granulometry (JL-1155, China) for grain size between 0.1 μm and 155 μm. LS has a mean particle size $D_{50} = 5\,\mu$m. The chemical composition of Portland cement and LS as given in Table 1 was determined by X-ray fluorescence (XRF). Phase identification of LS was performed at room temperature using X-ray diffraction (XRD, Bruker AXS D8-Focus, Germany) with Cu-Kα radiation in the range 2θ-scale from 5° to 65°. X-ray diffraction studies confirmed that the main phase of the LS is calcite (Figure 4).

2.2. Methods. To study the effects of NC/LS on the strength development of cement pastes and mortar, paste mixes were prepared by adopting a water-to-binder ratio of 0.4, and mortar mixes were prepared by adopting a binder-to-sand ratio of 1 : 3 and water-to-binder ratio of 0.5. Table 2 summarizes the proportions of cement paste and mortar. Specimens C-0 and M-0 are the control specimens of C-NC/LS and M-NC/LS, respectively. After 24 h, the specimens were demoulded and were immersed in water at 20°C until testing. The mechanical properties of these specimens were tested after 3 and 28 days. The test results of cement pastes are the average of three compression test. The test results of cement mortar are the average of three flexural specimens and six compression test. The amount of calcium silicate hydrates and calcium hydroxide present in specimens after 3 days was determined by thermogravimetric analysis (PerkinElmer Instrument, USA) from 100°C to 1000°C in a nitrogen atmosphere. Samples for DSC-TG analysis were obtained from 3-day-old specimens. Before grinding, specimens were kept in absolute ethyl alcohol up to test to stop the hydration process.

It is believed that pore structure has a direct effect on properties like strength, permeability, and durability of cementitious materials. Mercury intrusion porosimetry (MIP) is a widely used method to characterize the pore structure of cementitious materials. However, the technique has several limitations [29, 30], among which are the ink bottle effect and a cylindrical pore geometry assumption that lead to inaccurate pore size distribution curves [31]. Also, Lange et al. [32] compared MIP and image analysis. Results showed that the two methods generate PSD curves of similar shape, and the image-based PSD has the advantage of being able to better describe the large porosity in the microstructure. Therefore, researchers are interested in image analysis for a more complete picture of pore structure. According to the working principles of backscattered electron imaging in scanning electron microscope, the backscattered electrons refer to the incident electrons that are reflected by atoms of the solid sample and the contrast of images are related to the atomic number distribution of sample surface atoms. Areas with high average atomic numbers generate strong signals, resulting in a bright contrast in BSE images. Therefore, the relative atomic number of different areas can be determined based on BSE images contrasts [33].

The partially reacted cement grains, cement hydrates, and pores in concrete are particularly variable, but average features can be measured, which help to understand the microstructure of cementitious materials. A JSM-IT300 scanning electron microscope was used to observe the microstructure of the specimens based on backscattered electron imaging. A small fractured sample was soaked in acetone to stop hydration and dried at 80°C for 2 h. Then, the sample was coated with 20 nm of gold to make it conductive. A magnification of 500x was used for all images. At 500x, the

TABLE 2: Proportions of cement paste and mortar.

Constitute	C-0	C-NC/LS	M-0	M-NC/LS
Cement paste				
Water-to-binder ratio	0.4	0.4	—	—
NC/LS (%)	0	5.0	—	—
Cement mortar				
Water-to-binder ratio	—	—	0.5	0.5
NC/LS (%)	—	—	0	5.0
Binder-to-sand ratio	—	—	1 : 3	1 : 3

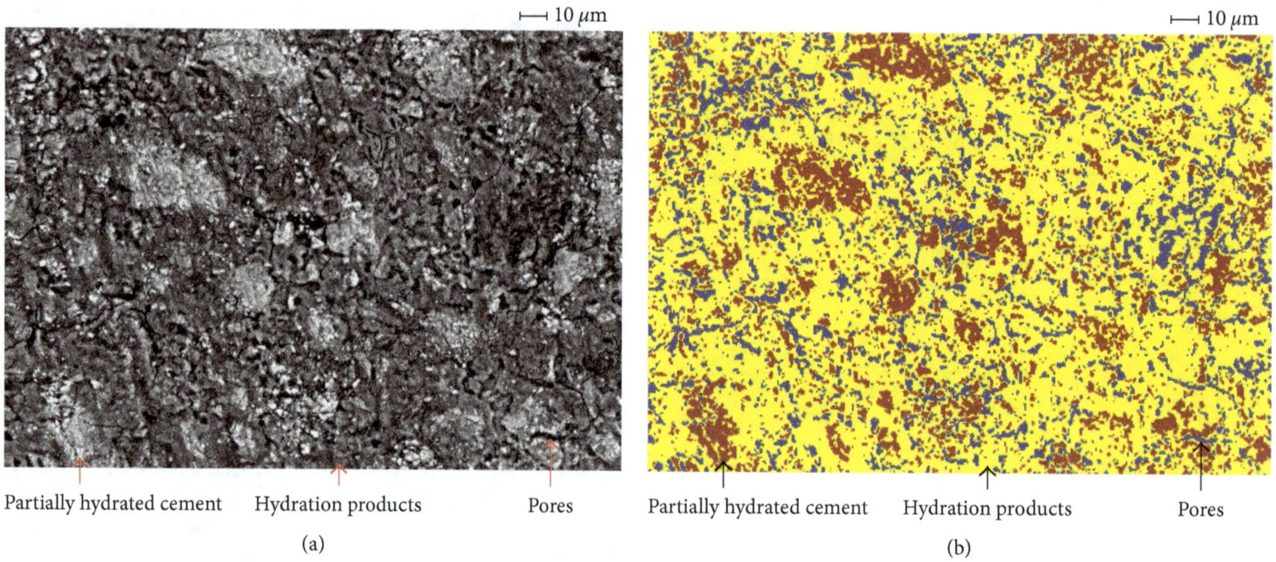

FIGURE 5: Cluster analysis of partially hydrated cement, hydration products, and pores. (a) A typical BSE image of C-0. (b) Results of cluster analysis.

resolution of the 1024×768 digital image (width of a single pixel) is 0.222 μm. Figure 5(a) shows a typical BSE image of specimen C-0. As can be seen, brighter areas correspond to partially reacted cement grains; less bright areas correspond to cement hydrates; and least bright areas correspond to pores. Cluster analysis was conducted on pretreated BSE images to separate partially reacted cement grains, cement hydrates, and pores, as shown in Figure 5(b). The effects of NC/LS on the pore structure and hydration products of cement specimens can be further investigated by calculating the respective areas of partially reacted cement grains, cement hydrates, and pores, as well as pore size distribution. Ten backscattered electron (BSE) images of random locations on the polished surface of each specimen were captured as image files in the image analysis computer.

3. Results and Discussion

3.1. Surface Structure of NC/LS. The micron-scale particles of LS and surface structure of NC/LS are shown in Figure 6(a) and (b), respectively. It is seen that LS is composed of micron-scale particles, which are characterized by smooth cleavage plane and sharp edges. However, the NC/LS, a composite of limestone powder and nanosized calcium carbonate, has very different particle morphology from that of raw limestone particles. As shown in Figure 6(b), nanosized calcium carbonate particles are attached to the surface of micron-scale limestone particles, resulting in a sophisticated surface structure of composite particles. The sharp-pointed edges of the micron-particles are not observed. Tanaka and Koishi [34] and Felekoglu [17] proposed that properties of cementitious materials are not solely related to the particle size distribution, also particle microshape and surface characteristics are important. Thus, the microshape and surface structure of particles have an important effect on the hydration and microstructure of cementitious materials. Furthermore, the most significant issue for nanoparticles is that of effective dispersion [35]. Nano-$CaCO_3$ on the surface of the micron-scale limestone particles will also significantly improve the dispersion of nano-$CaCO_3$ in cementitious materials.

3.2. Strength Development. Figure 7 shows the compressive strength of cement pastes and the flexural and compressive strength of cement mortar after 3 and 28 days. Cement pastes with NC/LS performed better in terms of compressive strength for all ages, probably due to the formation of calcium aluminate monocarbonate and the accelerating

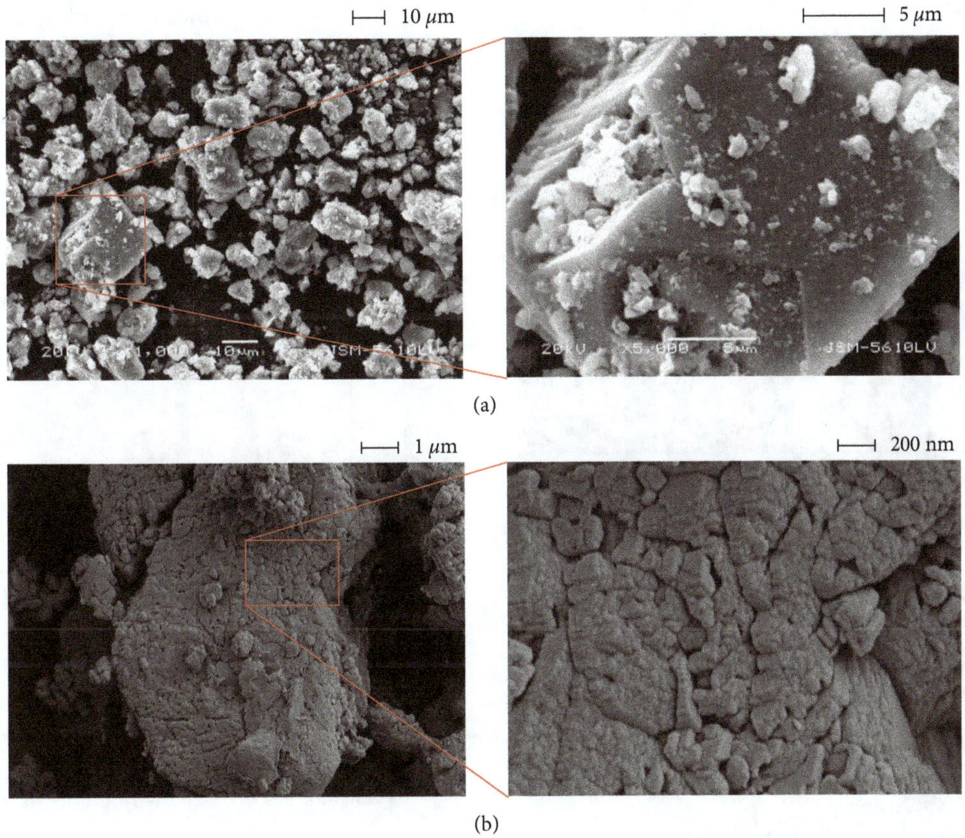

FIGURE 6: (a) Micron-scale particles of limestone powder. (b) Surface structure of NC/LS composite particles.

effect on cement hydration in the presence of calcium carbonate [11, 12, 14]. Similarly, the flexural and compressive strength of specimen M-NC/LS was higher than that of specimen M-0. These results suggest that NC/LS improved the mechanical properties of cement paste and mortar. On the other hand, the compressive strength of specimen C-NC/LS after 3 and 28 days was, respectively, 6.3% and 1.9% higher than that of specimen C-0. The compressive strength of M-NC/LS after 3 and 28 days was, respectively, 14.5% and 7.2% higher than that of M-0. These results indicate that NC/LS improves the cement paste and mortar mechanical properties significantly during the early stages.

As discussed that characteristics of NC/LS are quite different from that of LS and the surface of NC/LS is more favorable for the precipitation and further growth of calcium silicate hydrate gels. Therefore, the improved strength of specimens C-NC/LS and M-NC/LS can be attributed to the complex surface structure of NC/LS, which influenced the size and morphology of hydration products of the cement [16]. Furthermore, XRD results [16] have shown that the complex surface structure of NC/LS can disturb the nucleation and growth of calcium hydroxide, resulting in the reduction of particle size and crystal orientation degree. Furthermore, the nucleation of calcium silicate hydrate is accelerated by the presence of nano-$CaCO_3$ particles [21] because nano-$CaCO_3$ is finer than LS particles. Camiletti, Soliman, and Nehdi also believe that micro-$CaCO_3$ acted mainly as an inert filler, while nano-$CaCO_3$ accelerated the cement hydration process through nucleation, and also acted as an effective filling material [36]. Also, it is reported that the presence of nano-$CaCO_3$ may cause acceleration of early strength development [26], compensates the dilution effect of the binding material, and thus offsets the negative effects of fly ash on early-age properties to facilitate the development of a more environmentally fly ash concrete. Hence, through the use of NC/LS, more environmentally friendly cementitious materials can be produced by reducing its cement factor, while achieving enhanced engineering properties.

3.3. Thermal Gravimetric Analysis. Figure 8 shows the DSC-TG curves of specimen C-NC/LS and C-0 after 3 days. It shows the normalized mass loss in percent (current mass divided by initial mass) versus temperature. It can be seen that the two specimens are gradually losing mass with the increasing temperature. The tendency of each curve in varying with the increasing temperature is almost the same. Mass loss between 105°C and 420°C represents the loss of combined water due to calcium silicate hydrate, aluminate hydrate, and ferroaluminate hydrate, between 420°C and 540°C, it corresponds to the dehydration of calcium hydroxide, and decarbonation of well-crystalline calcium carbonate occurs at temperature between 540°C and 750°C. In Figure 8, three main rapid losses in weight exist in the vicinities of 200°C, 500°C, and 750°C for the two specimens.

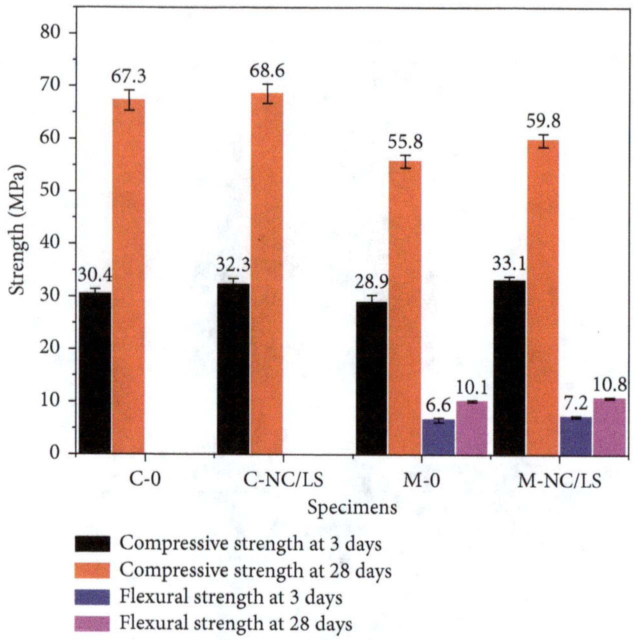

FIGURE 7: Strength of cement paste and mortar with NC/LS.

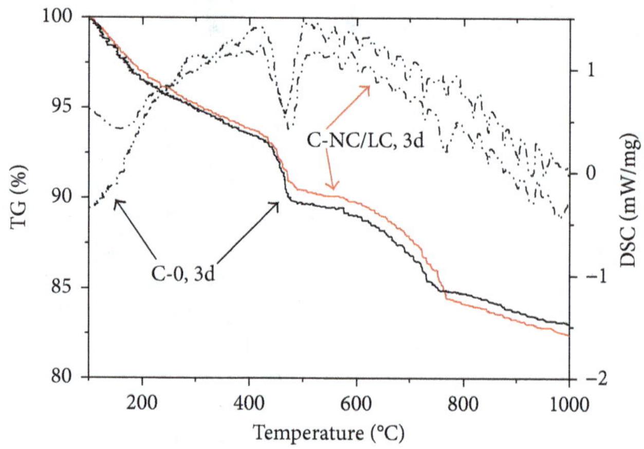

Specimen	C-0	C-NC/LS
Nonevaperable water/%	6.56	6.31
Calcium hydroxide/%	15.46	14.27

FIGURE 8: DSC-TG curves of specimens.

△W1 corresponded to the dehydration of calcium silicate hydrates, △W2 corresponded to the dehydration of calcium hydroxide, and △W3 corresponded to calcium carbonate decomposition. The △W3 of specimen C-NC/LS was higher than that of specimen C-0. This can be attributed to calcium carbonate decomposition in NC/LS. The △W1 and △W2 of specimen C-NC/LS is slightly lower than that of specimen C-0. This may be due to the part of Portland cement in specimen C-NC/LS which was substituted by NC/LS. On the other hand, the smaller weight loss of calcium hydroxide in specimen C-NC/LS compared to specimen C-0 indicates that the calcium hydroxide absorbs carbon dioxide to form more calcium carbonate probably. However, as shown in Figure 7, cement pastes and mortar containing NC/LS exhibited higher strength than that of control specimen at 3 days. The investigation of the phenomenon will be the subject of further research by BSE analysis.

TABLE 3: Porosity and contents of partially reacted cement grain and hydration products calculated based on BSE results.

Specimen	C-0	C-NC/LS
Partially reacted cement grain (%)	18.6	24.2
Hydration products (%)	63.5	58.5
Porosity (%)	17.9	17.3

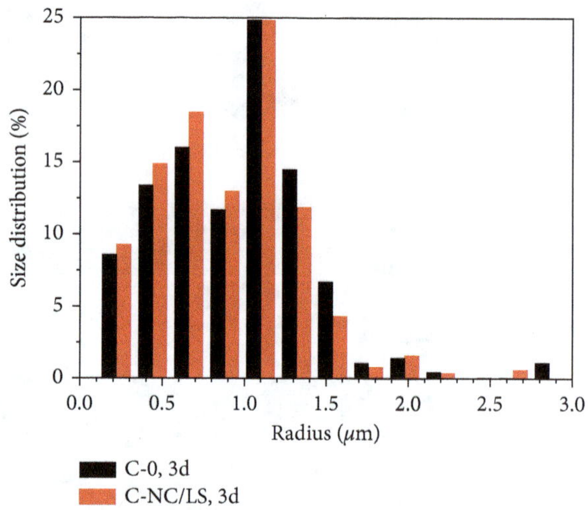

FIGURE 9: Pore size distributions of specimens.

3.4. BSE Image Analysis. Based on BSE images, the contents of partially reacted cement grains and hydration products of different specimens and their respective porosities were calculated and are shown in Table 3. After 3 days, specimen C-NC/LS had higher contents of partially reacted cement grains than that in specimen C-0, and specimen C-NC/LS had lower hydration products content than did specimen C-0. However, the porosity of specimen C-NC/LS was lower than that of specimen C-0. In other words, C-NC/LS had higher structure compactness than that of specimen C-0.

It can be observed from Figure 9 that the larger pores of pastes containing NC/LS are decreased, while the smaller pores are increased. The use of NC/LS leads to a more compact paste with a reduction of larger pores, indicating that the presence of NC/LS is advantageous in pores modification. It is noted that the sophisticated surface structure of NC/LS has a significant influence in decreasing the larger pores of cement pastes and further influences its macroscopic properties [16].

As indicated by both BSE images and DSC-TG results, specimen C-NC/LS had lower contents of calcium silicate hydrate and calcium hydroxide than that in specimen C-0. However, specimen C-NC/LS had higher compressive strength than that of specimen C-0, and specimen M-NC/LS had higher flexural and compressive strength than that of specimen M-0. According to further investigations of the porosity and pore structure of specimens obtained, the improved strength of specimens C-NC/LS and M-NC/LS can be attributed to their lower porosity and better pore structure. Hence, the microstructure of cementitious materials with NC/LS depends not only on the distribution of particles but also the morphology and surface structure of the particle. Rashad' research shows that the enhancement in strength of hardened paste is related to the refinement of calcium hydroxide grains, which occurred during the hydration of cement [37]. Thus, the growth of calcium hydroxide grains in cementitious materials may be affected by NC/LS, either.

4. Conclusions

Based on the results obtained in this study, the following conclusions can be drawn:

(1) Nano-$CaCO_3$/limestone composite particles were successfully prepared using the $Ca(OH)_2$-H_2O-CO_2 reaction system. The surface structure and morphology of micron-scale limestone particles have been altered. Nano-$CaCO_3$ scattered evenly on the surface of the micron-scale limestone particles. It is an effective method to solve the dispersion problem of nano-$CaCO_3$ particles in concrete.

(2) Adding nano-$CaCO_3$/limestone composite particles resulted in reduced porosity and enhanced pore structure for hardened cementitious materials. The microstructure of hardened cementitious materials was optimized by nano-$CaCO_3$/limestone composite particles. Therefore, the mechanical properties of cement paste and mortar were improved.

Conflicts of Interest

The authors declare that there are no conflicts of interest regarding the publication of this paper.

Acknowledgments

This work is supported by the National Natural Science Foundation of China under 51669004, and the Hubei Natural Science Foundation of China under 2014CFB705.

References

[1] P. C. Aitcin, "Cements of yesterday and today: concrete of tomorrow," *Cement and Concrete Research*, vol. 30, no. 9, pp. 1349–1359, 2000.

[2] E. Gartner, "Industrially interesting approaches to "low-CO_2" cements," *Cement and Concrete Research*, vol. 34, no. 9, pp. 1489–1498, 2004.

[3] P. K. Mehta and R. W. Burrows, "Building durable structures in the 21st century," *Concrete International*, vol. 23, no. 3, pp. 57–63, 2001.

[4] S. Kawashima, J. W. T. Seo, D. Corr, M. C. Hersam, and S. P. Shah, "Dispersion of $CaCO_3$ nanoparticles by sonication and surfactant treatment for application in fly ash-cement systems," *Materials and Structures*, vol. 47, no. 6, pp. 1011–1023, 2014.

[5] V. Bonavetti, H. Donza, G. Menendez, O. Cabrera, and E. F. Irassar, "Limestone filler cement in low w/c concrete: a rational use of energy," *Cement and Concrete Research*, vol. 33, no. 6, pp. 865–871, 2003.

[6] A. A. Elgalhud, R. K. Dhir, and G. Ghataora, "Limestone addition effects on concrete porosity," *Cement and Concrete Composites*, vol. 72, pp. 222–234, 2016.

[7] M. Heikal, H. El-Didamony, and M. S. Morsy, "Limestone-filled pozzolanic cement," *Cement and Concrete Research*, vol. 30, no. 11, pp. 1827–1834, 2000.

[8] S. Palm, T. Proske, M. Rezvani, S. Hainer, C. Muller, and C. A. Graubner, "Cements with a high limestone content-mechanical properties, durability and ecological characteristics of the concrete," *Construction and Building Materials*, vol. 119, pp. 308–318, 2016.

[9] S. Tsivilis, E. Chaniotakis, G. Kakali, and G. Batis, "An analysis of the properties of Portland limestone cements and concrete," *Cement and Concrete Composites*, vol. 24, no. 3-4, pp. 371–378, 2002.

[10] N. Voglis, G. Kakali, E. Chaniotakis, and S. Tsivilis, "Portland-limestone cements. Their properties and hydration compared to those of other composite cements," *Cement and Concrete Composites*, vol. 27, no. 2, pp. 191–196, 2005.

[11] V. L. Bonavetti, V. F. Rahhal, and E. F. Irasser, "Studies on the carboaluminate formation in limestone filler-blended cements," *Cement and Concrete Research*, vol. 31, no. 6, pp. 853–859, 2001.

[12] G. Kakali, S. Tsivilis, E. Aggeli, and M. Bati, "Hydration products of C_3A, C_3S and Portland cement in the presence of $CaCO_3$," *Cement and Concrete Research*, vol. 30, no. 7, pp. 1073–1077, 2000.

[13] B. Lothenbach, G. Le Saout, E. Gallucci, and K. Scrivener, "Influence of limestone on the hydration of Portland cements," *Cement and Concrete Research*, vol. 38, no. 6, pp. 848–860, 2008.

[14] J. Pera, S. Husson, and B. Guilhot, "Influence of finely ground limestone on cement hydration," *Cement and Concrete Composites*, vol. 21, no. 2, pp. 99–105, 1999.

[15] V. Rahhal and R. Talero, "Early hydration of Portland cement with crystalline mineral additions," *Cement and Concrete Research*, vol. 35, no. 7, pp. 1285–1291, 2005.

[16] H. Yang, Y. Che, and M. Zhang, "Effect of nano-$CaCO_3$/limestone powder composite on the early age cement hydration products," *Key Engineering Materials*, vol. 703, pp. 354–359, 2016.

[17] B. Felekoglu, "Effects of PSD and surface morphology of micro-aggregates on admixture requirement and mechanical performance of micro-concrete," *Cement and Concrete Composites*, vol. 29, no. 6, pp. 481–489, 2007.

[18] F. U. A. Shaikh and S. W. M. Supit, "Mechanical and durability properties of high volume fly ash (HVFA) concrete containing calcium carbonate ($CaCO_3$) nanoparticles," *Construction and Building Materials*, vol. 70, pp. 309–321, 2014.

[19] F. U. A. Shaikh and S. W. M. Supit, "Chloride induced corrosion durability of high volume fly ash concretes containing nano particles," *Construction and Building Materials*, vol. 99, pp. 208–225, 2015.

[20] J. Camiletti, A. M. Soliman, and M. L. Nehdi, "Effect of nano-calcium carbonate on early-age properties of ultra-high-performance concrete," *Magazine of Concrete Research*, vol. 65, no. 5, pp. 297–307, 2013.

[21] T. Sato and J. J. Beaudoin, "Effect of nano-$CaCO_3$ on hydration of cement containing supplementary cementitious materials," *Advances in Cement Research*, vol. 23, no. 1, pp. 33–43, 2011.
[22] S. Kawashima, P. K. Hou, D. J. Corr, and S. P. Shah, "Modification of cement-based materials with nanoparticles," *Cement and Concrete Composites*, vol. 36, pp. 8–15, 2013.
[23] J. M. Makar, G. W. Chan, J. J. Beaudoin, F. Torres, and K. Trischuk, "Effect of n-$CaCO_3$ and metakaolin on hydrated Portland cement," *Advances in Cement Research*, vol. 24, no. 4, pp. 211-219, 2012.
[24] G. C. Long, Y. Shi, K. L. Ma, and Y. J. Xie, "Reactive powder concrete reinforced by nanoparticles," *Advances in Cement Research*, vol. 28, no. 2, pp. 99–109, 2016.
[25] A. H. Korayem, N. Tourani, M. Zakertabrizi, A. M. Sabziparvar, and W. H. Duan, "A review of dispersion of nanoparticles in cementitious matrices: nanoparticle geometry perspective," *Construction and Building Materials*, vol. 153, pp. 346–357, 2017.
[26] T. Meng, Y. Yu, and Z. Wang, "Effect of nano-$CaCO_3$ slurry on the mechanical properties and micro-structure of concrete with and without fly ash," *Composites Part B: Engineering*, vol. 117, pp. 124–129, 2017.
[27] S. W. Tang, X. H. Cai, Z. He et al., "The review of pore structure evaluation in cementitious materials by electrical methods," *Construction and Building Materials*, vol. 117, pp. 273–284, 2016.
[28] S. Huang, J. C. Chen, C. W. Hsu, and W. H. Chang, "Effects of nano calcium carbonate and nano calcium citrate on toxicity in ICR mice and on bone mineral density in an ovariectomized mice model," *Nanotechnology*, vol. 20, no. 37, pp. 1-7, 2009.
[29] S. Diamond, "Mercury porosimetry: an inappropriate method for the measurement of pore size distributions in cement-based materials," *Cement and Concrete Research*, vol. 30, no. 10, pp. 1517–1525, 2000.
[30] K. Tanaka and K. Kurumisawa, "Development of technique for observing pores in hardened cement paste," *Cement and Concrete Research*, vol. 32, no. 9, pp. 1435–1441, 2002.
[31] K. L. Willis, A. B. Abell, and D. A. Lange, "Image-based characterization of cement pore structure using wood's metal intrusion," *Cement and Concrete Research*, vol. 28, no. 12, pp. 1695–1705, 1998.
[32] D. A. Lange, H. M. Jennings, and S. P. Shah, "Image-analysis techniques for characterization of pore structure of cement-based materials," *Cement and Concrete Research*, vol. 24, no. 5, pp. 841–853, 1994.
[33] K. L. Scrivener, "Backscattered electron imaging of cementitious microstructures: understanding and quantification," *Cement and Concrete Composites*, vol. 26, no. 8, pp. 935–945, 2004.
[34] I. Tanaka and M. Koishi, "Fundamental properties of powder, paste and mortar of surface modified cement and process of the surface modification," *Construction and Building Materials*, vol. 13, no. 5, pp. 285–292, 1999.
[35] F. Pacheco-Torgal, S. Miraldo, Y. Ding, and J. A. Labrincha, "Targeting HPC with the help of nanoparticles: an overview," *Construction and Building Materials*, vol. 38, pp. 365–370, 2013.
[36] J. Camiletti, A. M. Soliman, and M. L. Nehdi, "Effects of nano- and micro-limestone addition on early-age properties of ultra-high-performance concrete," *Materials and Structures*, vol. 46, no. 6, pp. 881–898, 2013.
[37] A. M. Rashad, "Effects of ZnO_2, ZrO_2, Cu_2O_3, CuO, $CaCO_3$, SF, FA, cement and geothermal silica waste nanoparticles on properties of cementitious materials–a short guide for civil engineer," *Construction and Building Materials*, vol. 48, pp. 1120–1133, 2013.

9

Experimental Characterization of Traditional Mortars and Polyurethane Foams in Masonry Wall

Dora Foti,[1] **Michela Lerna**,[2] **and Vitantonio Vacca**[1]

[1]*Department of Civil Engineering and Architecture, Polytechnic of Bari, Via Orabona 4, 70125 Bari, Italy*
[2]*Department of Innovation Engineering, University of Salento, Via per Monteroni, 73100 Lecce, Italy*

Correspondence should be addressed to Dora Foti; dora.foti@poliba.it

Academic Editor: Pietro Russo

Masonry is a composite material largely used in construction. It exhibits several advantages, including significant compressive strength, thermal inertia, and aesthetic beauty. A disadvantage of masonry is mainly related to the inadequate shear strength due to the poor capacity and ductility of the adopted mortar. This aspect is crucial in seismic areas. In this paper, the behavior of polyurethane foams, used as adhesives for the construction of thin joints brick masonry walls, has been investigated. First, the characterization of components was carried out, followed by laboratory uniaxial tests on masonry walls and shear tests on triplets. Moreover, a comparison of the behavior of the foam-brick walls with respect to the traditional mortars masonry was carried out, as the type of joints varies and the arrangement of the holes of the bricks varies with respect to the direction of the applied load. Results provide indications on which adhesive has to be adopted for masonry buildings in reference to the site of construction (i.e., seismic hazard).

1. Introduction

Masonry is one of the oldest building systems that characterized the history of the structures. Although it consists in the coworking of two constituents, such as the brick and the mortar, arranged in a more or less regular texture, the role of the mortar joints is currently slightly investigated in the scientific literature. An explanation for this is attributable to the common responsibility of the masonry structural members: the mechanical performance of the mortar is less important when compared to the compressive strength of the brick (natural or artificial). In the past centuries, masonry structures were designed by considering death loads. The choice of the type of brick and its dimensions were the main target of the structural calculus. Unfortunately, recent seismic events have shown that masonry elements hardly effectively react against horizontal forces (e.g., in earthquakes). This is principally devoted to the existing mortar in the masonry. In fact, brittle failure easily occurs within the mortar joints when an in-plane force is applied. In this scenario, more efforts need to be concentrated on the study of the mechanical performances of traditional mortars and even more on innovative once.

Similarly, since the twentieth century, the construction industry had been more prone towards the use of reinforced concrete for structures with infills. In addition, after the past seismic events, such as L'Aquila earthquake (2009), masonry showed a very brittle behavior and a diffused crack patterns even if included in reinforced concrete frames. As a consequence, many buildings were declared nonaccessible although the structure remained apparently intact. Today, new technologies have revived the masonry built with thin joints [1–6]. Moreover, further studies regarding the behavior of thin joints in presence of fiber-reinforced mortars have been developed. The mechanical characteristics of the masonry made with bricks and thin joints have been compared by means of laboratory tests on panel specimens [7, 8]. This kind of fiber-reinforced mortar has been also successfully utilized for the reinforcement of concrete columns [9] or other concrete members [10–13].

In seismic prone areas, thin joints become important elements in a masonry structure, being shear strength the main parameter connected with the earthquake resistance of a masonry. Increasing the shear strength of the masonry mainly concerns the execution of proper bed and head joints as this provides stiffness to the structure [14–16]. For this kind of joints, polyurethane foams can also be used to make up a wall. They are more deformable and ductile and show a higher damping capacity than traditional mortars. This aspect makes the foams more performable in seismic prone areas; however, this characteristic should be verified with respect to traditional mortars, as aimed in the present paper for thin joints brick masonry walls.

2. Background

Due to their versatility and ease of use, polyurethane foams are suitable to be effectively used in masonry walls; they present, in fact, a series of advantageous characteristics, such as excellent adhesion to substrates of various nature, extreme speed of installation, and strong reduction in processing costs and, consequently, in the total costs.

Several studies have been conducted on the use of foams as filler for sandwich panels [17–19]. The cores have been realized with different kinds of polyurethane foams and their mechanical characteristics have been compared.

Furthermore, the structural behavior of a prefabricated wall system made up of glass fiber-reinforced rigid polyurethane foam (PUF) and magnesium oxide (MgO) has been investigated. The experimental results showed a behavior governed by MgO, while a complete interaction between the fiber-reinforced rigid polyurethane foam and MgO board was achieved by using epoxy adhesives [20].

Taher et al. [21] assessed the crashworthiness characteristics of a composite glass fiber-reinforced plastic (GFRP) sandwich structure with the purpose of producing cost-effective sandwich structures. All the samples were based on the concept of the "double-layered" foam-filled block. Thus, two foam-core sheets are wrapped by reinforcement woven fabric; the latter acts as a reinforcement face and at the same time ties the core layers and faces together, thus preventing catastrophic failure under axial loading conditions. Experimental results indicated that a high crushing force efficiency was achieved.

Moreover, the metal foams have relatively recently found applications in construction, automotive, aerospace, and naval industries due to their unique properties, such as excellent stiffness-to-weight ratio, efficient energy absorption, shock wave attenuation, sound and vibration damping, and absorption and low thermal conductivity. The mechanical properties of the aluminum foam-polymer composites were also studied [22–25].

An interesting application regards the use of polyurethane within plaster by incorporating different proportions of the recycled polyurethane itself [26]. The objective of the conducted experimentation was to obtain a new lightweight plaster material with good thermal insulating properties for industrial applications. Also, the properties of the mix and the basic proportions to get the desiderate behavior have been determined. The experimental results showed that the quantity of polymeric foam present in the composite is the most important factor in explaining the variation in the properties of these compounds.

The behavior of masonry prisms under axial compression is influenced by the relative stiffness of masonry units and mortar, the thickness of mortar joints, and the shape of the units. Thamboo et al. [27] presented an experimental study carried out to examine the effects of the thickness of mortar joints, the type of mortar adhesives, and the presence of shells in the hollow concrete masonry prisms under axial compression test. It is found that the thinner and higher adhesive mortars increase the compressive strength and stiffness, while lowering Poisson's ratio.

A large experimental campaign regarded the investigation of the mortar composition effect, the dispersion methods, and the unit surface textures when flexural and shear bond characteristics is the main issue [28]. The results of the flexural and shear bond characteristics about the thin-layer polymer cement mortared concrete masonry evidenced the effectiveness of the polymers in significantly improving the mechanical performances of the masonry. A further study aimed to investigate the influence of the curing methods, and the age of the bond has been performed. As a result, the dry cured thin-layer mortared masonry exhibited higher bond strength and Young's and shear modules when compared to the wet cured specimens [29].

An experimental campaign has been conducted to analyze the behavior of a thin-layer polymer mortared concrete masonry under combined shear-compression stress state [30]. Similarly, to the conventional masonry, its strength and deformation were affected by the orientation of the bed joints to the direction of the applied loading. Nevertheless, the thin-layer mortared concrete masonry showed trending towards isotropic behavior compared with conventional masonry, which is distinctly orthotropic.

A finite element technique for high bond strength, thin-layer mortared masonry through material and interface modelling for simulating the behavior of the masonry was proposed in [31]. The developed interface model consistently simulates the behavior for masonry joints, which dominates many modes of failure of the masonry. In the study, the joint strength to shear and flexural actions were modelled and the agreement with the experimental results was accurate.

Finally, experimental and numerical investigations were conducted to evaluate the influence of the bond strength on the in-plane shear capacity of masonry walls [32]. The study figured out that the polymeric joints benefitted the masonry in favouring the failure in the brick units. Currently, very few studies specifically refer to the polyurethane foams utilized in bonding material for buildings' walls. Foams could be very useful in seismic areas; due to the ductility, they can assess to the joints and, therefore, to the entire masonry building. From this point of view, foams should be preferable with respect to ordinary bonding materials and adhesives. With this in mind, the present research consists in the characterization of the properties of the materials constituting the masonry made with thin joints from the experimental point of view by considering innovative and traditional adhesives, that is, polyurethane foams and mortar, respectively.

3. Experimental Program

After a preliminary study on the characteristics and classification of mortars according to the Italian Technical

Standards for Construction (DM 2008) [33], a series of laboratory tests have been performed aimed at highlighting the mortar-brick and foam-brick forces of adhesion and adherence [34]. The aim is a possible comparison between mortars and foams also based on the kind of joints and the different brick blocks used during the tests.

The specimens were assembled with rectified brick blocks (35 × 24.5 × 24 cm). The results have been discussed, highlighting the advantages and disadvantages of the two bonding solutions.

The main goal of the proposed study consists in evaluating the mortar role in the mechanical characterization of the masonry by varying the type of mortars. For this purpose, the adopted mortars itself have been tested, as well as the masonry in shear-stress state and the adhesion of the mortar with the brick. In order to clearly report the methodology, the experimental program has been arranged by reporting three different sections regarding the mortar, the block, and the masonry tests, respectively. Finally, the main results are presented and discussed by comparing the experienced performance of the traditional and innovative mortars. The mechanical characteristics have been computed according with Italian and International standard test methods.

3.1. Mortars. Foam can be described as a stable structure with low density and strong cohesion. In this study, only treated polyurethane foams, one-component based, usually available in spray cans have been considered. The polyurethane foam is a mixture of polyurethane prepolymer and propellant. Thus, polyurethane foams have a number of characteristics that make them suitable for many different applications, for example, in excellent adhesion and outstanding insulation ability against heat, cold, and noise. Furthermore, they present high durability when chemical agents are in contact, elasticity, and ability to absorb shocks. In addition, for their versatility and ease of use, polyurethane foams are well suited to be used in the building field. The main applications are the thermal insulation of window frames, the thermal insulation of water pipes and thermo-hydraulic installations, fixing insulating panels (polystyrene, polyester, cork, etc.), the consolidation of tiles, the bonding of metal sheets and tiles for roof covering, and so on.

The mortars utilized for comparison purposes are M1-mortar (i.e., mortar composed of cement and fine-graded mineral fillers, containing high quantities of synthetic resins and special additives; shear adhesion of 2.0 MPa), M2-mortar (i.e., mortar based on silicates and carbonates; shear adhesion of 2.5 MPa), and M3-mortar (i.e., special mortar for brick blocks with a good thermal insulation; shear adhesion of 2.7 MPa).

One of the fundamental properties of the mortar joints in masonry texture is the adhesion strength, which can be experimentally measured according to [35]. In particular, the adhesion strength is obtained by means of the pull-out method and specific tests on masonry samples. The sample is rigidly held and a clamp is fixed to the upper part. A bending moment is applied to the clamp with a lever until when the upper part is pulled out (Figure 1).

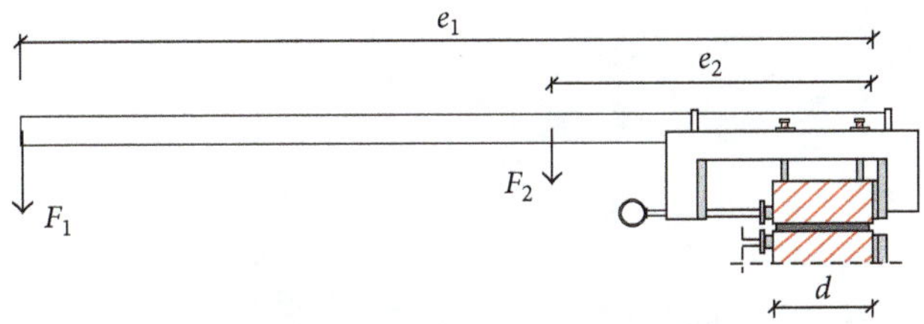

FIGURE 1: Sketch of the pull-out test method.

The nomenclature related to Figure 1 is listed as follows:

(i) e_1: distance between the point of application of the load and the connecting system (mm)

(ii) e_2: distance from the center of gravity of the lever and the connecting system (mm)

(iii) F_1: maximum applied load (N)

(iv) F_2: weight of the lever and the upper connecting system (N)

(v) d: average depth of the sample.

Here, the adhesion force is obtained according to the following equation:

$$f_{\mathrm{wi}} = \frac{F_1 e_1 + F_2 e_2 - (2/3) d \left(F_1 + F_2 + (W/4)\right)}{Z}, \qquad (1)$$

where

$$Z = \frac{bd^2}{6}, \qquad (2)$$

where b is the average width of the tested joint; d is the average depth of the specimen; and W is the weight of the masonry unit obtained from the specimen with the mortar.

The first series of laboratory tests consisted of two vertical compression tests on blocks joined in the direction parallel to the holes by means of polyurethane foam. Each specimen has been realized with 2 Module FV "MVI350" type brick blocks (Figure 2). By following the instructions about the use of the foam, two parallel curbs, of about 3 cm thick each (thin joints), were realized and then positioned at a distance of about 5 cm from the edge of the block (Figure 3). The hardening of the foam required only one hour; however, three days more have been spent before testing in order to be sure about the hardening of the bonding materials.

Table 1 reports the results for specimens made with two rectified bricks and foam.

Furthermore, a second series of tests have been done utilizing Gasbeton (i.e., autoclaved aerated concrete) elements and foam for comparison purposes (Table 2).

Figures 4 and 5 show, respectively, A4 failure and A7 failure occurred in this test.

Results show that higher performances are obtained in the case of specimens made with Gasbeton elements.

In addition, compression tests on a couple of bearing bricks, set in the direction of the holes, have been carried out. A comparison between the values obtained for different kinds of adhesives and joints has also been developed. In particular, 30 pairs of samples have been realized

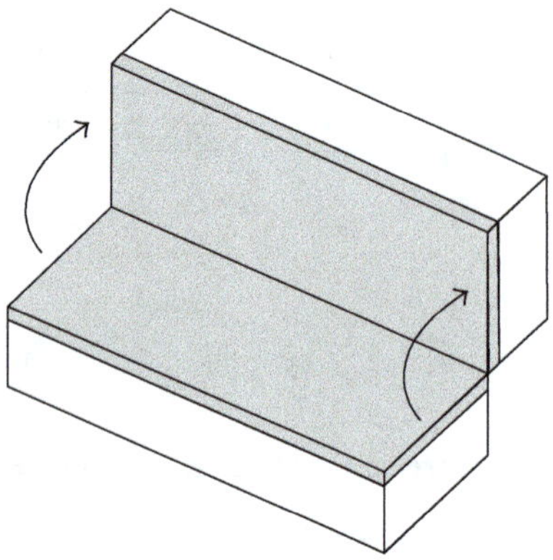

Figure 2: Failure A4: insufficient stress inside the joint.

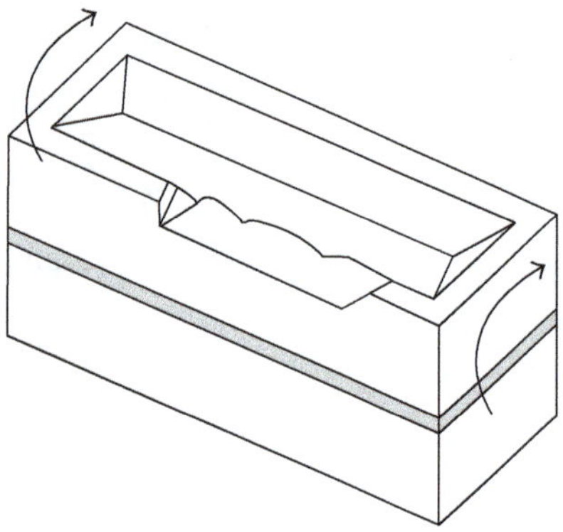

Figure 3: Failure A7: shear crash of the part of the specimen fixed to the clamp.

Table 1: Specimen 1: rectified brick 37 × 25 × 23.5 cm and foam for masonry.

Test number	F_1 (N)	e_1 (mm)	e_2 (mm)	F_2 (N)	f_{wi} (N/mm^2)	Type of collapse UNI EN 1052-5
1.1	1000	1100	300	365	0.30	A4
1.2	700	1100	300	365	0.21	A4
1.3	1100	1100	300	365	0.33	A4
1.4	710	1100	300	365	0.22	A4
1.5	400	1100	300	365	0.13	A4
2.1	800	1100	300	365	0.24	A4
2.2	840	1100	300	365	0.25	A4
2.3	670	1100	300	365	0.20	A4
2.4	615	1100	300	365	0.19	A4
2.5	865	1100	300	365	0.26	A4

f_{wi}: adhesion strength of the single specimen; f_{wmean}: mean adhesion strength; f_{wk}: characteristic adhesion strength.

and arranged into 3 groups of 10 each, one made with "ordinary" joints (maximum thickness of 5 mm), one with "thin" joints (maximum thickness of 3 mm), and one with "super-thin" joints (maximum thickness of 1.5 mm). The Module FV "MVI350" type brick blocks have dimensions 35 × 24 × 24 cm and percentage of holes ranging between 45% and 55%, with thin septa. All joints were made using M1-mortar. The specimens, perfectly rectified, do not need a "cover" in correspondence of the surfaces in contact with the press plates.

Table 2: Specimen 2: Gasbeton 63 × 25 × 24 cm and foam for masonry.

Test number	F_1 (N)	e_1 (mm)	E_2 (mm)	F_2 (N)	f_{wi} (N/mm^2)	Type of collapse UNI EN 1052-5
1.1	1750	1100	300	365	0.28	A4
1.2	2450	1100	300	365	0.47	A7
1.3	2600	1100	300	365	0.51	A7
1.4	2400	1100	300	365	0.45	A7
1.5	1950	1100	300	365	0.33	A7
2.1	2150	1100	300	365	0.39	A7
2.2	2200	1100	300	365	0.40	A7
2.3	2600	1100	300	365	0.51	A7
2.4	2500	1100	300	365	0.48	A7
2.5	2350	1100	300	365	0.44	A7

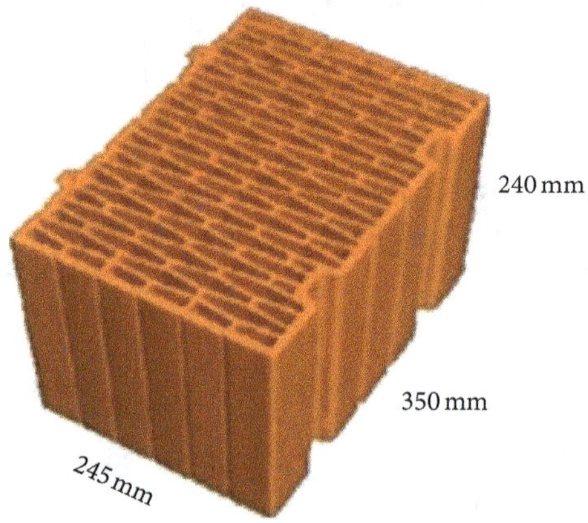

Figure 4: Module FV "MVI350" brick.

Figure 5: Placement of the foam.

The results are reported in Table 3, in which a significant dispersion of data, with a standard deviation of about 1.7 N/mm^2, is noticeable. The higher strength was obtained for masonry made with "thin" joints.

TABLE 3: Values of the ultimate load and compressive strength for different joints and mortar.

Specimens	M1-mortar ordinary joint	M1-mortar thin joint	M1-mortar super-thin joint	M2-mortar ordinary joint	M2-mortar thin joint	M2-mortar super-thin joint
Ultimate load $F_{i,\max}$ (kN)	741.35	818.6	751.375	1098.5	1235.625	1125.05
Compressive strength f_i (N/mm^2)	8.826	9.745	8.945	13.077	14.71	13.393

3.2. Block. The compressive strength, measured perpendicularly to the horizontal joints, is derived from the resistance of small specimens of masonry subjected to test up to collapse, according to UNI EN 1052-1:2001 [36]. The specimens are uniformly loaded in compression and the maximum load to collapse, $F_{\max}$, has been recorded. The characteristic nominal strength of a masonry is obtained from the strength of each specimen. Two specimens were loaded, and the force has been recorded by means of a loading cell, obtaining the ultimate compression load $F_{i,\max}$ and the compressive strength f_i by assuming

$$f_i = \frac{F_{i,\max}}{A_i}, \tag{3}$$

where $A_i = 350\,\text{mm} \times 240\,\text{mm} = 84000\,\text{mm}^2$ (area of the loaded transversal section) and $F_{i,\max}$ is the maximum reached load (kN).

Definitely, the two specimens exhibited ultimate compression loads of $F_{i,\max} = 430$ kN and 480 kN, corresponding to the compressive strengths of about $f_i = 5.714\,\text{N/mm}^2$ and $f_i = 5.119\,\text{N/mm}^2$, respectively. Thus, the mean values are: $F_{i,\max} = 455$ kN and $f_i = 5.4165\,\text{N/mm}^2$. Figure 6 shows the failure modes of the specimens consisting of vertical cracking close to the corners of the block.

The plot in Figure 7 shows the different values of the compressive strength in the case of mortar and foam. It is evidenced that the thickness of the joints does not affect M1-mortar much, contrary to M2-mortar.

In both tests, the results are lower than those obtained utilizing the mortar; taking as reference the higher result obtained for the sample made with M2-mortar with thin joints, it is possible to notice a reduction of about 67% with respect to the mean value obtained for the specimens realized with foam.

(a)

(b)

FIGURE 6: Details of the failure of the block in pure compression test.

3.3. Masonry (Adhesion). In the present section, the purpose of the test is the evaluation of the capacity of a masonry in resisting against the actions that impose shear stresses such as the laboratory slide strength of the foam-brick joints. The test consists in placing a masonry sample under a proper equipment capable of stressing the foam joint between two blocks, with a controlled increasing load up to failure. The tests were carried out on three brick blocks Module FV "MVI350" type with dimensions 35 × 24.5 × 24 cm (length, width, and height, respectively). The percentage of holes was $45\% < F \leq 55\%$. The blocks were jointed with polyurethane foam by running two parallel curbs of about 3 cm thick each, positioned at a distance of about 5 cm from the longitudinal edge of the block.

The first specimen was placed in between two plates of dimensions 30 × 30 × 3 cm, leaving the central block free to slide. The vertical displacement of the block with respect to the adjacent blocks was measured by four displacement transducers positioned on the right and left sides of the sample (Figures 8(a) and 8(b)). The load was applied by a hydraulic jack piston through the load cell connected to the

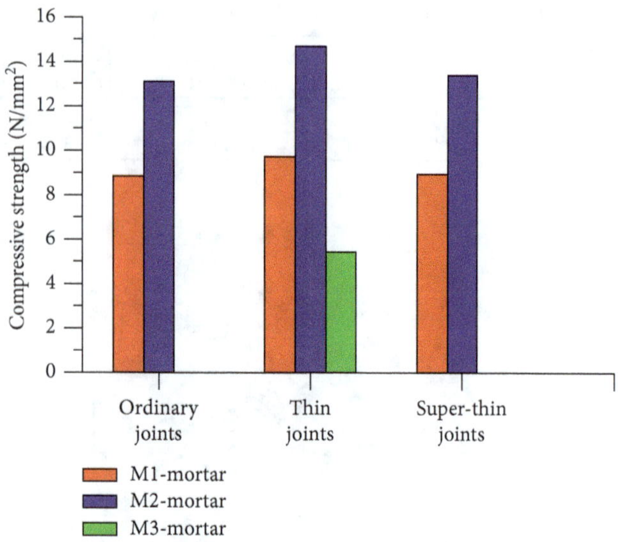

FIGURE 7: Values of the compressive strength as function of the bonding material and the joints.

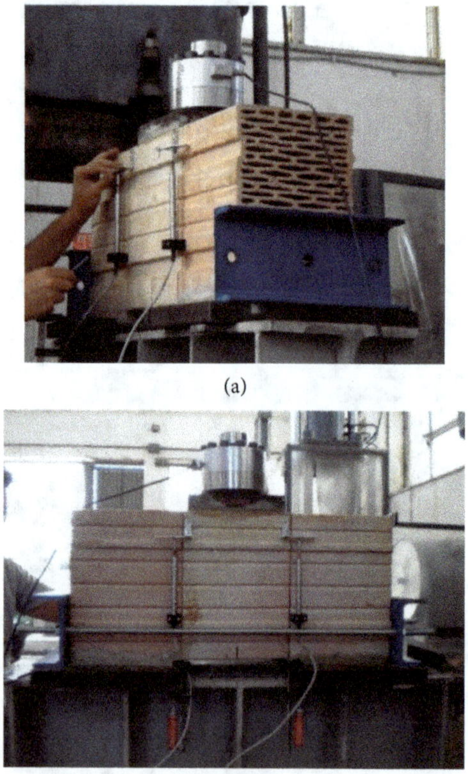

(a)

(b)

FIGURE 8: Details of the test setup: (a) positioning of the transducers for measurements; (b) specimen ready for the test.

computer and positioned on the central block by means of a $22 \times 22 \times 3.5$ cm steel plate.

Two external 16 cm diameter bars have to maintain the position of the sample, thus simulating the contiguity of the blocks in a wall (there was no precompression or presence of stresses normal to the direction of the holes of the bars

FIGURE 9: First specimen failure under the maximum load.

because the bolts have been tightened). The specimen was subjected to a crescent load up to the failure; the failure appeared in the central block under the loading plate (Figure 9). The maximum recorded load was 17.25 kN.

The displacements (Def) measured at the four transducers were as follows:

(i) Def1 = 0.369 mm

(ii) Def2 = 0.181 mm

(iii) Def3 = 0.244 mm

(iv) Def4 = 0.493 mm.

During the unloading phase, an increment of the displacements in nodes 1 and 2 towards the bottom was noticed, while a displacement towards the top occurred in nodes 3 and 4 (Figures 10(a) and 10(b)). However, the displacements obtained during the unloading phase were influenced by the collapse of the block and therefore irrelevant for the foam-block adherence test.

The load-displacement plots show an almost linear trend of displacements depending on the specimen's load up to failure (Figure 11).

The failure occurred in the central block (second specimen) under the loading plate is reported in Figure 12. The maximum recorded load was 14.80 kN, while the displacement recordings in the four transducers were as follows:

(i) Def1 = 0.237 mm

(ii) Def2 = 0.227 mm

(iii) Def3 = 0.206 mm

(iv) Def4 = 0.177 mm.

It is possible to notice that in this case, the displacements have more uniform values but a lower maximum failure load.

Similarly to the previous test, the specimen was then subjected to an unloading phase. From the plot in Figure 13,

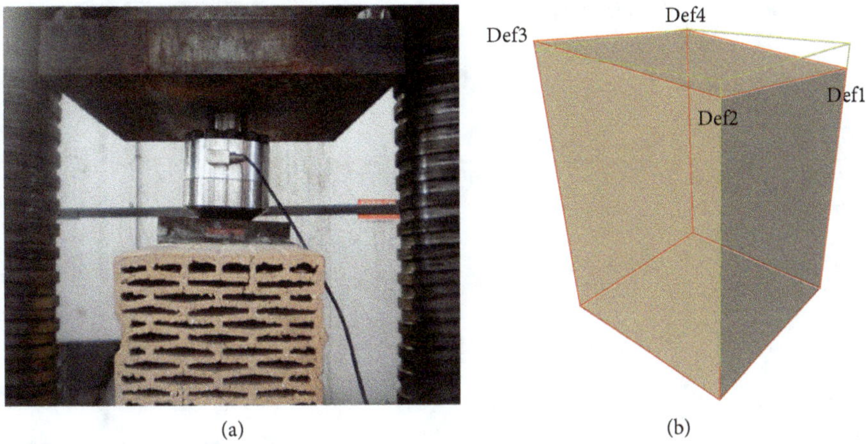

FIGURE 10: Details of the test on triplets: (a) specimen after the unloading; (b) 3D visual of the displacement at the end of the unloading tests.

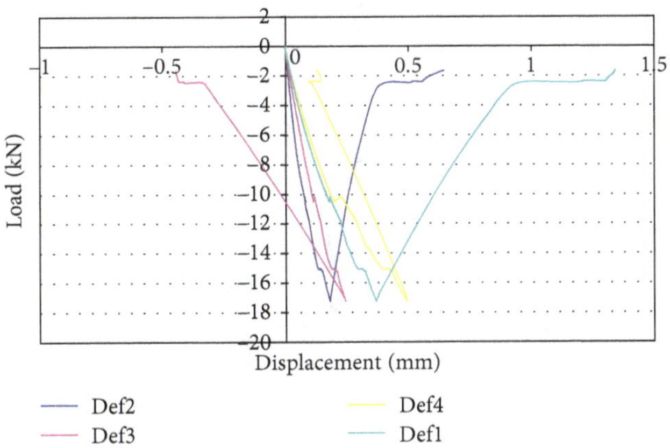

FIGURE 11: Load-displacement plot of the first triplet test.

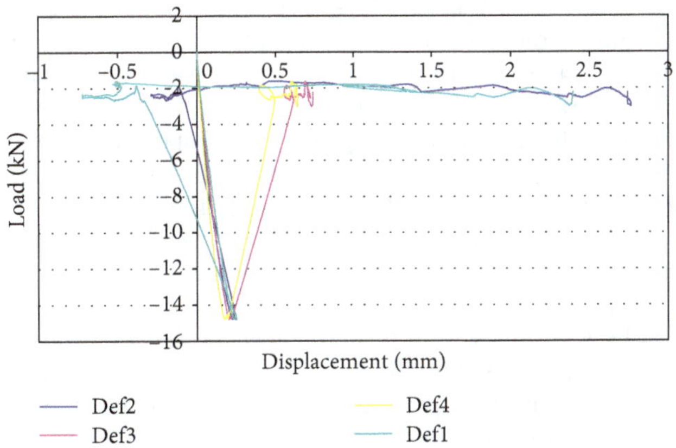

FIGURE 13: Load-displacement plot of the second triplet test.

FIGURE 12: Second specimen failure under the maximum load.

the trend of the displacements as function of the applied load can be observed. Also, in this case, it is possible to notice a linear trend of the displacements up to failure. By comparing the results obtained for the first specimen, the failure is obtained for lower values of the load, and the peak displacement is the same for all the transducers.

In addition, the initial characteristic shear strength without normal stresses has been experimentally determined according to [37]. The characteristic strength f_{vk0} has been deduced from the mean strength f_{vm}, obtained from the test results, by means of the following expression:

$$f_{vk0} = 0.7 f_{vm}. \quad (4)$$

In the present case, the mean strength on the two tests is equal to

$$f_{vm} = 0.09535 \text{ N/mm}^2, \quad (5)$$

and the experimental characteristic shear strength is

$$f_{vk0} = 0.7 f_{vm} = 0.066745 \text{ N/mm}^2. \quad (6)$$

The detailed results (6 specimens) are shown in Table 4, by reporting the geometrical dimensions of the six specimens, the failure loads, and the shear strength reached during the tests.

All samples showed a shear failure in the element/mortar connecting area on one side or divided between the two sides

TABLE 4: Results of the tests on the six specimens.

Specimen	Specimen's dimensions			Area of the transversal section A_i	Failure load $F_{i,\max}$	Shear strength f_{v0i}
	L (mm)	l (mm)	h (mm)	(mm^2)	(N)	(N/mm^2)
1	234.0	373.0	667.0	87282.0	55500	0.32
2	234.0	373.0	667.0	87282.0	58100	0.33
3	234.5	373.5	667.5	87212.3	57800	0.33
4	234.5	373.0	667.0	87468.5	45800	0.26
5	234.0	373.5	667.5	87165.0	50600	0.29
6	234.0	373.0	667.0	87282.0	54900	0.31
Total number of samples					n	6
Average shear strength f_{vm0}					N/mm^2	0.31
Characteristic value $f_{vk0} = 0.7 f_{vm0}$					N/mm^2	0.22

of the element (Figure 14) according to A1 type of failure (Appendix A of UNI EN 1052-3) [37].

After that, a total of 6 masonry samples have been realized assembling 3 brick blocks M3-mortar 38 × 25 × 21.9 cm, laid down in the direction of the vertical holes and joined together with M3-mortar with 0.1-0.2 cm thick continuous joints (Figure 15). The manufacturing of the specimens is of type I according to [37]. The test samples were imposed in a standard condition habitat ageing for at least 28 days at 20°C and relative humidity of 70% before the test. The shear strength test was conducted by applying the load according to type B procedure with zero prestressing (section 8.2 of EN 1052-3: 2007) [37].

The specimens were positioned in the testing machine following the prescriptions in section 8.1 of UNI EN 1052-3 [37] smoothing with the mortar the bearing surfaces at the fix connections (Figure 16).

Moreover, the loading speed was set in the range 0.1–0.4 N/mm^2 per minute. Again, the shear strength was obtained according to $f_{vi0} = F_{i,\max}/2A_i$, while the characteristic shear strength was obtained from $f_{vk0} = 0.7 f_{vm0}$.

The detected dispersions are related to the type of brick used in the experimental tests. In fact, the hollow bricks would not be the most suitable for this type of test because the brick's holes can be filled more or less by the foam in dependence of the thixotropy characteristics, the dimension of the holes, and the pressure exerted on the joints, which decreases passing from the lower joints of the wall to the top ones. With regard to the laboratory tests, both in the case of vertical compression and in the sliding of triplets, strength values lower with the foam than with the mortars have been obtained. Tables 5 and 6 show the results of the tests for the different cases.

The average value of the characteristic shear strength for the first two specimens is equal to 0.095 N/mm^2, lower than the value obtained for the specimens with mortar (0.263 N/mm^2).

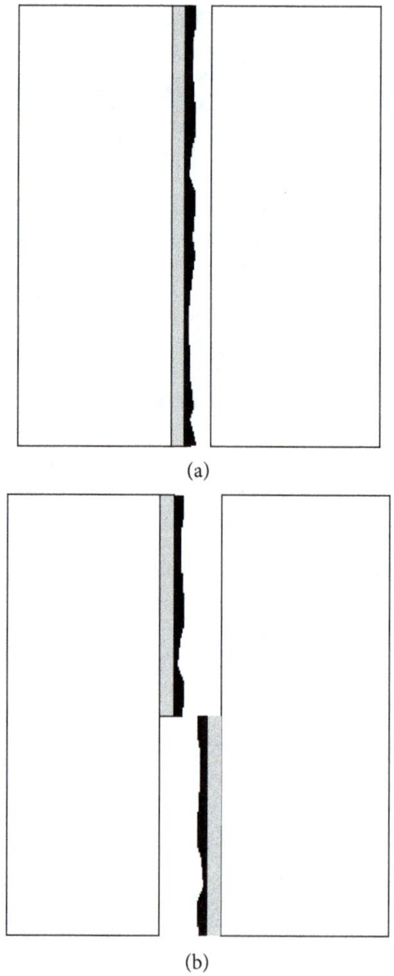

FIGURE 14: Shear failure in the element/mortar interface: (a) on one surface; (b) divided between two surfaces of the element.

FIGURE 15: Specimen constituted by three blocks.

3.4. Masonry (Shear). Substrates of rectified brick blocks must be sufficiently resistant, solid and smooth, clean and healthy, free from oils and grease, dust, loose and dirty material, and without any trace of old paint surfaces. They must also be sufficiently cured and free of significant shrinkage.

FIGURE 16: Positioning in the testing machine.

The foam is placed in curbs about 3 cm thick (Figure 17); in the case of rectified masonry blocks unperforated (cellular concrete, aerated, and/or autoclaved), the thickness of the foam curbs is reduced to 1.5 cm to avoid the possible phenomena of translational displacements and/or slight elevations of the block itself. Accompanying sheet reports useful date regarding the foam as listed down in Table 7.

The system was fast and easy to apply and allowed a reduction of the waiting time because the masonry was not wetted by the foam (contrary to what happens with mortar) and the higher speed of drying of the polyurethane foam itself. These elements lead to the final advantage to allow to continue with the subsequent processing steps in very short times.

The complete curing of the foam is achieved in a few hours after application. The final ageing and use takes place after 3-4 days approximately. In such a short time, it will already be possible to obtain the maximum performance of the wall. The thin-bed mortar was also characterized by a bonding with elastic characteristics, with all the consequent advantages. Consequently, the shear strength of the specimens is obtained from the following expression [38]:

$$f_{\mathrm{v0i}} = \frac{F_{i,\max}}{2A_i}, \quad (7)$$

where $F_{i,\max}$ is the maximum shear force (N) and A_i is the transversal section of a specimen parallel to the horizontal joints (mm^2). Two specimens have been prepared, one with foam and one with mortar.

The main results are listed as follows:

(i) Specimen 1: $37 \times 25 \times 23.5$ cm wall made of rectified bricks and foam for masonry. Mean value of the initial shear strength: $f_{\mathrm{v0m}} = 0.17\ \mathrm{N/mm}^2$.

(ii) Specimen 2: $37 \times 25 \times 23.5$ cm wall made of rectified bricks and mortar for masonry.

Mean value of the initial shear strength: $f_{\mathrm{v0m}} = 0.16\ \mathrm{N/mm}^2$.

The initial characteristic shear strength f_{v0k} is obtained as follows:

$$f_{v0k} = 0.8 f_{v0m} = 0.136\ \mathrm{N/mm}^2. \quad (8)$$

The mechanical shear strength of the masonry wall made with hollow bricks and one with polyurethane foam and one with M1-mortar has been investigated and then compared in diagonal compressive strength according to [38].

Specifically, seven wall specimens of nominal dimensions 1116×1245 mm and 240 mm thick have been realized for diagonal compression tests. In particular, the following tests have been carried out:

(i) 4 diagonal compression tests on walls made of hollow elements and mortar;

(ii) 3 diagonal compression tests on walls made of hollow elements and bedding foam.

Each sample was placed under a test setup properly designed and realized. It was made up of two steel caps; the caps were connected to each other by hauling systems consisting of chains and hydraulic jacks (Figure 18). The result of this imposed stress state is therefore a concentrated 45° oriented force with respect to the sides of the sample and central with respect to the section of the specimen. The wall specimen was instrumented with pressure gauges and strain bases placed along the diagonals. The mean shear strength is obtained from the following analytical expression:

$$S_{\mathrm{s}} = \frac{0.707P}{A_{\mathrm{n}}}, \quad (9)$$

where S_{s} is the mean shear strength along the sliding surface; P is the applied force along 45°; and A_{n} is the net area of the specimen, that is $(w + h)t/2$, in which w, h, and t are the dimensions of the specimen.

The results are shown in Table 8 for the specimens made with mortar and in Table 9 for the specimens made with foam. From the tables, the dissipative effect that the foams are able to provide to the masonry is clearly evidenced. In fact, a significantly higher value of the maximum deformation $\varepsilon_{\max}$ due to shear and energy dissipation G_{V} was reached.

It is possible to notice that in the second series of specimens, made with foam, the collapse was reached for higher values of the load, while deformations are comparable, except for 1-FOAM specimen.

4. Conclusions and Future Developments

The present study showed that a masonry element introduces some difficulties caused by the great variety of possible combinations of elements and binder and of the consequent difficulty in generalizing the information obtained on a limited number of combinations.

The phenomena of cracking and damage in the masonry evidence the need of the nonlinear analyses, carried out frequently in the case of materials not resistant to tensile forces. The results of the conducted experimental campaign were as follows:

(i) In the case of diagonal compression test for masonry with foam, the collapse is reached for higher values of the load, while deformations are comparable to the masonry with M2-mortar.

TABLE 5: Results of vertical compression tests.

	Test of vertical compression on blocks			
	Brick	Binder	Maximum load (KN)	Compressive strength (N/mm^2)
Specimen 1 (thin joint)	Module FV "MVI350" (35 × 25 × 24.5 cm)	Polyurethane foam	480	5.714
Specimen 2 (thin joint)	Module FV "MVI350" (35 × 25 × 24.5 cm)	Polyurethane foam	430	5.119
Specimen 3 (ordinary joint)	Iper (30 × 24 × 24 cm)	M1-mortar	741.35	8.513
Specimen 4 (thin joint)	Iper (30 × 24 × 24 cm)	M1-mortar	818.6	9.40
Specimen 5 (super-thin joint)	Iper (30 × 24 × 24 cm)	M1-mortar	751.37	8.629
Specimen 6 (ordinary joint)	Iper (30 × 24 × 24 cm)	M2-mortar	1098.5	12.615
Specimen 7 (thin joint)	Iper (30 × 24 × 24 cm)	M2-mortar	1235.62	14.19
Specimen 8 (Giunto Supersottile)	Iper (30 × 24 × 24 cm)	M2-mortar	1125.05	12.92

TABLE 6: Results of the sliding on triplet tests.

	Test of sliding on triplets			
	Brick	Binder	Maximum load (KN)	Shear strength (N/mm^2)
Specimen 1	Module FV "MVI350" (35 × 25 × 24.5 cm)	Polyurethane foam	17.236	0.102
Specimen 2	Module FV "MVI350" (35 × 25 × 24.5 cm)	Polyurethane foam	14.802	0.0881
Specimen 3	Porotherm Bio Plan (38 × 25 × 21.9 cm)	M3-mortar	55.5	0.274
Specimen 4	Porotherm Bio Plan (38 × 25 × 21.9 cm)	M3-mortar	58.1	0.283
Specimen 5	Porotherm Bio Plan (38 × 25 × 21.9 cm)	M3-mortar	57.8	0.283
Specimen 6	Porotherm Bio Plan (38 × 25 × 21.9 cm)	M3-mortar	45.8	0.223
Specimen 7	Porotherm Bio Plan (38 × 25 × 21.9 cm)	M3-mortar	50.6	0.248
Specimen 8	Porotherm Bio Plan (38 × 25 × 21.9 cm)	M3-mortar	54.9	0.265

FIGURE 17: Example of foam placement on the brick.

(ii) The higher compressive strengths of the tested specimens with M1-mortar and M2-mortar are obtained for masonry made with "thin" joints.

(iii) The compressive strength of the masonry specimens with foam joints is lower than those obtained utilizing the mortar.

(iv) Foam joints were able to improve the ductility and the energy dissipation of the masonry, which is a desirable capacity in seismic prone areas.

In conclusion, the following observations are proposed:

(i) All specimens show a failure due to crushing and/or split in the elements.

TABLE 7: Technical data of the foam.

Color	Light green
Exterior temperature during the application	From −5°C to +40°C
Temperature	From −40°C to +120°C
Action film surface (at 23°C and 50% of RU)	7–10 min
Traceability (20 mm diameter curb at 23°C and 50% of UR)	30–40 min
Bulk density (after opposed expansion)	13–18 kg/m^3
Dimensional variation (at 23°C and 50% of UR)	<3%
Fire behavior	B2
UV resistance	Scarce, tends to yellow
Impermeability/sealing pressure	Up to 0.5 bar

(ii) The blocks utilized in this study due to their geometry are not the most suitable for this type of testing, since the holes are insufficient to ensure appropriate stress distributions. However, these blocks meet the requirements in terms of insulation and thermal transmittance.

Then, the recorded shear strength values cannot be considered explanatory of the foam-brick adhesion phenomenon; they still apply the following properties:

(i) The excellent adhesion on different types of supports

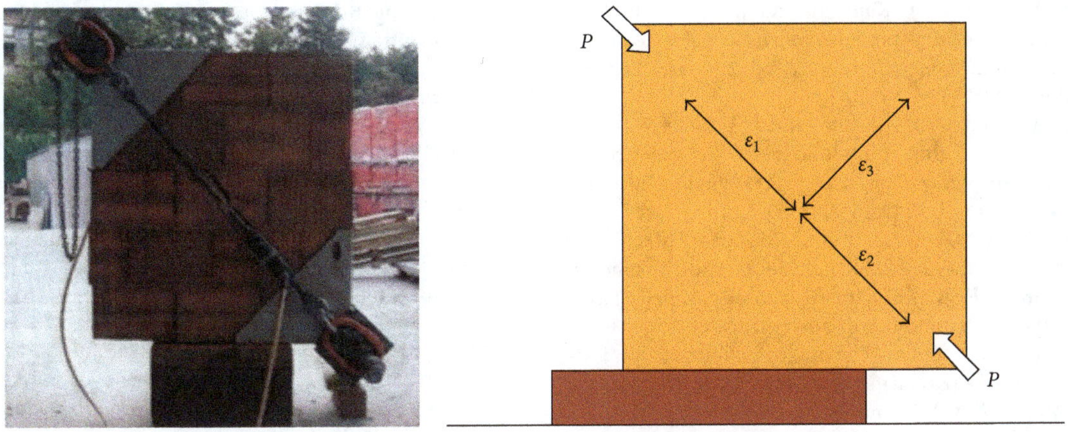

(a) (b)

FIGURE 18: Details of the diagonal compression test (a) and relative test setup (b).

TABLE 8: Mean shear strength in case of diagonal compression test for masonry with mortar.

Test	A_n(mm^2)	P (N)	S_s (N/mm^2)	ε_{max} (%)	G_V (MPa)	Type of collapse
1-MORTAR	283320	80000	0.20	0.13	0.017	Collapse for tensile cracking along the compressed diagonal.
2-MORTAR	283320	85000	0.21	0.14	0.018	Collapse for tensile cracking along the compressed diagonal.
3-MORTAR	283320	70000	0.17	0.07	0.006	Collapse for tensile cracking along the compressed diagonal.
4-MORTAR	283320	95000	0.24	0.16	0.023	Collapse for tensile cracking along the compressed diagonal.

TABLE 9: Mean shear strength in case of diagonal compression test for masonry with foam.

Test	A_n(mm^2)	P (N)	S_s (N/mm^2)	ε_{max} (%)	G_V (MPa)	Type of collapse
1-FOAM	283320	70000	0.17	0.23	0.026	Collapse for tensile cracking along the compressed diagonal and substantial collapse of the head brick in correspondence of the loading areas.
2-FOAM	283320	110000	0.27	0.14	0.025	Collapse for tensile cracking along the compressed diagonal.
3-FOAM	283320	110000	0.27	0.24	0.045	Collapse for tensile cracking along the compressed diagonal and corresponding to the loading areas and for shear sliding of the center of the specimen.

(ii) The high impermeability

(iii) The good chemical resistance

(iv) The extreme speed of installation with a reduction of up to 50% of the time compared to the use of thin-bed mortar

(v) The high cleanliness and the almost total reduction of tools

(vi) The sharp reduction in processing costs and the total costs.

In conclusion, the presented methodology can therefore be considered valid and can be used for such studies, considering an elastic-plastic behavior of the masonry, closer to its actual behavior.

Conflicts of Interest

The authors declare that they have no conflicts of interest.

References

[1] A. N. Fried, J. J. Roberts, M. C. Limbachiya, and A. Ahmed, "Thin joint masonry in the UK," *Masonry International Journal of British Masonry Society*, pp. 85–87, 2005.

[2] A. N. Fried, J. Ali, O. J. Kanyeto, and J. J. Roberts, "The properties of wet and dry thin layer mortar," in *Proceedings of 7th International Masonry Conference*, pp. 234–243, London, UK, October 2006.

[3] A. Jabbar, A. Fried, O. J. Kanyeto, and J. J. Roberts, "Compressive strength of thin joint solid dense concrete blockwork," in *Proceedings of 7th International Masonry Conference*, London, UK, October 2006.

[4] M. R. Valluzzi, F. Da Porto, and C. Modena, "Mechanical behaviour of masonry made of flat blocks and thin mortar joints," in *Proceedings of 6th International Masonry Conference*, pp. 503–511, London, UK, November 2002.

[5] O. J. Kanyeto and A. Fried, "Flexural behaviour of thin joint concrete blockwork: Experimental results," *Construction and Building Materials*, vol. 25, no. 8, pp. 3639-3647, 2011.

[6] N. Walliman, B. Baiche, and R. Ogden, "Thin-joint glued brickwork: Building in the British context," *Construction and Building Materials*, vol. 22, no. 6, pp. 1081-1092, 2008.

[7] D. Foti, "Shape optimization of rectified brick blocks for the improvement of the out-of-plane behavior of masonry," *International Journal of Mechanical Sciences*, vol. 7, pp. 417–424, 2013, ISSN: 1998-4448.

[8] D. Foti and A. Romanazzi, "Experimental analysis of fiber-reinforced mortar for walls in rectified brick blocks Analisi sperimentale di malte fibrorinforzate per pareti in blocchi di laterizio rettificati," *C and Ca*, vol. 41, no. 2, pp. 109–118, 2011, ISSN: 0045-6152.

[9] D. Foti and S. Vacca, "Comportamiento mecánico de columnas de hormigón armado reforzadas con mortero reoplástico/mechanical behavior of concrete columns reinforced with rheoplastic mortar," *Materiales de Construcción*, vol. 63, no. 310, pp. 267–282, 2013.

[10] D. Foti, "Preliminary analysis of concrete reinforced with waste bottles PET fibers," *Construction and Building Materials*, vol. 25, no. 4, pp. 1906–1915, 2011.

[11] D. Foti, "Use of recycled waste pet bottles fibers for the reinforcement of concrete," *Composite Structures*, vol. 96, pp. 396–404, 2013.

[12] D. Foti and F. Paparella, "Impact behavior of structural elements in concrete reinforced with PET fibers," *Mechanics Research Communications*, vol. 57, pp. 57–66, 2014.

[13] D. Foti, "Innovative techniques for concrete reinforcement with polymers," *Construction and Building Materials*, vol. 112, pp. 202–209, 2016.

[14] P. Medeiros, G. Vasconcelos, P. B. Lourenço, and J. Gouveia, "Numerical modelling of non-confined and confined masonry walls," *Construction and Building Materials*, vol. 41, pp. 968–976, 2013.

[15] R. Klingner, "Behavior of masonry in the northridge (US) and Tecomán–Colima (Mexico) earthquakes: lessons learned, and changes in US design provisions," *Construction and Building Materials*, vol. 20, no. 4, pp. 209–219, 2006.

[16] M. Sedlmajera, J. Zach, V. Novak, J. Hroudova, and A. Horsky, "Alternative technology of constructing masonry structures designed for areas with increased seismic activity," *Procedia Engineering*, vol. 151, pp. 177–182, 2016.

[17] H. Tuwair, M. Hopkins, J. Volz et al., "Evaluation of sandwich panels with various polyurethane foam-cores and ribs," *Composites Part B*, vol. 79, pp. 262–276, 2015.

[18] A. Shams, A. Stark, F. Hoogen, J. Hegger, and H. Schneider, "Innovative sandwich structures made of high performance concrete and foamed polyurethane," *Composite Structures*, vol. 121, pp. 271–279, 2015.

[19] G. Zhang, B. Wang, L. Ma, L. Wu, S. Pan, and J. Yang, "Energy absorption and low velocity impact response of polyurethane foam filled pyramidal lattice core sandwich panels," *Composite Structures*, vol. 108, pp. 304–310, 2014.

[20] A. Manalo, "Structural behaviour of a prefabricated composite wall system made from rigid polyurethane foam and Magnesium Oxide board," *Construction and Building Materials*, vol. 41, pp. 642–653, 2013.

[21] S. T. Taher, R. Zahari, S. Ataollahi, F. Mustapha, and S. Basri, "A double-cell foam-filled composite block for efficient energy absorption under axial compression," *Composite Structures*, vol. 89, no. 3, pp. 399–407, 2009.

[22] S. Kishimoto, Q. Wang, Y. Tanaka, and Y. Kagawa, "Compressive mechanical properties of closed-cell aluminum foam–polymer composites," *Composites: Part B*, vol. 64, pp. 43–49, 2014.

[23] M. Guden, S. Yüksel, A. Tasdemirci, and M. Tanoğlu, "Effect of aluminum closed-cell foam filling on the quasi-static axial crush performance of glass fiber reinforced polyester composite and aluminum/composite hybrid tubes," *Composite Structures*, vol. 81, no. 4, pp. 480–490, 2007.

[24] K. Stöbener and G. Rausch, "Aluminium foam–polymer composites: processing and characteristics," *Journal of Materials Science*, vol. 44, no. 6, pp. 1506–1511, 2009.

[25] S. Sharma, N. E. Rayess, and N. Dukhan, "Preliminary NVH characterization of metal foam–polymer interpenetrating phase composites," in *Proceedings of ASME International Mechanical Engineering Congress and Exposition*, pp. 445-451, Lake Buena Vista, FL, USA, November 2009.

[26] S. Gutiérrez-González, J. Gadea, A. Rodríguez, C. Junco, and V. Calderón, "Lightweight plaster materials with enhanced thermal properties made with polyurethane foam wastes," *Construction and Building Materials*, vol. 28, no. 1, pp. 653–658, 2012.

[27] J. A. Thamboo, M. Dhanasekar, and C. Yan, "Effects of joint thickness, adhesion and web shells to the face shell bedded concrete masonry loaded in compression," *Australian Journal of Structural Engineering*, vol. 14, no. 3, pp. 291–302, 2013.

[28] J. A. Thamboo, M. Dhanasekar, and C. Yan, "Flexural and shear bond characteristics of thin layer polymer cement mortared concrete masonry," *Construction and Building Materials*, vol. 46, pp. 104–113, 2013.

[29] J. A. Thamboo and M. Dhanasekar, "Characterisation of thin layer polymer cement mortared concrete masonry bond," *Construction and Building Materials*, vol. 82, pp. 71–80, 2015.

[30] J. A. Thamboo and M. Dhanasekar, "Behaviour of thin layer mortared concrete masonry under combined shear and compression," *Australian Journal of Structural Engineering*, vol. 17, no. 1, pp. 39–52, 2016.

[31] J. A. Thamboo and M. Dhanasekar, "Nonlinear finite element modelling of high bond thin-layer mortared concrete masonry," *International Journal of Masonry Research and Innovation*, vol. 1, no. 1, pp. 5–26, 2016.

[32] M. Dhanasekar, J. A. Thamboo, and S. Nazir, "On the in-plane shear response of the high bond strength concrete masonry walls," *Materials and Structures*, vol. 50, no. 5, p. 214, 2017.

[33] Consiglio Superiore dei Lavori Pubblici, "Approvazione delle nuove norme tecniche per le costruzioni, DM 14.01," *Gazzetta Ufficiale della Repubblica Italiana*, 2008, in Italian.

[34] Progetto di Ricerca LATEFICIO PUGLIESE SpA–MIUR, *Analisi Dell'adesione Malta-Laterizio in Funzione Dello Spessore Dei Giunti e Delle Modalità di Posa*, Dipartimento di Ingegneria Civile e Ambientale, Laboratorio Prove Materiali "M. Salvati", 2008.

[35] UNI EN 1052-5:2005, *Metodi di Prova Per Muratura–Parte 5: Determinazione Della Resistenza All'adesione Con il Metodo a Strappo*, Ente Nazionale Italiano di Unificazione, 2005.
[36] UNI EN 772-1:2002, *Metodi di Prova Per Elementi di Muratura–Parte 1: Determinazione Della Resistenza a Compressione*, Ente Nazionale Italiano di Unificazione, 2002.
[37] UNI EN 1052-3:2007, *Metodi di Prova Per Muratura–Parte 3: Determinazione Della Resistenza Iniziale a Taglio*, Ente Nazionale Italiano di Unificazione, 2007.
[38] ASTM E519, *Standard Test Method for Diagonal Tension (Shear) in Masonry Assemblages*, American Society for Testing and Materials, Baltimore, MD, USA, 2010.

10

Preparation and Properties of a Flexible Al_2O_3/Al/Al_2O_3 Composite

Yigang Tong,[1] Zhibin Zhou,[1] Hui Cai,[2] Xueliang Wang,[3] and Yaping Wang[1,4]

[1]*MOE Key Laboratory for Nonequilibrium Synthesis and Modulation of Condensed Matter, School of Science, Xi'an Jiaotong University, Xi'an 710049, China*
[2]*College of Materials Science and Engineering, Xi'an University of Science and Technology, Xi'an 710054, China*
[3]*MOE Key Laboratory of Thermo-Fluid Science and Engineering, School of Energy and Power Engineering, Xi'an Jiaotong University, Xi'an, Shaanxi 710049, China*
[4]*State Key Laboratory for Mechanical Behavior of Materials, Xi'an Jiaotong University, Xi'an 710049, China*

Correspondence should be addressed to Yaping Wang; ypwang@mail.xjtu.edu.cn

Academic Editor: Antonio Caggiano

A flexible $Al_2O_3/Al/Al_2O_3$ sandwich composite was prepared by microarc oxidation (MAO) on an aluminum foil with 50 μm thickness. The obtained ceramic layers with a thickness of 20–22 μm on the surface of the aluminum foil are mainly Al_2O_3. The composite, with exterior layers of Al_2O_3, can be bent more than 90° without observable fracture or delamination. The morphology of the MAO coatings, which has been characterized by scanning electronic microscopy (SEM), is porous and contains microcracks. The energy-dispersive spectrum (EDS) and the X-ray diffraction (XRD) results indicate that the MAO coatings mainly contain γ-Al_2O_3. Based on experimental observations, the flexibility mechanism was explained by the presence of the microcracks which improve the toughness and decrease the stiffness of the MAO coatings. The well-adhered layers of the sandwich structure can also prevent the fracture of the MAO coating. The thermal diffusivity of the composite is between that of aluminum and Al_2O_3 and the specific electrical resistance is 45.4% of that of Al_2O_3 under 50 V. This composite has the potential to be applied as tailorable low-voltage insulation material.

1. Introduction

Microarc oxidation (MAO), a novel technique to form oxide-based layers with special properties on a metal substrate by plasma discharging in aqueous solution under high voltage, is gaining increased attention [1, 2]. With the introduction of MAO Al_2O_3 ceramic coatings on Al-based materials, wear resistance, corrosion resistance, mechanical strength, thermal shock resistance, and electrical insulation of aluminum or its alloys can be efficiently increased [3, 4].

MAO ceramic layers grown on the surface of aluminum and its alloy are usually divided into two regions [5]. The compact region mainly contains α-Al_2O_3, and the loose region mainly contains γ-Al_2O_3. The compact region of MAO coatings possesses excellent wear, corrosion, and chemical resistance [6, 7], and the loose region is generally applied based on its large specific surface area, for example, improving the heat exchange of cases and heat sinks, as well as the capacity of electrode foils [8]. There have been many studies on the fabrication of the MAO coatings that are mainly composed by the compact region [5, 9]. However, all of these MAO coatings are a kind of naturally brittle material and are usually assumed to be unbendable. The application of material with such unbendable MAO coatings is limited, especially in ceramic biomaterial. Moreover, very few studies have explored deformable loose MAO coatings or their other physical properties [10–12].

In this paper, the flexible alumina ceramic layers are fabricated on the surface of the aluminum foil using the MAO method. The composition, phase structure, microstructure, and physical properties, as well as flexibility mechanisms of the alumina layers are investigated. This composite has the potential to be applied as tailorable low-voltage insulation material.

2. Materials and Methods

50 μm thick 1060 aluminum foil, which is essentially pure aluminum with a minimum 99 wt% aluminum content, was used to perform MAO in 15 g/L Na_2SiO_3 and 10 g/L $(NaPO_3)_6$ solution. The MAO equipment (240H-III) consisted of a potential adjustable pulsed DC source. The power source was set to manual mode during the treatment process. The electrical parameters of the treatment process are shown in Table 1. The obtained coatings were cleaned with distilled water and dried in air to remove the residual electrolyte from their surface.

The surface morphology, thickness, and microstructure of the MAO layers were observed in the JSM-7000F scanning electron microscope (SEM). The composition analysis was performed by energy-dispersive X-ray spectroscope (EDS) attached to another SEM (TESCAN MIRA3). Phase constitutes of the layers were examined by X-ray diffraction (XRD) with Cu k-α radiation.

To characterize the rapidity of heat propagation through the composite, the thermal diffusivity coefficient of the composite was measured using a LFA 457 laser flash apparatus. It had been set to measure three times under one temperature point (25°C) with the laser flash apparatus. Cowan model and pulse correction model have been applied by the accessional software named Netzsch Proteus to avoid the effect of the heat loss and that of the laser pulse width on the experimental results. The composite was cut into dimensions of 12.7 mm diameter. The thickness of the composite was measured by a micrometre. Three thickness measurements have been taken at three different dot positions of the composite, and the averaged result was used as the thickness of the measurement.

The resistance of the sample was measured using an insulation tester. The samples for testing had been washed by distilled water and dried at 30°C in the air overnight to avoid the unexpected effects on the resistance result.

Two bending tests have been carried out, bending by hand and by a rod (16 mm in diameter). All the bending processes have been recorded by a digital camera. A metalloscope (Olympus, BX61-32FAI-S09) has been used to observe the surface morphology of the bending sample on the rod.

Table 1: Electrical parameters of the treatment process.

Time (min)	Pulsed voltage (V)	Duty ratio (%)
0	500	12
5	520	12
10	530	12
15	540	10
20	550	10
25	560	10

3. Results and Discussion

The cross-sectional morphology of the as-prepared sample is shown in Figure 1(a), which displays a sandwich-type structure with two MAO and an inner Al foil layers. Two interfaces between the MAO layers and the Al substrate are found. These interfaces featured undulation and continuity, indicating that the specific area of the interface was expanded from that of the original Al substrate and the cohesion between the MAO layers and the Al substrate was good. It is observed that the average thickness of the MAO layers is about 20 μm. The MAO layers are fabricated into loose structure. Pores and microcracks are observed in the MAO layers. Microcracks have been circled in Figure 1(a).

The surface morphology of the MAO layer is presented in Figure 1(c). It can be seen that the surface of the MAO layer is uneven, and many micropores with various pore sizes distributing uniformly on the surface can be observed. Comparing with the cross-sectional morphology in Figure 1(a), it is found that the micropores mainly accumulate at the surface of the MAO layers; in other words, few micropores appear near the interface between the MAO layer and the Al substrate. The reason is that growth of both the inward and outward layers has taken place during the discharging process, and then the inner MAO layers are melted under high-temperature plasma discharging and consequently are ejected out through the original crater which are discharge channels. The ejected molten alumina is cooled by electrolyte and spread around the discharge channel. Finally, the volcano-like pores are formed by the solidification of the molten alumina [13]. These volcano-like micropores can be seen in Figure 1(c). Therefore, the formation of the micropores throughout the whole MAO process is attributed to the residual discharge channels, which are retained after the plasma discharge process in the interior material of the layers.

Figure 2 shows the X-ray diffraction profiles of the sandwich composites. The diffraction peaks corresponding to γ-Al_2O_3 and Al were clearly detected. The peaks of γ-Al_2O_3, which relate to the crystal faces including (222), (400), and (440), and that of Al, which include (111), (200), (220), and (311), confirm that the phase structure of the outer layers is Al_2O_3 ceramics. The possible reason that the XRD result, shown in Figure 2, contains the diffraction peaks of Al is that the thickness of the composite, less than 80 μm, is too thin to hinder the X-ray from penetrating the MAO layer on the surface of the aluminum substrate.

The EDS result for the MAO layer of the as-prepared sample is shown in Figure 3, which indicates that the composition in the outer surface mainly contains Al, O, and P elements, and the percentage of these three elements are about 72.23 at%, 27.03 at%, and 0.74 at%, respectively. The P element existing in the MAO layer stems from the $(NaPO_3)_6$ electrolyte which is used during the MAO process [14].

The thermal diffusivity coefficient of the composite is 21.5 mm^2/s (shown in Table 2), lower than that of Al (about 86.7 mm^2/s at room temperature) but higher than that of Al_2O_3 (about 5.7 mm^2/s at room temperature), which indicates that the transient thermal response of the Al foil has been passivized by MAO layers.

As shown in Figure 4, the specific electrical resistance has shown a rapid drop with the increase in applied voltage up to 50 V, and then it displayed a gradual drop in specific electrical resistance as the applied voltage is increased from 50 V to 250 V. The insulation properties are theoretically 45.4% of that of Al_2O_3 (10^{14} Ω·cm). The experimental results indicate

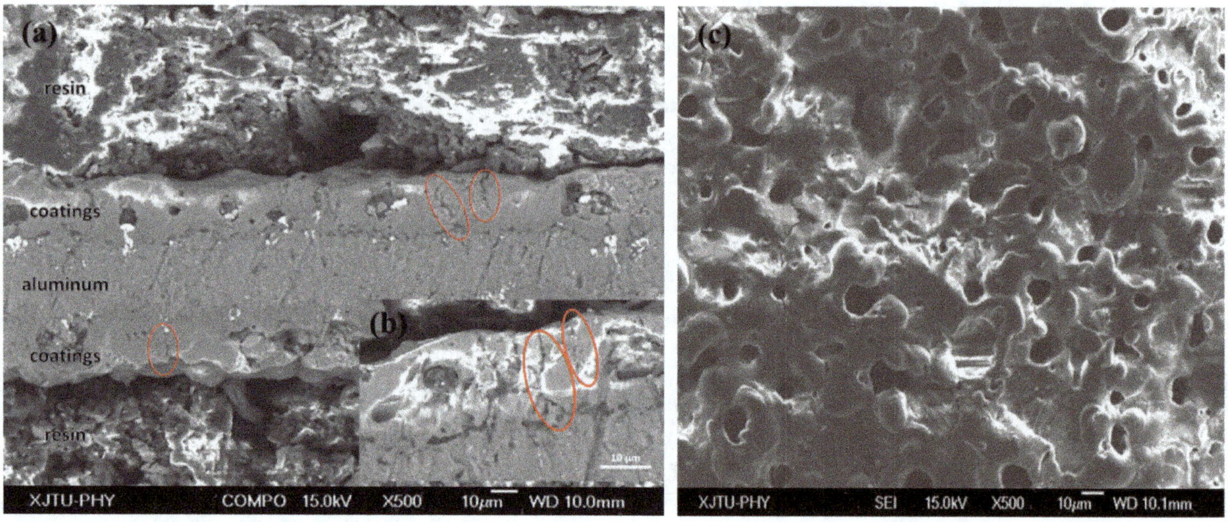

FIGURE 1: (a, b) The cross-sectional morphology of the specimen; microcracks are marked with red circle; pores accumulate at the surface of the MAO layer. (c) The surface morphology of the specimen.

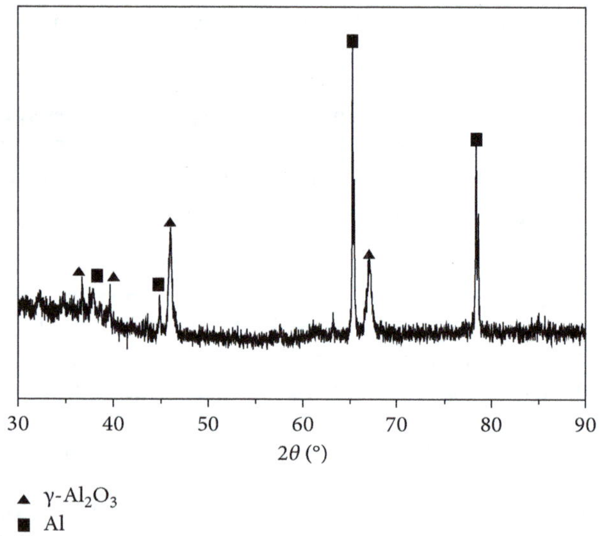

FIGURE 2: The XRD pattern of the as-prepared MAO sample.

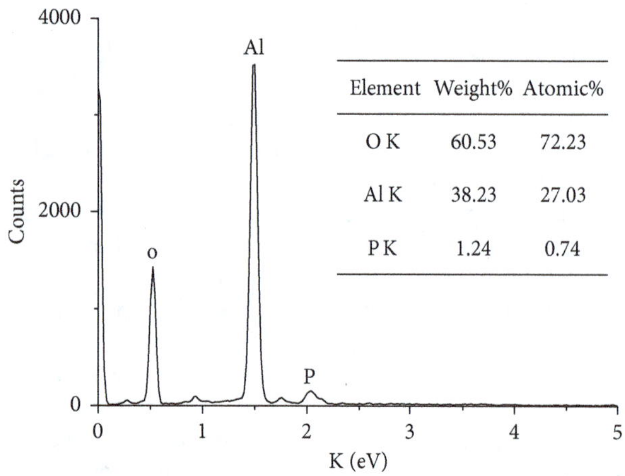

FIGURE 3: EDS test result for the MAO layers of the as-prepared sample.

TABLE 2: Thermal diffusivity result for the composite.

Number	Temperature (°C)	Thermal diffusivity (mm^2/s)
1	25.4	23.32
2	25.3	20.09
3	25.4	21.14
Mean value	25.4	21.5

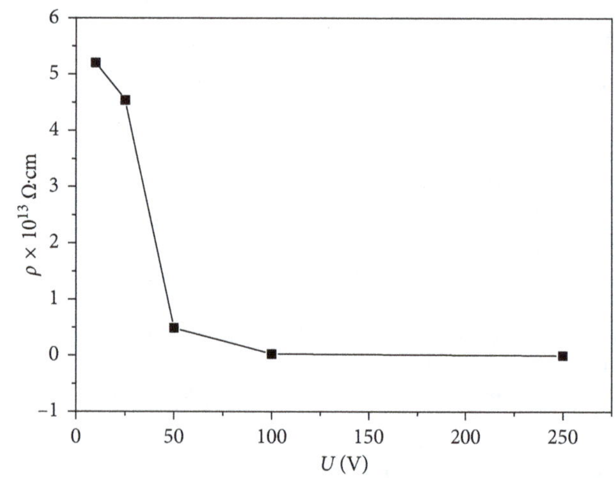

FIGURE 4: Specific electrical resistance of the specimen under different voltages.

that 50 V seems to be the breakdown voltage. Such a composite can be used as a low-voltage dielectric material [8]. This is related to the porous microstructure of the MAO layers and the irregular Al_2O_3/Al interface. A potential barrier is created at the interface between the insulator (Al_2O_3) and the conductor (Al) under external electric field, and this potential barrier can block the charge transportation and accumulate charges on both sides of the interface between two substrates. Once enough charges, which were induced by enough high external electric field, accumulated at the convexity of the interface, the potential barrier can be penetrated and the

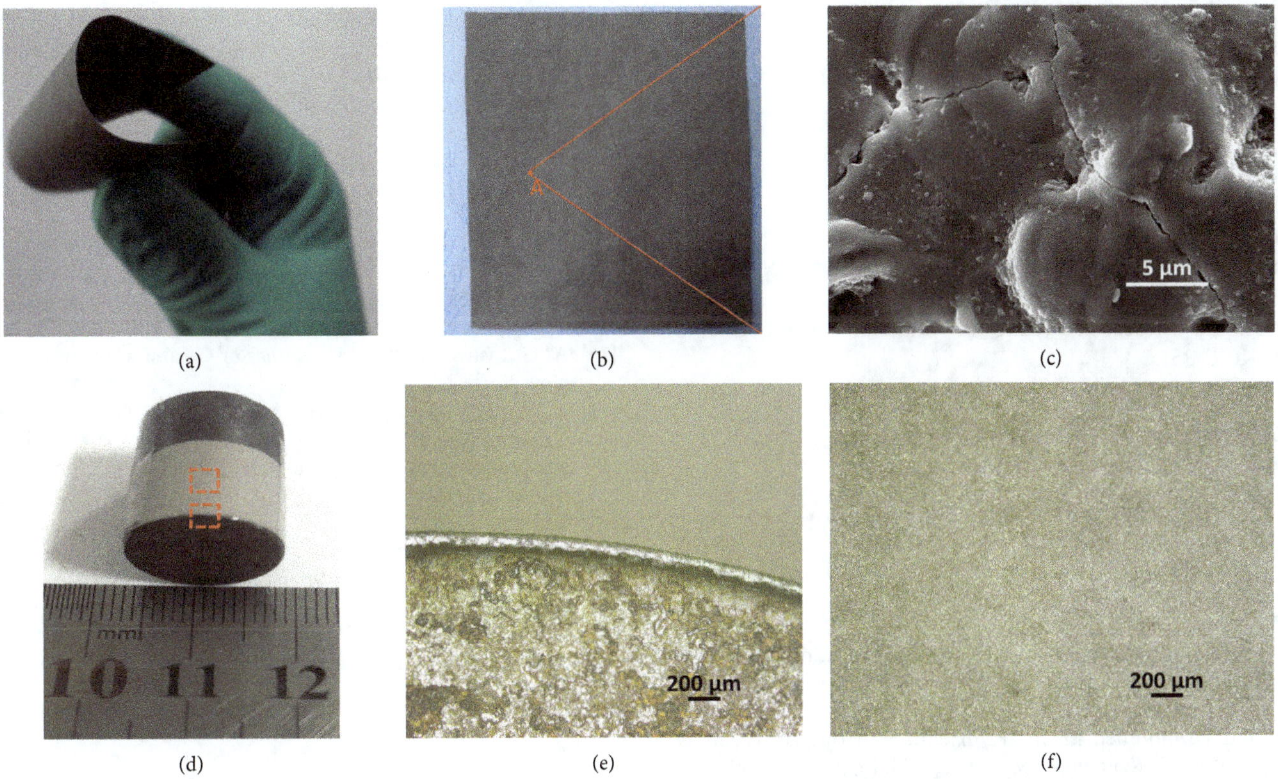

FIGURE 5: Bending test by hand: (a) bending test by hand; (b) the recovered composite; (c) SEM picture of the recovered composite, and the observed place has been marked as point A. Bending test by rod: (d) a tailored composite covering the rod of 16 mm diameter; (e) surface morphology of the tailored composite during the bending process on the rod; (f) cross-sectional view of the tailored composite during rod bending.

current would form; as a result, the macroscopic resistance declines. Moreover, porous structure can absorb moisture in a wet and hot environment, and thus the value of the real breakdown voltage of the MAO layers is lower than the theoretical one [15].

The flexibility of the composite is shown in Figure 5. When the composite was bent by hand, the MAO layers were still uniform without any observable fracture or falling-out piece, as shown in Figure 5(a). The composite can completely recover after releasing load as shown in Figure 5(b). The scanning electron micrograph (Figure 5(c)) at × 1000 shows the surface morphology of the recovered composite surface in the region which was under heavy deformation and is marked as point A. It shows that there has been no significant widening of microcracks after bending, suggesting the efficient recovery of MAO coatings. In another bending test, the composite has also been tailored and attached to a rod of 16 mm diameter, as shown in Figure 5(d). The surface and cross-sectional morphology at × 50 are shown in Figures 5(e) and 5(f), respectively, and no macrocracks can be found. The aforementioned performances reflect that MAO layers have good flexibility and good fracture toughness.

The flexibility of the MAO layers can be explained by two reasons: (1) the microcracks, which were observed in Figure 1, could effectively resist the propagation of the macrocracks of the MAO layers during the composite deformation, and (2) the strong adhesion of the interface made the aluminum substrate well support the MAO layers.

Firstly, the mechanism for resisting the cracks propagation in the coatings is that the microcracks can increase the energy for crack propagation by prolonging the crack paths and inducing the cracks branching [16, 17]. What is more, the presence of the microcracks and the pores in the layers can sharply reduce the stiffness of the MAO layer [18, 19].

Secondly, it appears that MAO coatings can exhibit excellent interfacial cohesion [2, 3]; meanwhile, the specific area of the interface between the MAO layers and Al substrate is large, so the stress that can make MAO layers fall out from substrate has been enlarged. As a result, the MAO layers can hardly fall out or be fractured during the deformation of the composite.

4. Conclusions

The flexibility around the 20 μm thick MAO layer, which mainly consists of γ-Al_2O_3, grows on and closely compacts the Al substrate. The surface of the MAO layers is uniformly porous, and there are microcracks in layer structure. The transient thermal response of the Al foil has been passivized by MAO layers. With the high electrical resistant, the composite has the potential to be used as a low-voltage dielectric material.

Conflicts of Interest

The authors declare that there are no conflicts of interest regarding the publication of this paper.

Acknowledgments

The authors acknowledge the financial support of the National Science Foundation of China (Nos. 51101177 and 51607132) and the Program for Key Science and Technology Innovative Research Team of Shaanxi Province (No. 2013KCT-05).

References

[1] P. Wang, T. Wu, Y. T. Xiao et al., "Effect of Al2O3 micropowder additives on the properties of micro-arc oxidation coatings formed on 6061 aluminum alloy," *Journal of Materials Engineering and Performance*, vol. 25, no. 9, pp. 1–5, 2016.

[2] P. Wang, T. Wu, H. Peng et al., "Effect of $NaAlO_2$ concentrations on the properties of micro-arc oxidation coatings on pure titanium," *Materials Letters*, vol. 170, pp. 171–174, 2016.

[3] S. Ji, Y. Weng, Z. Wu et al., "Excellent corrosion resistance of P and Fe modified micro-arc oxidation coating on Al alloy," *Journal of Alloys and Compounds*, vol. 710, pp. 452–459, 2017.

[4] X. Guo, K. Du, Y. Huang et al., "Application of a composite electrolyte in a solid-acid fuel cell system: A micro-arc oxidation alumina support filled with CsH_2 PO_4," *International Journal of Hydrogen Energy*, vol. 38, no. 36, pp. 16387–16393, 2013.

[5] T. Wei, F. Yan, and J. Tian, "Characterization and wear- and corrosion-resistance of microarc oxidation ceramic coatings on aluminum alloy," *Journal of Alloys and Compounds*, vol. 389, no. 1-2, pp. 169–176, 2005.

[6] W. Liu, W. Liu, and A. Bao, "Microstructure and properties of ceramic coatings on 7N01 aluminum alloy by Micro-Arc oxidation," *Procedia Engineering*, vol. 27, pp. 828–832, 2012.

[7] O. Tazegul, F. Muhaffel, O. Meydanoglu, M. Baydogan, E. S. Kayali, and H. Cimenoglu, "Wear and corrosion characteristics of novel alumina coatings produced by micro arc oxidation on AZ91D magnesium alloy," *Surface and Coatings Technology*, vol. 258, pp. 168–173, 2014.

[8] A. L. Yerokhin, X. Nie, A. Leyland, A. Matthews, and S. J. Dowey, "Plasma electrolysis for surface engineering," *Surface and Coatings Technology*, vol. 122, no. 2–3, pp. 73–93, 1999.

[9] J. Tian, Z. Luo, S. Qi, and X. Sun, "Structure and antiwear behavior of micro-arc oxidized coatings on aluminum alloy," *Surface and Coatings Technology*, vol. 154, no. 1, pp. 1–7, 2002.

[10] K. H. Dittrich, W. Krysmann, P. Kurze, and H. G. Schneider, "Structure and properties of ANOF layers," *Crystal Research and Technology*, vol. 19, no. 1, pp. 93–99, 1984.

[11] W. Araki and J. Malzbender, "Ferroelastic deformation of La0.58Sr0.4Co0.2Fe0.8O3–δ under uniaxial compressive loading," *Journal of the European Ceramic Society*, vol. 33, no. 4, pp. 805–812, 2013.

[12] Z. Y. Deng, T. F. Fukasawa, M. Ando, G.-J. Zhang, and T. Ohji, "Microstructure and mechanical properties of porous alumina ceramics fabricated by the decomposition of aluminum hydroxide," *Journal of the American Ceramic Society*, vol. 84, no. 11, pp. 2638–2644, 2001.

[13] M. M. S. Al Bosta and K.-J. Ma, "Suggested mechanism for the MAO ceramic coating on aluminium substrates using bipolar current mode in the alkaline silicate electrolytes," *Applied Surface Science*, vol. 308, pp. 121–138, 2014.

[14] G. Lv, W. Gu, H. Chen et al., "Characteristic of ceramic coatings on aluminum by plasma electrolytic oxidation in silicate and phosphate electrolyte," *Applied Surface Science*, vol. 253, no. 5, pp. 2947–2952, 2006.

[15] K. H. Lin, Z. H. Xu, and S. T. Lin, "A study on microstructure and dielectric performances of alumina/copper composites by plasma spray coating," *Journal of Materials Engineering and Performance*, vol. 20, no. 2, pp. 231–237, 2011.

[16] A. G. Evans, *Fracture Mechanics of Ceramics*, Plenum Press, New York, NY, USA, 1978.

[17] Y. Wang, T. Lei, B. Jiang, and L. Guo, "Growth, microstructure and mechanical properties of microarc oxidation coatings on titanium alloy in phosphate-containing solution," *Applied Surface Science*, vol. 233, no. 1, pp. 258–267, 2004.

[18] J. Dean, T. Gu, and T. W. Clyne, "Evaluation of residual stress levels in plasma electrolytic oxidation coatings using a curvature method," *Surface and Coatings Technology*, vol. 269, pp. 47–53, 2015.

[19] J. A. Curran and T. W. Clyne, "Porosity in plasma electrolytic oxide coatings," *Acta Materialia*, vol. 54, no. 7, pp. 1985–1993, 2006.

11

Behaviors of Oxide Film during Semisolid Brazing of SiC_p/6063Al Composite Materials

Jing Xiao [1], **Shun Li** [1], **Shuxin Bai** [1], **Jiuchun Yan**, [2] **and Degan Xiong** [1]

[1] *Department of Materials Science and Engineering, National University of Defense Technology, Changsha 410073, China*
[2] *State Key Laboratory of Advanced Welding and Joining, Harbin Institute of Technology, Harbin 150001, China*

Correspondence should be addressed to Shuxin Bai; 13272093806@163.com

Academic Editor: Xinquan Zhang

The semisolid brazing of SiC_p/6063Al under an applied pressure using Zn-Al-Cu filler metal was investigated. The samples to be joined were heated from 380°C to 382°C, 386°C, 392°C, and 410°C under a constant pressure of 10 MPa, respectively. Effects of the temperature on microstructural evolution and deformation behavior of the filler metal, interfacial structure, and shear strength of the bonded joint were discussed, and the disruption behavior of the surface oxide film was studied. The results show that, after heating, the solid grains of the filler metal transform into a globular structure surrounded by liquid. The degree of sphericity and the liquid fraction tend to improve with increasing temperature. During the heating process, the deformation of the filler metal is first accomplished by plastic deformation of solid grains and then by intergrain sliding and liquid flow. The surface oxides are broken and stripped by a cocontribution of compressive and shear stress which is caused by depressing and sliding of solid grains along the composites. It is found that the heating of 380°C to 392°C under pressure is the optimum condition to disrupt the surface oxide films and obtain sound bonds. The mechanical test results show that the maximum shear strength of the bond joints is as high as 105 MPa, reaching 78% that of the parent materials.

1. Introduction

High volume fraction SiC-reinforced aluminum metal matrix composites (Al-MMCs) is considered as an ideal electronic packaging material due to its attractive properties (e.g., high thermal conductivity, low tunable coefficient of thermal expansion, high modulus, and low density). The associated welding technologies thus attract intense attentions. Among the current joining methods for Al-MMCs, brazing is a competitive candidate because of its flexibility, reliability, and low bonding temperature. However, the presence of the surface oxides is the main barrier to successful bonding. Therefore, conventional brazing methods are normally conducted in vacuum [1, 2] or with flux to reduce the high-temperature oxidation of the faying surfaces. However, joining processes in vacuum increase the overall bonding time and reduce the design flexibility of the parts to be joined; besides, the residual flux is hard to be removed, which could result in a severe corrosive problem. Therefore, the vacuum-free bonding processes which are assisted by physical energy to remove the oxides have been developed increasingly.

Ultrasonic brazing is an elegant method to join Al-MMCs in air with the physical assistance. Due to the cavitation effect caused by the ultrasonic wave, not only the surface oxides are disrupted but also the wettability is enhanced [3–9]. However, higher ultrasonic wave energy may aggravate dissolution of the base metal, which leads to degradation of the substrate properties [10].

Semisolid joining techniques have been developed recently based on semisolid metal processing. The results show that the bonding of the substrate could be realized using the semisolid joining technique with constant pressure. Kiuchi et al. first performed semisolid joining for the aluminum alloy with glass or stone and confirmed the excellent joining ability of the semisolid alloys [11]. Cheng and Zhang [12] carried out semisolid joining for SiC_p/2024Al composites and the 2024 aluminum alloy under high pressure. The bond strength was increased due to the fine microstructure which arose from the spherical semisolid billet. To overcome the oxide problem and improve the bond interface structure, the semisolid joining technique with motion energy has been recently developed. Narimannezhad et al. [13] and Hosseini et al. [14, 15] developed semisolid stirring joining in the zinc alloy and AZ91 magnesium alloy, respectively. Ideal and nonlinear bond interface were observed under the action of stirring. Yan et al. [16] and Xu et al. [17, 18] carried out semisolid brazing of SiC_p/A356 composites with the assistance of vibration and stirring, respectively. It was found that the mechanical assistance aids the disruption of the oxide films. However, the current methods are often relatively complicated to operate and usually involved with the porosity and crack problem in the joining region [19, 20]. What is more, the intermechanisms including the disruption mechanism of the surface oxides in the processes are not clear.

In this work, a semisolid brazing of SiC_p/6063Al composites under applied pressure was investigated, aiming at simplifying the operation, lowering the cost, and retaining the potential of the semisolid joining. Effects of the brazing temperature on the interfacial structure and shear strength of the joint were probed. Additionally, the disruption behavior of the surface oxide film was also studied by analyzing the microstructural evolution and deformation behavior of the filler metal during the brazing process.

2. Materials and Methods

2.1. Materials. The base materials used in this work were the 6063Al alloy that was reinforced with 63 vol.% SiC particles with a diameter range of 10–100 μm. The materials were produced by vacuum pressure infiltration with shear strength of ~135 MPa at room temperature. The filler metals used in this work were Zn-Al-Cu alloys, whose chemical composition and solidus (T_S) and liquidus (T_L) temperatures are shown in Table 1. Before joining, as-cast Zn-Al-Cu alloys were homogenized at 350°C for 12 h and rolled at 300°C from 30 mm to a thickness of 0.3 mm trip (99% reduction) in several passes. The base materials and filler metal were processed into plates with dimensions of 30 mm × 30 mm × 3 mm and 30 mm × 30 mm × 0.3, respectively. All samples were mechanically polished to a 600-grit finish and then ultrasonically degreased in acetone prior to heating.

2.2. Brazing Process. The filler metal was preset between the base materials. The thermocouple used to measure the temperature was fixed near the bonding line, and an accuracy of the temperature control was ±1°C.

The specimens were heated up to 380°C (T_S) at a heating rate of 20°C·min^{-1} and held for 10 min to homogenize the samples. Then, an axial pressure of 10 MPa was imposed on the upper specimen. Afterwards, the specimens were further heated to four different temperatures of 382, 386, 392, and 410°C at a low rate of 0.5°C·min^{-1} with the pressure, held at the corresponding temperature for 5 min and then cooled down to 200°C at a rate of 5°C·min^{-1}. To ensure adequate solidification of the filler metal, the loading pressure was not removed until the temperature went down to 200°C. The whole brazing process as described in Figure 1 was conducted under argon atmosphere.

2.3. Characterization Test and Observation. To investigate the effect of temperature on the microstructure of the filler metal, the filler metals with dimensions of 30 mm × 30 mm × 0.3 mm were heated to 382°C–410°C without compression and with compression between two SiC_p/Al composites plates (similar to brazing in a furnace), respectively. Then, the specimens were quickly quenched with water.

The filler metals and cross sections of the bonded joints were etched with a 2% solution of nitric acid after mechanical grinding and polishing with 0.5 μm diamond paste. The microstructures were examined by HIROX 200 optical microscope (OM) and SIRION-type field emitting scanning electron microscope (SEM) produced by Japan FEI Corporation. The chemical composition was examined by energy dispersive spectroscopy (EDS). The average grain size, shape factor, and the liquid volume fraction were measured using Leica Q550MW image analysis software. The shear strength tests of the brazed joints were evaluated on the WDW-100 electromechanical universal testing machine with a constant speed of 0.5 mm·min^{-1} at room temperature. In order to guarantee the reliability, a designed jig was applied for the test, as shown in Figure 2(a). In particular, a notch with specific dimension was created on the top of the specimen beside the bond line (a distance of 0.15 mm from the bond line as shown in Figure 2(b)) to accommodate the load. At least five independent tests were conducted under each experiment condition.

3. Results and Discussion

3.1. Effect of Temperature on the Interfacial Structure of the Joint. The interfacial structure of the joint brazed at different temperatures is shown in Figure 3. A continuous oxide film is observed at the interface at 382°C, and a slight zinc diffusion layer appears at the side of the base material (as shown by arrow in Figure 3(a)). As reported by Xu et al. [3], there exist many fractures in the oxide inclusion zone, and zinc is inclined to diffuse into the matrix through the fractures to form a penetration zone. At 386°C, the oxide layer becomes discontinuous and a wider diffusion layer is formed in the composites side (Figure 3(b)). When heated to 392°C, it can be found that the oxide films were broken into small fragments, and a sharply increasing diffusion layer is observed (Figure 3(c)). However, with further increasing of temperature, no visible removal of the oxide layer is obtained. Discrete oxide films are still contained at the bond line, despite the metallic bonding between them (Figure 3(d)).

3.2. Microstructure Evolution and Deformation Behavior of the Filler Metal. The disruption of the surface oxide film

Table 1: Chemical composition and solidus (T_S) and liquidus temperatures (T_L) of the filler metals.

Composition (wt.%)			Solidus and liquidus temperature (°C)	
Zn	Al	Cu	T_S	T_L
Balance	12	4.5	380	420

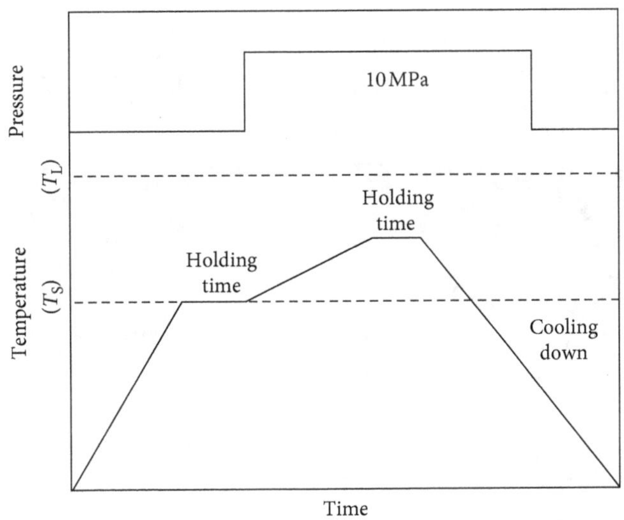

Figure 1: Schematic diagram of the brazing process.

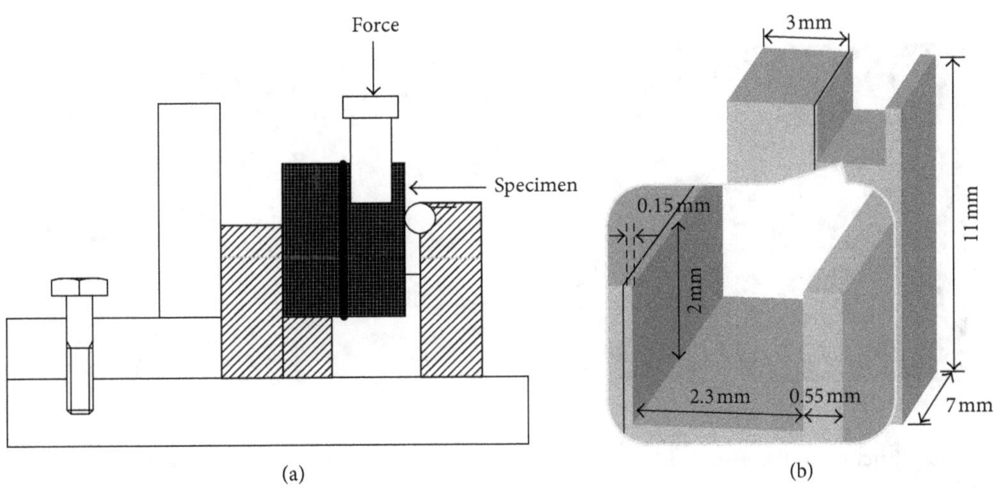

Figure 2: Schematic diagram of (a) shear test arrangement and (b) dimension of the shear specimen (bond line is shown by a thick line).

during the brazing process is considered to be related to the microstructure, compression behavior of the filler metal, and the behavior against the base metal. Therefore, it is very important to verify the microstructural evolution and compression behavior of the filler metal during the heating process. The microstructure of the as-rolled Zn-Al-Cu alloy is shown in Figure 4.

Optical micrograph shows that the microstructure consists of plastically deformed grains aligned in the rolling direction. The backscattered electron image and spot analysis as shown in Table 2 reveal that the microstructure of the Zn-Al-Cu alloy consists of the proeutectic β phase (marked "A"), aluminum-rich α phase (marked "B"), and Zn-rich η phase (marked "C"). Copper additions less than 5% had no significant effect on the microstructure of this alloy. During solidification of the Zn-Al-Cu alloy, proeutectic β forms first and the remaining liquid transforms into the $\beta + \eta$ mixture by the eutectic reaction. During cooling to room temperature, the proeutectic and eutectic β phases both decompose into an alternating layer of α and η [21]. Due to the rolling process, all grains orient themselves in the rolling direction.

The microstructure of the as-rolled Zn-Al-Cu alloy filler metal heated to different temperatures without compression is shown in Figures 5(a)–5(d). It can be seen that no obvious

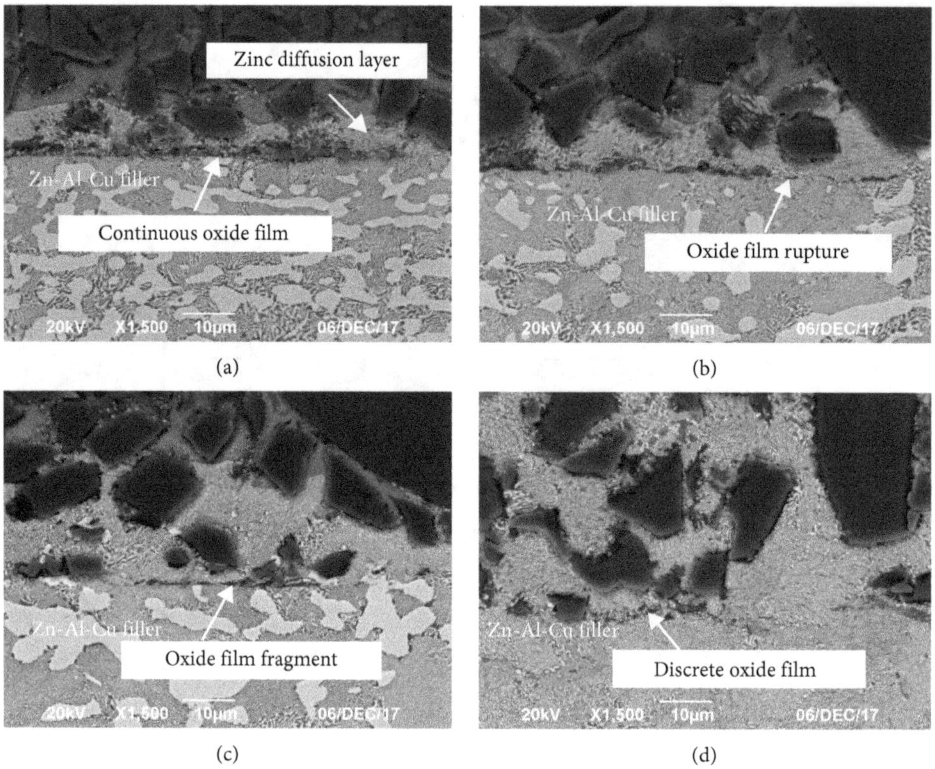

FIGURE 3: The interfacial structure of the joint brazed at different temperatures: (a) 382°C; (b) 386°C; (c) 392°C; (d) 410°C.

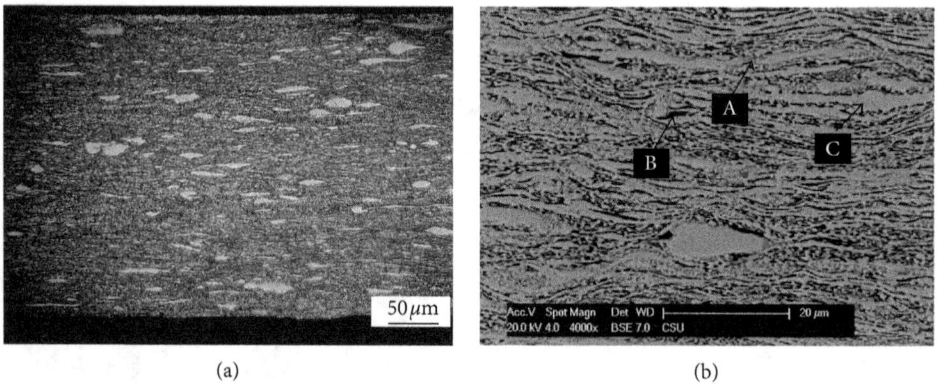

FIGURE 4: Microstructure of the as-rolled Zn-Al-Cu alloy: (a) optical micrograph; (b) backscattered electron image.

TABLE 2: EDS analysis results of regions A, B, and C in Figure 4.

Region	Zn	Al	Cu
A	73.31	23.58	3.11
B	61.15	36.86	1.97
C	84.43	1.38	14.19

change can be observed in the microstructure of the Zn-Al-Cu alloy after being heated to 382°C as shown in Figure 5(a). When heated to 386°C, both deformed η-Zn grains and β-Al grains have transformed to near-spherical grains with local liquid at the grain boundary (Figure 5(b)). With the increase of the temperature, most of the η-Zn and β-Al grains become more spheroidal and uniformly distributed in a liquid matrix as shown in Figure 5(c). When the temperature increases to 410°C, coalescence occurs between the adjoining grains and liquid pool forms due to the increasing liquid fraction (Figure 5(d)).

It can also be found that the amount of η-Zn grains tends to decrease with increasing temperature from 382°C to 410°C and they almost disappear after being heated to 410°C.

Figure 6 shows the variation in the shape factor of the solid grains and the liquid fraction as a function of the temperature. As shown in Figure 6, the shape factor increases to a maximum at 392°C and then decreases slightly with further increasing temperature. The liquid fraction measure implies that the liquid fraction is relatively low at

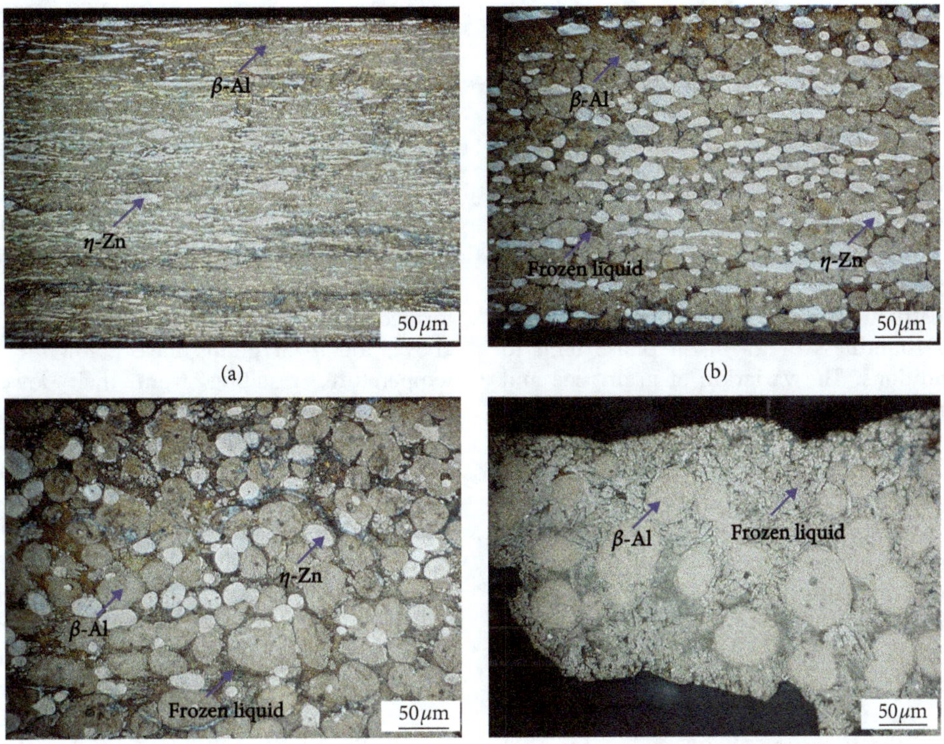

FIGURE 5: The microstructure of the filler metal heated to different temperatures without compression: (a) 382°C; (b) 386°C; (c) 392°C; (d) 410°C.

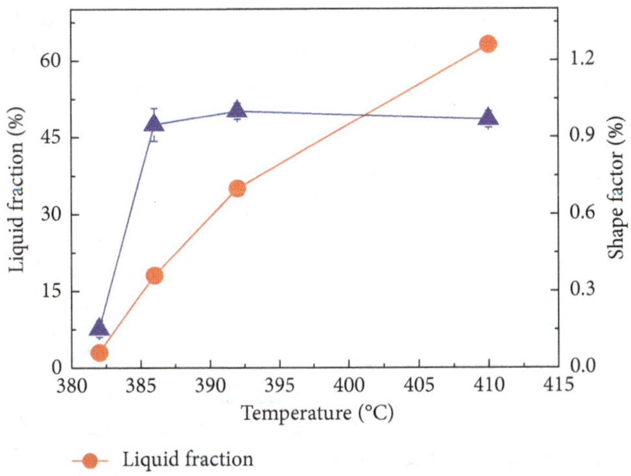

FIGURE 6: The variation in the shape factor of the solid grains and the liquid fraction as a function of the temperature.

382°C ($f_L = 3\%$), and it increases markedly with increasing temperature.

It is well known that the plastic deformation of the alloy supplied by hot or cold work ultimately results in sufficient stored strain energy. The strain energy then activates the recrystallization and spheroidization of the solid grains during the reheating process [22–25]. Recrystallization is diffusion controlled, and sufficient temperature is thus necessary for diffusion. At temperature below 386°C, thermal energy for recrystallization is insufficient, and recrystallization is thus difficult to be induced. Therefore, there seems to be no detectable recrystal grain appearing in the microstructure at 382°C, as shown in Figure 5(a). With the increase of the temperature, the diffusion rate increases and hence, recrystal grains are able to formed, as illustrated in Figure 5(b). These recrystal grains are, however, partial spheroidal. By continuing the heating process, the grains spheroidize and coarsen due to the minimum interface energy principle. However, spheroidization and coarsening are codominant and compete with each other. During the heating period of 382°C to 392°C, small grains which have a lower melting point are melted in favor of the larger grains and the number is reduced. This process is less effective in particle growth but has a great influence on the spheroidization of solid grains [26], so the shape factor of the solid grains increase obviously in this stage, as shown in Figure 6. During the heating period of 392°C to 410°C, the majority of the solid grains have gained uniform surface curvature, and then growth of spherical grains commences. Hence, obvious coarsening of the grains and decrease of the shape factor are observed as shown in Figures 5 and 6. In addition, according to Figures 5 and 6, it is seen that the increase of the temperature results in the increase of the liquid fraction and the decrease of the number of the η-Zn phase. It is known that when temperature is higher than the eutectic temperature, intergranular eutectic phases ($\alpha + \eta$) melt first to form liquid. This induces a thin liquid layer at the grain boundaries and a little decrease of the η-Zn phase as shown in Figure 5(b).

With the increase of the temperature, large η-Zn particles start to melt, resulting from their low melting point. The higher the temperature, the more the melting of the η-Zn particles, and consequently, the higher the liquid fraction.

Figures 7(a)–7(d) shows the microstructure of the as-rolled Zn-Al-Cu samples heated to different temperatures with compression. The microstructure which underwent both thermal and thermal-compression treatment is somewhat different to that without compression. Firstly, by comparing Figures 5 and 7, β-Al grains in all thermal-compression microstructure have no obvious change, but the grain size and roundness of the η-Zn phase tend to decrease in all conditions. The variation of grain size and roundness of the η-Zn phase with different microstructures is shown in Figure 8. It can be seen that the grain size and roundness of the η-Zn phase in the noncompression microstructure are always higher than that in the thermal-compression microstructure. Moreover, the grain size of η-Zn grains in the noncompression microstructure increases apparently with increasing temperature. However, the trend in the thermal-compression microstructure is quite the reverse (Figure 8(a)). Secondly, the thermal-compression microstructure at the center region is similar to that at the edge region at 382°C as shown in Figures 7(a_1) and 7(a_2). With the increase of the temperature, it can be seen that strong compression makes the liquid squeeze out, which shows a lower fraction of the liquid in the center and higher fraction in the edge (Figures 7(b)–7(d)). Thirdly, the thickness of the filler in the thermal-compression microstructure decreases apparently with increasing temperature. As shown in Figure 7, the thickness of the filler metal is around 290 μm at 382°C, which is a little lower than the initial thickness of the filler metal. As temperature increased from 382°C to 392°C and deformation proceeded, the thickness decreases obviously from 290 μm to around 85 μm. Finally, the interface structure between the filler metal and composites improves with increasing temperature. Unjointed gaps are observed on the interface at 382°C, which is probably due to no disruption of the oxide film in this region and huge residual thermal stress resulted from quenching. The gaps disappear at 386°C, and metallurgical bonding is achieved at 392°C and 410°C, as shown by arrows in Figures 7(c_1) and 7(d_1).

It is well known that the high temperature strength of metals is related to the diffusion ability of metal atoms. The lower the diffusion ability is, the greater the high temperature strength will be. On the contrary, the higher the melting point of metal, the stronger the bonding force of metal atoms and, consequently, the lower the diffusion coefficient of the atoms. In this case, η-Zn grains with the lower melting point will have a lower strength at the elevated temperature. The grains are thus deformed and broken easily under the applied pressure at elevated temperature compared to the β-Al grains with the higher melting point. This can explain the lower roundness and grain size of the η-Zn grains in the thermal-compression microstructure as shown in Figure 8.

As reported by Chen and Tsao [27], four deformation mechanisms have been proposed in the semisolid deformation of nondendritic structures. Liquid flow (LF) and flow of liquid incorporating solid particles (FLS) mechanisms are dominant when the solid particles are surrounded by the liquid phase; plastic deformation of solid particles (PDS) and sliding between solid particles (SS) mechanisms become dominant when the solid particles are in contact with each other [28]. Considering the high volume of solid particles and the irregular geometry as shown in Figure 5(a), the PDS mechanism is dominant initially during the heating period of 380°C to 382°C under pressure. As a result, the deformation of the filler metal is formed by plastic deformation of solid grains. As mentioned above, the η-Zn grains have a lower strength at elevated temperature, resulting from their lower melting point. Therefore, the deformation of the filler metal is formed mainly by plastic deformation of the η-Zn solid grains. By comparing with Figure 5(a), the thickness of the filler metal in the thermal-compression microstructure decreases thus with the deformed η-Zn grains as shown in Figures 7(a_1) and 7(a_2). During the heating period of 382°C to 386°C, both of the β-Al grains and deformed η-Zn grains tend to become spherical, resulting from the recrystallization and the effect of interface curvature. In addition, the β-Al particles with the higher melting point have a higher degree of sphericity than the η-Zn grains as shown in Figure 7(b_1), which leads to the SS mechanism to respond. As a result, the intergranular sliding of the spherical β-Al particles occurs. Along with the intergranular sliding, liquid is expelled towards the free surface of the filler metal. That is why tightly connected grains with less liquid appear in the center of the filler metal, while more liquid are present at the edge of the filler metal as shown in Figures 7(b_1) and 7(b_2). Therefore, the deformation of the filler metal is formed mainly by intergranular sliding of β-Al grains, rarely by deformation of the η-Zn grains and fluidity. The thickness of the filler metal decreases thus with sliding grains, deformed grains, and less fluidity as well. During the heating period of 386°C to 392°C, on the one hand, the η-Zn grains become spheroidized grains of the lowest surface energy. On the other hand, the η-Zn grains are liable to break or fused in the neck shrinkage for a lower free energy. So many near-spherical and small sizes of η-Zn grains can be observed in the filler metal as shown in Figures 7(c_1) and 7(c_2). However, the liquid fraction and the roundness of the β-Al solid grains increase with increasing temperature. The results are consistent with the noncompression microstructure as shown in Figure 5(c). Therefore, the FLS mechanism is activated. That is why the liquid incorporating solid grains can be observed at the edge of the filler metal as shown in Figure 7(c_2). Accordingly, the deformation of the filler metal is achieved mainly by the flow of the liquid incorporating solid grains. The thickness of the filler metal hence decreases with sliding grains and increasing fluidity by comparing with Figure 5(c).When the temperature increases to 410°C, the microstructure consists of coarsen β-Al solid grains with liquid pools as shown in Figure 5(d), and the LF mechanism is hence activated. Due to the nature of the LF mechanism, larger fraction of the liquid is expelled towards the edge. This leads to a high degree segregation of the spheroidal β-Al particles in the center and an amount of fine dendritic grains at the edge as shown in Figures 7(d_1) and 7(d_2). The fine dendritic grains are induced by water quenching

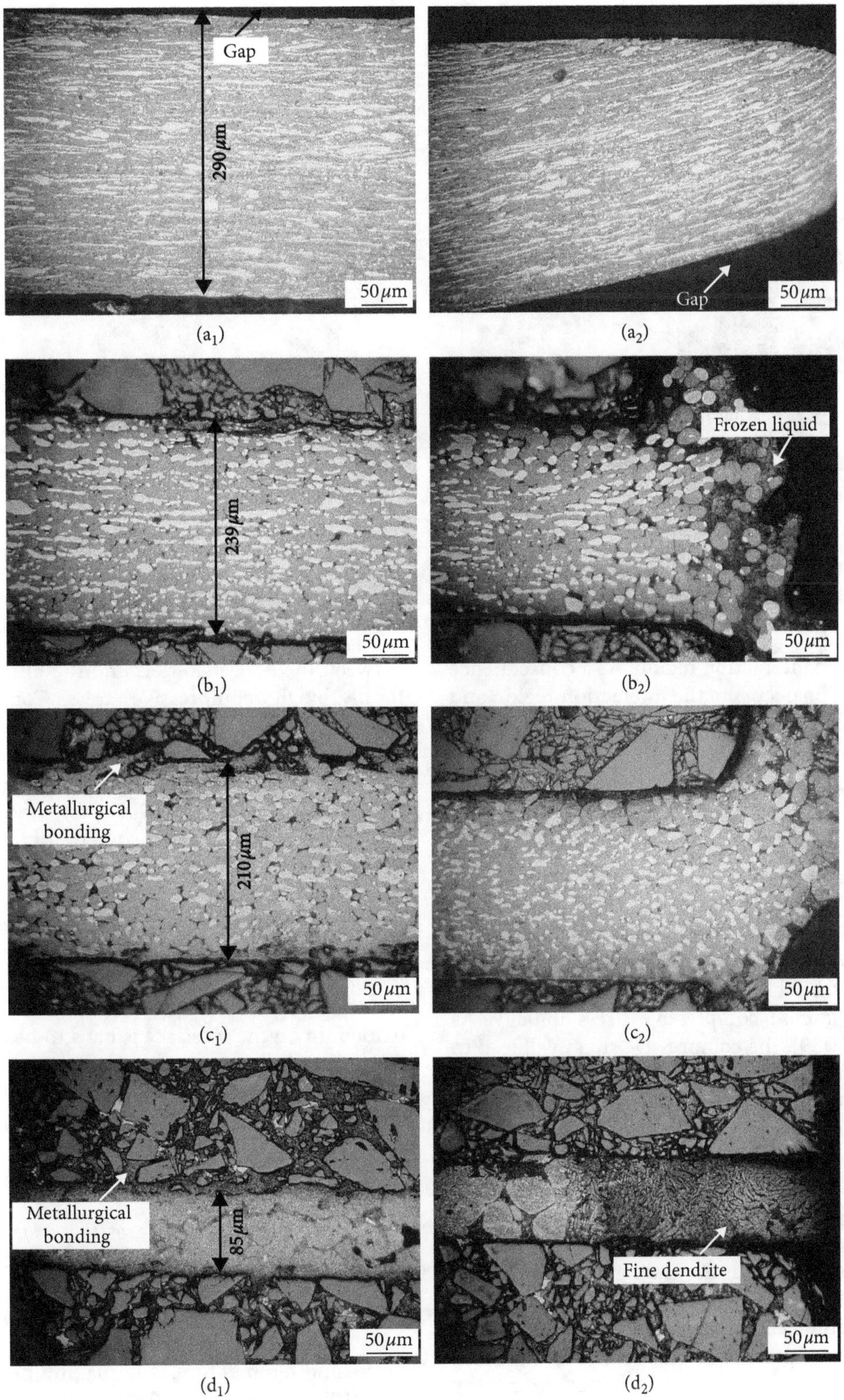

FIGURE 7: The microstructure of the as-rolled filler metal heated to different temperatures with compression between the SiC_p/Al plates: (a_1)–(d_1) center region and (a_2)–(d_2) edge region of the filler metal heated to 382°C, 386°C, 392°C, and 410°C.

of the extrusive liquid. As expected, the deformation of the filler metal is formed by the higher fluidity. The thickness of the filler metal is relatively lower because of the larger fraction of the liquid and higher fluidity.

It is worth highlighting that the increase of the temperature and the disruption of the oxides could aid the diffusion of the Zn-Al-Cu alloy into the base materials. The diffusion will then lower the local melting point and provide

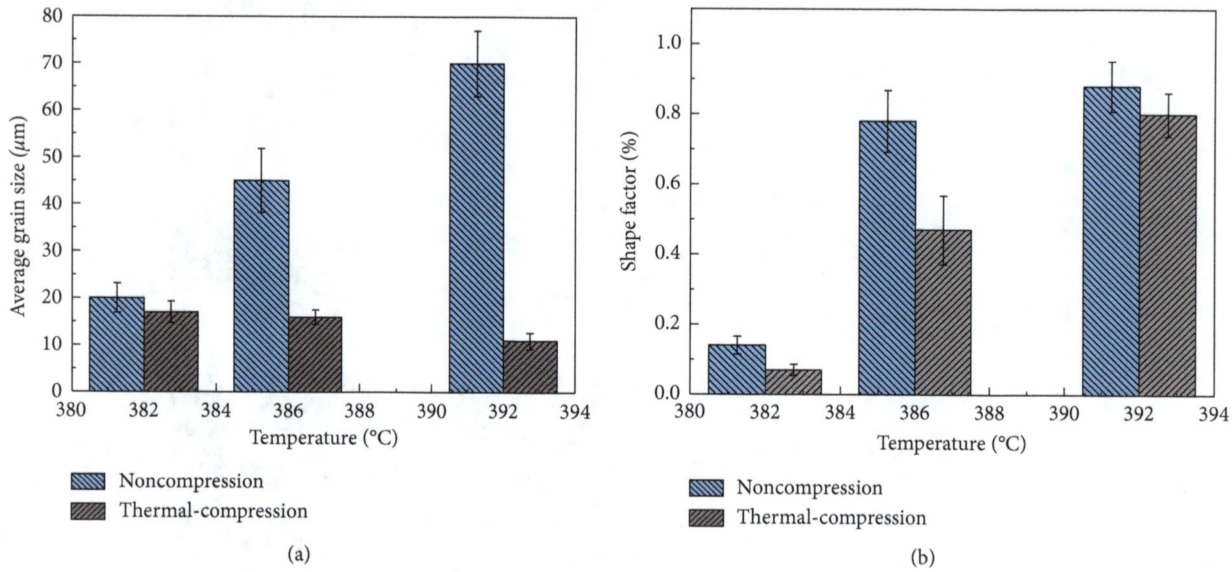

FIGURE 8: The variation of grain size and shape factor of the η-Zn phase with different microstructures.

a partial melting of the interaction region. As a consequence of cooling, solidification occurs in the interaction region and hence, metallurgical bonding forms.

3.3. Interaction of the Filler Metal with Base Materials on the Interface. Considering the deformation behavior of the filler metal, it can be deduced that the behavior of the filler against the base materials are different during compression. As the PDS mechanism is dominant, the solid grains of the filler metal are deformed under a compressive stress, which, in turn, creates a larger compressive resistance to the interface. So the interaction of the solid grains with base materials at the interface shows an intense compressive stress initially. As reported by Shi et al. [29], the compressive stress will lead to a microplastic deformation to the faying surface. Due to the protruding SiC particles on the surface of the composites, the microplastic deformation is higher with the asperities. When the SS mechanism is dominant, in addition to the compressive stress on the interface, a shear action is exerted by sliding of solid grains and less fluidity of the liquid. The interaction on the interface is thus shear stress and compressive stress. As the FLS mechanism starts to set in, the sliding and fluidity becomes more active. So, the interaction on the interface is higher shear stress combined with compressive stress. Finally, LF becomes dominant. According to the results of Chen and Tsao [27], deformation in this stage is achieved mainly by lateral flow of the liquid. Consequently, the interaction on the interface shows a smaller compressive stress and higher shear stress mainly by fluidity and less by sliding.

3.4. Disruption Mechanism of the Oxide Film. According to the above observations and analysis, the disruption of the oxide film during the brazing process could be divided into four stages. As schematically depicted in Figure 9, when heated from 380°C to 382°C (the first stage), the faying face between the base materials and the filler metal is deformed locally by the compressive stress. Consequently, the continuous oxide films on the faying face crack due to their lower ductility. As the temperature is increased from 382°C to 386°C under pressure (the second stage), the crack extends by substantial compressive stress, and meanwhile, oxide films are scraped as a consequence of the shear stress. With further increase of the temperature (the third stage), more oxide films are broken and stripped away from the faying surface by codominant compressive stress and shear stress. Finally (the last stage), the relative sliding and compressive resistance of the solid grains to the interface diminish, resulting from the larger fraction of the liquid. Therefore, the oxide film on the interface is hard to be broken further with the decreasing compressive and shear stress (Figure 9). In this regard, the heating period of 380°C to 392°C which includes the fore three stages can be considered as an optimum stage for the disruption of the oxide film.

3.5. Mechanical Properties of the Brazed Joint. Figure 10 shows the variation of the shear strength of the joint with the brazing temperature. As shown in Figure 10, the shear strength increases with increasing temperature from 382°C to 392°C and then keeps a constant state value with further increasing temperature. The maximum shear strength reaches 105 MPa which is 78% of the shear strength of the parent material. Figure 11 shows the typical fracture surfaces of the joint brazed at 382°C and 392°C, respectively. A grinding trace with no evidence of bonding is observed at 382°C (Figure 11(a)), while a nonplanar fracture is presented at 392°C as shown in Figure 11(b). The results indicate that, with the effective disruption of the oxide film at 392°C, a sound bonding of the Al matrix and Zn-Al-Cu filler metal is obtained and a high improvement in the shear strength of the joint is achieved.

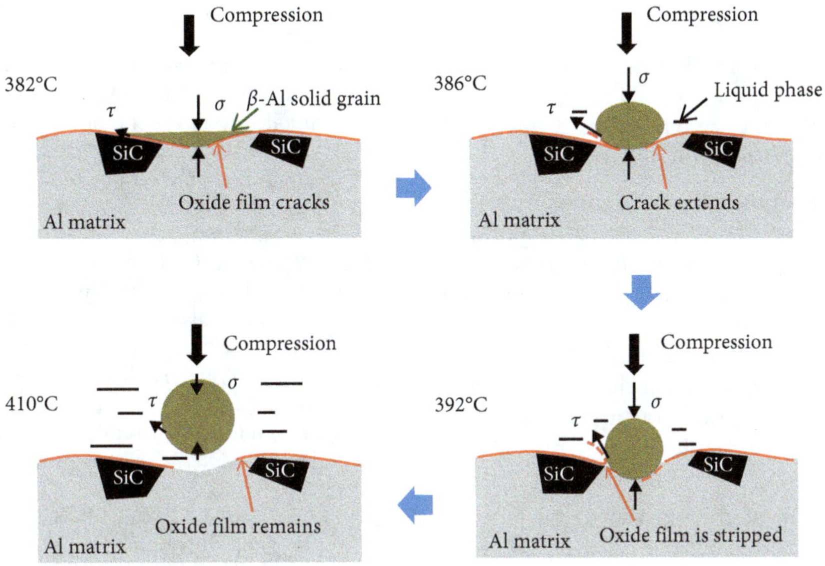

FIGURE 9: Model of the disruption of the surface oxide film during brazing.

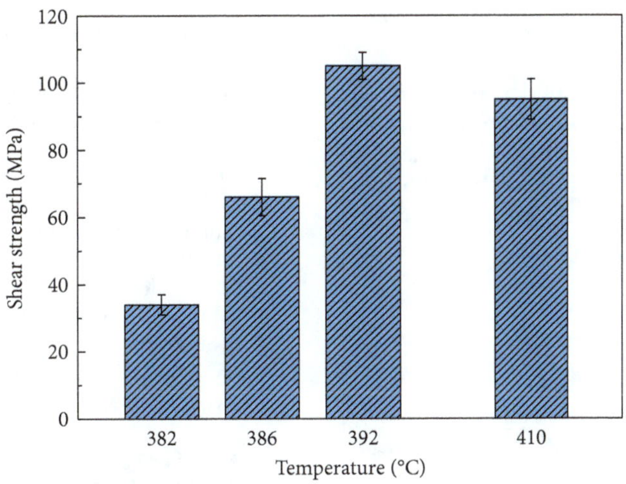

FIGURE 10: Relationship of shear strength of the joint with the brazing temperature.

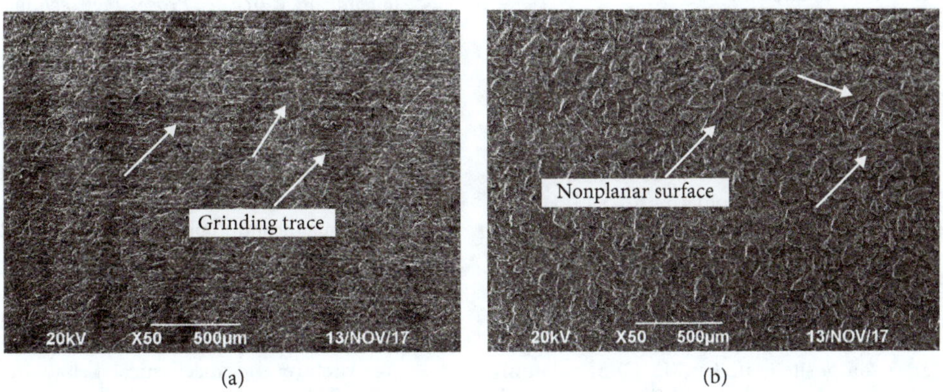

FIGURE 11: The fracture surface of the joint brazed at different temperatures: (a) 382°C; (b) 392°C.

4. Conclusions

Semisolid brazing of 63% SiC_p/6063Al composites under an action of applied pressure using Zn-Al-Cu filler metal was performed successfully. The surface oxide film is disrupted sufficiently, and metallurgical bonding is formed.

(1) The microstructure of the filler metal transforms into the semisolid microstructure consisting of globular grains surrounded by liquid after heating treatment. The degree of sphericity and the liquid fraction tend to improve with increasing temperature. Under the applied pressure, the deformation of the filler metal is accomplished by plastic deformation of solid grains when heated to 382°C and then mainly by intergrain sliding and liquid flow with further heating up to 410°C.

(2) Disruption behavior of the surface oxide layer is related to the deformation behavior of the filler metal and the interaction of solid grains to the base material surface during compression. The oxide films are first broken by compressive stress that came from depressing of the solid grains. The oxide films are then stripped away by shear stress which is caused by intergranular sliding of the solid grains and fluidity of the liquid.

(3) After being heated to 392°C under pressure, the surface oxides film can be disrupted significantly and high strength bonds can be obtained. The maximum shear strength of the brazed joints reached 105 MPa. This achieves 78% of the shear strength of the parent material (135 MPa).

Conflicts of Interest

The authors declare that there are no conflicts of interest regarding the publication of this paper.

Acknowledgments

This project was supported by the National Natural Science Foundation of China (No. 51435004).

References

[1] A. Urena, L. Gil, E. Escriche, J.-M. G. de Salazar, and M.-D. Escalera, "High temperature soldering of SiC particulate aluminum matrix composites (series 2000) using Zn-Al filler alloys," *Science and Technology of Welding and Joining*, vol. 6, no. 1, pp. 1–11, 2001.

[2] X.-P. Zhang, G.-F. Quan, and W. Wei, "Preliminary investigation on joining performance of SiCp-reinforced aluminum metal matrix composite (Al/SiCp-MMC) by vacuum brazing," *Composites Part A: Applied Science and Manufacturing*, vol. 30, no. 6, pp. 823–827, 2009.

[3] Z.-W. Xu, L. Ma, J.-C. Yan, S.-Q. Yang, and S.-Y. Du, "Wetting and oxidation during ultrasonic soldering of an alumina reinforced aluminum-copper-magnesium (2024 Al) matrix composite," *Composites Part A: Applied Science and Manufacturing*, vol. 43, no. 3, pp. 407–414, 2012.

[4] L. Ma, Z.-W. Xu, K. Zheng, J.-C. Yan, and S.-Q. Yang, "Vibration characteristics of aluminum surface subjected to ultrasonic waves and their effect on wetting behavior of solder droplets," *Ultrasonics*, vol. 54, no. 3, pp. 929–937, 2014.

[5] Q. Wang, X.-G. Chen, L. Zhu et al., "Rapid ultrasound-induced transient-liquid-phase bonding of Al-50Si alloys with Zn interlayer in air for electrical packaging application," *Ultrasonics Sonochemistry*, vol. 34, pp. 947–952, 2017.

[6] L. Zhu, Q. Wang, L. Shi et al., "Ultrarapid formation of multi-phase reinforced joints of hypereutectic Al-Si alloys via an ultrasound-induced liquid phase method using Sn-51In interlayer," *Materials Science and Engineering A*, vol. 711, pp. 94–98, 2018.

[7] X.-G. Chen, J.-C. Yan, F. Gao, J.-H. Wei, Z.-W. Xu, and G.-H. Fan, "Interaction behaviors at the interface between liquid Al-Si and solid Ti-6Al-4V in ultrasonic-assisted brazing in air," *Ultrasonics Sonochemistry*, vol. 20, no. 1, pp. 144–154, 2013.

[8] Z. Huiwen, C. Wei, H. Jingshan, Y. Jiuchun, and Y. Shiqin, "Formation and evolution of intermetallic compounds at interfaces of Cu/Al joints by ultrasonic-assisted soldering," *Journal of Materials Processing Technology*, vol. 223, pp. 1–7, 2015.

[9] W.-B. Guo, T.-M. Luan, J.-S. He, and J.-C. Yan, "Ultrasonic-assisted soldering of fine-grained 7034 aluminum alloys using ZnAl filler metals," *Materials and Design*, vol. 125, pp. 85–93, 2017.

[10] T. Watanabe, K. Noguchi, A. Yanagisawa, and S. Satoh, "Soldering of aluminum with the aid of ultrasonic vibration (5th Report). The effect of ultrasonic vibration on soldering Al-Mg alloy," *Quarterly Journal of the Japan Welding Society*, vol. 14, pp. 260–266, 1996.

[11] S. Sugiyama, M. Kiuchi, and J. Yanagimoto, "Application of semisolid joining—Part 4 glass/metal, plastic/metal, or wood/metal joining," *Journal of Materials Processing Technology*, vol. 201, no. 1–3, pp. 623–628, 2008.

[12] Y.-S. Cheng and X.-H. Zhang, "Interfacial strength and structure of joining between 2024 aluminum alloy and SiC_p/2024 Al composite in semi-solid state," *Materials and Design*, vol. 65, pp. 7–11, 2015.

[13] A. Narimannezhad, H. Aashuri, A.-H. Kokabi, and A. Khosravani, "Microstructural evolution and mechanical properties of semisolid stir welded zinc AG40A die cast alloy," *Journal of Materials Processing Technology*, vol. 209, no. 8, pp. 4112–4121, 2009.

[14] V.-A. Hosseini, H. Aashuri, and A.-H. Kokabi, "Mechanisms of joint formation throughout semisolid stir welding of AZ91 magnesium alloy," *Transactions of Nonferrous Metals Society of China*, vol. 23, no. 9, pp. 2585–2590, 2013.

[15] V.-A. Hosseini, H. Aashuri, and A.-H. Kokabi, "Characterization of newly developed semisolid stir welding method for AZ91 magnesium alloy by using Mg-25% Zn interlayer," *Materials Science and Engineering A*, vol. 565, pp. 165–171, 2013.

[16] J.-C. Yan, H.-B. Xu, L. Shi, X.-L. Wang, and S.-Q. Yang, "Vibration assisted brazing of SiCp/A356 composites: microstructure and mechanical behavior," *Science and Technology of Welding and Joining*, vol. 13, no. 8, pp. 760–764, 2008.

[17] H.-B. Xu, Q.-X. Xing, Y.-L. Zeng, Y. Luo, and C.-H. Du, "Semisolid stirring brazing of SiC_p/A356 composites with Zn27Al filler metal in air," *Science and Technology of Welding and Joining*, vol. 16, no. 6, pp. 483–487, 2011.

[18] H.-B. Xu, B.-F. Zhou, C.-H. Du, Q.-X. Luo, and H.-Y. Chen, "Microstructure and properties of joint interface of semisolid stirring brazing of composites," *Journal of Materials Science and Technology*, vol. 28, no. 12, pp. 1163–1168, 2012.

[19] H.-B. Xu, Q.-X. Luo, B.-F. Zhou, Y.-L. Zeng, and C.-H. Du, "The effect of stirring rate on semisolid stirring brazing of SiC_p/A356 composites in air," *Materials and Design*, vol. 34, pp. 452–458, 2012.

[20] S.-M. J. Alvani, H. Aashuri, A. Kokabi, and R. Beygi, "Semisolid joining of aluminum A356 alloy by partial remelting and mechanical stirring," *Transactions of Nonferrous Metals Society of China*, vol. 20, no. 9, pp. 1792–1798, 2010.

[21] T. Savaşkan and A.-P. Hekimoğlu, "Microstructure and mechanical properties of Zn−15Al-based ternary and quaternary alloys," *Materials Science and Engineering A*, vol. 603, pp. 52–57, 2014.

[22] A. Bolouri, M. Shahmiri, and E.-N. H. Cheshmeh, "Microstructural evolution during semisolid state strain induced melt activation process of aluminum 7075 alloy," *Transactions of Nonferrous Metals Society of China*, vol. 20, no. 9, pp. 1663–1671, 2010.

[23] E. Parshizfard and S.-G. Shabestari, "An investigation on the microstructural evolution and mechanical properties of A380 aluminum alloy during SIMA process," *Journal of Alloys and Compounds*, vol. 509, no. 40, pp. 9654–9658, 2011.

[24] L. P. Wang, W.-Y. Jiang, T. Chen et al., "Spheroidal microstructure formation and thixoforming of AM60B magnesium alloy prepared by SIMA process," *Transactions of Nonferrous Metals Society of China*, vol. 22, pp. 435–444, 2012.

[25] B. Behzad and A.-K. Mehrdad, "Phase evolution and mechanical behavior of the semi-solid SIMA processed 7075 aluminum alloy," *Metals*, vol. 6, no. 3, pp. 1–23, 2016.

[26] T.-A. Witten and L.-M. Sander, "Diffusion-limited aggregation, a kinetic critical phenomenon," *Physical Review Letter*, vol. 47, no. 19, pp. 1400–1408, 1981.

[27] C.-P. Chen and C.-Y. A. Tsao, "Semi-solid deformation of non-dendritic structures-I. Phenomenological behavior," *Acta Metallurgica*, vol. 45, no. 5, pp. 1955–1968, 1997.

[28] C.-P. Chen and C.-Y. A. Tsao, "Semisolid deformation of aluminium alloy 356 with non-dendritic structure," *Materials Science and Technology*, vol. 15, no. 9, pp. 981–985, 1999.

[29] L. Shi, J.-C. Yan, Y.-F. Han, and B. Peng, "Behaviors of oxide layer at interface between semi-solid filler metal and aluminum matrix composites during vibration," *Journal of Materials Science and Technology*, vol. 27, no. 8, pp. 746–752, 2011.

12

Finite Element Analysis of Bend Test of Sandwich Structures Using Strain Energy Based Homogenization Method

Hassan Ijaz,[1] Waqas Saleem,[1] Muhammad Zain-ul-Abdein,[1] Tarek Mabrouki,[2] Saeed Rubaiee,[3] and Abdullah Salmeen Bin Mahfouz[4]

[1]*Mechanical Engineering Department, University of Jeddah, Jeddah, Saudi Arabia*
[2]*Mechanical Engineering Department, University of Tunis El Manar, ENIT, Tunis, Tunisia*
[3]*Industrial Engineering Department, University of Jeddah, Jeddah, Saudi Arabia*
[4]*Chemical Engineering Department, University of Jeddah, Jeddah, Saudi Arabia*

Correspondence should be addressed to Hassan Ijaz; hassan605@yahoo.com

Academic Editor: Frederic Dumur

The purpose of this article is to present a simplified methodology for analysis of sandwich structures using the homogenization method. This methodology is based upon the strain energy criterion. Normally, sandwich structures are composed of hexagonal core and face sheets and a complete and complex hexagonal core is modeled for finite element (FE) structural analysis. In the present work, the hexagonal core is replaced by a simple equivalent volume for FE analysis. The properties of an equivalent volume were calculated by taking a single representative cell for the entire core structure and the analysis was performed to determine the effective elastic orthotropic modulus of the equivalent volume. Since each elemental cell of the hexagonal core repeats itself within the in-plane direction, periodic boundary conditions were applied to the single cell to obtain the more realistic values of effective modulus. A sandwich beam was then modeled using determined effective properties. 3D FE analysis of Three- and Four-Point Bend Tests (3PBT and 4PBT) for sandwich structures having an equivalent polypropylene honeycomb core and Glass Fiber Reinforced Plastic (GFRP) composite face sheets are performed in the present study. The authenticity of the proposed methodology has been verified by comparing the simulation results with the experimental bend test results on hexagonal core sandwich beams.

1. Introduction

Applications of composite sandwich structures are continuously increasing in the recent times due to their excellent out-of-plane shear and compressive properties. Modern manufacturing aerospace, ship building, and automotive industries are the main users of sandwich structures. A typical sandwich structure consists of a central core material covered by top and bottom face sheets. Generally honeycomb, truss, and foam are used as the inner cores in sandwich structures. Nomex, Polypropylene, and Aluminium are regularly used core materials, whereas, Glass Fiber Reinforced Plastic (GFRP), Carbon Fiber Reinforced Plastic (CFRP), and Aluminium are commonly used face sheet materials.

Composite sandwiches are used as substantial structures in cars, ships, beams, and so forth. Complex and large hexagonal honeycomb core shapes are not only difficult to model but also are computationally expensive. Due to these limitations, an alternative strategy is desired in which a complex shaped core material may be replaced by a simple equivalent volume having elastic orthotropic properties.

Many authors attempted to determine the equivalent properties of central core by separately modeling the honeycomb core sheet with hexagonal unit cells [1–6]. Each cell of the honeycomb sheet periodically repeats itself within in-plane direction. The required equivalent core properties are the in-plane (E_{11}, E_{22}) and out-of-plane (E_{33}) elastic moduli, in-plane (G_{12}) and out-of-plane (G_{13}, G_{23}) shear moduli, and the Poisson ratios (υ_{12}, υ_{13}, and υ_{23}). The homogenization method is based on the mechanical equivalence between the microstructures of RVE (representative volume element) and a similar homogeneous macroscopic volume.

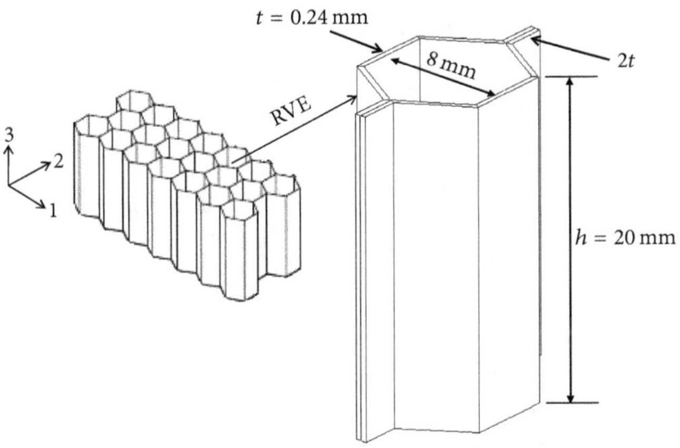

FIGURE 1: Representative volume element (RVE).

Different authors like Kelsey et al. [7], Gibson and Ashby [8], Zhang and Ashby [9], Klintworth and Stronge [10], Hohe and Becker [11], and Yang and Becker [12] developed the analytical schemes to estimate the equivalent properties of cellular core. Normally analytical theories are shape dependent and also not able to predict all the 9 engineering constants of the elastic orthotropic properties. In recent years, finite element analysis (FEA) has become a popular tool to determine the equivalent elastic properties of any cellular structure of periodic nature [13–17].

In the present study only a single hexagonal element with periodic boundary conditions has been modeled to determine the equivalent properties. This approach to determine the equivalent properties is based on the work of Gornet et al. [13, 15]. The required orthotropic properties in terms of 9 engineering constants are determined using the strain energy based homogenization method. Subsequently, a 3D sandwich structure with face sheets and simple inner core volume was modeled along with the calculated equivalent properties and analyzed for Three-Point Bend Test (3PBT) and Four-Point Bend Test (4PBT). For the 3PBT and 4PBT finite element (FE) simulations, a complex hexagonal honeycomb is replaced by the equivalent 3D core material. Finally, the results were compared with the experimental data.

This article is organized as follows: representative volume element (RVE) for honeycomb core material is defined and described in Section 2. Mathematical formulation is explained in Section 3. FEA results and its comparison with experimental results are discussed in Section 4. Finally, some concluding remarks are made in Section 5.

2. Representative Volume Element (RVE)

Figure 1 shows a representative volume element (RVE) for overexpanded hexagonal honeycomb material along with necessary orthotropic basis, that is, axes 1, 2, and 3. Now by successive in-plane translation of RVEs, one can obtain a larger volume of the honeycomb cellular structure. RVE for honeycomb core material and corresponding equivalent homogenized volume is shown in Figure 2. Once the orthotropic properties of homogenized volume are known, one can model the large and complex sandwich structures with simple homogenized volume for different FE analyses.

3. Formulation

By considering Y to be the RVE and $|Y|$ its volume measurement, the average of the second-order tensor X_{IJ} may be represented by the following relation:

$$\langle X_{IJ} \rangle = \frac{1}{|Y|} \int_Y X_{IJ}(Y)\, dY. \tag{1}$$

One can express the average stress and strain tensors as $\Sigma_{IJ} = \langle \sigma_{IJ} \rangle$ and $E_{IJ} = \langle \varepsilon_{IJ} \rangle$, respectively. Periodic homogenization is based upon the data average over the strain field and then one determines the compliance operator C^{RVE} of the RVE by imposing displacement boundary conditions [13]. Here, Hill-Mandel [18, 19] energy approach is used to calculate the equivalent homogenized components of 6×6 compliance matrix C_{IJ}^{RVE}. Now homogenized stress and strain tensors are related to each other with the help of the following relation:

$$\begin{aligned} \Sigma &= C^{\text{RVE}} E \\ E &= S^{\text{RVE}} \Sigma. \end{aligned} \tag{2}$$

Here S^{RVE} represents the 6×6 stiffness matrix. The homogenized mechanical stiffness characteristics (C^{RVE}) of the RVE can be derived with the Hill-Mandel theorem [18, 19]. The Hill-Mandel theorem equates the RVE energy to the average of the energies of its components:

$$\varepsilon_I C_{IJ}^{\text{RVE}} \varepsilon_J = \frac{1}{|Y|^{\text{RVE}}} \int_Y \text{Tr}[\sigma\varepsilon]\, dY. \tag{3}$$

The 21 components of the orthotropic compliance matrix C_{IJ}^{RVE} can be determined by applying 21 linear combinations of applied loading on RVE and the resultant set of the corresponding equations may be expressed as follows [13, 15]:

$$\begin{aligned}
&\varepsilon_I C_{II}^{\text{RVE}} \varepsilon_I = \frac{1}{|Y|^{\text{RVE}}} \int_Y \text{Tr}[\sigma\varepsilon]\, dY \quad I \in [1,3] \\
&\sqrt{2}\varepsilon_I C_{II}^{\text{RVE}} \sqrt{2}\varepsilon_I = \frac{1}{|Y|^{\text{RVE}}} \int_Y \text{Tr}[\sigma\varepsilon]\, dY \quad I \in [4,6] \\
&\varepsilon_I C_{II}^{\text{RVE}} \varepsilon_I + 2\varepsilon_I C_{IJ}^{\text{RVE}} \varepsilon_J + \varepsilon_J C_{JJ}^{\text{RVE}} \varepsilon_J \\
&\quad = \frac{1}{|Y|^{\text{RVE}}} \int_Y \text{Tr}[\sigma\varepsilon]\, dY \\
&\qquad (I,J) \in [1,3]^2 \text{ with } (I \neq J) \\
&\varepsilon_I C_{II}^{\text{RVE}} \varepsilon_I + 2\varepsilon_I C_{IJ}^{\text{RVE}} \sqrt{2}\varepsilon_J + \sqrt{2}\varepsilon_J C_{JJ}^{\text{RVE}} \sqrt{2}\varepsilon_J \\
&\quad = \frac{1}{|Y|^{\text{RVE}}} \int_Y \text{Tr}[\sigma\varepsilon]\, dY \quad (I,J) \in [1,3] \times [4,6] \\
&\sqrt{2}\varepsilon_I C_{II}^{\text{RVE}} \sqrt{2}\varepsilon_I + 2\sqrt{2}\varepsilon_I C_{IJ}^{\text{RVE}} \sqrt{2}\varepsilon_J + \sqrt{2}\varepsilon_J C_{JJ}^{\text{RVE}} \sqrt{2}\varepsilon_J \\
&\quad = \frac{1}{|Y|^{\text{RVE}}} \int_Y \text{Tr}[\sigma\varepsilon]\, dY \\
&\qquad \text{with } (I,J) \in [4,6]^2 \text{ with } (I \neq J).
\end{aligned} \tag{4}$$

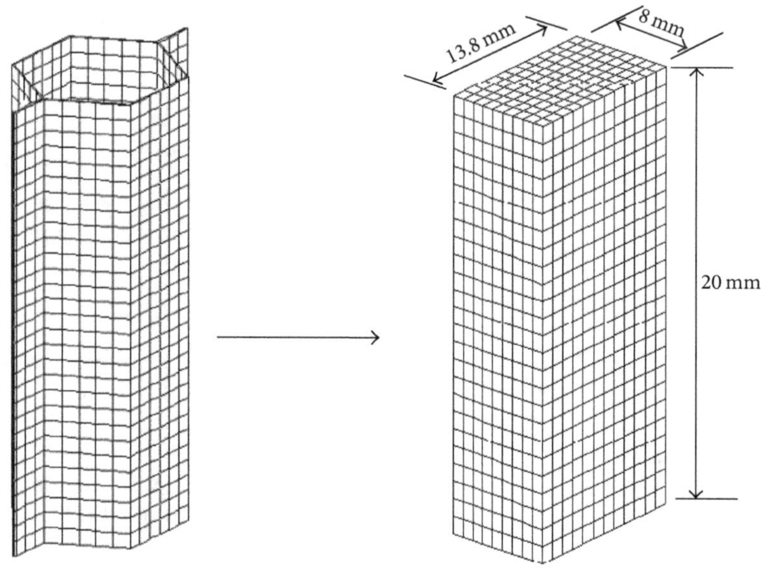

FIGURE 2: Honeycomb core RVE.

Equations (4) are used to find the components of compliance matrix C_{IJ}^{RVE} and components of the stiffness matrix S_{IJ}^{RVE} can be determined by using the relation $S_{IJ}^{\mathrm{RVE}} = [C_{IJ}^{\mathrm{RVE}}]^{-1}$. The selected RVE is loaded with the imposed strain for different loading conditions corresponding to different coefficients of the compliance matrix. The resultant strain energy of the RVE can be calculated in the postprocessing after the FEA is completed. The calculated strain energy is related to the coefficients of the compliance matrix (C_{IJ}) by the Hill-Mandel theorem, (3). The coefficients of the compliance matrix (C_{IJ}) corresponding to different loading conditions are represented by (4). The orthotropic material parameters (E_{11}, E_{22}, E_{33}, G_{12}, G_{13}, G_{23}, υ_{12}, υ_{13}, and υ_{23}) can be determined from the stiffness matrix S_{IJ}^{RVE}. See the Appendix for detailed mathematical relations.

4. 3D FEA and Results

In this section, FE analysis and results are presented and discussed in detail. Two different FE models were developed and analyzed. First, FEA of a single cell of hexagonal honeycomb representing the RVE is performed and the equivalent orthotropic properties are determined using the methodology explained in previous section. Then a sandwich beam is modeled and FEA (3PBT and 4PBT) is performed using the determined equivalent properties.

4.1. Determination of Equivalent Properties. FEA to determine the equivalent properties was done using the commercially available software CAST3M [20]. This code was developed by the "Département Mécanique et Technologie (DMT) du Commissariat français à l'Energie Atomique (CEA), France." The purpose of the development of CAST3M is to provide high level support for the design and analysis of structures and components in the research field. In this context, CAST3M not only incorporates the process of calculations but also serves as preprocessor for the modeling of structural parts and analysis of the final results (posttreatment).

The honeycomb core is modeled with three-dimensional brick elements having 20 nodes (CU20) [20]. The core material is polypropylene with a density of 80.0 Kg/m^3 and a sheet thickness of 0.24 mm. The unit cell is 8.0 mm wide and 20 mm high (see Figure 1).

As adopted by Gornet et al. [13, 15], FEA was done for different loading conditions to determine the orthotropic coefficients of the compliance matrix. The principal orthotropic directions are shown in Figure 1. The boundary conditions for five different loadings are shown in Figure 3. Von Mises stress profiles of 20 mm thick core for three tensile loading conditions are shown in Figure 4. The determined orthotropic properties and their comparison with analytical results as proposed by Gibson and Ashby [8] and Chamis et al. [6] are given in Table 1.

Honeycomb cores are usually designed for out-of-plane normal (*E*33) and shear (*G*13, *G*23) loading conditions. Gibson and Ashby [8] presented the solution for orthotropic properties of honeycomb cores; however, they did not consider the effects of double wall thickness and periodic nature of core media. Due to this reason, the values for *E*11 and *E*22 are not only similar to each other but also lower than those depicted by FE results. Likewise, the values of *G*13 and *G*23 are also identical for the same reasons. The out-of-plane modulus predicted by the FEA is in close approximation to the Gibson and Ashby results. Still the value of *E*33 is towards the lower side for the Ashby and Gibson due to constant wall approximation. Chamis et al. [6] also proposed approximated formulas to calculate the equivalent orthotropic properties of honeycomb structures.

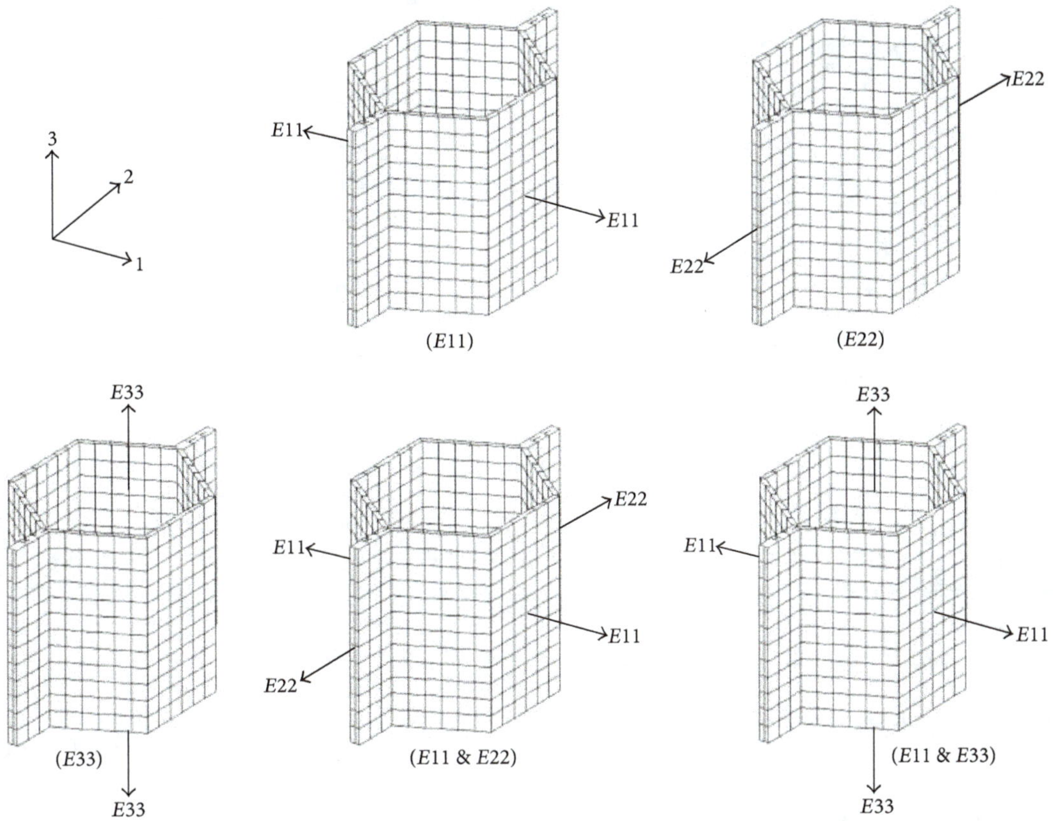

FIGURE 3: Boundary conditions for five different loadings.

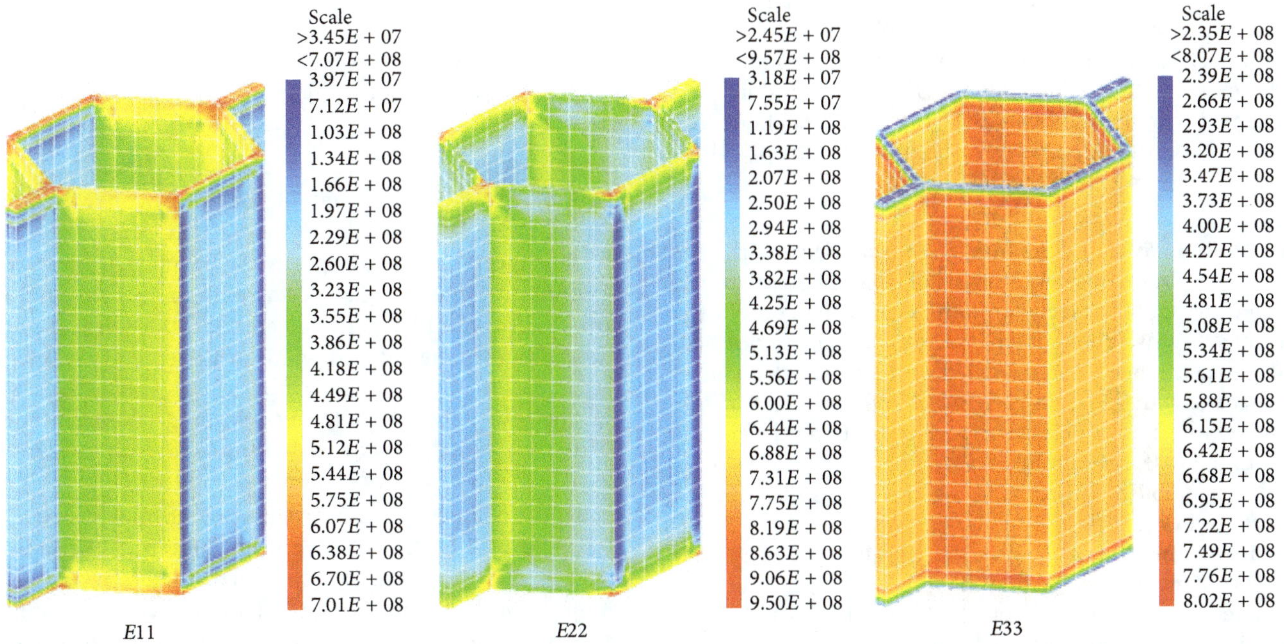

FIGURE 4: Von Mises stress profile for different loading conditions on hexagonal core.

TABLE 1: Comparison of FEA results with published [6, 8] data.

Property	3D FEA results	Gibson and Ashby [8]	Chamis et al. [6]
Unit cell	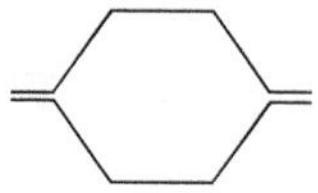	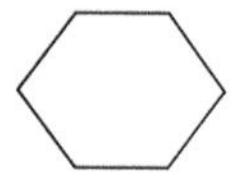	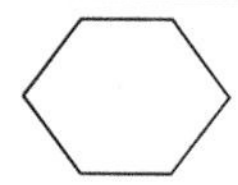
E_{11} (N/m^2)	5.8E6	4.2E5	5.85E7
E_{22} (N/m^2)	6.9E6	4.2E5	3.9E7
E_{33} (N/m^2)	10.2E7	7.8E7	11.7E7
G_{12} (N/m^2)	1.97E6	1.05E5	1.03E7
G_{13} (N/m^2)	1.44E7	1.37E7	1.37E7
G_{23} (N/m^2)	2.2E7	1.37E7	2.06E7
υ_{12}	0.81	0.99	0.42
υ_{13}	0.024	0.031	0.42
υ_{23}	0.028	0.0054	0.42

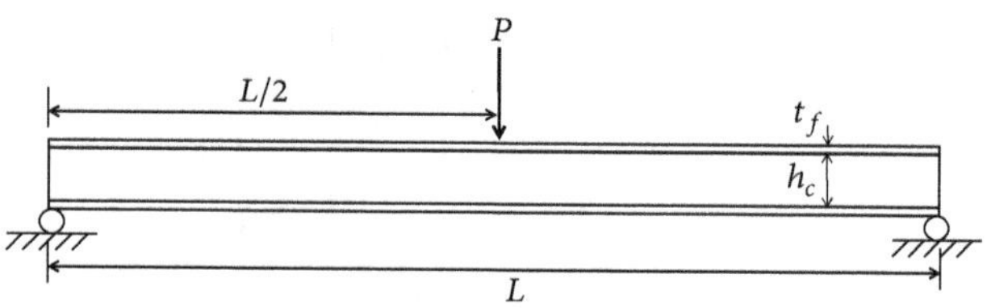

FIGURE 5: Three-point bend test (3PBT) specimen.

The formulations proposed by Chamis et al. overestimated the calculated values. Readers may consult the work of Gibson and Ashby and Chamis et al. for more details [6, 8].

In the present work, only one honeycomb core cell with actual double wall thickness is modeled and results are obtained by applying periodic displacement boundary conditions. The predicted out-of-plane elastic modulus $E33$ and the two shear moduli $G13$ and $G23$ were found in good agreement with the values obtained by Chamis et al. [6]. The identified equivalent properties were used for the 3D 4PBT and 3PBT FE analyses in the next section.

4.2. 3D FE Analysis of Three- and Four-Point Bend Tests. The sandwich structure usually consists of central core materials covered by the face sheets on both sides of the core. In the present study, two different types of analyses were performed, where the face sheets were modeled with the shell elements in the first and with the 3D solid elements in the second analysis. However, the core material is modeled by 3D solid elements in both the cases. The authenticity of the performed analysis has been established by comparing the FE results with the reported work [2, 21].

The geometry and boundary conditions for 3PBT and 4PBT specimens are shown in Figures 5 and 6, respectively. Considering the symmetry of 4PBT, only half of the specimen is modeled and analysis is performed by employing symmetric boundary conditions. The specimen length is L = 300 mm, the face sheet thickness is t_f = 1 mm, and the width is b = 35 mm for all the specimens. A core height (hc) of 20 mm is used for both 3PBT and 4PBT analyses. The 3D FE analyses are performed using the commercial software Abaqus/Standard. C3D8R solid elements are used for core material modeling. S4R shell and C3D8R solid elements are used for face sheet material modeling.

Figures 7 and 8 present the FE results of force versus deflection curve for 3PBT and 4PBT, respectively. Note that changing the type of elements from solid to shell did not reflect any noticeable difference in the FE results. Nevertheless, the simulation results are in close agreement with the experimental ones.

The above simulation results on 3PBT and 4PBT indicate that the adopted strategy to model the hexagonal honeycomb material by equivalent orthotropic material gives good results. Hence, instead of modeling a large number of hexagonal cells, one can determine the equivalent properties by modeling a single hexagonal cell as an RVE only.

Figures 9 and 10 present the Von Mises and normal stress (σ_{33}) profile on the 3PBT specimen for both core and face sheet materials, respectively.

Similarly Figures 11 and 12 show the Von Mises and normal stress (σ_{33}) profile on the 4PBT specimen for core and face sheet materials, respectively. The normal stress (σ_{33}) distribution/trend along a specific path of core material for both 3PBT and 4PBT is discussed here. The selected nodal paths along the height and width of the core are shown in Figure 13 for 3PBT and in Figure 14 for 4PBT. The stress distribution is studied at the nodes along width direction, where the loading is imposed and the same path is extended along the height for the creation of second set of nodes.

Figure 15 presents the variation of normal stress (σ_{33}) along the core height, starting from the top to the bottom node for 3PBT (Figure 13). The normal stress produced is compressive in nature because of the direction of applied load at the selected nodes. The top node stress value is observed maximum and it decreases gradually along the path towards the bottom node. Figure 16 shows the variation of normal compressive stress along the width for 3PBT. The stress value increased as it moves from corner node and then

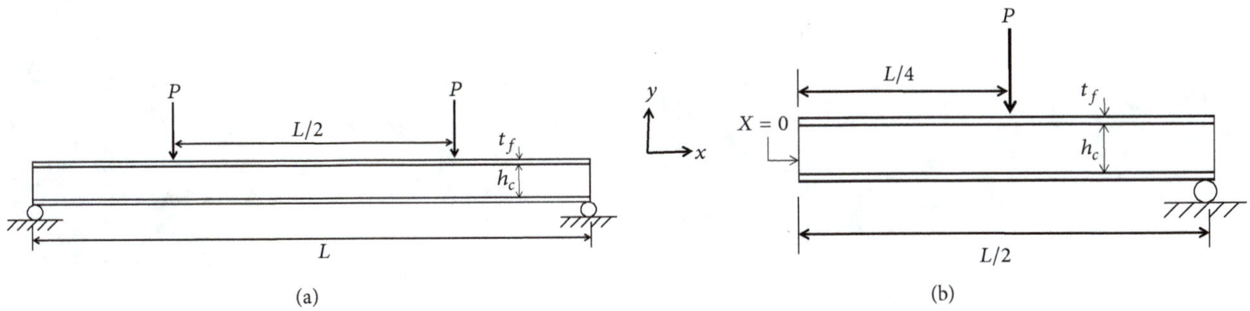

FIGURE 6: Four-point bend test (4PBT) specimen: (a) full length; (b) symmetric.

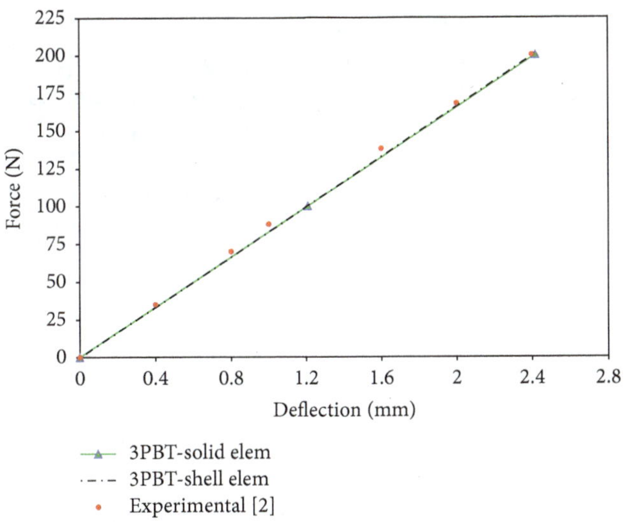

FIGURE 7: Force versus deflection curve (3PBT).

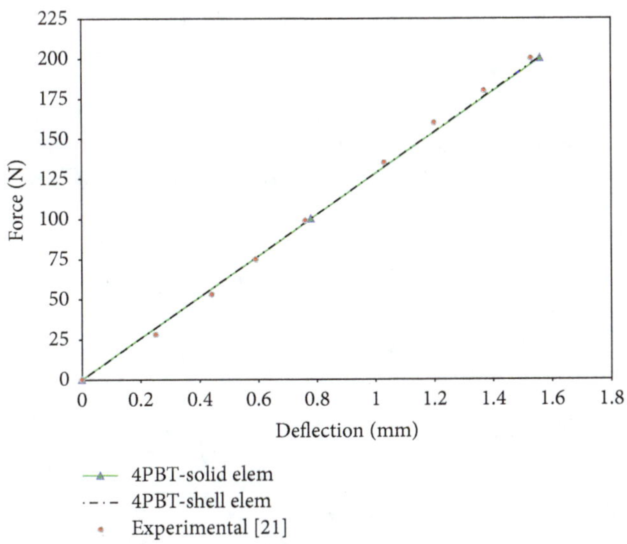

FIGURE 8: Force versus deflection curve (4PBT).

tends to stabilize as it reaches the middle node; after reaching the middle node the exact symmetry is observed for the remaining half width.

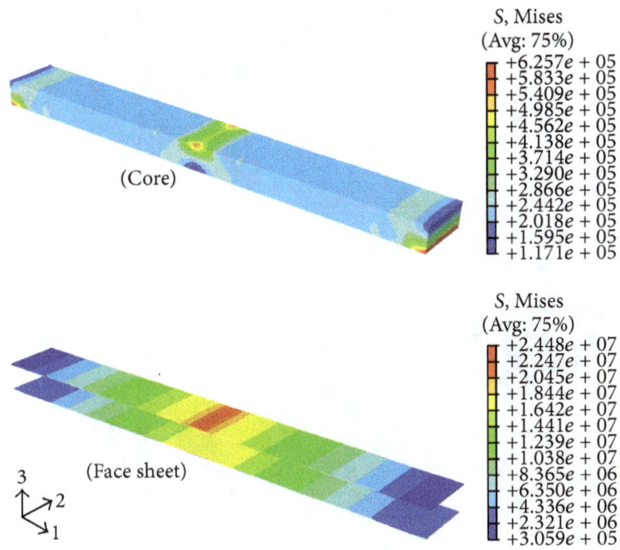

FIGURE 9: Von Mises stress profile (3PBT).

Similarly, Figures 17 and 18 present the normal compressive stress variation along the selected nodal paths for height and width directions for 4PBT. Although the stress magnitudes observed in the 4PBT are lesser as compared to those of 3PBT, the trends are quite similar for both the cases. The lower values of normal stress for 4PBT can be attributed to the small bending arm's length for the given loading in comparison to 3PBT; see Figures 5 and 6.

5. Conclusion

In this article, the strain energy based homogenization methodology is explained to determine the equivalent orthotropic properties of honeycomb sandwich structures. The obtained properties are further successfully employed to simulate the 3PBT and 4PBT. A 3D representative volume element (RVE) of one cell of honeycomb is selected to apply the homogenization technique rather than selecting a continuous structure having multiple cells, hence considerably reducing the computational time and effort. In the first part, homogenization is performed for honeycomb core material and determined orthotropic properties are compared with the properties based on available theories of different authors.

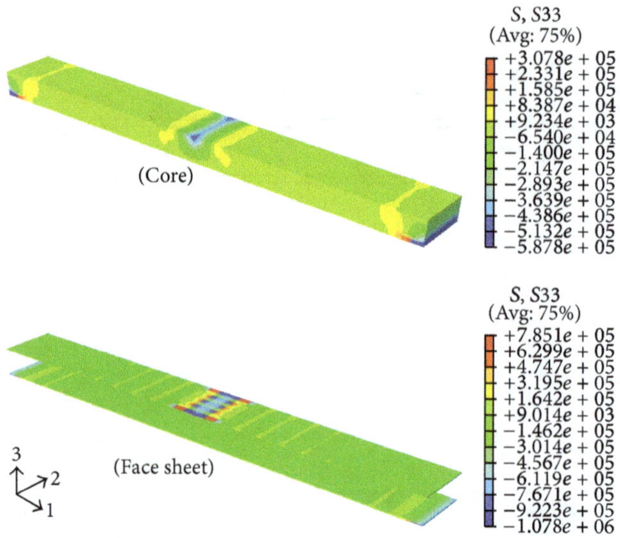

FIGURE 10: Normal stress (σ_{33}) profile (3PBT).

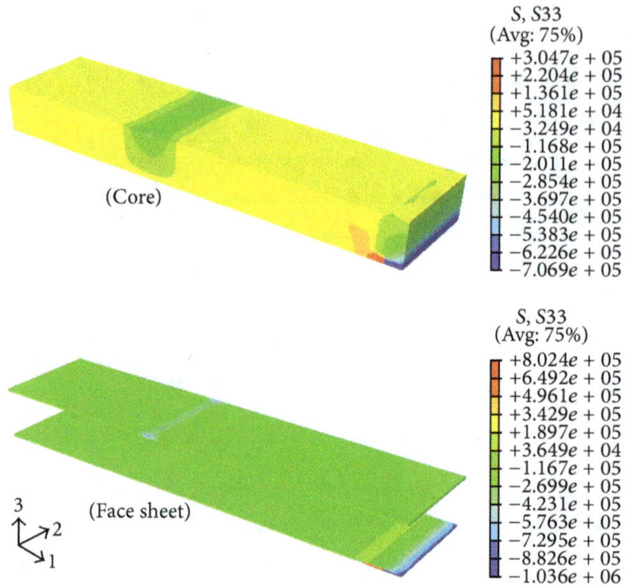

FIGURE 12: Normal stress (σ_{33}) profile (4PBT).

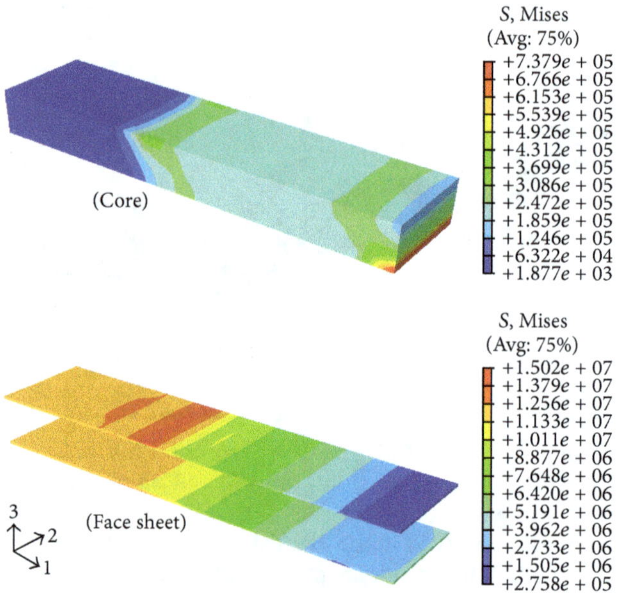

FIGURE 11: Von Mises stress profile (4PBT).

In the second part, 3D FEA of 3PBT and 4PBT are performed based on the determined equivalent orthotropic properties. The bend test results for both analyses are compared with experimental data and are found in good agreement.

Appendix

The constitutive equations representing the 3D stress-strain relation in terms of 6×6 compliance matrix, C_{IJ} (Herakovich, 1997) [17], are

$$\sigma_1 = C_{IJ}\varepsilon_J \tag{A.1}$$

$$\begin{bmatrix}\sigma_{11}\\ \sigma_{22}\\ \sigma_{33}\\ \sqrt{2}\sigma_{23}\\ \sqrt{2}\sigma_{13}\\ \sqrt{2}\sigma_{12}\end{bmatrix} = \begin{bmatrix} C_{11} & C_{12} & C_{13} & C_{14} & C_{15} & C_{16}\\ & C_{22} & C_{23} & C_{24} & C_{25} & C_{26}\\ & & C_{33} & C_{34} & C_{35} & C_{36}\\ & & & C_{44} & C_{45} & C_{46}\\ & \text{symmetry} & & & C_{55} & C_{56}\\ & & & & & C_{66}\end{bmatrix} \begin{bmatrix}\varepsilon_{11}\\ \varepsilon_{22}\\ \varepsilon_{33}\\ \sqrt{2}\varepsilon_{23}\\ \sqrt{2}\varepsilon_{13}\\ \sqrt{2}\varepsilon_{12}\end{bmatrix} \tag{A.2}$$

The inverse relationship in terms of 6×6 stiffness matrix, S_{IJ}, may be expressed as follows (Herakovich, 1997) [17]:

$$\varepsilon_1 = S_{IJ}\sigma_J$$

$$\begin{bmatrix}\varepsilon_{11}\\ \varepsilon_{22}\\ \varepsilon_{33}\\ \sqrt{2}\varepsilon_{23}\\ \sqrt{2}\varepsilon_{13}\\ \sqrt{2}\varepsilon_{12}\end{bmatrix} = \begin{bmatrix} S_{11} & S_{12} & S_{13} & S_{14} & S_{15} & S_{16}\\ & S_{22} & S_{23} & S_{24} & S_{25} & S_{26}\\ & & S_{33} & S_{34} & S_{35} & S_{36}\\ & & & S_{44} & S_{45} & S_{46}\\ & \text{symmetry} & & & S_{55} & S_{56}\\ & & & & & S_{66}\end{bmatrix} \begin{bmatrix}\sigma_{11}\\ \sigma_{22}\\ \sigma_{33}\\ \sqrt{2}\sigma_{23}\\ \sqrt{2}\sigma_{13}\\ \sqrt{2}\sigma_{12}\end{bmatrix}, \tag{A.3}$$

where $S_{IJ} = C_{IJ}^{-1}$.

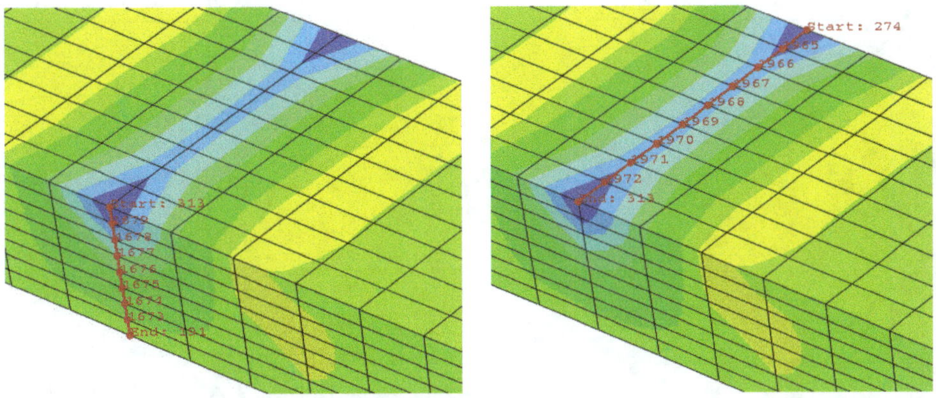

(a) (b)

FIGURE 13: Selected nodal path direction for 3PBT: (a) height; (b) width.

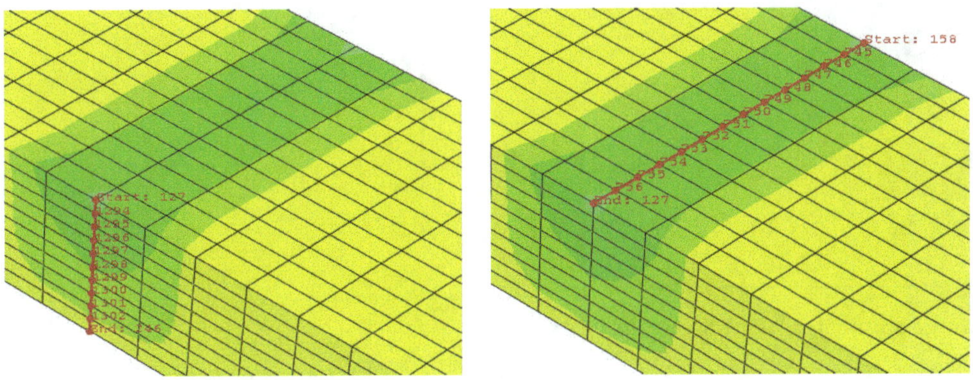

(a) (b)

FIGURE 14: Selected nodal path direction for 4PBT: (a) height; (b) width.

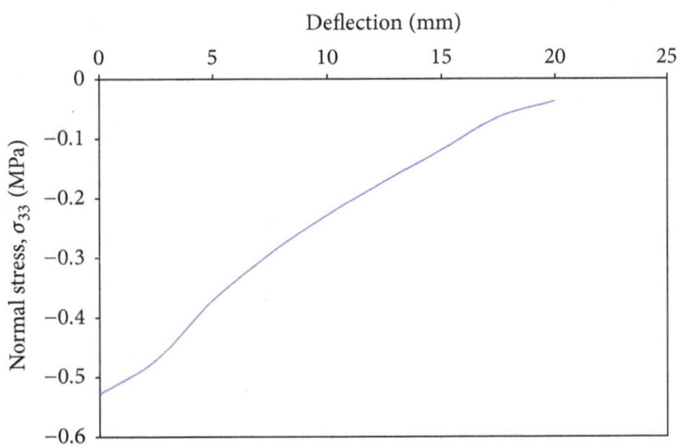

FIGURE 15: Normal stress distribution along the height direction (3PBT).

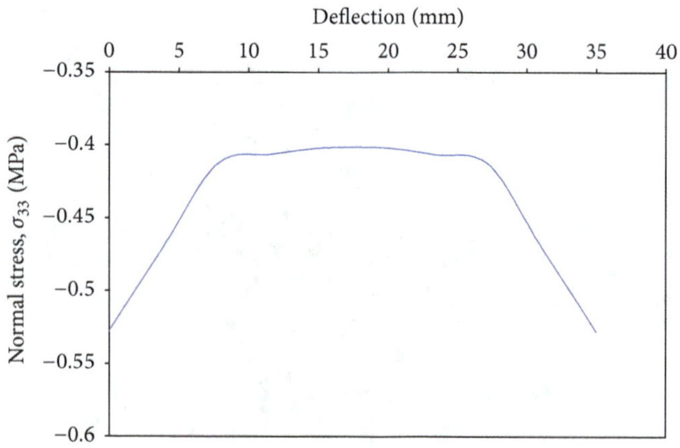

FIGURE 16: Normal stress distribution along the width direction (3PBT).

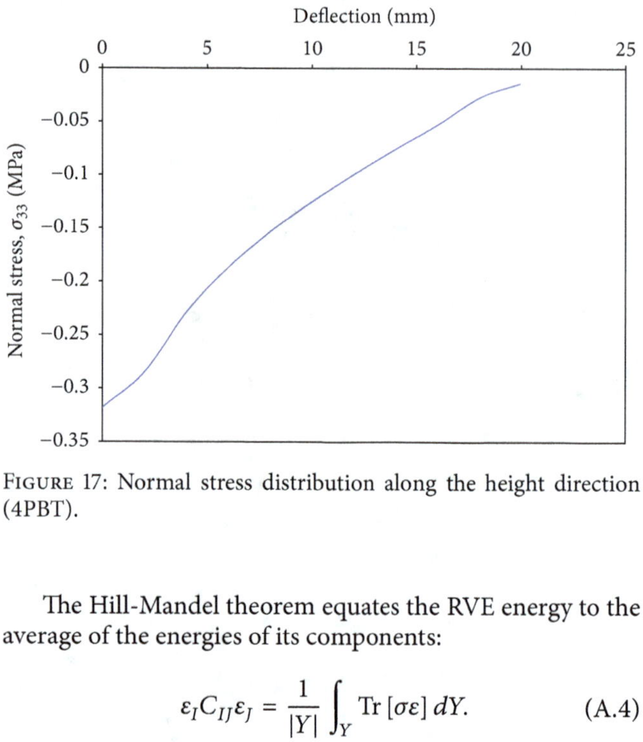

FIGURE 17: Normal stress distribution along the height direction (4PBT).

The Hill-Mandel theorem equates the RVE energy to the average of the energies of its components:

$$\varepsilon_I C_{IJ} \varepsilon_J = \frac{1}{|Y|} \int_Y \mathrm{Tr}\,[\sigma\varepsilon]\, dY. \tag{A.4}$$

Components of the 6×6 compliance matrix, C_{IJ}, are given in (A.2). The coefficients of the compliance matrix are calculated using (4) for different imposed strain loading conditions. The components of the stiffness matrix may be calculated using relation $S_{IJ} = C_{IJ}^{-1}$. Stiffness matrix for the orthotropic material is represented by (A.5) where some coefficients become zero due to symmetry of three mutually orthogonal planes (Herakovich, 1997) [17].

$$\begin{bmatrix} S_{11} & S_{12} & S_{13} & 0 & 0 & 0 \\ S_{12} & S_{22} & S_{23} & 0 & 0 & 0 \\ S_{13} & S_{23} & S_{33} & 0 & 0 & 0 \\ 0 & 0 & 0 & S_{44} & 0 & 0 \\ 0 & 0 & 0 & 0 & S_{55} & 0 \\ 0 & 0 & 0 & 0 & 0 & S_{66} \end{bmatrix}. \tag{A.5}$$

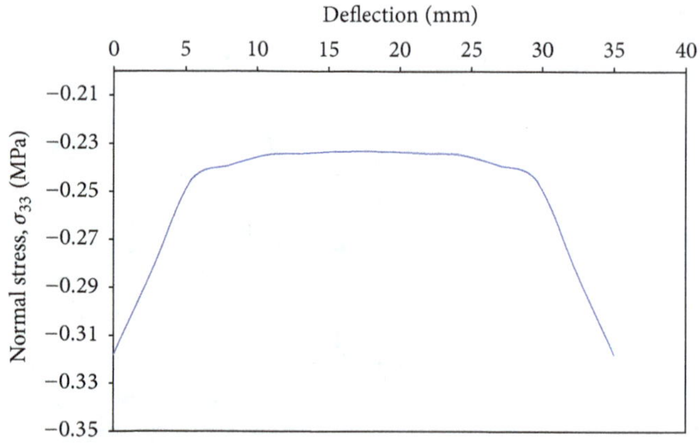

FIGURE 18: Normal stress distribution along the width direction (4PBT).

The orthotropic material parameters (E_{11}, E_{22}, E_{33}, G_{12}, G_{13}, G_{23}, υ_{12}, υ_{13}, and υ_{23}) can be determined from stiffness matrix S_{IJ}; see (A.6) (Herakovich, 1997) [17].

$$\begin{bmatrix} S_{11} & S_{12} & S_{13} & 0 & 0 & 0 \\ S_{12} & S_{22} & S_{23} & 0 & 0 & 0 \\ S_{13} & S_{23} & S_{33} & 0 & 0 & 0 \\ 0 & 0 & 0 & S_{44} & 0 & 0 \\ 0 & 0 & 0 & 0 & S_{55} & 0 \\ 0 & 0 & 0 & 0 & 0 & S_{66} \end{bmatrix} = \begin{bmatrix} \frac{1}{E_1} & \frac{-\nu_{12}}{E_1} & \frac{-\nu_{13}}{E_1} & 0 & 0 & 0 \\ \frac{-\nu_{21}}{E_2} & \frac{1}{E_2} & \frac{-\nu_{23}}{E_2} & 0 & 0 & 0 \\ \frac{-\nu_{31}}{E_3} & \frac{-\nu_{32}}{E_3} & \frac{1}{E_3} & 0 & 0 & 0 \\ 0 & 0 & 0 & \frac{1}{2G_{23}} & 0 & 0 \\ 0 & 0 & 0 & 0 & \frac{1}{2G_{13}} & 0 \\ 0 & 0 & 0 & 0 & 0 & \frac{1}{2G_{12}} \end{bmatrix}. \tag{A.6}$$

Conflicts of Interest

The authors declare that there are no conflicts of interest regarding the publication of this paper.

References

[1] C. Xu, H. Yanjiao, W. Shou, L. Chun, and F. Luping, "Stress distribution on sandwich structure with triangular grid cores suffered from bending load," *International Journal of Aerospace Engineering*, vol. 2015, Article ID 723487, 8 pages, 2015.

[2] J. Arbaoui, Y. Schmitt, J. L. Pierrot, and F. X. Royer, "Numerical simulation and experimental bending behaviour of multi-layer sandwich structures," *Journal of Theoretical and Applied Mechanics*, vol. 52, no. 2, pp. 431–442, 2014.

[3] A. Zinno, A. Prota, E. Di Maio, and C. E. Bakis, "Experimental characterization of phenolic-impregnated honeycomb sandwich structures for transportation vehicles," *Composite Structures*, vol. 93, no. 11, pp. 2910–2924, 2011.

[4] B. P. Russell, T. Liu, N. A. Fleck, and V. S. Deshpande, "Quasi-static three-point bending of carbon fiber sandwich beams with square honeycomb cores," *Journal of Applied Mechanics, Transactions ASME*, vol. 78, no. 3, Article ID 031008, 2011.

[5] M. Giglio, A. Gilioli, and A. Manes, "Numerical investigation of a three point bending test on sandwich panels with aluminum skins and Nomex™ honeycomb core," *Computational Materials Science*, vol. 56, pp. 69–78, 2012.

[6] C. Chamis, R. Aiello, and P. Murthy, "Fiber composite sandwich thermostructural behavior: computational simulation. Structures," in *Proceeding of the Structural Dynamics And Materials Conference*, pp. 370–381, American Institute of Aeronautics and Astronautics, San Antonio, TX, USA, 1986.

[7] S. Kelsey, R. A. Gellatly, and B. W. Clark, "The shear modulus of foil honeycomb cores: a theoretical and experimental investigation on cores used in sandwich construction," *Aircraft Engineering and Aerospace Technology*, vol. 30, no. 10, pp. 294–302, 1958.

[8] L. J. Gibson and M. F. Ashby, *Cellular Solids: Structures and Properties*, Cambridge University Press, Cambridge, UK, 2nd edition, 1997.

[9] J. Zhang and M. F. Ashby, "The out-of-plane properties of honeycombs," *International Journal of Mechanical Sciences*, vol. 34, no. 6, pp. 475–489, 1992.

[10] J. W. Klintworth and W. J. Stronge, "Elasto-plastic yield limits and deformation laws for transversely crushed honeycombs," *International Journal of Mechanical Sciences*, vol. 30, no. 3-4, pp. 273–292, 1988.

[11] J. Hohe and W. Becker, "Effective stress-strain relations for two-dimensional cellular sandwich cores: homogenization, material models, and properties," *Applied Mechanics Reviews*, vol. 55, no. 1, pp. 61–87, 2002.

[12] Q. S. Yang and W. Becker, "A comparative investigation of different homogenization methods for prediction of the macroscopic properties of composites," *Computer Modeling in Engineering and Sciences*, vol. 6, no. 4, pp. 319–332, 2004.

[13] L. Gornet, G. Marckmann, and M. Lombard, "Failure and elastic properties of Nomex® honeycombs: periodic homogeneization and mechanical simulation," *Mecanique et Industries*, vol. 6, no. 6, pp. 595–604, 2005.

[14] Y. Zhang and Z. Xia, "Micromechanical analysis of interface damage for fiber reinforced composite laminates," *Computers Materials & Continua*, vol. 38, pp. 3689–3717, 2005.

[15] L. Gornet, S. Marguet, and G. Marckmann, "Numerical modelling of Nomex® honeycomb cores: failure and effective elastic properties," in *Proceeding of the III European Conference on Computational Mechanics, 509*, vol. 509, 2006.

[16] H. Ijaz, M. Asad, A. Memon, KB. Ahmed, H. Abbasi, and AN. Laghari, "Strain energy based homogenization method to find the equivalent orthotropic properties of sandwich structures," *Sindh University Research Journal (Science Series)*, vol. 46, no. 1, pp. 93–98, 2014.

[17] C. T. Herakovich, *Mechanics of Fibrous Composites*, John Wiley and Sons Ltd, New York, NY, USA, 1st edition, 1997.

[18] X. Li, Q. Liu, and J. Zhang, "A micro-macro homogenization approach for discrete particle assembly—Cosserat continuum modeling of granular materials," *International Journal of Solids and Structures*, vol. 47, no. 2, pp. 291–303, 2010.

[19] S. Forest, "Mechanics of generalized continua: construction by homogenizaton," *Journal de Physique IV*, vol. 18, pp. 39–48, 1998.

[20] P. Verpeaux, T. Charras, and A. Millard, "Castem 2000: Une approche moderne du calcul des structures," in *Calcul des Structures et Intelligence Artificielle*, J. M. Fouet, P. Ladevèze, and R. Ohayon, Eds., vol. 2, pp. 227–261, 1988.

[21] J. Arbaoui, Y. Schmitt, J.-L. Pierrot, and F.-X. Royer, "Effect of core thickness and intermediate layers on mechanical properties of polypropylene honeycomb multi-layer sandwich structures," *Archives of Metallurgy and Materials*, vol. 59, no. 1, pp. 11–16, 2014.

13

An Analysis of the Thermal Conductivity of Composite Materials (CPC-30R/Charcoal from Sugarcane Bagasse) Using the Hot Insulated Plate Technique

René Salgado-Delgado,[1] Alfredo Olarte-Paredes,[1] Areli Marlen Salgado-Delgado,[1] Zully Vargas-Galarza,[1] Teresa Lopez-Lara,[2] Juan Bosco Hernández-Zaragoza,[2] Israel Rico-Rodríguez,[1] and Gonzalo Martínez-Barrera[3]

[1]*Division de Estudios de Posgrado e Investigación del Instituto Tecnológico de Zacatepec, Calzada Tecnológico No. 27, Col. Centro, 62780 Zacatepec, MOR, Mexico*
[2]*Division de Estudios de Posgrado, Facultad de Ingeniería, Universidad Autónoma de Querétaro, Cerro de las Campanas S/N, Col. de lãs Campanas, 76010 Querétaro, QRO, Mexico*
[3]*Laboratorio de Investigación y Desarrollo de Materiales Avanzados, Facultad de Química, Universidad Autónoma del Estado de México, Km 12 Toluca-Atlacomulco, 50200 San Cayetano, MEX, Mexico*

Correspondence should be addressed to René Salgado-Delgado; renesalgado@hotmail.com

Academic Editor: Fernando Lusquiños

The production of new thermally insulating composite materials from solid residues such as charcoal from sugarcane bagasse (CSB) is of great importance because it takes advantage of materials that might otherwise pollute the environment. Therefore, for this study, we obtained composite materials based on a portland cement (CPC-30R) matrix and CSB particles with a 4 : 1 water-cement ratio and CSB concentrations of 5%, 10%, and 15% by weight. The thermal conductivities of these materials were characterized following ASTM guideline C177, their CSB morphological properties were analyzed using scanning electron microscopy (SEM), and their compositions were determined using energy-dispersive spectrometer (EDS). The results show that the composite materials have reduced thermal conductivities. The metallic oxide percentage composition of the CSB was also determined. It was observed that there was a reduction in thermal conductivity when CSB was used as compared to 100% CPC-30R, and the influence of the CSB concentration on thermal conductivity was measured.

1. Introduction

The increasing concern about industrial pollution has forced the construction and manufacturing industries to seek innovative materials that are trustworthy and sustainable to replace the conventional synthetic fibers used as reinforcements in structural materials. Natural fibers such as sisal, jute, cotton, linen, hemp, and kenaf have been considered as possible alternatives because of their environmental friendliness and their availability in fibrous form and because they can be obtained from plant leaves at a low cost [1, 2].

Charcoal from sugarcane bagasse (CSB) is a residue generated by the bagasse burning process that has low pozzolanic reactivity due to its high porosity. This material is frequently used as a fertilizer even though it is inadequate for this use due to its high silica content (SiO_2) of more than 60% of the total metallic oxide mass. CSB has a broad and variable particle size distribution in addition to a low nutrient content. Therefore, it is important to promote other uses of CSB that generate an added value and synergistic properties for the material.

The locality of *Zacatepec*, in the Mexican state of Morelos, has a sugarcane mill. The residues generated by this industry accumulate every year, generating a series of negative consequences for both human health and the environment. In addition, this is a waste of a potential source of products. In this study, a composite material was obtained from the agroindustrial residue CSB and used as a reinforcement and matrix for portland cement (CPC-30R).

Various studies on the characterization of composite materials based on cement and various inorganic loads that establish synergistic properties due to the volume and characteristics of the loads are reported. The fillers confer properties such as thermal insulation, light weight, and impact resistance [3, 4].

Other studies involve the use of sugarcane bagasse ash, but these differ from the charcoal from sugarcane bagasse. The use of ash does not improve the mechanical strength of the material and water permeability decreases [5].

Our material intends to use the charcoal from sugarcane bagasse to achieve thermal insulation. Bagasse coal has completely different properties to the ash. In ash the organic part is removed leaving only metal oxides.

This method is commonly employed to explain the properties of compounds even though more precise models that include the impact of the shape, size, and distribution of the particles on the elastic properties of the material have been developed [6, 7].

The thermal conductivity (λ) is a measure of a material's ability to conduct heat. It is most commonly used in Fourier's law of heat conduction. This property is very sensitive to the material's microstructure, and there are thermally conducting and insulating materials.

Fourier's law of heat transfer by conduction states that heat conduction occurs due to a temperature difference between two bodies and obeys the following equation [8]:

$$\lambda \equiv -\frac{q_x \Delta x}{A \Delta T}, \quad (1)$$

where λ is the thermal conductivity in watts per meter-degree Celsius (W/m°C), q_x is the heat flow in the x direction in watts, Δx is the thickness of the sample in meters, ΔT is the temperature difference in °C, and A is the area of the sample in m^2.

The method is valid within the linear regime and allows for several material properties, including the thermal conductivity, to be calculated.

2. Experimental

2.1. Materials. We used charcoal from sugarcane bagasse (CSB) collected from the "*Emiliano Zapata Salazar*" sugarcane mill in Zacatepec, Morelos, fast-drying composite portland cement (CPC-30R, "*Tolteca*" brand), and distilled water from the company "*Químicos Farmacéuticos Industriales.*"

2.2. Equipment. The following devices were used: a Denver Instrument digital scale with a limit of 320 g, certified to meet ISO quality guideline 9001, a Barnant Series 10 mixer, model number 700-5400, granulometry equipment from FIIC, SA de CV A CRODE SCH 0001 97 drying oven, an insulated hot plate with K-type thermocouples and an Arduino open-source hardware platform, a Thermo Haake K20 water circulation system, a PID temperature controller with a temperature range of 20 to 300°C, and JEOL/JSM-6010LA scanning electron microscope (SEM) and energy-dispersive spectroscope (EDS).

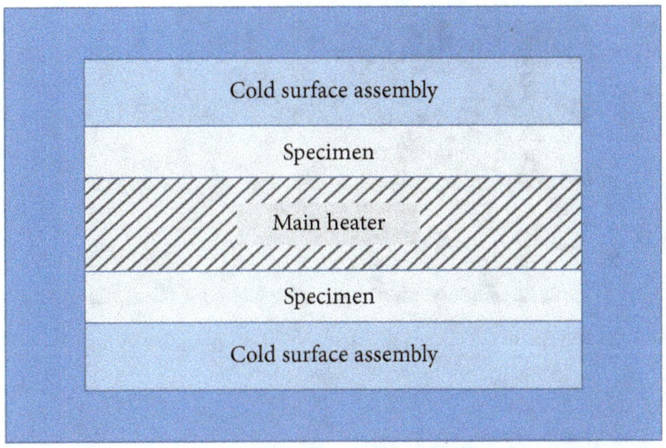

FIGURE 1: System mounting hot isolated plate by ASTM C177-97 standard.

2.3. Methods

2.3.1. Preparation of the Charcoal from Sugarcane Bagasse (CSB). To eliminate any traces of organic fiber, 250 g of CSB was placed in the drying oven for 2 hours at 400°C and then allowed to cool down. Subsequently, different sizes of CSB particles were separated using the granulometry equipment (mesh 100: 0.150 mm and mesh 200: 0.075 mm of hole opening. It is relative with ASTM E11-87).

2.3.2. Preparation of the Composite Materials. The composite materials were prepared by mixing different weight concentrations of CSB (5, 10, and 15 g) with cement (100 g) at a rate of 140 ± 5 rpm for 5 minutes. Then, water was added (40 g) and mixed at a rate of 285 ± 10 rpm for 5 minutes. The mixtures obtained were poured into 10 × 2.5 × 2.5 cm metallic molds. The molds containing the mixtures were left to dry at room temperature for 72 hours. Then, the composite materials were extracted from the molds for characterization.

2.4. Characterization

2.4.1. Measurement of Thermal Conductivity. For this study, a hot plate system based on ASTM guideline C177-97 (Standard Test Method for Steady State Heat Flux Measurements and Thermal Transmission Properties by Means of the Guarded Hot Plate Apparatus) was used. Figures 1 and 2 show the experimental setup, which consists of the following elements: a hot plate, samples of the materials, two cold plates, one on each side of the system, a mounting system, an isolation chamber, and the necessary mechanisms for tuning and measurement [9].

Some of the most important specifications in the guideline are as follows: the method must be applied to materials with thermal conductivities of up to 0.62 kcal/mh°C; the temperature during the test must be kept between −45 and

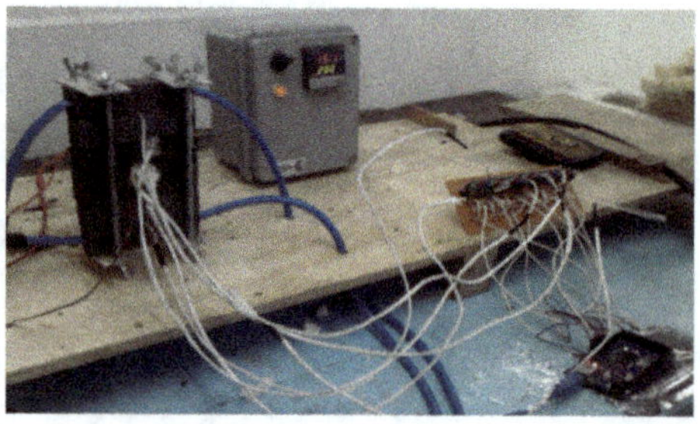

Figure 2: System hot insolated plate technique.

Table 1: Composition of composite materials (CBC/CPC-30R/H_2O).

Mesh number	CBC (g)	CPC-30R (g)	H_2O (g)
M-100	15	100	40
M-100	10	100	40
M-100	5	100	40
M-200	15	100	40
M-200	10	100	40
M-200	5	100	40

760°C; the external insulation surrounding the guard ring must have a thermal resistance of at least twice that of the material being tested; the thermocouples must be made from wire with a diameter no greater than 0.57 mm; and finally, the temperature difference between the hot and cold faces of the sample must be at least 22°C [10].

2.4.2. Scanning Electron Microscopy (SEM) and Energy-Dispersive Spectroscopy (EDS) Analyses. The morphology and load distribution of the composite materials were studied using the desktop SEM described in the equipment section. It was set to 3 kV and to a magnification of 1000x. The EDS analysis was performed simultaneously under the same conditions to determine the elemental content and chemical composition of the CSB.

3. Results and Discussion

The composite materials were used to create a sample matrix (2 sizes × 3 weight concentrations), as shown in Table 1.

To measure the thermal conductivity, the samples were placed on the hot plate apparatus, which maintained a temperature between 30 and 230°C for 4 hours, with a flow temperature of 10°C at the cold plate. The thermal conductivities of the composite are shown in Figure 3.

The results show the behavior of the thermal conductivity as a function of temperature and the influence of the particle size and CSB concentration. At 122°C, the thermal conductivity of the pure cement matrix (CPC-30R) was high at 0.358 W/m°C.

At this temperature, all of the composite materials tested had lower thermal conductivities than the pure matrix; the values were between 0.350 and 0.300 W/m°C. Between 122 and 180°C, a decrease in the conductivity due to the CSB load was observed. For one load size (M-200 or M-100), the 10 g weight concentration marked the threshold for a minimal response because the thermal conductivity increased when the load was decreased (to 5 g) and increased (to 15 g). This may have been due to its distribution within the cement matrix. When the results of the M-200-15 g and M-100-15 g samples were compared, they showed that a larger particle size results in lower conductivity values when the load is constant. The same result was found when comparing the M-200-5 g and M-100-5 g samples. From this, we concluded that a smaller load leads to more efficient heat conduction, a finding that explains the higher conductivities obtained when the CSB particle size was increased [8].

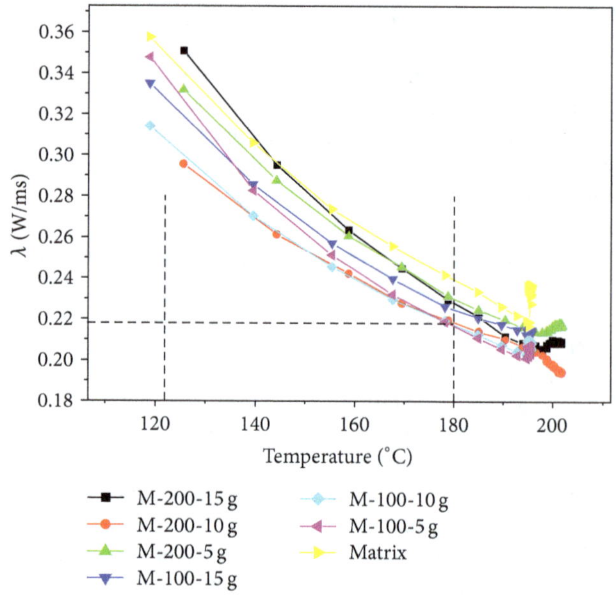

Figure 3: Thermal conductivity of composite materials with two sizes (M-100 and M-200) and three concentrations (15, 10, and 5 g) in the range of 120–180°C.

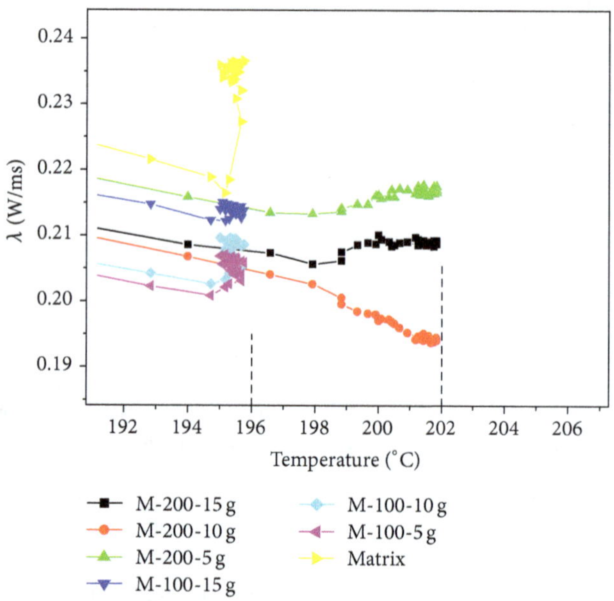

Figure 4: Thermal conductivity of composite materials with two sizes (M-100 and M-200) and three concentrations (15, 10, and 5 g) in the range of 194–202°C.

TABLE 2: EDS: elemental analysis of the CBC.

Chemical formula	C	Na_2O	MgO	Al_2O_3	SiO_2	P_2O_5	SO_3	K_2O	CaO	PbO
% mol	70	0.32	1.58	2.31	15.89	0.66	1.53	1.93	5.57	0.21

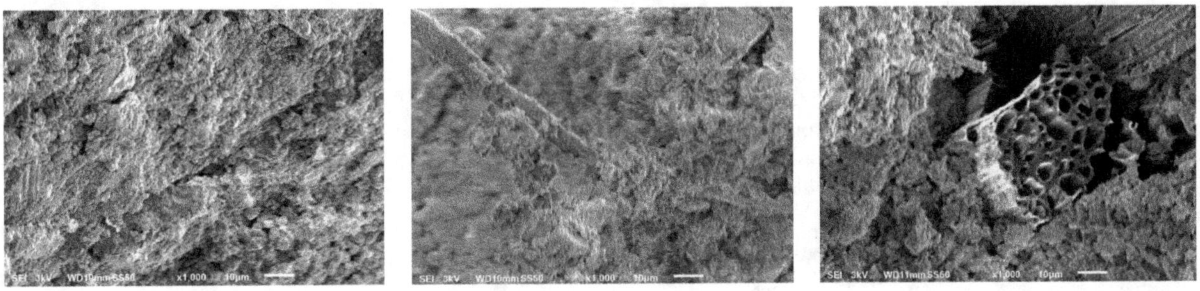

(a) (b) (c)

FIGURE 5: SEM micrographs (1000x) of composite materials: (a) matrix (CPC-30R/H_2O), (b) M-200-10 g, and (c) M-100-10 g.

Figure 4 shows that at 195.5°C a maximum conductivity of 0.235 W/m°C for the sample was obtained (matrix: portland cement). Samples containing charge (CSB) showed a lower thermal conductivity in the range of 0.215 to 0.205 W/m°C at this temperature (the latter being shown for M-200-10 g).

In the temperature range of 196 to 199°C the thermal conductivity decreases for all samples with CSB.

The sample M-200-10 g shows the lowest thermal conductivity of 0.194 W/m°C at a temperature of 202°C. The CSB is contributing to having a low thermal conductivity. With a concentration of 10 g CBS has the lowest income considering that this concentration is critical because by using 200-mesh distributes the load better generating sufficient space for the thermal conductivity decreases. This demonstrates that a good load distribution and a small particle size favor insulating material properties, as explained by Peña-Rodríguez et al., 2008 [8], where the thermal resistivity obtained polystyrene thermal clay loads and the same behavior is observed in the distribution and particle size.

SEM micrographs of the analysis showed that CPC-30R/H_2O ratio allows a hardening material without leaving a broken surface (Figure 5(a)). Likewise in the sample M-200-10 g good wetting of the matrix on the load can be demonstrated (see Figure 5(b)).

Figure 5(c) shows a particle load in the fracture of the composite material after undergoing strain. It can be seen that part of the load fell off the matrix suggesting a weak interaction between the filler and the matrix. It notes that the porosity of the charge also contributes to the thermal conductivity of composites decreases.

It is important to establish the composition of the CSB. We used EDS analysis, which provided evidence for the amount of carbon and oxides present in the material for that purpose. Table 2 shows the chemical composition of the CSB surface. Oxides contributed to the decrease in the thermal conductivity of the material, and here, SiO_2 was present in a greater proportion.

4. Conclusions

Composite materials were created using portland cement (CPC-30R) and charcoal from sugarcane bagasse with varying load particle sizes and concentrations. From studying these materials, we concluded that loading diminishes a composite material's thermal conductivity and that increasing the particle size reduces the thermal conductivity. This is because an increase in the particle size affects the heat conduction by decreasing the contact area. Similarly, an increase in the particle size increases the pore size, which may increase the volume of air inside the load and, therefore, reduce the thermal conductivity. The EDS analysis revealed the presence of oxides as an additional factor contributing to the diminished thermal response. The composite material with the lowest conductivity was the M-200-10 g sample (0.194 W/m°C), which behaved as a thermal insulator; it had a reduced heat transfer rate in comparison to the rest of the samples. The conductivity of the M-200-10 g sample was less than the conductivities reported by other authors for ceramic bricks and asbestos plates (0.87 and 0.36 W/m°C, resp.). The micrographs show both the porosity of the load and the load-matrix interaction at the point of fracture. In future studies, it is recommended that the effects of thermal shock on the mechanical properties of the materials should be investigated.

Competing Interests

The authors declare that they have no competing interests.

References

[1] J. Wei and C. Meyer, "Degradation mechanisms of natural fiber in the matrix of cement composites," *Cement and Concrete Research*, vol. 73, pp. 1–16, 2015.

[2] F. D. A. Silva, N. Chawla, and R. D. D. T. Filho, "Tensile behavior of high performance natural (sisal) fibers," *Composites Science and Technology*, vol. 68, no. 15-16, pp. 3438–3443, 2008.

[3] M. Salas, R. Luco, and R. Ojeda, "Análisis Estructural de un Camarán de materiales compuestos," *Anales de Mecánica de la Fractura*, vol. 20, 2003.

[4] C. S. Smith, *Design of Marine Structures in Composite Materials*, Elsevier Science, Amsterdam, The Netherlands, 1990.

[5] Marcos De Paula Oliveira et al., "La caña de azúcarbagazo ash como material parcial- portland-cemento-reemplazo," *Dyna, Revista de la Facultad de Minas*, vol. 77, no. 163, pp. 47–54, 2010.

[6] N. Martínez-Mateos, D. Busquets-Mataix, M. D. Salvador-Moya, and V. Amigó-Borrás, "Propiedades mecánicas de compuestos de matriz de aluminio reforzados con partículas cerámicas, obtenidos por extrusión de polvos," in *Proceedings of the VIII Congreso Nacional de Propiedades Mecánicas de Sólidos*, pp. 215–224, 2002.

[7] G. Bao, J. W. Hutchinson, and R. M. McMeeking, "Particle reinforcement of ductile matrices against plastic flow and creep," *Acta Metallurgica et Materialia*, vol. 39, no. 8, pp. 1871–1882, 1991.

[8] G. Peña-Rodríguez, J. Sánchez-Molina, and R. Monroy, "Efecto de la concentración de poliestireno expandido en la conductividad térmica efectiva de bloques de termoarcilla," *Revista Colombiana de Física*, vol. 40, no. 2, pp. 285–288, 2008.

[9] M. M. Pérez-Sánchez, R. Centeno-Lara, and F. Lazcano-Serrano, "Desarrollo de un prototipo para la caracterización térmica de los materiales de construcción regionales," *Ingeniería*, vol. 6, no. 2, pp. 13–22, 2002.

[10] ASTM C 177-97(Standard Test Method for Steady State Heat Flux Measurements and Thermal Transmission Properties by Means of the Guarded Hot plate Apparatus).

14

Effect of the Yield Criterion of Matrix on the Brittle Fracture of Fibres in Uniaxial Tension of Composites

Sergei Alexandrov,[1] Yaroslav Erisov,[2] and Fedor Grechnikov[2]

[1] *Institute for Problems in Mechanics, Russian Academy of Sciences, 101-1 Prospect Vernadskogo, Moscow 119526, Russia*
[2] *Samara State Aerospace University, 34 Moskovskoe Shosse, Samara 443086, Russia*

Correspondence should be addressed to Yaroslav Erisov; yaroslav.erisov@mail.ru

Academic Editor: Filippo Berto

This paper examines the effect of the yield criterion of the matrix on brittle fracture of the fibre in continuous fibre reinforced metal matrix composites subjected to tension in the direction parallel to the fibres. It is assumed that the matrix obeys quite a general isotropic yield criterion. An approximate approach to predicting the tensile load at which the fibre breaks previously proposed in the literature is adopted. It is shown that this tensile load is practically independent of the yield criterion of the matrix. This is a great advantage for engineering applications since an analytic solution is available in the case of Tresca yield criterion. This solution can be used for a wide range of matrix materials with no loss of the accuracy of the prediction of the tensile load at which the fibre breaks.

1. Introduction

There are many applications in which continuous and short fibre reinforced metal matrix composites are required. There is a class of such composites in which the fibre is made of brittle material and the matrix of ductile material [1–3]. Failure of such composites involves various mechanisms such as fibre fracture, interfacial debonding, and matrix plasticity. The effect of short fibre reinforcement on the fracture toughness of metal matrix composites has been studied in [4, 5]. Interfacial debonding has been investigated in [6–8]. Computational methodologies for modeling fracture in continuous fibre reinforced metal matrix composites and in laminated composites at the micromechanical level have been proposed in [7–10]. It has been shown in [11] that the interfacial conditions strongly affect tensile fracture characteristics of a boron-fibre-reinforced aluminum composite.

In the present paper the brittle fracture of fibres in continuous fibre reinforced metal matrix composites subjected to uniaxial tension is predicted using the approach proposed in [1]. In the latter work, this approach has led to a simple analytic solution for the axial force at which the fibre breaks. In particular, the solution predicts the influence of the volume fraction of fibres on the strength of composites, which is well known from experiment [12]. Tresca's yield criterion has been adopted for the matrix in [1]. However, it is known that the strength of metal matrix composites is influenced by many factors including matrix strength [13]. Moreover, the in situ flow properties of the matrix of metal matrix composites differ from the properties of the matrix bulk material and the experimental scatter is much larger for experiments in the matrix than in the corresponding bulk material [14]. It is therefore of importance to extend the approach proposed in [1] to generalized yield criteria. Such criteria have been proposed, for example, in [15, 16]. In the present paper, the criterion proposed in [15] is used. It is shown that the predicted magnitude of the axial force at which the fibre breaks is almost independent of the yield criterion of the matrix. Therefore, the analytical expression derived in [1] can be used for a large class of continuous fibre reinforced metal matrix composites. The main result of the present paper can be used in conjunction with methods developed to study the behaviour of composites in the presence of fractured fibres (e.g., [17]).

The rigid plastic solution found in the present paper is of academic interest as well. To the best of authors' knowledge,

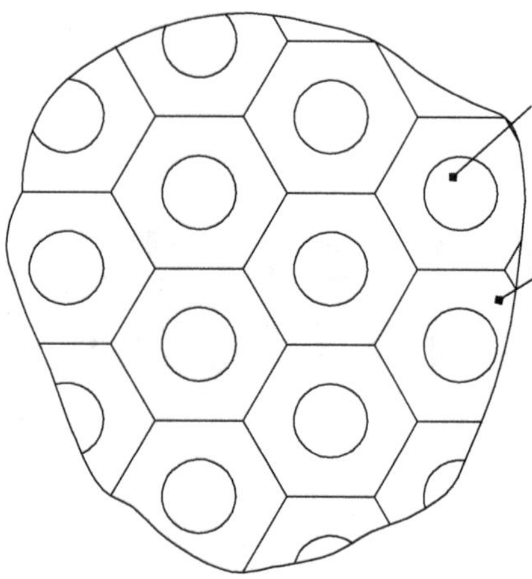

FIGURE 1: Typical cross section of the idealized composite.

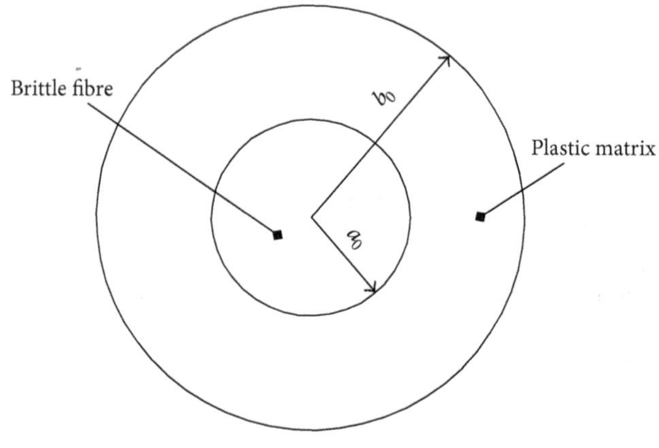

FIGURE 2: Cross section of a typical idealized cell.

the only available nontrivial axisymmetric semianalytic rigid plastic solution for the generalized yield criterion has been outlined in [18]. In this paper, flow through an infinite conical channel has been analyzed. An alternative solution to this problem and quantitative results have been presented in [19], where the yield criterion proposed in [15] was adopted. The present paper provides another solution. It is also worthy to note that the solution given in [1] has been extended to the double shearing model in [20] and to a viscoplastic model in [21]. The fracture of fibres has not been considered in these works. A description of the double shearing model can be found in [22].

2. General Approach

An elegant theory of the failure of ductile materials reinforced by elastic fibres has been proposed in [1]. This theory is outlined in this section to formulate the boundary value problem to be solved. It is assumed that the composite consists of a large number of equal cells, each of which is a hexagonal cylinder of the matrix material containing a concentric circular fibre. A typical cross section of the idealized composite is illustrated in Figure 1. These cells are further idealized replacing the hexagonal cylinders with circular cylinders of the same cross-sectional area. Figure 2 illustrates the cross section of a typical cell that will be used in the mathematical formulation. In particular, the radius of the fibre is denoted by a_0 and the radius of the cell by b_0. The length of the cell will be denoted by $2L$.

It is supposed that the composite is subjected to uniaxial tension applied in the direction parallel to the fibres. The equations for the cell shown in Figure 2 are referred to a cylindrical polar coordinate system (r, θ, z) whose origin is located at the centre of the cell and z-axis lies along the axis of the fibre (Figure 3). The plane $z = 0$ coincides with the plane of symmetry of the cell. It is assumed that the state of stress is approximately axially symmetric about the z-axis. Therefore, the nonzero stress components referred to the cylindrical coordinate system are σ_{rr}, $\sigma_{\theta\theta}$, σ_{zz}, and σ_{rz}. The material of the matrix is isotropic. Therefore, the nonzero strain rate components referred to the cylindrical coordinate system are ξ_{rr}, $\xi_{\theta\theta}$, ξ_{zz}, and ξ_{rz}. Moreover, the circumferential velocity vanishes everywhere. The radial and axial velocities will be denoted by u_r and u_z, respectively.

Since the boundary value problem is symmetric about the plane $z = 0$, it is sufficient to consider the region $z \geq 0$. Based on the assumptions made the following boundary conditions have been formulated in [1]:

$$u_r = -U \tag{1}$$

for $r = b_0$,

$$u_r = 0 \tag{2}$$

for $r = a_0$,

$$u_z = 0 \tag{3}$$

for $z = 0$,

$$\sigma_{rz} = 0 \tag{4}$$

for $r = b_0$,

$$\sigma_{rz} = \tau_f \tag{5}$$

for $r = a_0$, and

$$\int_0^L \sigma_{rr}\big|_{r=b_0} dz = 0. \tag{6}$$

In (5), τ_f is the shear yield stress of the matrix. The boundary conditions (1) to (6) should be used to solve the plasticity boundary value problem in the matrix. It is also assumed that the ends of the fibres do not support tensile load (Figure 3). This load is transferred to the fibre only through the shear stress on the contact surface. The resulting force acting in the axial direction at the cross section $z = L$ is

$$P = 2\pi \int_{a_0}^{b_0} \sigma_{zz}\big|_{z=L} r\, dr. \tag{7}$$

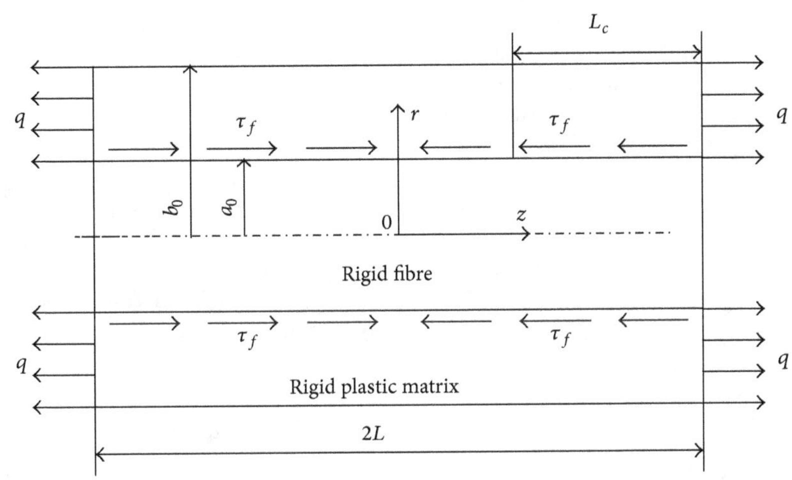

FIGURE 3: Configuration and coordinate system for a typical cell.

Let T be the mean tensile stress at which the fibre breaks. It has been shown in [1] that failure occurs by plastic flow of the matrix if $L < L_c$ and by fracture of the fibre if $L > L_c$ where

$$L_c = \frac{a_0 T}{2\tau_f}. \tag{8}$$

The approach to determining the tensile load at which the fibre breaks is as follows. It is assumed that in the matrix

$$\frac{\sigma_{zz}}{\sigma_0} = 1 + C,$$
$$\frac{\sigma_{rr}}{\sigma_0} = \frac{\sigma_{\theta\theta}}{\sigma_0} = C, \tag{9}$$
$$\sigma_{rz} = 0$$

in the range $0 \le z \le L - L_c$ (Figure 3). Here σ_0 is the yield stress in tension and C is an arbitrary constant. It is evident that the state of stress given by (9) satisfies any yield criterion. The state of stress in the matrix in the range $L - L_c \le z \le L$ should be found from the solution satisfying the boundary conditions (1) to (5). It is then assumed that

$$\left[\int_{a_0}^{b_0} \sigma_{zz} r\, dr\right] = 0 \tag{10}$$

at $L = L_c$. Here [ ] denotes the amount of jump in the quantity enclosed in the brackets. Equations (6) and (10) allow the axial stress in the matrix to be determined. Then, (7) can be used to find the tensile load at which the fibre breaks.

The approach outlined in this section has been adopted in [1] in conjunction with Tresca's yield criterion in the matrix. As a result, the dimensionless tensile load at which the fibre breaks has been found as

$$\frac{P}{\pi\sigma_0 b_0^2} = q_S = \left(1 - a^2\right)\left[1 - \frac{(2-n)}{2}\frac{L_c}{L}\right] + \frac{Ta^2}{\sigma_0}\left(1 - \frac{L_c}{2L}\right), \tag{11}$$

where $a = a_0/b_0$ and n is a numerical coefficient. This coefficient is given by

$$n = -\frac{1}{2} + \frac{a^2}{(1-a^2)}\ln a - \frac{\left(1-a^2\right)}{a^2} - \frac{\left(1+a^2\right)}{(1-a^2)}\left[a^2 F(\gamma,\phi) - \frac{E(\gamma,\phi)}{a^2}\right]. \tag{12}$$

Here $F(\gamma,\phi)$ and $E(\gamma,\phi)$ are the incomplete elliptic integrals of the first and second kinds, respectively, and

$$\gamma = \sqrt{1-a^4},$$
$$\phi = \arctan\left(\frac{1}{a}\right). \tag{13}$$

Many materials satisfy other yield criteria compared to that of Tresca. Therefore, it is of interest to correct (11) taking into account a generalized yield criterion or to show that this simple equation is accurate enough for predicting the tensile load at which the fibres breaks. In the present paper, the yield criterion proposed in [15] is adopted.

3. Rigid Plastic Solution in the Matrix

Let σ_1, σ_2, and σ_3 be the principal stresses. With no loss of generality it is possible to assume that

$$\sigma_1 \ge \sigma_2 \ge \sigma_3. \tag{14}$$

Then, the yield criterion proposed in [15] can be written as

$$\left[\frac{(\sigma_1-\sigma_2)^m + (\sigma_2-\sigma_3)^m + (\sigma_1-\sigma_3)^m}{2}\right]^{1/m} = \sigma_0. \tag{15}$$

The associated flow rule is

$$\xi_1 = \lambda\left[(\sigma_1-\sigma_2)^{m-1} + (\sigma_1-\sigma_3)^{m-1}\right],$$
$$\xi_2 = \lambda\left[(\sigma_2-\sigma_3)^{m-1} - (\sigma_1-\sigma_2)^{m-1}\right], \tag{16}$$
$$\xi_3 = -\lambda\left[(\sigma_2-\sigma_3)^{m-1} + (\sigma_1-\sigma_3)^{m-1}\right],$$

where ξ_1, ξ_2, and ξ_3 are the principal strain rates and λ is a nonnegative multiplier. Equations (15) and (16) are supplemented with the condition that the principal stress and strain rate directions coincide. In the case under consideration the equilibrium equations in the cylindrical coordinate system are

$$\frac{\partial \sigma}{\partial r} + \frac{\partial \tau_{rr}}{\partial r} + \frac{\partial \tau_{rz}}{\partial z} + \frac{\tau_{rr} - \tau_{\theta\theta}}{r} = 0, \quad \frac{\partial \tau_{rz}}{\partial r} + \frac{\partial \sigma}{\partial z} + \frac{\partial \tau_{zz}}{\partial z} + \frac{\tau_{rz}}{r} = 0, \tag{17}$$

where

$$\sigma = \frac{\sigma_1 + \sigma_2 + \sigma_3}{3}, \quad \tau_{rr} = \sigma_{rr} - \sigma, \quad \tau_{\theta\theta} = \sigma_{\theta\theta} - \sigma, \quad \tau_{zz} = \sigma_{zz} - \sigma, \quad \tau_{rz} = \sigma_{rz}. \tag{18}$$

It is reasonable to assume that

$$\sigma_2 = \sigma_{\theta\theta}. \tag{19}$$

This assumption should be verified a posteriori. It follows from (15) and (16) that

$$\tau_f = \frac{\sigma_0}{\sqrt[m]{1 + 2^{m-1}}}. \tag{20}$$

The basic assumption concerning the velocity field is [1]

$$\frac{u_r}{U} = -\eta(\rho), \tag{21}$$

where $\eta(\rho)$ is a function of ρ and $\rho = r/b_0$. Using (21) the radial and circumferential strain rates are found as

$$\xi_{rr} = -\frac{U}{b_0}\frac{d\eta}{d\rho}, \quad \xi_{\theta\theta} = -\frac{U}{b_0}\frac{\eta}{\rho}. \tag{22}$$

Then, the axial strain rate is determined from the incompressibility equations as

$$\xi_{zz} = \frac{U}{b_0 \rho}\frac{d(\eta\rho)}{d\rho}. \tag{23}$$

Since $\xi_{zz} = \partial u_z/\partial z$ and the right hand side of (23) is independent of z, this equation can be immediately integrated to give

$$\frac{u_z}{U} = \frac{\zeta}{\rho}\frac{d(\eta\rho)}{d\rho} + \mu(\rho), \tag{24}$$

where $\zeta = z/b_0$ and $\mu(\rho)$ is an arbitrary function of its argument. It follows from (21) and (24) that the shear strain rate in the cylindrical coordinate system is

$$\xi_{rz} = \frac{U}{2b_0}\frac{d}{d\rho}\left[\frac{d(\eta\rho)}{\rho d\rho}\right]\zeta + \frac{U}{2b_0}\frac{d\mu}{d\rho}. \tag{25}$$

Let φ be the inclination of the σ_1 principal stress direction to the r-axis, measured anticlockwise. Then,

$$\tan 2\varphi = \frac{2\tau_{rz}}{\tau_{rr} - \tau_{zz}} = \frac{2\xi_{rz}}{\xi_{rr} - \xi_{zz}}. \tag{26}$$

It follows from (22), (25), and (26) that

$$\tan 2\varphi = -\frac{(d/d\rho)\left[d(\eta\rho)/\rho\, d\rho\right]\zeta + d\mu/d\rho}{\left[d\eta/d\rho + (1/\rho)(d(\rho\eta)/d\rho)\right]}. \tag{27}$$

The solution derived in [1] suggests that φ is independent of z. Then, it follows from (27) that $d(\eta\rho)/d\rho = 2C\rho$ where C is constant. Integrating this equation gives

$$\eta = C\rho + \frac{C_1}{\rho}, \tag{28}$$

where C_1 is constant of integration. It is seen from (21) that the boundary conditions (1) and (2) are equivalent to the conditions $\eta = 1$ for $\rho = 1$ and $\eta = 0$ for $\rho = a$, respectively. Using these boundary conditions the constants C and C_1 are determined from (28) as

$$C = \frac{1}{(1 - a^2)}, \quad C_1 = -\frac{a^2}{(1 - a^2)}. \tag{29}$$

Substituting (29) into (28) results in

$$\eta = \frac{(\rho^2 - a^2)}{(1 - a^2)\rho}. \tag{30}$$

Substituting (30) into (27) and using (26) result in

$$\frac{2\tau_{rz}}{\tau_{rr} - \tau_{zz}} = \frac{\rho^2 (a^2 - 1)}{(a^2 + 3\rho^2)}\frac{d\mu}{d\rho}. \tag{31}$$

It is convenient to introduce the stress variables

$$s = \frac{\sigma_1 + \sigma_3}{2\sigma_0},\quad t = \frac{\sigma_1 - \sigma_3}{2\sigma_0},\quad s_{\theta\theta} = \frac{\sigma_{\theta\theta}}{\sigma_0}. \tag{32}$$

It is evident from (14) that $t > 0$. Using (32) it is possible to express the stress components in the cylindrical coordinate system as

$$\frac{\sigma_{rr}}{\sigma_0} = s + t\cos 2\varphi,\quad \frac{\sigma_{zz}}{\sigma_0} = s - t\cos 2\varphi,\quad \frac{\sigma_{rz}}{\sigma_0} = t\sin 2\varphi. \tag{33}$$

The direction of flow requires that $\sigma_{rz} > 0$ (Figure 3). Assume that $\sigma_{zz} > \sigma_{rr}$. This assumption should be verified a posteriori. Then, it is possible to find from (32) and (33) that

$$\frac{\pi}{4} \le \varphi \le \frac{\pi}{2}. \tag{34}$$

Equation (19) dictates that $\xi_{\theta\theta} = \xi_2$. Therefore,

$$\frac{\xi_{rr} - \xi_{zz}}{\xi_{\theta\theta}} = \frac{(\xi_1 - \xi_3)}{\xi_2}\cos 2\varphi. \tag{35}$$

Substituting (16), (22), and (28) into this equation and using (32) and (33) yield

$$\frac{(3\rho^2 + a^2)}{(\rho^2 - a^2)} = \frac{\left[(s + t - s_{\theta\theta})^{m-1} + 2(2t)^{m-1} + (s_{\theta\theta} + t - s)^{m-1}\right]\cos 2\varphi}{\left[(s_{\theta\theta} - s + t)^{m-1} - (s - s_{\theta\theta} + t)^{m-1}\right]}. \tag{36}$$

It follows from (19) and (32) that the yield criterion (15) can be rewritten as

$$\left[\frac{(s + t - s_{\theta\theta})^m + (s_{\theta\theta} - s + t)^m + (2t)^m}{2}\right]^{1/m} = 1. \tag{37}$$

Substituting (33) into the equilibrium equations (17) and assuming that t is independent of z yield

$$\frac{\partial s}{\partial \rho} + \frac{d(t\cos 2\varphi)}{d\rho} + \frac{s - s_{\theta\theta} + t\cos 2\varphi}{\rho} = 0,\quad \frac{\partial s}{\partial \zeta} + \frac{d(t\sin 2\varphi)}{d\rho} + \frac{t\sin 2\varphi}{\rho} = 0. \tag{38}$$

These equations have a solution if and only if

$$s = C_2(\rho) + C_3\zeta,\quad s - s_{\theta\theta} = \beta(\rho), \tag{39}$$

where C_3 is constant and $C_2(\rho)$ and $\beta(\rho)$ are arbitrary functions of ρ. Substituting (39) into (38) results in

$$\frac{d(C_2 + t\cos 2\varphi)}{d\rho} + \frac{\beta(\rho) + t\cos 2\varphi}{\rho} = 0, \tag{40a}$$

$$\frac{d(t\sin 2\varphi)}{d\rho} + \frac{t\sin 2\varphi}{\rho} = -C_3. \tag{40b}$$

Equation (40b) can be immediately integrated to give

$$t\sin 2\varphi = -\frac{C_3\rho}{2} + \frac{C_4}{\rho}, \tag{41}$$

where C_4 is constant of integration. Using (33) the boundary conditions (4) and (5) are transformed to $t\sin 2\varphi = 0$ for $\rho = 1$ and $t\sin 2\varphi = \tau_f/\sigma_0$ for $\rho = a$, respectively. These boundary conditions and (41) combine to give

$$C_3 = \frac{2a}{\sqrt[m]{1 + 2^{m-1}}(1 - a^2)},\quad C_4 = \frac{a}{\sqrt[m]{1 + 2^{m-1}}(1 - a^2)}. \tag{42}$$

Therefore, (41) becomes

$$t = \frac{1}{2}\left(\frac{1}{\rho} - \rho\right)\frac{C_3}{\sin 2\varphi}. \tag{43}$$

Eliminating t and $s - s_{\theta\theta}$ in (36) and (37) by means of (39) and (43) yields

$$\frac{(3\rho^2 + a^2)}{(\rho^2 - a^2)} = \frac{\left\{\left[2\beta\rho\sin 2\varphi + C_3(1 - \rho^2)\right]^{m-1} + 2^m C_3^{m-1}(1 - \rho^2)^{m-1} + \cdots + \left[C_3(1 - \rho^2) - 2\beta\rho\sin 2\varphi\right]^{m-1}\right\}\cos 2\varphi}{\left[C_3(1 - \rho^2) - 2\beta\rho\sin 2\varphi\right]^{m-1} - \left[2\beta\rho\sin 2\varphi + C_3(1 - \rho^2)\right]^{m-1}},$$
$$\left[2\beta\rho\sin 2\varphi + C_3(1 - \rho^2)\right]^m + \left[C_3(1 - \rho^2) - 2\beta\rho\sin 2\varphi\right]^m + \cdots + 2^m C_3^m(1 - \rho^2)^m = 2^{m+1}\rho^m\sin^m 2\varphi. \tag{44}$$

Equations (44) should be solved for $\beta(\rho)$ and $\varphi(\rho)$ numerically. C_3 should be eliminated by means of (42). Once these equations have been solved, t can be found from (42) and (43) and, then, C_2 is determined from (40a) as

$$C_2 = -t\cos 2\varphi + \int_a^\rho \frac{(\beta + t\cos 2\varphi)}{\chi} d\chi + C_5. \quad (45)$$

Here C_5 is constant of integration and χ is a dummy variable of integration. β, t, and φ in the integrand are understood to be functions of χ. The state of stress in the matrix is given by (33), (39), and (45). Then, C_5 and C involved in (9) can be determined by means of the approach outlined in Section 2.

4. Effect of the Yield Criterion of the Matrix on Brittle Fracture of the Fibre

It follows from (33), (39), and (45) that

$$\frac{\sigma_{rr}}{\sigma_0} = C_3\zeta + \int_a^\rho \frac{(\beta + t\cos 2\varphi)}{\chi} d\chi + C_5,$$
$$\frac{\sigma_{zz}}{\sigma_0} = C_3\zeta - 2t\cos 2\varphi + \int_a^\rho \frac{(\beta + t\cos 2\varphi)}{\chi} d\chi + C_5. \quad (46)$$

These equations are valid in the range $L - L_c \leq z \leq L$ (Figure 3). Substituting (9) and (46) into (6) and (10) yields

$$C\left(\frac{l}{l_c} - 1\right) + C_5 = \frac{C_3}{2}(l_c - 2l) - \int_a^1 \frac{(\beta + t\cos 2\varphi)}{\rho} d\rho, \quad (47)$$

$$C - C_5 = C_3(l - l_c) - 1 + \frac{2}{(1-a^2)}\int_a^1 \rho \int_a^\rho \frac{(\beta + t\cos 2\varphi)}{\chi} d\chi\, d\rho - \frac{4}{(1-a^2)}\int_a^1 \rho t \cos 2\varphi\, d\rho, \quad (48)$$

respectively. Here $l = L/b_0$ and $l_c = L_c/b_0$. Equations (47) and (48) constitute a linear system for C and C_5. The solution to this system can be found with no difficulty. Once the value of C_5 has been found, P and its dimensionless representation, q, are determined from (7) and (46) as

$$\frac{P}{\pi\sigma_0 b_0^2} = q = (C_3 l + C_5)(1 - a^2) - 4\int_a^1 \rho t\cos 2\varphi\, d\rho + 2\int_a^1 \rho \int_a^\rho \frac{(\beta + t\cos 2\varphi)}{\chi} d\chi\, d\rho. \quad (49)$$

In order to evaluate the effect of the yield criterion on the tensile force at which the fibre breaks, it is convenient to introduce the parameter δ as

$$\delta = \frac{|q - q_S|}{q_S} \cdot 100\%. \quad (50)$$

Equations (44) have been solved numerically in the range $1.05 \leq m \leq 20$. Condition (19) has been verified using (32), (33), (39), and (45). Then, q has been calculated by means of (42), (47), (48), and (49). It has been found by means of (11) and (50) that $\delta < 1.5\%$ for typical values of a, l, and T/σ_0 provided in [1]. Therefore, (11) derived in that paper is a very good approximation for the tensile load at which the fibre breaks independently of the yield criterion of the matrix.

5. Conclusions

Brittle fracture of fibres in continuous fibre reinforced composites subjected to tensile loading in the direction parallel to the fibres has been predicted by means of the theory developed in [1]. A distinguished feature of this new solution is that the generalized yield criterion proposed in [15] has been adopted for the matrix. The solution has been compared to the analytic solution for Tresca's yield criterion derived in [1]. It has been found that the magnitude of the tensile load at which the fibre breaks is practically independent of the yield criterion of the matrix. In particular, formula (11) obtained in [1] predicts this magnitude for any isotropic pressure-independent yield criterion with a very high accuracy. In general, this feature of the solution is capable of experimental verification and hence provides a means for testing the theory. However, such verification requires the determination of the yield criterion of the matrix before testing the composite. To the best of authors' knowledge, no research that includes both testing the matrix material and testing the composite has been reported in the literature. Therefore, the present paper may be considered as an encouragement to experimentalists to attempt to verify the theoretical predictions made.

A new result of academic interest is the solution for the constitutive equations (15) and (16) supplemented with the equilibrium equations and satisfying the boundary conditions (1)–(5). This solution can be regarded as a generalization of the famous Prandtl's solution for plane strain compression of a plastic layer between parallel, rough plates (see, e.g., [23]). In contrast to the plane strain problem, the solution to the axisymmetric problem depends on the yield criterion chosen. A distinguished feature of the present solution is that the generalized yield criterion has been adopted.

Competing Interests

The authors declare that they have no competing interests.

Acknowledgments

The reported study was funded by RFBR according to the research project no. 16-58-52051.

References

[1] A. J. M. Spencer, "A theory of the failure of ductile materials reinforced by elastic fibres," *International Journal of Mechanical Sciences*, vol. 7, no. 3, pp. 197–209, 1965.

[2] G. J. Dvorak and Y. A. Bahei-El-Din, "Plasticity analysis of fibrous composites," *Journal of Applied Mechanics*, vol. 49, no. 2, pp. 327–335, 1982.

[3] C. Cheng and N. Aravas, "Creep of metal-matrix composites with elastic fibers—part I: continuous aligned fibers," *International Journal of Solids and Structures*, vol. 34, no. 31-32, pp. 4147–4171, 1997.

[4] H. Naji, S. M. Zebarjad, and S. A. Sajjadi, "The effects of volume percent and aspect ratio of carbon fiber on fracture toughness of reinforced aluminum matrix composites," *Materials Science and Engineering A*, vol. 486, no. 1-2, pp. 413–420, 2008.

[5] I. T. Lee, Y. Q. Wang, Y. Ochi, S. I. Bae, K. S. Han, and J. I. Song, "Effect of short fiber reinforcement on the fracture toughness of metal matrix composites," *Advanced Composite Materials*, vol. 19, no. 1, pp. 41–53, 2010.

[6] S. Ghosh, Y. Ling, B. Majumdar, and R. Kim, "Interfacial debonding analysis in multiple fiber reinforced composites," *Mechanics of Materials*, vol. 32, no. 10, pp. 561–591, 2000.

[7] A. Caporale, R. Luciano, and E. Sacco, "Micromechanical analysis of interfacial debonding in unidirectional fiber-reinforced composites," *Computers & Structures*, vol. 84, no. 31-32, pp. 2200–2211, 2006.

[8] S. Li and S. Ghosh, "Modeling interfacial debonding and matrix cracking in fiber reinforced composites by the extended voronoi cell FEM," *Finite Elements in Analysis and Design*, vol. 43, no. 5, pp. 397–410, 2007.

[9] D. H. Allen, R. H. Jones, and J. G. Boyd, "Micromechanical analysis of a continuous fiber metal matrix composite including the effects of matrix viscoplasticity and evolving damage," *Journal of the Mechanics and Physics of Solids*, vol. 42, no. 3, pp. 505–529, 1994.

[10] C. R. Ananth, S. R. Voleti, and N. Chandra, "Effect of fiber fracture and interfacial debonding on the evolution of damage in metal matrix composites," *Composites Part A: Applied Science and Manufacturing*, vol. 29, no. 9-10, pp. 1203–1211, 1998.

[11] Z. P. Luo and C. Y. Sun, "Effect of the interfacial bonding status on the tensile fracture characteristics of a boron-fiber-reinforced aluminum composite," *Materials Characterization*, vol. 50, no. 1, pp. 51–58, 2003.

[12] S. Fukumoto, A. Hirose, and K. F. Kobayashi, "Evaluation of the strength of diffusion bonded joints in continuous fiber reinforced metal matrix composites," *Journal of Materials Processing Technology*, vol. 68, no. 2, pp. 184–191, 1997.

[13] T. Pacheco, H. Nayeb-Hashemi, and H. E. M. Sallam, "The effects of matrix and fiber properties on the mechanical behavior and acoustic emission in continuous fiber reinforced metal matrix composites," *Materials Science and Engineering A*, vol. 247, no. 1-2, pp. 88–96, 1998.

[14] J. L. Bucaille, A. Rossoll, B. Moser, S. Stauss, and J. Michler, "Determination of the matrix in situ flow stress of a continuous fibre reinforced metal matrix composite using instrumented indentation," *Materials Science and Engineering A*, vol. 369, no. 1-2, pp. 82–92, 2004.

[15] W. F. Hosford, "A generalized Isotropic Yield Criterion," *International Journal of Applied Mechanics*, vol. 39, no. 2, pp. 607–609, 1972.

[16] W. H. Yang, "A generalized von Mises criterion for yield and fracture," *Journal of Applied Mechanics*, vol. 47, no. 2, pp. 297–300, 1980.

[17] S. R. Voleti, C. R. Ananth, and N. Chandra, "Effect of fiber fracture and matrix yielding on load sharing in continuous fiber metal matrix composites," *Journal of Composites Technology & Research*, vol. 20, no. 4, pp. 203–209, 1998.

[18] R. T. Shield, "Plastic flow in a converging conical channel," *Journal of the Mechanics and Physics of Solids*, vol. 3, no. 4, pp. 246–258, 1955.

[19] S. Alexandrov and F. Barlat, "Modeling axisymmetric flow through a converging channel with an arbitrary yield condition," *Acta Mechanica*, vol. 133, no. 1–4, pp. 57–68, 1999.

[20] S. E. Aleksandrov and R. V. Goldstein, "Generalization of the prandtl solution to the case of axisymmetric deformation of materials obeying the double shear model," *Mechanics of Solids*, vol. 47, no. 6, pp. 654–664, 2012.

[21] S. Alexandrov and Y.-R. Jeng, "A generalization of Prandtl's and Spencer's solutions on axisymmetric viscous flow," *Archive of Applied Mechanics*, vol. 81, no. 4, pp. 437–449, 2011.

[22] A. J. M. Spencer, "Deformation of ideal granular materials," in *Mechanics of Solids: The Rodney Hill 60th Anniversary Volume*, H. G. Hopkins and M. J. Sewell, Eds., pp. 607–652, Pergamon Press, Oxford, UK, 1982.

[23] R. Hill, *The Mathematical Theory of Plasticity*, Clarendon Press, Oxford, UK, 1950.

15

Development and Characterization of Norite-Based Cementitious Binder from an Ilmenite Mine Waste Stream

Mahmoud Khalifeh,[1] Arild Saasen,[1] Helge B. Larsen,[2] and Helge Hodne[1]

[1] *Department of Petroleum Engineering, University of Stavanger, 4036 Stavanger, Norway*
[2] *Department of Mathematics and Natural Sciences, University of Stavanger, 4036 Stavanger, Norway*

Correspondence should be addressed to Mahmoud Khalifeh; mahmoud.khalifeh@uis.no

Academic Editor: Kedsarin Pimraksa

Norite is a major type of solid waste generated during the production of ilmenite. This study focuses on the usability of norite as a solid precursor of alkali-activated cement. Sensitivity analysis was performed to find the influence of different alkali types and particle sizes of the source material on polycondensation. Norite was ground and mixed with sodium hydroxide and potassium hydroxide solutions to produce a binder. Potassium-containing systems were more effective compared to sodium-containing systems with respect to strength development. The X-ray diffraction patterns indicated formation of zeolites, albite, and oligoclase based on the type of activator and used additives. The patterns also revealed formation of an amorphous phase in the matrices of the binder that was cured at 87°C. Microstructure analysis revealed some degree of crystallization with different Si : Al ratios, which indicated heterogeneity of the binder matrices.

1. Introduction

Alkali-activated binders are a class of inorganic alkaline polymers, which are produced during the alkaline activation of silica-rich and alumina-rich materials. The reaction shows a complex process. Generally, in an alkaline medium, the bonds of Si-O-Si are broken and aluminum (Al) atoms penetrate into the original Si-O-Si structure. Consequently, aluminosilicate gels (oligomers) are formed and finally polycondensation takes place. Cations (e.g., Na^+, K^+, and Ca^+) must be present in the framework cavities to balance the negative charges of ions [1, 2]. The result is a cementitious phase with high mechanical strength and high fire and acid resistance. Polycondensation depends on parameters such as chemical and mineralogical composition, particle size, and surface area of the source materials. In addition, curing pressure and temperature, Si : Al ratio of the used substances (the active silica content), alkali type (e.g., NaOH, KOH), alkali concentration, liquid-to-solid ratio, and types of additives significantly influence the binder synthesis [3–8].

Based on the source used, different types of binders can be produced such as fly ash-based and blast furnace slag-based binders. Utilization of fly ashes in the production of binders has been extensively studied [9, 10]. However, few studies have been carried out on the usability of natural rocks to make binders in alkaline media.

One of the world's largest ilmenite producers has its quarry on the southwest coast of Norway. The mine is operated by Titania AS and the facility uses different methods to separate ilmenite from ore. The facility produces 850,000 metric tons of ilmenite concentrate per year. Ilmenite is a mineral chemically described as $FeTiO_3$. When the iron is removed, the final product becomes TiO_2 which is known as "titanium white," a widely used white pigment. Normally, ilmenite is found in large quantities in the anorthosite or norite rocks. Thus, norite or anorthosite must be separated from the valuable ilmenite. This can be done by conventional separation techniques that leave large amounts of ground norite and anorthosite particles as a waste stream. Norite is one of the main waste tailings during the production of ilmenite at Titania AS. The norite used in this study was supplied by Titania AS.

Norite is an intrusive igneous rock and is predominantly composed of orthopyroxene and plagioclase. Orthopyroxene

TABLE 1: Chemical composition of norite (wt.%).

SiO_2	TiO_2	Al_2O_3	Fe_2O_3	MgO	CaO	Na_2O	MnO	K_2O	P_2O_5	Cr_2O_3	SO_3	LOI
43	8.2	15.5	12.5	6.8	6.6	3.4	0.1	0.9	0.3	0.03	0.1	2.57

TABLE 2: Traced minerals in the norite.

Minerals	Mass%
Plagioclase [(Na, Ca, Al)-silicates]	64
Hyperstone [(Ca, Mg, Fe, Al)-silicates]	10
Biotite [(K, Mg, Al)-silicates]	7
Other silicates	5
Ilmenite [$FeOTiO_2$]	13
Sulfides	0.5
Apatite [$Ca(PO_4)_3$(F, Cl, OH)]	0.5

TABLE 3: Chemical composition of GGBFS (wt.%).

SiO_2	Al_2O_3	CaO	MgO	TiO_2	Mn_2O_3	S^{2-}	Na_2O
34	13	31	17	2.4	0.6	1.1	0.9

is an inosilicate and therefore has interlocking chains of silicate tetrahedra. Plagioclase is a common series of aluminosilicate (tectosilicate) minerals within the feldspar family. In the Streckeisen classification system, norite is in the group with gabbro and anorthosite. Similar to gabbro and anorthosite, norite is very rich in plagioclase compared to K-feldspar, feldspathoids, and quartz [11, 12]. Therefore, norite has the potential to be utilized as a solid precursor in the development of a new cementitious binder. The current study presents a fundamental procedure for developing a norite-based binder in alkali media.

2. Experimental Program

2.1. Materials. Titania AS provided the norite; its chemical analysis and the traced minerals are presented in Tables 1 and 2. The laboratory at Titania AS reported a pH value of 6.5 for the norite [13]. Sodium hydroxide (NaOH) and potassium hydroxide (KOH) pellets used for the preparation of alkali solutions were caustic soda and caustic potash with 99% purity. Sodium silicate solution (Na_2SiO_3) and potassium silicate solution (K_2SiO_3) were used in some mix designs. The Na_2SiO_3 (supplied by Merck KGaA in Germany) was reported to contain 28.5% SiO_2, 8.5% Na_2O, and 63% H_2O; the K_2SiO_3 (supplied by Univar AS, Norway) was reported to contain 38% potassium silicate and 62% H_2O. An aluminum- and calcium-rich blast furnace slag (product name "Merit 5000" from Merit 5000, Sweden) was used as an additive (Table 3). Distilled water was used throughout the experiments.

2.2. Test and Analysis Methods. As the norite particle size was high, 0.4–2 mm, the specific surface area of the norite was increased to enhance its reactivity by milling. Therefore, every 100 g of the norite was ground by using a Retsch PM100 ball mill at 356 rpm for 10 minutes. Particle size distribution (PSD) of the ground norite, Figure 1, was measured using a Sympatec HELOS laser diffraction particle size analyzer. The analysis was done at Tel-Tek National Research Institute, Norway. Sauter mean diameter (SMD) and volume mean diameter (VMD) were reported to be 6.43 and 47.16 μm, respectively. The particle density of norite was estimated to be 2.00 g/cc.

A Zeiss Supra 35VP model scanning electron microscopy-energy dispersive X-ray spectroscopy (SEM-EDXS) analyzer was used. The analysis was performed at ambient temperature and under high vacuum conditions.

An Agilent Cary 630 FTIR (Fourier transform infrared) spectrometer was used to investigate the structure of the norite and the investigation was carried out in the 1200–400 cm^{-1} region. The KBr (potassium bromide) wafer technique was used for the spectra determination [2].

X-ray powder diffraction (XRD) analysis of the norite and the binder was performed using X-ray synchrotron radiation with the wavelength of 0.6888 Å. The XRD measurements were performed with a PILATUS2M pixel detector-based diffractometer at the Swiss-Norwegian beamline (SNBL-BM01A), at the European Synchrotron Radiation Facility, ESRF, Grenoble, France. The two-dimensional diffraction patterns spanned a maximal diffraction angle, 2θ, of 46 degrees. One-dimensional diffraction patterns were obtained by azimuth integration using the program Fit2D [14]. The mineral analysis program, Match!, and Crystallography Open Database (COD) were used for phase identification [15].

To measure the compressive strength of the binder, two different uniaxial compressive strength (UCS) mechanical testers were utilized: a Zwick/Z020 for specimens with compressive strength lower than 10 MPa and a Toni Technik-H for specimens with compressive strength higher than 10 MPa. The Zwick/Z020 and Toni Technik-H testers applied TestXpert II v3.2 and TestXpert v7.11 testing software, respectively, to calculate UCS of the specimens.

2.3. Specimen Preparation. In order to produce a pure norite-based binder, the ground norite was mixed with an alkali solution for 2 minutes. To investigate the influences of ground granulated blast furnace slag (GGBFS) and type of alkali silicate solution, the GGBFS and alkali silicate solution were introduced into some of the mixes as presented in Tables 4 and 5. When an alkali silicate solution was required, the alkali solution and the alkali silicate solution were mixed prior to adding the solid phase. Whenever the GGBFS was necessary, it was entered into the blend during the final part of the mixing. A Hamilton Beach blender was used for the mixing.

The slurries were poured in cylindrical plastic molds with 52 mm diameter and 100 mm length. The specimens were cured at ambient pressure and different temperatures: ambient temperature and 87°C. Tables 4 and 5 present the mix designs.

Table 4: Compressive strength of the binders without any additives. The specimens were cured at 87°C and ambient pressure.

Specimen 1					
Norite (g)	Alkali sol.	Wt. of alkali sol. (g)	L/S ratio	Curing time (days)	Estimated compressive strength (±2 MPa)
300	8 M NaOH	125	0.42	5	2.5
300	8 M NaOH	125	0.42	7	3
300	8 M NaOH	125	0.42	14	23
300	8 M NaOH	125	0.42	28	24

Table 5: Influence of GGBFS on the compressive strength of the binder. The specimens were cured at ambient pressure.

Specimen	Norite (g)	GGBFS (g)	Alkali sol.	Wt. of alkali sol. (g)	K_2SiO_3 (g)	Na_2SiO_3 (g)	L/S ratio	Curing temp. (°C)	Curing time (days)	Estimated compressive strength (±2 MPa)
2	183	91	8 M NaOH	52	0	72	0.45	23	7	20
3	183	91	4 M KOH	53	72	0	0.46	23	7	26
4	248	40	10 M NaOH	61	61	0	0.42	87	1	5.0
4	248	40	10 M NaOH	61	61	0	0.42	87	5	9.0
4	248	40	10 M NaOH	61	61	0	0.42	87	7	9.0

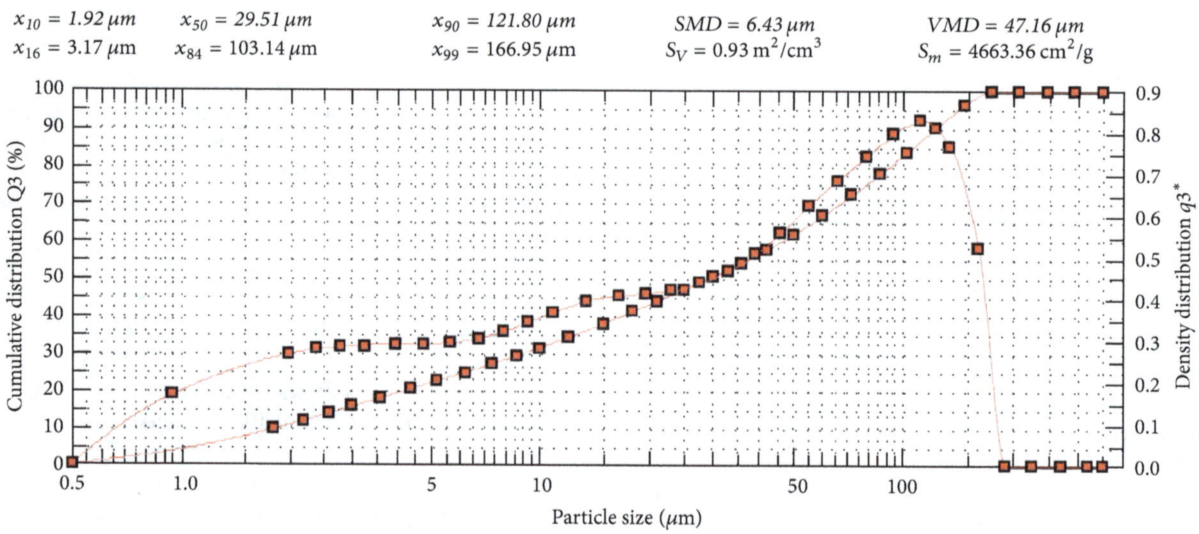

Figure 1: Graphical representation of particle size distribution analysis of the analyzed norite.

3. Results and Discussion

For a better understanding of the binder synthesis, sets of analytical experiments were conducted. For each test, only one parameter was changed while the others were kept constant. The compressive strength of the different specimens was measured, besides the XRD and microstructure analyses.

3.1. Effect of Particle Size. The norite particle size could be important for understanding its physical and chemical properties. So, two specimens were prepared to evaluate the influence of the norite particle size on its reactivity. The specimens were prepared by using the nonground and the ground norite. The maximum particle size of the nonground norite was reported to be 2 mm. Both specimens were mixed by using 8 M NaOH and cured at ambient pressure and temperature for 7 days. A liquid-to-solid mass ratio of 0.42 was selected to maintain fluidity of the slurries; however, generally, a liquid-to-solid mass ratio of 0.35 is recommended [16]. The prepared slurry with the nonground norite never solidified while the ground norite solidified (Figure 2). Although the larger precursor particles have lower surface-to-volume ratio and the slurry requires a lower activator content, the formation of gels was not observed. The smaller the particle sizes, the higher the reactivity of the norite. Likely, the large particles do not participate in the reaction to form the binder but they act as an aggregate (Figure 4(a)). It could be said that milling activates the material mechanically.

FIGURE 2: Effect of particle size on the reactivity at 87°C and ambient pressure: (a) nonground norite and (b) ground norite.

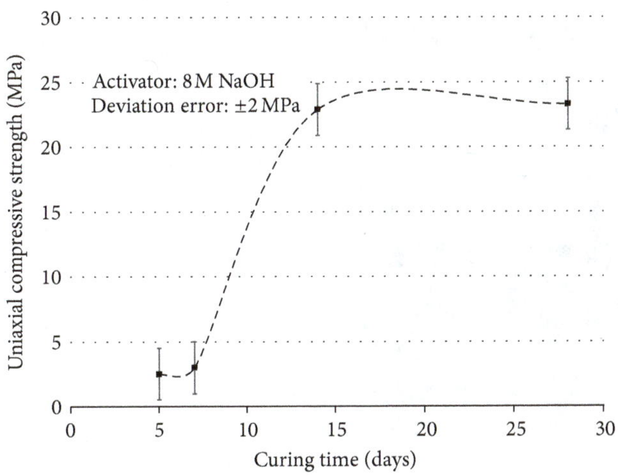

FIGURE 3: Strength development of specimen 1 at 87°C and ambient pressure.

Figure 3 presents the compressive strength development of the norite-based binder, which was cured at 87°C and ambient pressure. A radical strength development was observed after 7 days of curing. The specimens were prepared by mixing the norite with 8 M solution of sodium hydroxide (Table 4).

3.2. Microstructure Analysis of the Binder

3.2.1. Scanning Electron Microscopy (SEM). Figure 4 presents the SEM images of specimen 1, which was cured at 87°C and ambient pressure for 7 days. The crystallization is shown through Figures 4(a)–4(c). Two crystals were selected for element analysis to confirm the polycondensation. The crystals are marked with a green cross (Figures 4(b)-4(c)). The resulting crystals mainly contain Si, Al, O, and Na. The element analysis of the marked crystals resembled erionite and oligoclase. From microstructural observations, it could be said that the selected matrices are not homogeneous (Figures 4(b)-4(c)). One crystal has a Si : Al ratio of 1.5 (Figure 4(b)), while the other crystal has a Si : Al ratio of 2 (Figure 4(c)). Higher magnifications on the crystals showed existence of some fissures as shown in Figure 4(b).

As shown in Figure 4(b), some rod-like species have been identified in some parts of the binder. The influence of these rod-like species on the compressive strength could not be figured out and, in this regard, further studies should be carried out in the future.

3.2.2. X-Ray Diffraction Analysis. Figure 5 shows the XRD patterns of the binder cured for 7 days at 87°C and ambient pressure. The XRD analysis shows that the starting material mainly consists of anorthite sodian ($Al_{1.52}Ca_{0.52}Na_{0.48}O_8Si_{2.48}$). But erionite, albite, and stilbite-Na seem to be synthesized as a result of polycondensation. In addition, the formation of an amorphous phase was noticed as a broad diffuse halo, or a "hump," in the 2θ range between 5 and 15°. Nonetheless, it has been shown that amorphous inorganic polymers are formed in the 2θ range between 20 and 39 degrees. The identification of the amorphous phases requires further study.

The XRD patterns show that when the norite was mixed with pure 8 M NaOH solution, the content of the amorphous phase increased (Figure 5(a)). He et al. [17] reported that curing alkali-activated based slurries at elevated temperatures leads to a higher degree of crystallization. It should be mentioned that, for the K-containing system, the pattern showed lower content of the amorphous phase compared to the Na-containing system. However, the GGBFS used as an additive was almost amorphous.

The oligoclase determined by the XRD patterns ($Al_{1.277}Ca_{0.277}Na_{0.723}Si_{2.723}$) with low content of erionite-Ca ($Al_{3.954}Ca_{1.22}H_{72.096}K_6Mg_{0.26}O_{49.2}Si_{14.046}$) was the reaction product of a Na-containing system (Figure 5(a)). Oligoclase is a high sodium content feldspar which could be reasonable due to the use of Na-containing solution [12]. Erionite-Ca is a member of the zeolite minerals. Gougazeh and Buhl [18] studied the synthesis of zeolite A by treating the activated metakaolin from natural kaolin with various concentrations of NaOH at 100°C. Their result shows that zeolite A was the major constituent phase, while quartz and hydroxysodalite were the minor constituents of the final product.

Further, the patterns show that albite ($AlNaO_8Si_3$) and stilbite-Na ($Al_{2.42}Ca_{0.84}H_{18}Na_{0.75}O_{24.12}Si_{6.64}$) are the reaction products for the K-containing system (Figure 5(b)). Albite is a plagioclase feldspar mineral and stilbite is a series of

FIGURE 4: SEM images of specimen 1 cured for 7 days at 87°C and ambient pressure; (a) pores are visible; (b) needle structures; (c) elemental analysis of a selected point.

tectosilicate minerals of the zeolite group. Kawano and Tomita [19] reported the synthesis of zeolites from obsidian in various concentrations of NaOH and KOH solutions at 150 and 200°C. They showed that smectite, phillipsite, and rhodesite formed in an NaOH solution as pH increased and smectite, merlinoite, and sanidine were produced in KOH solutions as pH increased. The pH and Si : Al and Na : K ratios of the reacting solutions were important parameters determining the nature of the products formed from obsidian. Formation of the zeolite could be due to the quantity of silica and grain size as investigated by Prudhomme et al. [20].

3.2.3. Infrared Analysis (IR). The infrared spectra of the norite and the norite-based binder are given in Figure 6. The binder was cured for 7 days at 87°C and ambient pressure. As shown in Figure 6, the chemical shifts of main IR characteristic bands and appearance of new peaks were observed. The IR for the solid precursors (norite) and the binder consists of the strongest vibrations at 980–1020 cm^{-1} and 420–560 cm^{-1}. Comparing the IR spectra of the norite and norite-based binder shows that, during polycondensation, the outstanding band at 1008 cm^{-1} shifted towards a lower wavenumber (1000 cm^{-1}). It seems that 1008 cm^{-1} band is caused by asymmetrical vibration of Si(Al)-O bonds and 540 cm^{-1} band by bending vibration of Si-O-Al bonds [2, 21]. The appearance of new peaks at 881 and 577 cm^{-1} is due to reaction. Mucsi et al. [22] studied the formation of fly ash-based geopolymer by using Fourier transformed infrared (FTIR) spectroscopy and suggested that the appearance of a new peak at 881 cm^{-1} is related to Si-O stretching and the bending of O-H in the Si-OH bond.

3.3. Influence of the GGBFS on the Compressive Strengths. Inclusion and influence of GGBFS on fly-ash have been studied on the setting time and strength development of fly ash-based binder. It is believed that addition of an amorphous silica phase leads to a shorter setting time and higher compressive strength. It has also been reported that the addition of GGBFS to concrete enhances setting time of concrete at ambient temperature. However, it may affect the

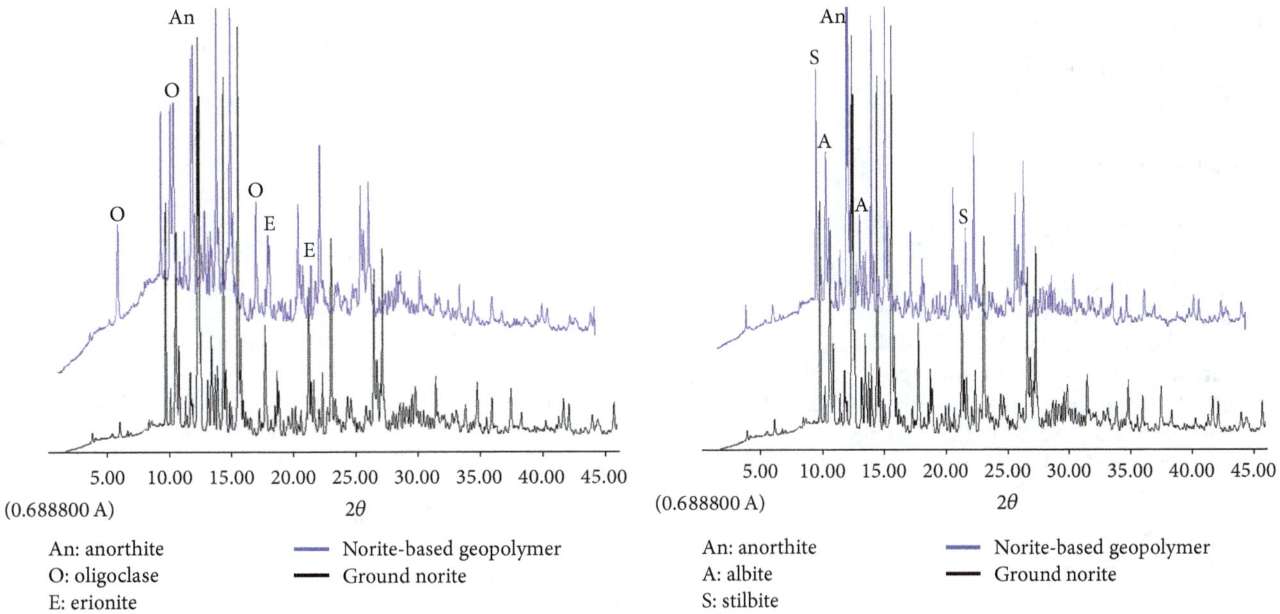

(a) Specimen 1: Na-containing system, amorphous phase as a broad diffuse halo in the 2θ range between 5 and 15°

(b) Specimen 4: K-containing system with GGBFS, lower content of amorphous phase in the 2θ range between 5 and 15°

FIGURE 5: X-ray patterns of the norite-based binders cured for 7 days at 87°C and ambient pressure.

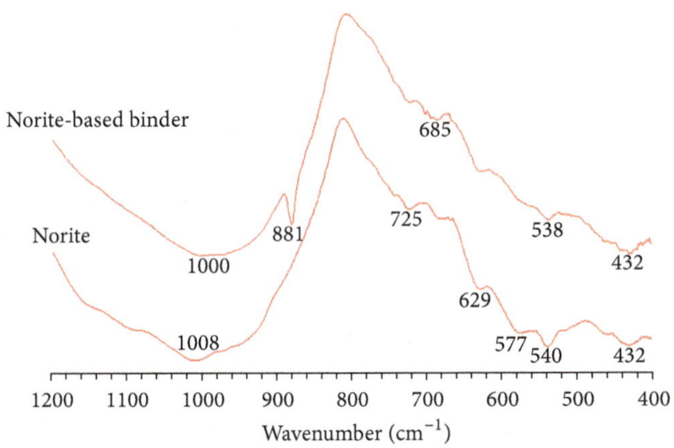

FIGURE 6: FTIR of aluminosilicates of the solid precursor and the norite-based binder, which was cured for 7 days at 87°C and ambient pressure.

final compressive strength of the slurry due to incomplete reaction [16, 23].

As the measured 7-day compressive strength of the binder was low, the influence of GGBFS was studied. As shown in Table 5, addition of the GGBFS to the slurries strongly influenced the compressive strength of the specimens, which were cured at ambient pressure and temperature. However, there was no supplementary information about the hydraulic reactivity relation for the used GGBFS.

3.4. Effect of Alkali Solution. The influence of alkali type was studied in this work. The obtained results show higher compressive strength for the specimens prepared using K-containing systems than with sodium (Table 5). It has been reported that dissolution of aluminosilicate in a sodium solution is more effective than in a potassium solution. However, the compressive strength of the binder is higher for the K-containing systems than for the Na-containing systems [2, 24, 25]. This phenomenon is explained by the size of K^+ cations and the formation of zeolite in the geopolymer structure [26]. Therefore, to get the same strength as the K-containing system, the amount of sodium solution may be increased. The used concentrations are classified as corrosive in the criteria for classification of substances; therefore, the K-containing system may have a lower risk of hazard, which makes it user-friendly. Hence, it could be said that K-containing systems are more user-friendly in comparison to Na-containing systems. However, a combination of potassium and sodium solutions substantially reduces the compressive strength (Table 5).

4. Conclusion

A norite-based binder has been successfully produced from the waste tailing of an ilmenite mine consisting mainly of anorthosite. The particle size of the source material significantly affected the reactivity. The IR spectra of norite-based binder showed a shift towards a lower wavenumber. The maximum compressive strength was obtained for the fully potassium-containing system while a combination of sodium and potassium reduced the strength development. The XRD patterns showed formation of oligoclase, albite, and zeolites. Furthermore, the patterns indicated amorphous phases for the Na-containing systems at 87°C of curing; however, crystallization was expected at elevated temperatures. The SEM images showed some degree of crystallization and heterogeneity in the binder matrices.

Conflicts of Interest

The authors declare that there are no conflicts of interest regarding the publication of this paper.

Acknowledgments

The authors acknowledge the Research Council of Norway, ConocoPhillips, AkerBP, Statoil, and Wintershall for financing the work through the research center DrillWell (Drilling and Well Centre for Improved Recovery), research cooperation between IRIS, NTNU, SINTEF, and UiS. The authors are also grateful to the management of Titania AS, Norway, for their support. Special thanks are due to the Swiss-Norwegian beamlines (SNBL) at the European Synchrotron Radiation Facility (ESRF) for providing beamtime and technical support during the experiment.

References

[1] F. Pacheco-Torgal, J. A. Labrincha, C. Leonelli et al., *Handbook of Alkali-activated Cements, Mortars and Concretes*, Oxford: Woodhead Publishing, Oxford, UK, 5th edition, 2014.

[2] J. Davidovits, *Geopolymer Chemistry and Application*, Saint-Quentin: Institut Geopolymere, Saint-Quentin, France, 3rd edition, 2011.

[3] I. Garcia-Lodeiro, A. Fernández-Jimenez, P. Pena, and A. Palomo, "Alkaline activation of synthetic aluminosilicate glass," *Ceramics International*, vol. 40, no. 4, pp. 5547–5558, 2014.

[4] D. L. Y. Kong, J. G. Sanjayan, and K. Sagoe-Crentsil, "Factors affecting the performance of metakaolin geopolymers exposed to elevated temperatures," *Journal of Materials Science*, vol. 43, no. 3, pp. 824–831, 2008.

[5] E. I. Diaz, E. N. Allouche, and S. Eklund, "Factors affecting the suitability of fly ash as source material for geopolymers," *Fuel*, vol. 89, no. 5, pp. 992–996, 2010.

[6] J. Nemecek, V. Smilauer, and L. Kopecky, "Nanoindentation characteristics of alkali-activated aluminosilicate materials," *Cement and Concrete Composites*, vol. 33, pp. 163–170, 2011.

[7] J. Stark, "Recent advances in the field of cement hydration and microstructure analysis," *Cement and Concrete Research*, vol. 41, pp. 666–678, 2011.

[8] D. Ravikumar, S. Peethamparan, and N. Neithalath, "Structure and strength of NaOH activated concretes containing fly ash or GGBFS as the sole binder," *Cement and Concrete Composites*, vol. 32, no. 6, pp. 399–410, 2010.

[9] M. Khalifeh, A. Saasen, T. Vralstad, and H. Hodne, "Potential utilization of class C fly ash-based geopolymer in oil well cementing operations," *Cement and Concrete Composites*, vol. 53, pp. 10–17, 2014.

[10] M. C. M. Nasvi, P. G. Ranjith, and J. Sanjayan, "The permeability of geopolymer at down-hole stress conditions: application for carbon dioxide sequestration wells," *Applied Energy*, vol. 102, pp. 1391–1398, 2013.

[11] H. Pichler and C. Schmitt-Riegraf, *Rock-Forming Minerals in Thin Section*, Chapman and Hall, london, UK, 2nd edition, 1993.

[12] W. D. Nesse, *Introduction to Mineralogy*, Oxford University Press, Inc, Oxford, UK, 1st edition, 2000.

[13] P. I. Norkyn, *Laboratory Analysis of Norite*, Kronos Titania AS, 2004.

[14] A. P. Hammersley, S. O. Svensson, M. Hanfland, A. N. Fitch, and D. Häusermann, "Two-dimensional detector software: from real detector to idealised image or two-theta scan," *High Pressure Research*, vol. 14, no. 4-5, pp. 235–248, 1996.

[15] Match!, Phase Identification from Powder Diffraction, Version 2, Crystal Impact, Dr. H. Putz and Dr. K. Brandenburg GbR, Kreuzherrenstr, 102, 53227 Bonn, Germany, http://www.crystal-impact.com/match.

[16] P. S. Deb, P. Nath, and P. K. Sarker, "The effects of ground granulated blast-furnace slag blending with fly ash and activator content on the workability and strength properties of geopolymer concrete cured at ambient temperature," *Materials & Design*, vol. 62, pp. 32–39, 2014.

[17] P. He, D. Jia, M. Wang, and Y. Zhou, "Thermal evolution and crystallization kinetics of potassium-based geopolymer," *Ceramics International*, vol. 37, no. 1, pp. 59–63, 2011.

[18] M. Gougazeh and J.-C. Buhl, "Synthesis and characterization of zeolite A by hydrothermal transformation of natural Jordanian kaolin," *Journal of the Association of Arab Universities for Basic and Applied Sciences*, vol. 15, no. 1, pp. 35–42, 2014.

[19] M. Kawano and K. Tomita, "Experimental study on the formation of clay minerals from obsidian by interaction with acid solution at 150° and 200°C," *Clays and Clay Minerals*, vol. 43, no. 2, pp. 212–222, 1995.

[20] E. Prudhomme, P. Michaud, E. Joussein, and S. Rossignol, "Influence of raw materials and potassium and silicon concentrations on the formation of a zeolite phase in a geopolymer network during thermal treatment," *Journal of Non-Crystalline Solids*, vol. 358, no. 16, pp. 1908–1916, 2012.

[21] Z. Yunsheng, S. Wei, C. Qianli, and C. Lin, "Synthesis and heavy metal immobilization behaviors of slag based geopolymer," *Journal of Hazardous Materials*, vol. 143, no. 1-2, pp. 206–213, 2007.

[22] G. Mucsi, S. Kumar, B. Csoke et al., "Control of geopolymer properties by grinding of land filled fly ash," *International Journal of Mineral Processing*, vol. 143, pp. 50–58, 2015.

[23] H. Xu, W. Gong, L. Syltebo, K. Izzo, W. Lutze, and I. L. Pegg, "Effect of blast furnace slag grades on fly ash based geopolymer waste forms," *Fuel*, vol. 133, pp. 332–340, 2014.

[24] L. Xin, J.-Y. Xu, W. Li, and E. Bai, "Effect of alkali-activator types on the dynamic compressive deformation behavior of geopolymer concrete," *Materials Letters*, vol. 124, pp. 310–312, 2014.

[25] G. S. Ryu, Y. B. Lee, K. T. Koh, and Y. S. Chung, "The mechanical properties of fly ash-based geopolymer concrete with alkaline activators," *Construction and Building Materials*, vol. 47, pp. 409–418, 2013.

[26] M. Khalifeh, A. Saasen, T. Vrålstad, H. B. Larsen, and H. Hodne, "Experimental study on the synthesis and characterization of aplite rock-based geopolymers," *Journal of Sustainable Cement-Based Materials*, vol. 5, no. 4, pp. 233–246, 2016.

16

Study of Tribological Behavior of Silicon Carbide Based Aluminum Metal Matrix Composites under Dry and Lubricated Environment

Kalyan Kumar Singh,[1] Saurabh Singh,[1] and Anil Kumar Shrivastava[2]

[1]*Department of Mechanical Engineering, Indian School of Mines, Dhanbad 826004, India*
[2]*Department of Mechanical Engineering, Galgotias College of Engineering & Technology, Greater Noida 201306, India*

Correspondence should be addressed to Saurabh Singh; saurabh.091638.bme@gmail.com

Academic Editor: Guru P. Dinda

Friction and wear behavior of silicon carbide based aluminum metal matrix composite and aluminum matrix alloy have been studied for sliding speeds of 3.14 m/s and 3.77 m/s and load range from 10 N to 30 N under dry and lubricated environment, respectively. The experiments were performed on pin on disk tribometer (Make: DUCOM). The composite was fabricated by stir casting process and has several challenges like inferior bonds and interfacial reaction products which will deteriorate the mechanical and tribological properties. Therefore, addition of reactive metal like magnesium (Mg) should be done which will lead to reduced solidification shrinkage, lower tendency towards hot tearing, and faster process cycles. Results have revealed that the developed composites have lower coefficient of friction and wear rates when compared with aluminum matrix alloy under dry and lubricated environment. Experimental results show that under dry condition coefficient of friction of both the matrix alloy and the composite decreases with increase in load, whereas it increases with increase in sliding speeds; on the other hand wear rates of both aluminum matrix alloy and the composites increase with increase in load as well as with sliding speeds. FESEM of worn surfaces are also used to understand the wear mechanisms.

1. Introduction

Aluminum alloy possesses excellent properties like high strength-to-weight ratio, corrosion resistance, good ductility, low density, high thermal conductivity, and so forth, thus making it suitable for wide range of applications in several industries like automobile, aerospace, structural applications, and so forth [1]. In recent times, the uses of high strength aluminum alloys are focused on structural applications in aerospace as well as on general engineering sectors. However aluminum alloys have not been well known for high performance tribological application because of their low hardness, poor wear resistance, and poor weldability. Thus, there is a greater possibility of oscillation since these materials are joined by riveting, bolting, and so forth and due to this it may often lead to sliding wear in dry environmental conditions. Therefore, to overcome this problem reinforcement of particulates, fiber, or whiskers can be used; these reinforcements are very well known for their high-specific strength and uniform distribution [2] which leads to use of metal matrix composites to complete the demand for light weight, high performance, environment friendly, wear and corrosion resistance material. Generally, Ag, Al, Be, Co, Cu, Fe, Mg, Ni, and Ti are produced and used as matrices in metal matrix composites. Due to excellent physical, mechanical, and tribological properties and good dimensional stability of aluminum matrix composite it has high commercial application compared to other available metals. The demand for aluminum composite in automobile sector has been increased drastically; this composite is used for the manufacturing of pistons, brake drums, cylinder liners, and pushrods [3, 4]. A commercial Al 7075 aluminum alloy is one of the highest strength aluminum alloys available and is having excellent strength-to-weight ratio and is formed in annealed conditions and subsequently heat treated. But due to its low wear resistance and poor tribological behavior it cannot be widely used; therefore

Table 1: Chemical composition (weight %) of aluminum matrix alloy (Al 7075).

Material	Si	Fe	Cu	Cr	Zn	Ti	Mn	Mg	Al	Others
Content	Max 0.4	0.5	1.2–2	0.18–0.28	5.1–6.1	0.2	0.3	2.1–2.9	87.1–91.4	0.15

reinforcement of particles has been done to improve its tribological properties. There are various particles available which are used for reinforcement such as B_4C, SiC, TiC, B, C, and Al_2O_3. With the reinforcement of high modulus ceramics particulates such as silicon carbide (SiC) having properties like more stability and chemical compatibility with aluminum matrix and low cost, wide range of available grade made aluminum alloys are suitable candidate for wider spectrum of applications [5]. The hard reinforcement of silicon carbide particle improved the properties like hardness and wear resistance of composite by acting as load bearing components [6, 7]. Many investigators have studied both parameters, that is, reinforcement size and volume fraction; Aiguo and Rack [8] observed that abrasive wear resistance in unreinforced 7091 alloys is less in comparison with both silicon carbide particle and silicon carbide whiskers. Wear resistance increases and plastic deformation decreases with the increase of reinforcement [9, 10]. Kwok and Lim [11] observed the size of silicon carbide reinforcement at high speed and suggested that when the particulate size becomes small the wear resistance rapidly deteriorates. Therefore, silicon carbide based aluminum metal matrix composites are mainly preferred for low speed applications; silicon carbide is commonly used among the reinforcement particles like carbides, nitrides, borides, and oxides, because it has several advantages like high melting point, high elastic modules, hardness, good thermal conductivity, high strength, excellent wear resistance, good availability, and low cost. Silicon carbide based aluminum metal matrix composites are widely used in automotive, marine, and other sectors because of its excellent properties such as good fluidity and low coefficient of thermal expansion and its quality depends on chemical composition, melting process, casting process, and solidification rate. As the volume fraction addition of silicon carbide in aluminum alloys increases, the value of ultimate tensile strength (UTS) increases and low degree of porosity is obtained.

There is not much improvement in strength and porosity of material, if the volume fraction of silicon carbide exceeds 10%; therefore in the present analysis 10% silicon carbide has been added in aluminum alloys as reinforcement. Due to low degree of porosity and uniform distribution of silicon carbide particles there is enhancement in tensile strength of composites and as a result it leads to effective transfer of tensile loads to the uniformly distributed silicon carbide particulates. In silicon carbide based aluminum metal matrix composites, casting properties increased by adding silicon whereas magnesium (Mg) made its heat treatable which greatly influenced the mechanical properties of composites.

There are several manufacturing techniques that are available for particle reinforcement in metal matrix composites such as liquid infiltration, spray decomposition, squeeze casting, compocasting, powder metallurgy, and mechanical alloying. Among all these available methods, stir casting method is attractive method for manufacturing of metal matrix composites (MMCs) because of low cost, near-net shape formation, and wide range of processing applications [1, 5, 12]. The mechanical strength of aluminum metal matrix composites is greatly influenced by particles size reinforcement; if size of particles ranges from micrometer to nanometric scale there is drastic increase in mechanical strength of composites but it will increase chance of agglomeration and particle clustering which will reduce the strength of composites. Therefore, it is important that reinforcement of particles should be homogeneous to achieve the improved properties of composites [13]. In stir casting method, ceramics reinforcement (i.e., silicon carbide) and metal matrix which is melted during casting are mixed homogeneously by mechanical stirrer. Fabrication of aluminum metal matrix composites has several challenges like inferior bonds and interfacial reaction products and it will directly deteriorate the mechanical properties and tribological properties of composites, which make them unsuitable for industrial application. Therefore, to overcome this problem, addition of reactive metal like magnesium has been done and the addition of this metal results in reduced solidification shrinkage, lower tendency towards hot tearing, suppression of segregation, settling or agglomeration, and faster process cycles [12].

Thus by analyzing various parameters and their effects and utilization this experiment was carried out under various environmental conditions to produce a material which can sustain a longer time. Composite fabricated has much better properties as compared with its base alloys (details and validation of which are shown in the experimental analysis). The severity of worn-out surfaces (i.e., wear scars, change of microstructure) was also analyzed using field emission scanning electron microscopes (FESEM).

2. Experimental Procedure

2.1. Fabrication of Silicon Carbide Based Aluminum Metal Matrix Composite. The composite was prepared by commercially available aluminum (Al 7075) matrix alloy and silicon carbide (10% by weight) which is used as reinforcement. The average size of aluminum matrix alloy was 150 microns to 160 microns and that of silicon carbide was 40 microns to 50 microns. In the present investigation, aluminum matrix alloy was chosen as a base matrix since it is having very high tensile strength of about 275 MPa. The addition of silicon carbide will enhance the wear resisting property of silicon carbide based aluminum metal matrix composite. The chemical composition of aluminum matrix alloys is presented in Table 1.

2.2. Composite Preparation. In order to achieve high level of mechanical properties of the composite, a good interfacial

TABLE 2: Chemical composition of silicon carbide based aluminum metal matrix composite.

Material	C	O	Mg	Al	Si	Fe	Ag	Zn	Cu
Content	24.27	20.61	1.13	36.43	10.62	1.96	0.84	3.34	0.80

bonding (wetting) between the dispatched phase and the liquid matrix has to be obtained. Fabrication of composite with discontinuous fiber can be economically done using liquid metallurgy route/technique [10]. Stir casting is a liquid state method which is the simplest and is cost effective to fabricate metal matrix composite. Care was taken to maintain optimum casting parameters of pouring temperature (670°C) stirring speed of 500 rpm, stirring time 20 minutes. The reinforcement was preheated at 600°C to the addition in aluminum matrix alloy (Al 7075). To avoid moisture from the particles, preheating of the reinforced particle should be done; otherwise, there is a chance of agglomeration of particulate which occurs due to moisture and gasses present in the particles. Degassing agent (hexachloroethane and magnesium) was used to reduce the gas porosities. The addition of magnesium will also lead to reduced solidification shrinkage, lower tendency towards hot tearing, suppression of segregation, settling or agglomeration, and faster process cycles. The depth of immersion of stirrer was maintained at about two-thirds of the depth of molten metal. The molten metal was then poured into a permanent mould preheated at 300°C of 9 mm diameter and 55 mm length. Specimen was left for curing and die was released after 6 hours; finally, on the lathe cast specimens were parted out in 2 pieces, that is, 25 mm length and 8 mm diameter, respectively. The chemical compositions of silicon carbide based aluminum metal matrix composite are presented in Table 2.

2.3. Composite and Aluminum Matrix Alloy Testing Procedure. The experiment was conducted at room temperature on a pin on disk wear and friction maintaining machine (Make: DUCOM instruments private limited, Bangalore diameter of 165 mm, thickness of 8 mm and made of material EN31 hardened to 60 HRC ground to 1.6 Ra surface roughness. Figure 1 shows the photographic view of disk). The size of cylindrical specimen which was used for testing is 8 mm diameter as well as 25 mm length. Before performing every experiment disk surface was cleaned by rubbing 600-grade Si6 emery paper and washed with acetone. Three different loads, that is, 10 N, 20 N, and 30 N, have been selected for wear and friction behavior of specimen with track diameter of 100 mm and 120 mm used for both aluminum matrix alloy and silicon carbide based aluminum metal matrix composite. The test duration was 10 minutes at a constant speed of 600 rpm for the entire test in dry and oil lubricated (SAE 20 with kinematic viscosity 25 to 30 centistokes at 50°C and oil has been used for lubrication) conditions. For oil lubricated experiments, initially, 2 drops of lubricating oil were placed over sliding surface after that oil was poured at 0.02 mL/min. All sets of experiments were repeated three times in the same manner for dry and oil lubricated conditions, respectively. Frictional forces and coefficient of friction were measured directly from the machine. For the calculation of wear loss

TABLE 3: Variation of coefficient of friction with load for aluminum matrix alloy and silicon carbide based aluminum metal matrix composite at $V = 3.14$ m/sec.

Sliding condition & load	COF for aluminum matrix alloy	COF for silicon carbide based aluminum metal matrix composite
Dry 10 N	0.65	0.42
Dry 20 N	0.57	0.37
Dry 30 N	0.47	0.30
Lubricated 10 N	0.46	0.32
Lubricated 20 N	0.43	0.25
Lubricated 30 N	0.36	0.19

Track diameter = 100 mm, $N = 600$ rpm, sliding speeds = 3.14 m/sec, and ΔT = 600 sec.

FIGURE 1: DUCOM Machine.

in terms of weight, weights of specimen before and after the experiment were taken.

The sliding speed of the test specimen is calculated by using the following formula:

$$V = \frac{\pi * D * N}{60000}. \tag{1}$$

The following calculation is used to obtain wear rate:

$$K_0 = \frac{\Delta m}{\rho * F * d}. \tag{2}$$

3. Results and Discussion

3.1. Coefficient of Friction. The variations of coefficient of friction for aluminum matrix alloy (Al 7075) and composites under dry and lubricated condition at different normal loads and two sliding speeds (i.e., 3.14 m/s and 3.77 m/s) have been shown in Tables 3 and 4, respectively. Table 3 shows that there is significant decrease in the value of coefficient of friction for composite compared with aluminum matrix alloy under dry and lubricated condition. Under dry condition, 30%–55%

TABLE 4: Variation of coefficient of friction with load for aluminum matrix alloy and silicon carbide based aluminum metal matrix composite at $V = 3.77$ m/sec.

Sliding condition & load	COF for aluminum matrix alloy	COF for silicon carbide based aluminum metal matrix composite
Dry 10 N	0.72	0.56
Dry 20 N	0.63	0.43
Dry 30 N	0.50	0.35
Lubricated 10 N	0.58	0.44
Lubricated 20 N	0.47	0.32
Lubricated 30 N	0.37	0.25

Track diameter = 120 mm, $N = 600$ rpm, sliding speeds = 3.77 m/sec, and $\Delta T = 600$ sec.

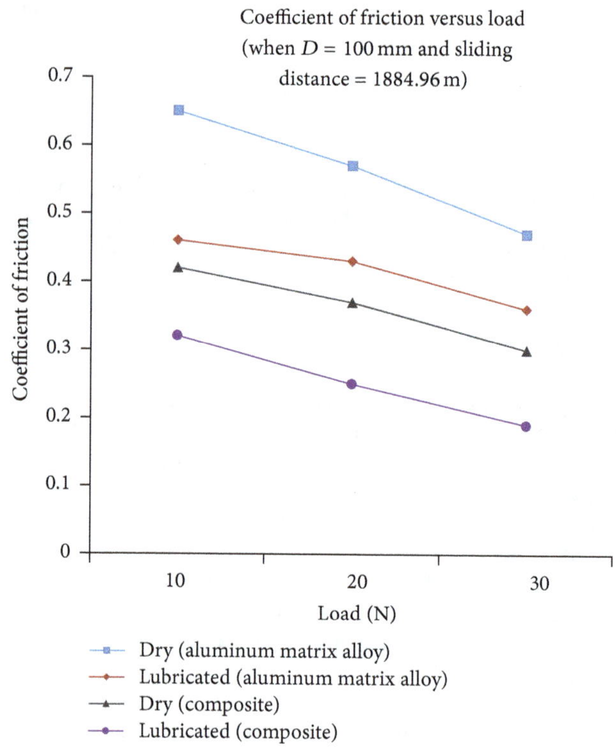

FIGURE 2: Variation of COF at $V = 3.14$ m/s.

reductions in value of coefficient of friction for load range 10 N–30 N have been observed; this is because of uniform mixing of silicon carbide with aluminum matrix alloy and results in excellent bond between matrix alloy and silicon carbide particle. These reinforced silicon carbide particles act as load bearing members as the loads increase and reduce the touch area between specimen and rotating disk's surfaces [1] which leads to lower coefficient of friction. Similar observations have been observed by other investees [14, 15].

However, under lubricated condition, the value of coefficient of friction reduced for both aluminum matrix alloy and composites compared with dry condition; this is because of thin lubricating film formation between specimen and rotating disk surface. This thin lubricating film contains lumps of particles and as temperature and normal loads increase these particles break into a small number of particles and reorient themselves along the sliding plane and this leads to lower coefficient of friction.

Table 4 shows variation of coefficient of friction at increased sliding distance for both aluminum matrix alloy and composites. As the sliding distance increases the value of coefficient of friction increases for both aluminum matrix alloy and composites because of temperature rise of rubbing surfaces for aluminum matrix alloy and reinforcement of silicon carbide into aluminum matrix alloys makes the matrix alloy plastically constrained and it can sustain high temperature compared with aluminum matrix alloy. Silicon carbide also protects the matrix alloy from severe contact with counter surface and leads to lower coefficient of friction. Similar observations have been made by other investees [16]. As sliding distance increases, the thin lubricating film between specimen and rotating disk surface may get thickened and be responsible for breakage of small particles from specimen and these broken particles may be clogged between the rubbing surfaces and this leads to higher coefficient of friction.

Figures 2 and 3 show the variation of coefficient of friction with increasing normal loads under dry and lubricating condition at sliding speeds 3.14 m/s and 3.77 m/s, respectively. It has been observed that temperature rise is greater in case of unreinforced aluminum matrix alloys compared to those in composites irrespective of the applied load and surface conditions. The rise in temperature leads to frictional heating in between the rubbing surfaces and this frictional heating is continuous because there is much less time for heat dissipation that causes higher coefficient of friction for aluminum matrix alloy compared to composites. Initially, aluminum alloys and composites particles are stronger and sharper which causes higher frictional heating and leads to higher coefficient of friction but as the time increases, there is increase in flow ability of material on specimen surface that leads to increase in slipping action and results in reduced frictional heating that leads to lower coefficient of friction at higher loads [1]. The composite shows lower coefficient of friction compared with aluminum matrix alloy because of presence of secondary phase of silicon carbide particles in aluminum matrix alloy which will restrict the flow of metal during sliding [17]. The uniform distribution of silicon carbide particle in aluminum matrix alloy would lead to sustaining higher loads with less plastic deformation and give us good load bearing capacity of composite at higher loads. The minimum values of coefficient of friction for both aluminum matrix alloy and composite have been observed at lower speeds; the same observations have been made by other investees [11].

3.2. Wear Studies

3.2.1. Weight Loss and Wear Rates. Effects of increasing loads (i.e., 10 N, 20 N, and 30 N) with sliding distance 1884.96 m (numerical values shown in Table 5) and 2262 m (numerical values in Table 6) on weight loss and wear rates for both aluminum matrix alloy and composites under dry and lubricated condition have been shown in Figures 4 and 5, respectively.

TABLE 5: Variation of weight loss and wear rate for aluminum matrix alloy and silicon carbide based aluminum metal matrix composite when sliding speeds = 3.14 m/s.

Sliding condition & load	$(\text{Weight loss})_{\text{alloy}}$ (mg)	$(\text{Wear rate})_{\text{alloy}}$ (mm^3/N-m)	$(\text{Weight loss})_{\text{composite}}$ (mg)	$(\text{Wear rate})_{\text{composite}}$ (mm^3/N-m)
Dry 10 N	3.8	4.662×10^{-5}	2.1	2.349×10^{-5}
Dry 20 N	8.6	5.275×10^{-5}	4.5	2.516×10^{-5}
Dry 30 N	15.2	6.216×10^{-5}	8.5	3.168×10^{-5}
Lubricated 10 N	3.0	3.680×10^{-5}	1.5	1.677×10^{-5}
Lubricated 20 N	6.5	3.987×10^{-5}	3.1	1.733×10^{-5}
Lubricated 30 N	13.2	5.398×10^{-5}	6.2	2.311×10^{-5}

$\rho_{\text{alloy}} = 4.324 \times 10^{-6}$ kg/mm^3, $\rho_{\text{composite}} = 4.744 \times 10^{-6}$ kg/mm^3, sliding distance (d) = 1884.96 m, and ΔT = 600 sec.

TABLE 6: Variation of weight loss and wear rate for aluminum matrix alloy and silicon carbide based aluminum metal matrix composite when sliding speeds = 3.77 m/s.

Sliding condition & load	$(\text{Weight loss})_{\text{alloy}}$ (mg)	$(\text{Wear rate})_{\text{alloy}}$ (mm^3/N-m)	$(\text{Weight loss})_{\text{composite}}$ (mg)	$(\text{Wear rate})_{\text{composite}}$ (mm^3/N-m)
Dry 10 N	6.7	6.85×10^{-5}	2.6	2.423×10^{-5}
Dry 20 N	13.7	7.003×10^{-5}	5.9	2.750×10^{-5}
Dry 30 N	22.4	7.634×10^{-5}	11.1	3.448×10^{-5}
Lubricated 10 N	4.9	5.009×10^{-5}	1.9	1.770×10^{-5}
Lubricated 20 N	11.2	5.725×10^{-5}	4.2	1.957×10^{-5}
Lubricated 30 N	17.5	5.964×10^{-5}	7.9	2.454×10^{-5}

$\rho_{\text{alloy}} = 4.324 \times 10^{-6}$ kg/mm^3, $\rho_{\text{composite}} = 4.744 \times 10^{-6}$ kg/mm^3, sliding distance (d) = 2262 m, and ΔT = 600 sec.

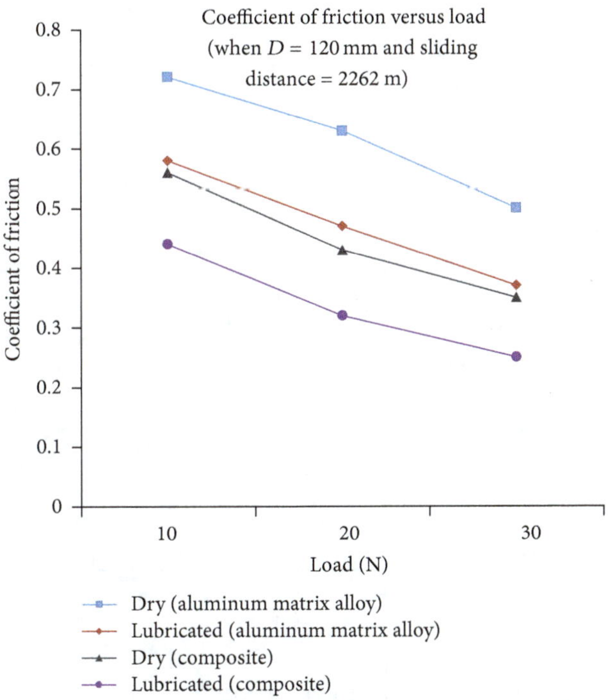

FIGURE 3: Variation of COF at V = 3.77 m/s.

Figure 4 shows the variation of weight loss and wear rates at sliding distance 1884.96 m with increasing loads. Mazahery and Shabani [18] observed that wear rates vary with normal rate and significantly it is low for composites compared to unreinforced alloys and it is indicative of Archard's law. In dry condition, rate of weight loss increases with increase in loads for both aluminum matrix alloy and composites because, with increasing loads, the rate of abrasion also increases and it will lead to detachment of metals from surfaces and leads to more loss of materials. But the rates of weight loss in composites are lower compared with aluminum matrix alloy because the presence of silicon carbide in aluminum matrix alloy controls the deformation and leads to less weight loss. The worn surfaces of composites are smoother and also shallower groove formations as the time increases which lead to higher wear resistance for composites compared with unreinforced aluminum matrix alloys [5]. As earlier we have seen that wear rates are directly proportional to weight loss, therefore wear rates will also be lower for composites compared with aluminum matrix alloy.

In lubricated condition, wear rates are lower compared with dry condition for both aluminum matrix alloy and composites because of thin lubricating film formation over rotating disk surfaces, and this thin lubricating film decreases the adhesion of small broken particles of aluminum matrix alloy and composites and leads to lower wear rates although it increases with increasing loads and if an applied load exceeds critical values, reinforced particle will break and loses its load carrying capacity properties and leads to more wear rates [5]. Critical loads basically mean that wear rates increase marginally with linear trends but suddenly these trends will

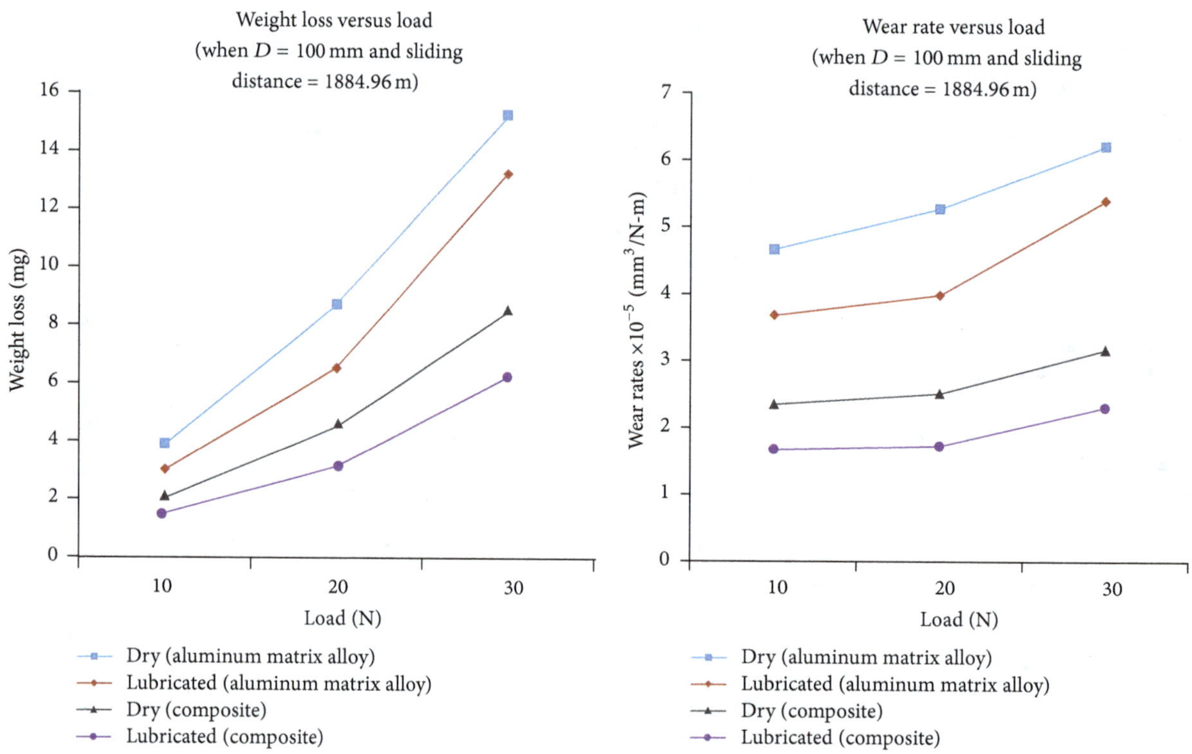

FIGURE 4: Variation of weight loss and wear rates at sliding distance 1884.96 m with increasing loads.

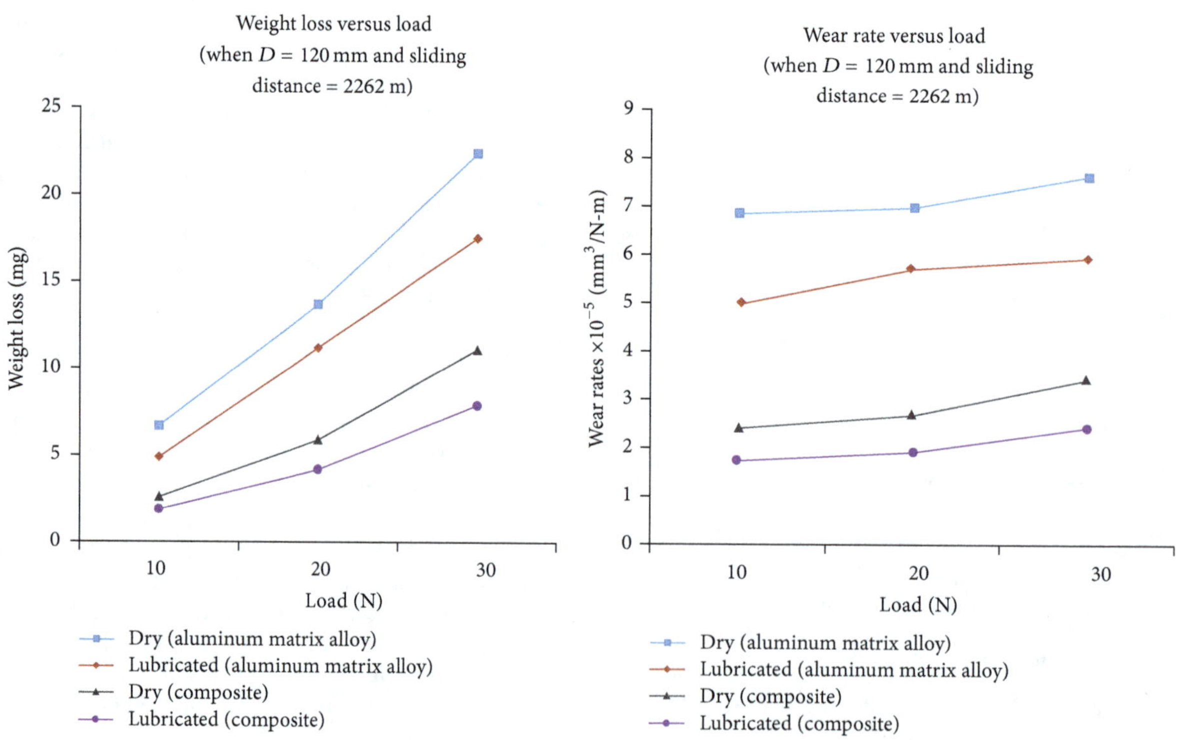

FIGURE 5: Variation of weight loss and wear rate at sliding distance 2262 m with increasing loads.

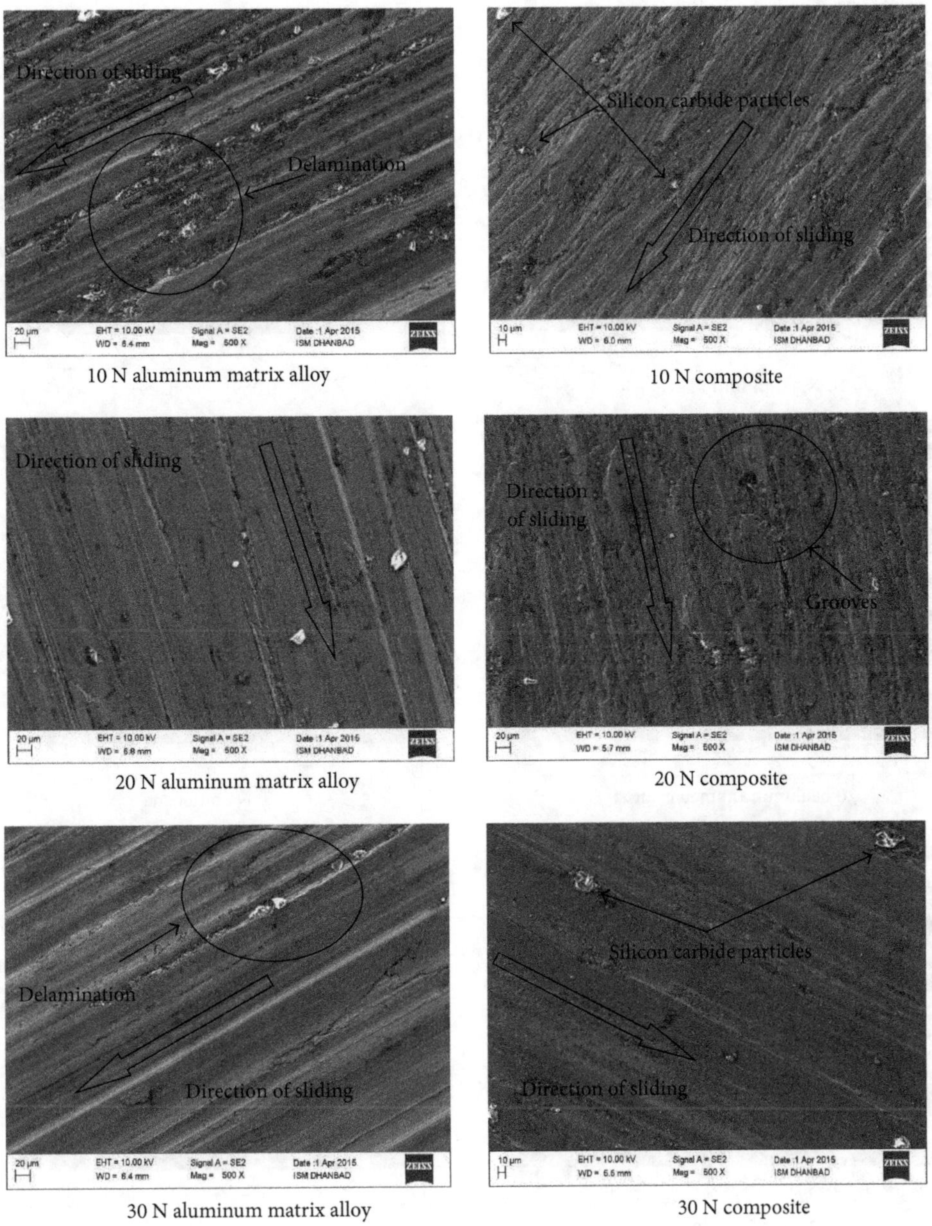

FIGURE 6: FESEM of worn surfaces at different loads with sliding speeds 3.14 m/s under dry condition.

not follow the loads up to which these trends that follow are known as critical loads.

Figure 5 shows the variation of weight loss and wear rates with increased sliding distance 2262 m with increasing loads. The increase of sliding distance leads to increase in weight loss and wear rates for both aluminum matrix alloy and composites because as sliding distance increases, the contact between rotating disk surface and sliding surface of specimen becomes more familiar with elapse of time; because of this the temperature between the rotating disk surface and sliding surface of specimen increases and leads to softening of materials and plastic state of materials occurs. Similar observations have been made by other investees [19, 20].

3.2.2. FESEM Analysis. FESEM micrographs of all worn surfaces for both aluminum matrix alloy and composites under dry and lubricated condition at sliding speeds 3.14 m/s and 3.77 m/s have been shown at magnification rate (×500).

Figure 6 shows the FESEM analysis of all worn surfaces for both aluminum matrix alloy and composites at different loads with sliding speeds 3.14 m/s under dry condition. It is clearly observed that, because of plastic flow of material, the wear rates for both aluminum matrix alloy and composites increase with increase of loads and severity of plastic flow increases at higher loads. At higher loads, the wear debris is mixture of fine and coarse size particles; this is because of breakdown of large size particles. As the load increases

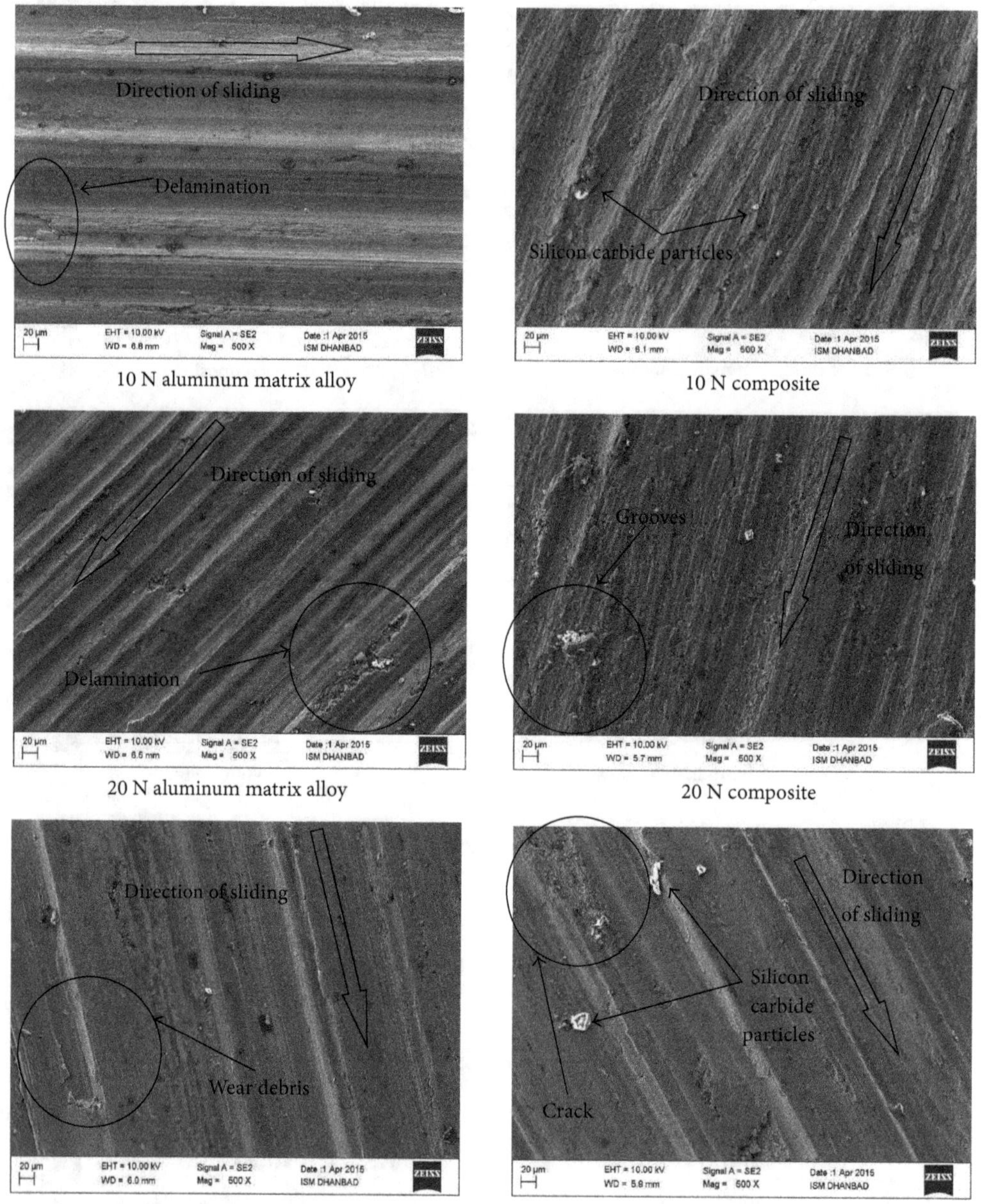

FIGURE 7: FESEM of worn surfaces at different loads with sliding speeds 3.14 m/s under lubricated condition.

the intersection between the sliding surface of specimen and rotating disk surface increases and leads to breakage of large particles into smaller number of particles. At lower loads, worn surfaces are smooth and grooves are small in size compared with higher loads; with increase in loads these grooves become severe and lead to formation of fracture spots. At these conditions reinforced particles cannot be in stable equilibrium and this leads to increase in wear rates.

Figure 7 shows the FESEM analysis of all worn surfaces for both aluminum matrix alloy and composites at different loads with sliding speeds 3.14 m/s under lubricated condition. In lubricated condition lower wear rates have been observed compared with dry condition because of thin lubricating film formation between sliding surface of specimen and rotating disk surface and it decreases the ploughing action and leads to decrease in breakage of small particles from aluminum matrix alloy and composites; therefore, lower wear rates have been observed in lubricated condition compared with dry condition.

Figure 8 shows the FESEM analysis of all worn surfaces for both aluminum matrix alloy and composites at different loads with sliding speeds 3.77 m/s under dry condition. As the sliding distance increases, there is gradual increase in wear rates for both aluminum matrix alloy and composites. The above results can be observed because of the following reason:

(i) With increase in sliding distance, the temperature at specimen interface increases and leads to softening the materials and plastic flow of materials.

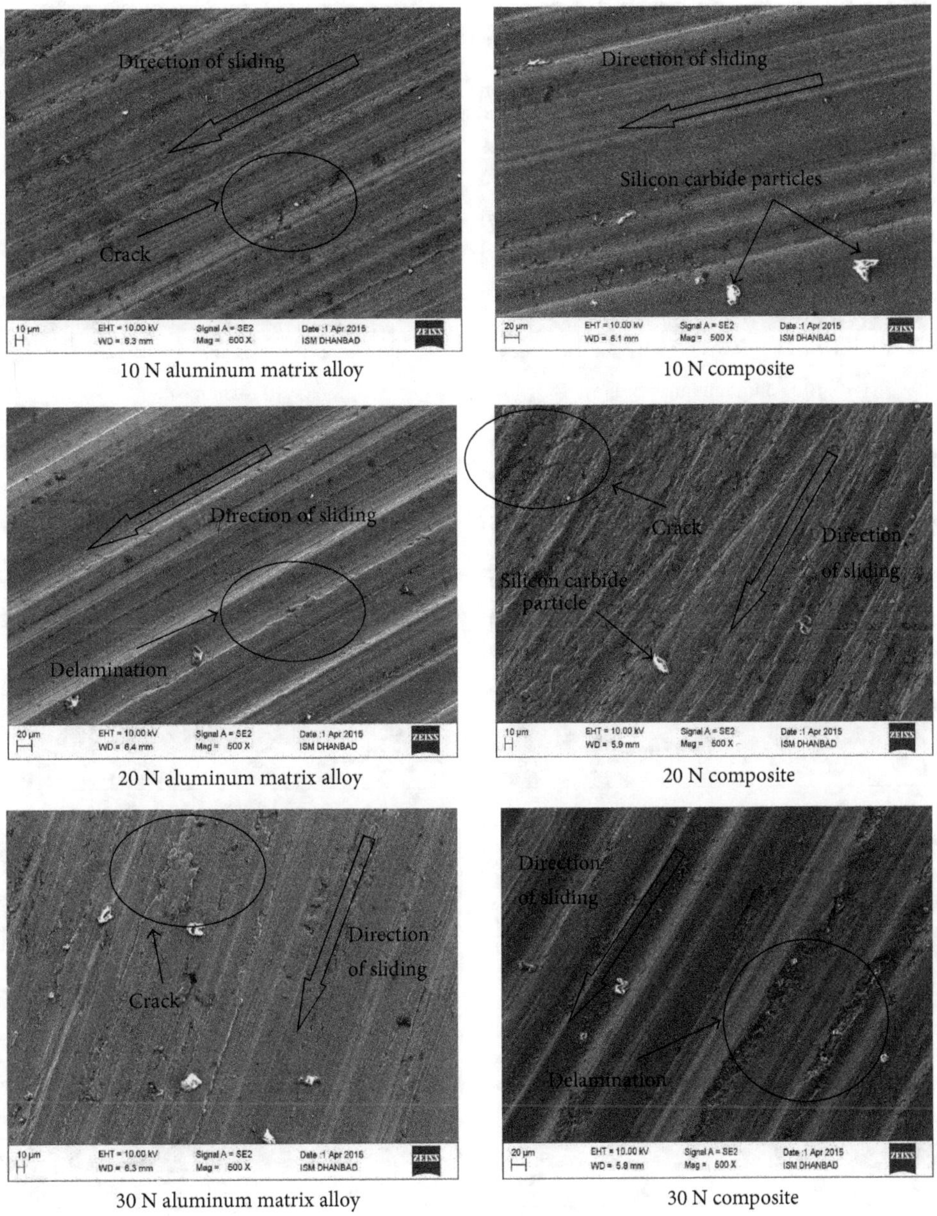

FIGURE 8: FESEM of worn surfaces at different loads with sliding speeds 3.77 m/s under dry condition.

(ii) With increase in sliding distance, the contact between sliding surface of specimen and rotating disk surface becomes intimate as time increases resulting in higher wear rates.

However, the wear rates are lower for composites compared with aluminum matrix alloy because some particles of silicon carbide break into small pieces and crushed as the disk rotates and act as abrasive medium. These powdered silicon carbide particles along with aluminum small particles made a layer which is called mechanically mixed layer (MML). These layers prevent the plastic flow and reduce the wear rates compared with aluminum matrix alloy although plastic flow of composites occurs at later stage.

Figure 9 shows the FESEM analysis of all worn surfaces for both aluminum matrix alloy and composites at different loads with sliding speeds 3.77 m/s under lubricated condition. At higher sliding speeds, wear rates are lower in lubricated condition because of thin lubricating film formation and another mechanically mixed layer (MML) formation delays the plastic flow of material and leads to lower wear rates in lubricated condition compared with dry condition.

4. Conclusions

After observing the condition, that is, dry and oil lubricated condition of aluminum matrix alloy and silicon carbide based aluminum metal matrix composite, the following conclusion can be drawn:

(1) The values of coefficient of friction are found to be maximum in case of dry condition in matrix alloy

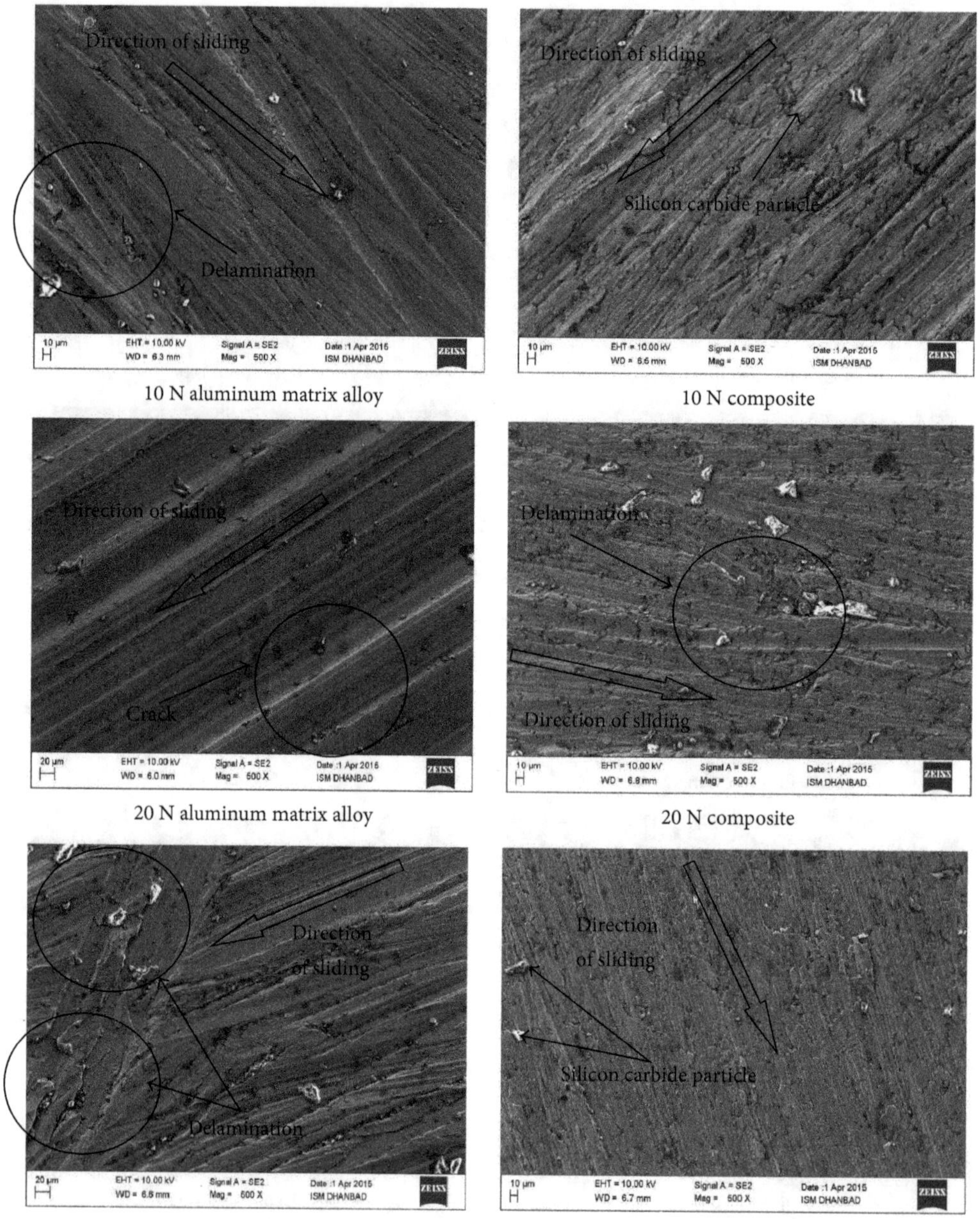

10 N aluminum matrix alloy — 10 N composite

20 N aluminum matrix alloy — 20 N composite

30 N aluminum matrix alloy — 30 N composite

FIGURE 9: FESEM of worn surfaces at different loads with sliding speeds 3.77 m/s under lubricated condition.

at sliding speed 3.77 m/s and minimum in case of lubricated condition in composite at sliding speed 3.14 m/s. The range of coefficient of friction lies from 0.19 to 0.72.

(2) The values of coefficient of friction and wear rates are minimum for both aluminum matrix alloy and silicon carbide based aluminum metal matrix composites under lubricated condition compared with dry condition.

(3) As we increase normal load and sliding speed, the wear rate increases and it is found to be maximum in dry condition of matrix alloy at 30 N.

(4) Uniform distribution of silicon carbide in aluminum matrix alloy reduces the values of coefficient of friction and wear rates for composites compared with aluminum matrix alloy under dry and lubricated condition which prolongs the life of composites for longer duration.

(5) The reinforcements of silicon carbide in aluminum matrix alloy significantly improve the strength of material in dry and lubricated condition and act as load bearing members as the loads and sliding speeds increase and give the best results in lubricated condition at low sliding speeds and loads.

(6) From the above analysis and discussions we conclude that silicon carbide based aluminum metal matrix composite should be preferred for low speed applications as the same results are also observed by other researchers [21].

Nomenclature

V: Sliding speeds (m/sec)
D: Diameter of track (mm)
N: Rotating disk speed (rpm)
F: Normal load (N)
K_0: Wear rate (mm^3/N-m)
d: Sliding distance in meter
ρ: Density of specimen (kg/mm^3)
Δm: Weight loss of specimen (mg).

Competing Interests

The authors declare that they have no competing interests.

Acknowledgments

The authors are grateful to the Advanced Composite Lab of Department of Mechanical Engineering of Indian School of Mines, Dhanbad, India, for providing them with the experimental support for carrying out the laboratory work.

References

[1] A. Mazahery and M. O. Shabani, "Study on microstructure and abrasive wear behavior of sintered Al matrix composites," *Ceramics International*, vol. 38, no. 5, pp. 4263–4269, 2012.

[2] C. G. E. Mangin, J. A. Isaacs, and J. P. Clark, "MMCs for automotive engine applications," *JOM*, vol. 48, no. 2, pp. 49–51, 1996.

[3] G. Straffelini, F. Bonollo, A. Molinari, and A. Tiziani, "Influence of matrix hardness on the dry sliding behaviour of 20 vol.% Al_2O_3-particulate-reinforced 6061 Al metal matrix composite," *Wear*, vol. 211, no. 2, pp. 192–197, 1997.

[4] H.-S. Chu, K.-S. Liu, and J.-W. Yeh, "An in situ composite of Al (graphite, Al4C3) produced by reciprocating extrusion," *Materials Science and Engineering A*, vol. 277, no. 1-2, pp. 25–32, 2000.

[5] A. Mazahery and M. O. Shabani, "The effect of primary and secondary processing on the abrasive wear properties of compocast aluminum 6061 alloy matrix composites," *Protection of Metals and Physical Chemistry of Surfaces*, vol. 50, no. 6, pp. 817–824, 2014.

[6] S. C. Lim, M. Gupta, L. Ren, and J. K. M. Kwok, "The tribological properties of Al–Cu/SiCp metal-matrix composites fabricated using the rheocasting technique," *Journal of Materials Processing Technology*, vol. 80–90, pp. 591–596, 1999.

[7] C. García-Cordovilla, J. Narciso, and E. Louis, "Abrasive wear resistance of aluminium alloy/ceramic particulate composites," *Wear*, vol. 192, no. 1-2, pp. 170–177, 1996.

[8] W. Aiguo and H. J. Rack, "Abrasive wear of silicon carbide particulate- and whisker-reinforced 7091 aluminum matrix composites," *Wear*, vol. 146, no. 2, pp. 337–348, 1991.

[9] P. Rodrigo, M. Campo, B. Torres, M. D. Escalera, E. Otero, and J. Rams, "Microstructure and wear resistance of Al–SiC composites coatings on ZE41 magnesium alloy," *Applied Surface Science*, vol. 255, no. 22, pp. 9174–9181, 2009.

[10] Y. Sahin and V. Kilicli, "Abrasive wear behaviour of SiCp/Al alloy composite in comparison with ausferritic ductile iron," *Wear*, vol. 271, no. 11-12, pp. 2766–2774, 2011.

[11] J. K. M. Kwok and S. C. Lim, "High-speed tribological properties of some Al/SiCp composites. I. Frictional and wear-rate characteristics," *Composites Science and Technology*, vol. 59, no. 1, pp. 55–63, 1999.

[12] A. Mazahery and M. O. Shabani, "Existence of good bonding between coated B4C reinforcement and al matrix via semisolid techniques: enhancement of wear resistance and mechanical properties," *Tribology Transactions*, vol. 56, no. 3, pp. 342–348, 2013.

[13] M. O. Shabani and A. Mazahery, "Application of finite element model and artificial neural network in characterization of al matrix nanocomposites using various training algorithms," *Metallurgical and Materials Transactions A*, vol. 43, no. 6, pp. 2158–2165, 2012.

[14] Z. F. Zhang, L. C. Zhang, and Y.-W. Mai, "Particle effects on friction and wear of aluminium matrix composites," *Journal of Materials Science*, vol. 30, no. 23, pp. 5999–6004, 1995.

[15] S. Wilson and A. T. Alpas, "Wear mechanism maps for metal matrix composites," *Wear*, vol. 212, no. 1, pp. 41–49, 1997.

[16] F. Rana and D. M. Stefanescu, "Friction properties of Al-1.5% Mg/SiC particulate metal-matrix composites," *Metallurgical and Materials Transactions A*, vol. 20, pp. 1564–1566, 1989.

[17] S. Kumar and V. Balasubramanian, "Effect of reinforcement size and volume fraction on the abrasive wear behaviour of AA7075 Al/SiCp P/M composites-a statistical analysis," *Tribology International*, vol. 43, no. 1-2, pp. 414–422, 2010.

[18] A. Mazahery and M. O. Shabani, "Mechanical properties of squeeze-cast A356 composites reinforced with B_4C particulates," *Journal of Materials Engineering and Performance*, vol. 21, no. 2, pp. 247–252, 2012.

[19] I. M. Hutchings, "Tribological properties of metal matrix composites," *Materials Science and Technology*, vol. 10, no. 6, pp. 513–517, 1994.

[20] T. Ma, H. Yamaura, D. A. Koss, and R. C. Voigt, "Dry sliding wear behavior of cast SiC-reinforced Al MMCs," *Materials Science and Engineering A*, vol. 360, no. 1-2, pp. 116–125, 2003.

[21] K. R. Brown, M. S. Venie, and R. A. Woods, "The increasing use of aluminum in automotive applications," *JOM*, vol. 47, no. 7, pp. 20–23, 1995.

17

An Investigation on Self-Compacting Concrete Using Ultrafine Natural Steatite Powder as Replacement to Cement

P. Kumar,[1] K. Sudalaimani,[2] and M. Shanmugasundaram[3]

[1]*Department of Civil Engineering, P.S.R. Engineering College, Sivakasi, Tamil Nadu 626140, India*
[2]*Department of Civil Engineering, Thiagarajar College of Engineering, Thiruparankundram, Madurai, Tamil Nadu 625015, India*
[3]*Structural Engineering Division, School of Mechanical and Building Sciences, VIT University, Chennai Campus, Chennai, Tamil Nadu 600127, India*

Correspondence should be addressed to P. Kumar; kumarponnudurai.civil@gmail.com

Academic Editor: Prinya Chindaprasirt

An experimental investigation was made on flow properties and compressive strength of self-compacting concrete (SCC) with ultrafine natural steatite powder (UFNSP) as replacement to cement. The tests were conducted on specimens with 5%, 10%, 15%, 20%, and 25% of replacement of UFNSP to the weight of cement and compared to the control specimens. The flow properties of all specimens were tested and checked for their limit with the existing guidelines. The compressive strength test was done on all specimens for strength of 7 days, 14 days, 28 days, and 56 days. The hardened samples were tested for their microstructural behavior and the elements Mg, Ca, and Si were mapped. Through mapping, the formations of M-S-H along with C-S-H are observed. The results show that the addition of UFNSP influences the flow property, by reducing the flow, and increases the compressive strength till 20% replacement. Further the addition of UFNSP increases the denseness of microstructure of the specimens thus resulting in the strength increment.

1. Introduction

The self-compacting concrete (SCC), also known as self-consolidating concrete, is in the limelight for the last two decades in construction industry. Self-compacting concrete (SCC) is a concrete which can be placed and compacted into every corner of formwork purely by means of its self-weight by eliminating the need of external energy. So it does not require compaction at site or concrete plants. It has been developed in Japan to improve the durability and uniformity of concrete [1]. The mix composition is chosen to satisfy all performance criteria for the concrete in both the fresh and hardened states. To achieve this, fly ash and silica fume were used as mineral admixtures, and super plasticizers were used in mix as chemical admixtures for design of concrete. In this regard, mass of fine aggregate is typically equal or greater when compared to coarse aggregate. Selection of coarse aggregate size also has an impact on requirement of self-compacting concrete. The development of SCC and its results on theoretical and experimental aspects were reviewed in various studies [2–4]. The methods of manufacturing of SCC initially had high content of paste; later on, researches were made to reduce the paste content and achieve strength [5, 6]. Due to the high content of water and voids, it has become important to look into the workability and segregation behavior of SCC [7, 8]. The important property of a SCC is to have workable and segregation resistant concrete which can flow through the reinforcement without any external vibrations or compactions [9]. The bond strength of reinforced bars and the SCC are higher when compared to that of normal concrete which also made the SCC more predominant material [10]. There were two types of usage of materials in SCC; one is as a pozzolanic replacement and the other one is filler. In this study ultrafine natural steatite powder (UFNSP) is used as cementitious replacement to cement. The steatite was used as insulators in ceramics, due to its high thermal resistivity. Its other major uses are in carving and sculpture industry. It is a proven durable raw material in refractory industry. In

TABLE 1: Properties of UFNSP.

	UFNSP	Cement
Physical properties		
Blaine surface area (m^2/Kg)	750	380
Particle mean Dia (μm)	<5	<32
Density	2.7	3.1
Loss of ignition	3.33%	2%
Chemical properties		
SiO_2	62.67%	23%
Al_2O_3	0.24%	4.20%
MgO	33.26%	0.20%
Fe_2O_3	0.30%	1.20%
CaO	0.20%	62%

terms of steatite the grinded powder was already proved to be good replacement to cement until 15% [11, 12]. Durability and strength of concrete increase with replacement of steatite powder to certain limit [12–14]. The phase of steatite is crystalline which belongs to magnesium metasilicate from talc mineral based on the following reaction [15]:

$$3MgO \cdot 4SiO_2 \cdot H_2O \longrightarrow 3MgSiO_3 + SiO_2 + H_2O \quad (1)$$

And this reaction supports the M-S-H formation. Generally the earlier studies show that the M-S-H gel has amorphous phase [11, 12]. The magnesium-rich binders can lead to low pH binding system which has wide range of applications in adverse concreting condition, and also this magnesium richness leads to early strength development and durability [16]. The present proposal will cover the area on development of SCC based on magnesium-rich steatite powder replacement and its mechanical behavior.

2. Materials, Mix Proportions, and Experimental Investigation

2.1. Materials Used in This Study. Ordinary Portland cement of grade 43 conforming to Indian Standard (IS) 12269-1987 was used and the properties were listed in Table 1. The UFNSP of less than 5 microns similar to that of earlier studies [11–14] is used and the physical and chemical properties were presented in Table 1. UFNSP was used as pozzolanic replacement in this study. The river sand procured from Tamil Nadu Minerals Limited (TAMIN) with the specific gravity of 2.65 and passing through 2.36 mm sieve was used. Crushed granite was used as coarse aggregate with a maximum nominal size of 12 mm and bulk density of 1640 kg/m^3. The normal potable water was used for mixing and curing process. Polycarboxylic ether based superplasticizer sky 8233 and viscosity modifier, Mastermatrix VMA 362, were used in this study as admixtures to maintain the workability.

2.2. Mix Proportion. The mix proportion was finalized based on trial and error process conforming to the EFNARC guidelines [17], and the target strength to be achieved was fixed as 25 MPa. The ratio of powder : fine aggregate : coarse aggregate was maintained as 1 : 1.8 : 1.9. The water/powder ratio was fixed as 0.8 [17]. To achieve the workability 2% superplasticizer and 0.4% viscosity modifier were added, and the addition was carried out with respect to weight of cement.

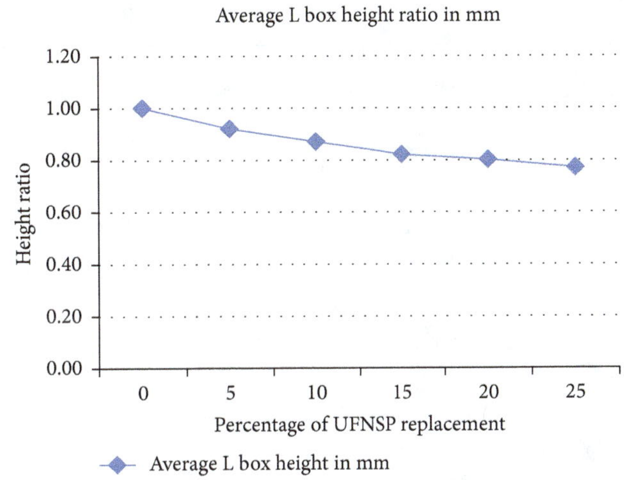

FIGURE 1: Results on *L* box test.

2.3. Experimental Investigation. The workability and flow properties of the fresh concrete were examined by L box test, J-ring test, V-funnel flow test, slump flow test, and T50 flow timing conforming to EFNARC guidelines [17]. The compressive strength of the hardened specimens was tested on cubes of size 150 mm × 150 mm × 150 mm conforming to specification of Indian Standard (IS) 10086: 1982 and the tests were carried out in accordance with IS 516: 1959 (reaffirmed 2004) in hydraulic compression testing machine of 2000 kN capacity. The microstructural behavior of the specimens is studied by scanning electron microscopy. Samples for scanning electron microscopy (SEM) analysis are taken near the surface (0-1 mm depth) of specimens. Microstructural studies utilized SEM (ZEISS) equipped with EDAX analyzer for microstructural observations of the surfaces, which is coated with evaporated copper for examination. SEM analysis is done at a maximum magnification of 1000x with energy 20 keV and a high resolution of 3.5 nm. For this analysis, samples of size 10 mm cubes were cut with a saw cutter.

3. Results and Discussion

The results were observed on the fresh and hardened state of concrete. The tests were carried out to ascertain the flow properties of fresh concrete, compressive strength on hardened concrete, and its respective microstructural behavior.

3.1. Workability and Flow Properties of Fresh Concrete. The results for "L" box test were presented in Table 2 and Figure 1. The results show that the addition of UFNSP as replacement to cement increases the water requirement by reducing the height ratio. The average value of "L" box height for 25% UFNSP replacement crosses the safe limit of EFNARC permissible limit. In comparison with the control

Table 2: Results on flow properties of fresh concrete.

S. number	Mix Ids	Percentage replacement of UFNSP	Average L box height ratio	Average J-ring height (mm)	Average V-funnel flow time in sec	Average slump flow diameter in mm	Average T50 cm slump flow in sec
1	SCCCS	0	1.00	7.88	7.90	710.00	2.60
2	SCC5	5	0.92	7.93	9.20	700.00	2.70
3	SCC10	10	0.87	8.23	9.50	680.00	3.20
4	SCC15	15	0.82	8.45	10.40	672.00	4.00
5	SCC20	20	0.80	8.73	11.20	650.00	4.50
6	SCC25	25	0.77	9.33	11.80	630.00	5.00

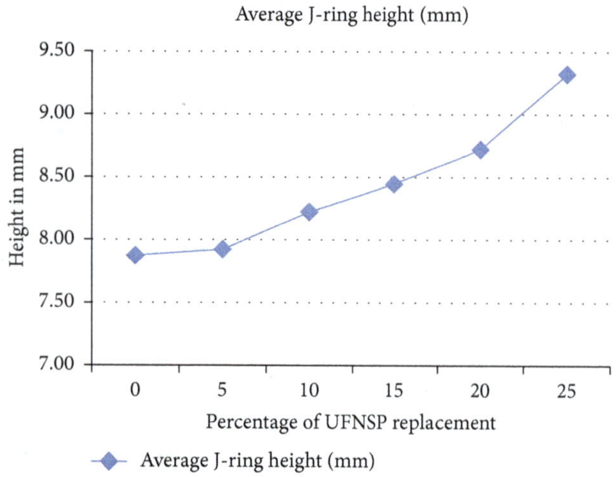

Figure 2: Results on J-ring test.

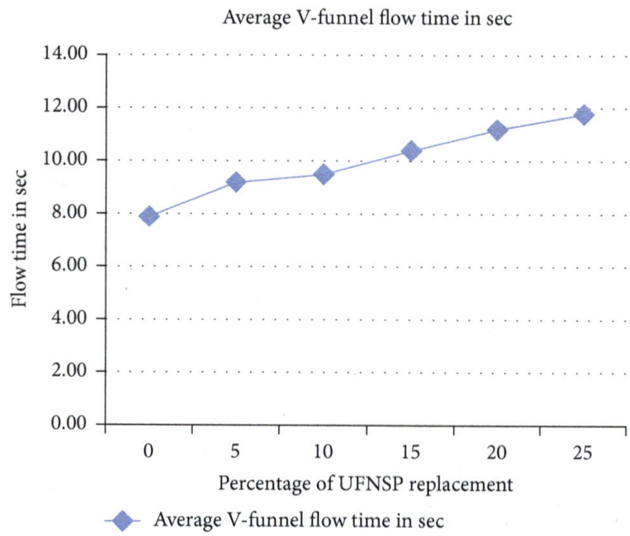

Figure 3: Results on V-funnel flow test.

specimen (SCCCS), specimens SCC5, SCC10, SCC15, SCC20, and SCC25 lose their height ratio by 8%, 13%, 18%, 20%, and 23%, respectively. The results show that there was a sudden reduction in height ratio between 0 and 15% replacements and then the ratio gets stabilized and this is due to the fact that after certain percentage of replacement the cement UFNSP reaction was lesser.

The results on J-ring test on fresh self-compacting concrete were presented in Table 2 and Figure 2. The test results shows that the height of fresh concrete in J-ring rises by 0.63%, 4.4%, 7.3%, 10.79%, and 18.41% for the specimens SCC5, SCC10, SCC15, SCC20, and SCC25, respectively, in comparison with SCCCS. From Figure 2 it was observed that height increment is not notable for specimen with 5% of UFNSP replacement. From 5% to 20% of UFNSP replacement the height gradually increases from 7.93 mm to 8.73 mm; with the replacement of 25% the height gets a steep increase to 9.33 mm. Even though the height rises the values were within the permissible limit which is 0–10 mm.

The results on "V"-funnel test on fresh self-compacting concrete specimens were presented in Table 2 and Figure 3. The results show that the time period for flow of fresh concrete through the "V"-funnel increases as the replacement percentage of UFNSP increases. The minimum time is observed at SCCCS specimen with flow period as 7.90 seconds and maximum time period for flow was observed at SCC25 with 11.80 seconds. The flow timing increases almost linearly which can be observed from Figure 3. The flow period for SCC5, SCC10, SCC15, SCC20, and SCC25 increases by 16.46%, 20.25%, 31.65%, 41.77%, and 49.37% in comparison to SCCCS specimen. The values lie well within the permissible limit for a typical self-compacting concrete. The actual permissible limit given by EFNARC is 0–25 Sec.

The slump flow values and the corresponding T50 cm test results were presented in Table 2 and Figures 4 and 5. The results show that the slump value reduces as the percentage replacement of UFNSP increases, which shows the UFNSP has an impact on the workability of SCC. The SCCCS specimen has the maximum value as observed by the other tests earlier and the lowest value was observed at SCC25 measuring 630 mm. The flow of SCC5, SCC10, SCC15, SCC20, and SCC25 reduces by 1.41%, 4.23%, 5.35%, 8.45%, and 11.27% with reference to SCCCS. The time period of flow for T50 test was presented in Figure 5. The time for spreading increases and the time for SCC25 is longer with its value as 5 sec. As the addition of UFNSP reduces the workability, the flow gets affected and hence the delay in spreading occurs. The spreading time increases as 3.85%, 23.08%, 53.85%, 73.08%, and 92.31% for SCC5, SCC10, SCC15, SCC20, and SCC25 in comparison with SCCCS. Even though

TABLE 3: Compressive strength of hardened concrete cubes.

S. number	Mix Ids	Average compression strength in MPa			
		7 days	14 days	28 days	56 days
1	SCCCS	21.80	25.27	26.60	26.90
2	SCC5	27.95	31.20	32.50	32.65
3	SCC10	31.50	33.61	34.30	34.40
4	SCC15	34.20	35.90	36.20	36.30
5	SCC20	29.07	31.01	32.30	32.37
6	SCC25	22.80	25.44	26.50	26.80

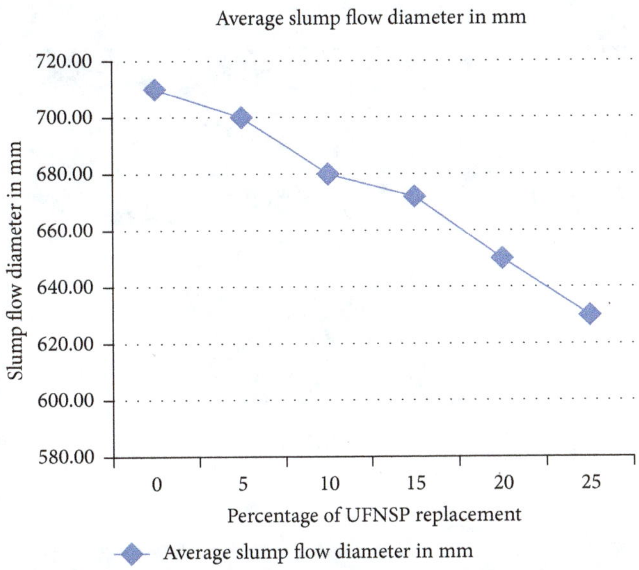

FIGURE 4: Results on slump flow test.

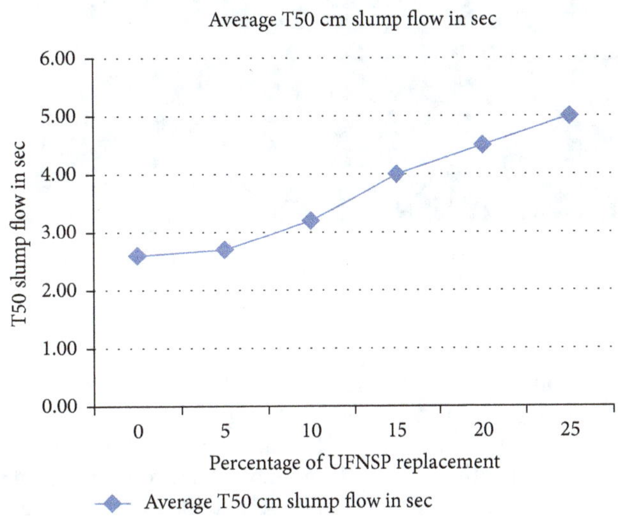

FIGURE 5: Results on T50 slump flow test.

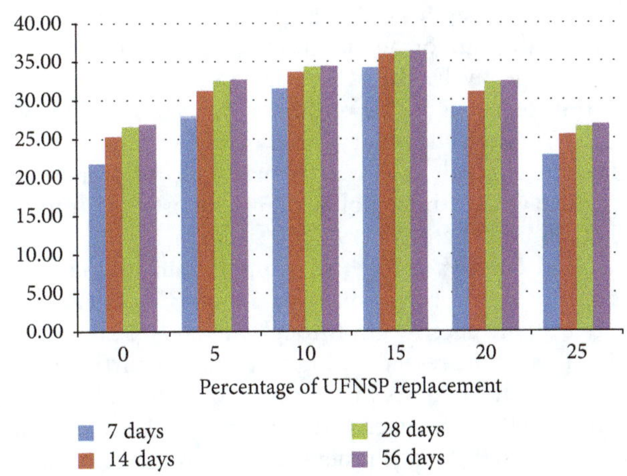

FIGURE 6: Results on average compressive strength test on 28 days.

the slump value decreases and the T50 test time increases the values are all well within the limit and show typical self-compacting concrete characteristics as indicated by EFNARC guidelines.

The observations shows that the water requirement was more in samples with UFNSP as replacement, and this is due to the fact that the M-S-H binder matrix with MgO presence has large water requirement due to texture, size, and shape of the particles [18]. This higher water requirement is also due to the reactive magnesia [19], a part in the UFNSP. From the tests on workability, even though it was observed that the replacement of UFNSP reduces the flow, almost all specimen lies within the typical limit as specified by EFNARC, and hence the replacement can be done until 25% without any effect on SCC limiting values.

3.2. Compressive Strength on Hardened Concrete at Various Ages. The compressive strength is tested on hardened concrete and the results are presented in Table 3 and Figure 6. The results show that at early ages all the replacement specimens show higher strength gain than the control specimens. The strength attainment of SCCCS, SCC5, SCC10, SCC15, SCC20, and SCC25 was 82%, 86%, 92%, 94%, 90%, and 86%, respectively, on 7 days and 95%, 96%, 98%, 99%, 97%, and 95% on 14 days. Through this, the SCC5 and SCC10 specimens are attaining 90% and above of designed strength on 7 days, which may be a notable parameter. The 7 days' test shows that the SCC5, SCC10, SCC15, SCC20, and SCC25 has strength increment of about 28.21%, 44.50%, 56.88%, 33.35%, and 4.59%, respectively, when compared to SCCCS. From the

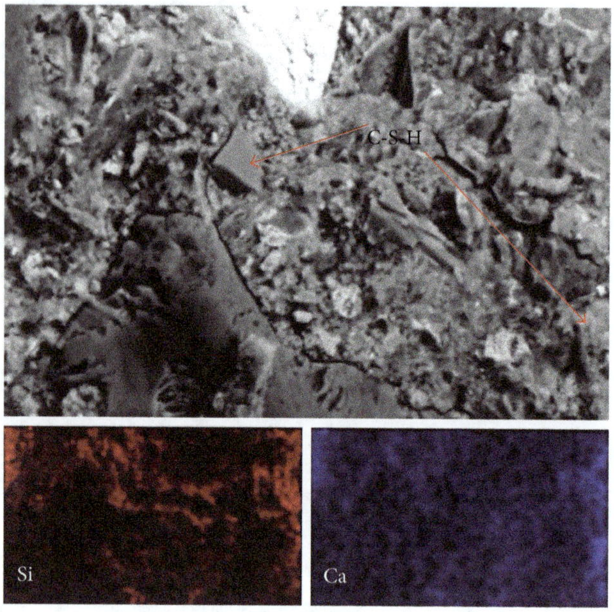

Figure 7: SEM image and its EDAX mapping on Si, Mg, and Ca of control specimen (CS).

results on 14 days' test it was evident that the SCC5, SCC10, SCC15, SCC20, and SCC25 have strength increment of about 23.47%, 33.30%, 42.07%, 22.71%, and 0.67%, respectively, in comparison with SCCCS. The 28 days' compressive strength for SCC5, SCC10, SCC15, and SCC20 has strength gain of about 22.18%, 28.95%, 36.09%, and 21.43%, respectively. The SCC25 specimens have a strength reduction of about 0.38% on 28 days in comparison with SCCCS. The 56 days' compressive strength for SCC5, SCC10, SCC15, and SCC20 has strength gain of about 21.38%, 27.88%, 34.94%, and 20.33%, respectively. The SCC25 specimens have a strength reduction of about 0.37% on 56 days in comparison with SCCCS. Through the observations it was evident that until 14 days the strength of all replacement specimens increases. From 28 days the strength of SCC25 is almost similar to that of SCCCS and there is very little strength loss for SCC25 on 28 and 56 days' tests. The strength gain is due to the presence of right proportion of M-S- and C-S-H [20]. The concrete attains its strength through the UFNSP; the reason behind this is the presence of Mg and reduced particle size of UFNSP, which facilitates the intrusion of particle in cement matrix [11–14]. The higher hydration rate is the main cause of early age strength attainment [21]. From the result it was clear that the strength attainment can be achieved until 20% replacement. The specimen with 25% UFNSP replacement is almost equal to and little lesser than SCCCS specimens on 28 days and 56 days.

3.3. Microstructure and Elemental Mapping on 56 Days' Hardened Specimens. The scanning electron microscopy analysis on the microstructure of the concrete specimen at 56 days and its relevant mapping for silica (Si), magnesium (Mg), and Calcium (Ca) of those specimens are presented through Figures 7–12. The elemental EDAX mapping is carried out

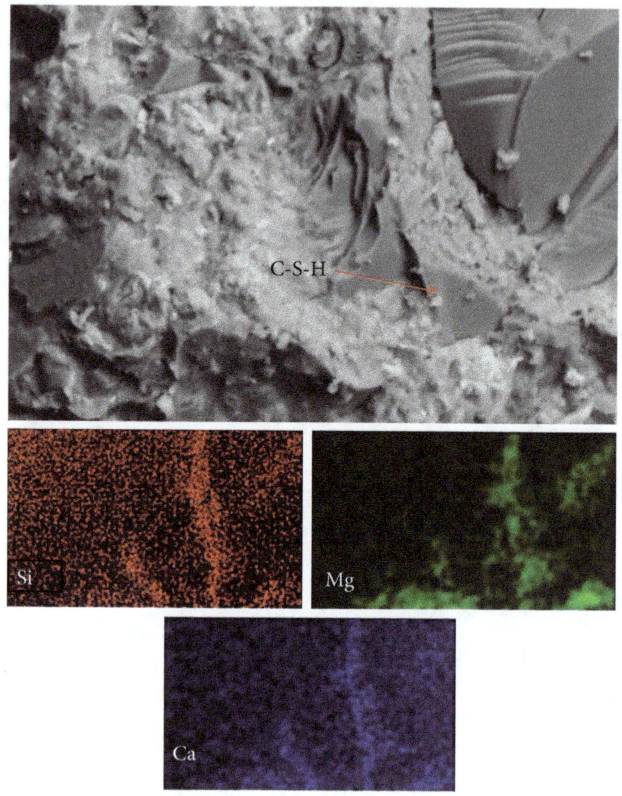

Figure 8: SEM image and its EDAX mapping on Si, Mg, and Ca of SCC-5.

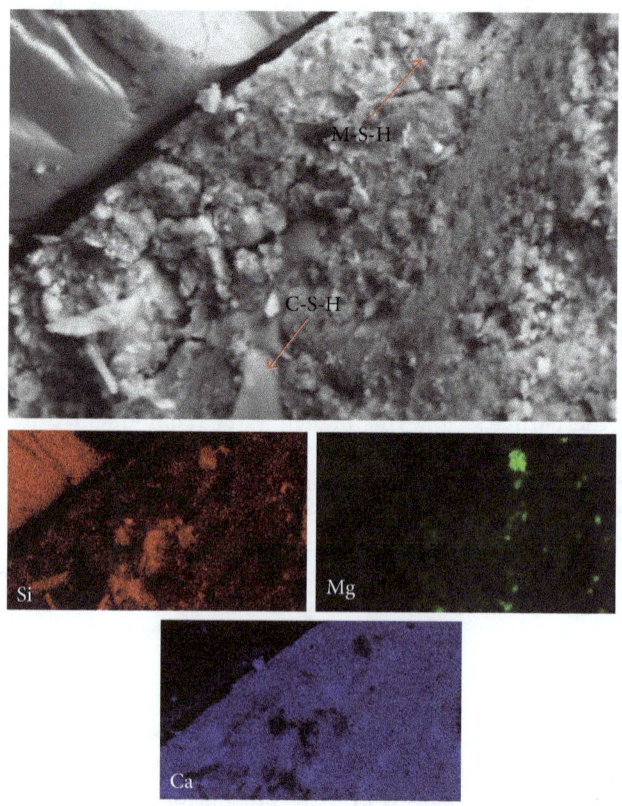

Figure 9: SEM image and its EDAX mapping on Si, Mg, and Ca of SCC-10.

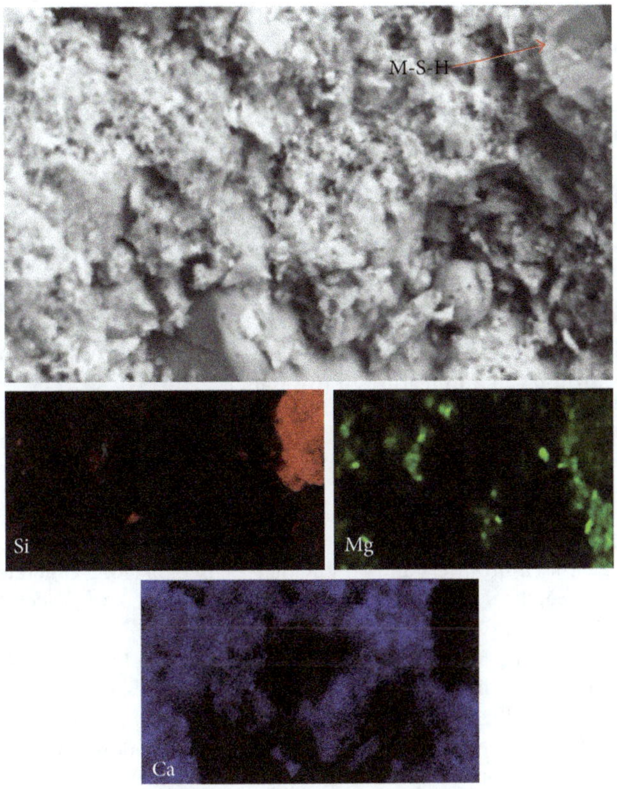

FIGURE 10: SEM image and its EDAX mapping on Si, Mg, and Ca of SCC-15.

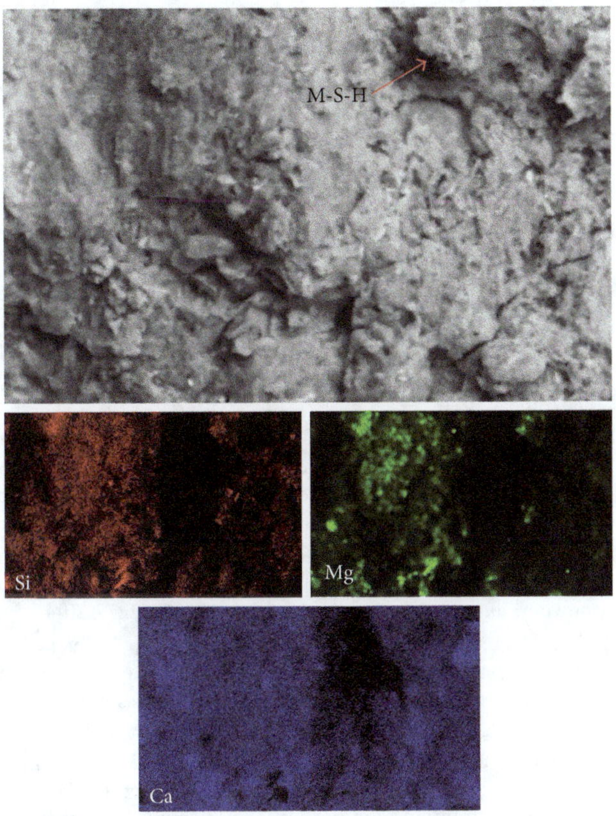

FIGURE 11: SEM image and its EDAX mapping on Si, Mg, and Ca of SCC-20.

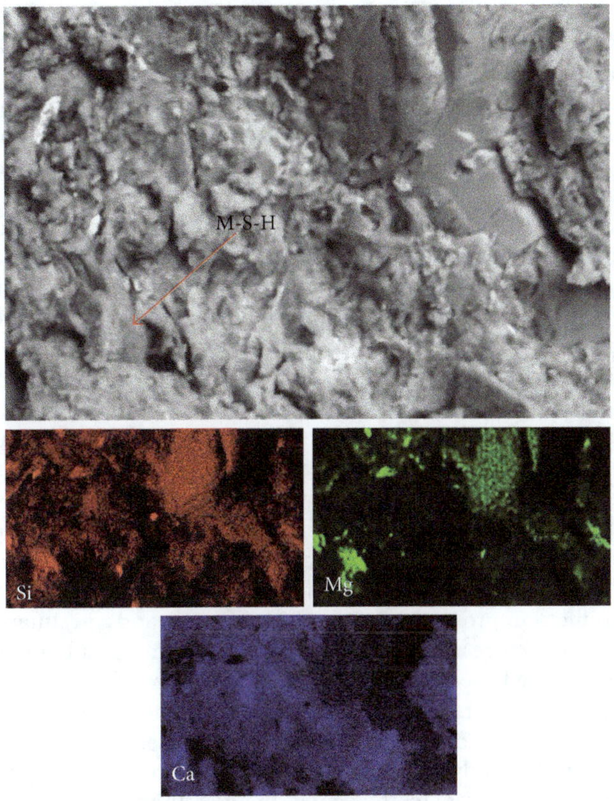

FIGURE 12: SEM image and its EDAX mapping on Si, Mg, and Ca of SCC-25.

to ascertain whether the addition of UFNSP has genuine effect on the strength development. The microstructural evolution was observed to be similar on all samples. Figure 7 shows the microscopy image of control specimen and its relevant mapping. This shows conventional specimen has no traces of magnesium elements. The formation of calcium silicate hydrates is noted. Figure 8 shows the images of SCC5 specimen has the traces of C-S-H along with M-S-H. The presence of magnesium is very limited when compared to calcium and this is due to the lesser replacement of UFNSP. The microscopy and mapping images were shown in Figure 9, and it is observed that the denseness of M-S-H formation is increasing and it is validated through the mapping, and a possible trace of magnesium hydroxide is also noted. Figure 10 shows the results of SCC15, and it is observed that the formation of M-S-H is more prominent and denser. The solidity of the magnesium silicate has increased widely in comparison with the specimens earlier, and also few spots of free calcium are traced which may cause a break point on strength gain. The maximum strength is observed at SCC15 which is to be noted at this point. Figure 11 shows the result images for SCC20 specimen and it shows large formation of magnesium silicate along with the calcium silicates, and the microstructure indicates positive attainment of microstructural evolution of control specimen, but not in comparison with SCC15 specimen. Figure 12 shows the microscopy and mapping images of SCC25 specimens; the microstructure of SCC25 was almost similar to that of SCCCS specimens.

In SCC25 it is noted that the magnesium silicates tend to break down to tiny M-S-H crystal which is identified through mapping, and this phenomenon causes the major strength reduction which is not observed in earlier specimens. In all the specimens it is noted that both magnesium based hydration products and calcium based hydration products have a vital role in establishing the strength gain.

From the above observations it can be ascertained that magnesium and magnesium based products are stable in all specimens, and it is also shown that when water exhausts the reaction of components end thus stabilizes its microstructure and attains early age strength in case of replacement up to 20%, this reaction product takes up space around magnesium compounds thus giving denser structure [22]. The magnesium based hydration products were found to be present in regions devoid of other elements, suggesting the presence of Mg crystals [23]. The observation shows that the M-S-H was less crystalline as observed and when this becomes more crystalline then the strength gain is reduced [24]. The process of increasing the $Mg(OH)_2$ and other magnesium hydration products improve the mechanical strength and specimen's microstructure as it is observed [21]. The addition of UFNSP has induced the magnesium based hydration products, and the results were little bit contrary to earlier studies [12, 13]. The contradiction is due to the increased water usage in current study through which the brucite development is accommodated in the pores. In this study as there are larger voids due to the higher water content and other admixtures, the $Mg(OH)_2$ which is higher in density than that of $Ca(OH)_2$ gets enough space to accommodate itself hence paving the way to good strength and microstructure, even at 25% replacement, which was not so in earlier studies [11, 12]. From the above study it can be ascertained that the usage of UFNSP will enhance the strength parameters and denser microstructural bond. The replacement percentage of UFNSP shall be maintained at 15% for maximum strength attainment, but still the 25% replacement gives strength similar to control specimen which shows that the replacement shall be carried out up to 25%.

4. Conclusions

From the present study it can be concluded that the replacement of UFNSP in self-compacting concrete system can have an influence on the workability, flow of fresh concrete, compressive strength of hardened concrete, and microstructural properties. There was decrease in flow properties of SCC with increase in addition of UFNSP. The entire replacement percentage exhibits safer limit for the SCC, but the specimens with 25% of UFNSP replacement reach the limit where the typical limit ends, which confirms that further replacement is not possible in terms of flow properties. The early age strength attainment is seen in samples SCC5, SCC10, SCC15, and SCC20 even on 7 days' test where more than the target strength is achieved. The maximum strength is achieved in SCC15 specimens. The strength enhancement is seen on all replacement specimens, wherein the strength of SCC25 specimens is almost equal to that of SCCCS. The microstructures of SCC10, SCC15, and SCC20 are denser and show denser magnesium hydroxide which has an effect on strength improvement. Further magnesium and silicate are mapped, which shows the dispersion over the surface and forms denser structures. The reduced strength of SCC25 is due to the higher magnesium content. From the investigation and discussion it is concluded that the replacement of UFNSP should be maintained below 20% and the UFNSP will enhance the strength parameters of the SCC.

Conflicts of Interest

The authors declare that there are no conflicts of interest regarding the publication of this paper.

References

[1] H. Okamura and M. Ouchi, "Self-compacting concrete," *Journal of Advanced Concrete Technology*, vol. 1, no. 1, pp. 5–15, 2003.

[2] H. J. H. Brouwers and H. J. Radix, "Self-compacting concrete: Theoretical and experimental study," *Cement and Concrete Research*, vol. 35, no. 11, pp. 2116–2136, 2005.

[3] P. L. Domone, "Self-compacting concrete: An analysis of 11 years of case studies," *Cement and Concrete Composites*, vol. 28, no. 2, pp. 197–208, 2006.

[4] P. L. Domone, "A review of the hardened mechanical properties of self-compacting concrete," *Cement and Concrete Composites*, vol. 29, no. 1, pp. 1–12, 2007.

[5] N. Su, K.-C. Hsu, and H.-W. Chai, "A simple mix design method for self-compacting concrete," *Cement and Concrete Research*, vol. 31, no. 12, pp. 1799–1807, 2001.

[6] N. Su and B. Miao, "A new method for the mix design of medium strength flowing concrete with low cement content," *Cement and Concrete Composites*, vol. 25, no. 2, pp. 215–222, 2003.

[7] M. F. Granata, "Pumice powder as filler of self-compacting concrete," *Construction and Building Materials*, vol. 96, pp. 581–590, 2015.

[8] EFNARC, *Specifications and Guidelines for Self-compacting Concrete*, Association House, Surrey, UK, 2002.

[9] E. Güneyisi, M. Gesoğlu, S. Al-Rawi, and K. Mermerdaş, "Effect of volcanic pumice powder on the fresh properties of self-compacting concretes with and without silica fume," *Materials and Structures/Materiaux et Constructions*, vol. 47, no. 11, pp. 1857–1865, 2014.

[10] N. Verma and A. K. Misra, "Bond characteristics of reinforced TMT bars in self compacting concrete and normal cement concrete," *Alexandria Engineering Journal*, vol. 54, no. 4, pp. 1155–1159, 2015.

[11] M. Shanmugasundaram and K. Sudalaimani, "An investigation on high performance concrete with ultra fine natural steatite powder," *Information*, vol. 17, no. 6 A, pp. 2267–2277, 2014.

[12] K. Sudalaimani and M. Shanmugasundaram, "Influence of ultrafine natural steatite powder on setting time and strength development of cement," *Advances in Materials Science and Engineering*, vol. 2014, Article ID 532746, 6 pages, 2014.

[13] M. Shanmugasundaram, S. Karthiyaini, and K. Sudalaimani, "Influence of ultrafine natural steatite powder on strength and permeability of high performance concrete," *International Journal of Applied Engineering Research*, vol. 10, no. 18, pp. 38967–38971, 2015.

[14] M. Shanmugasundaram, R. Premkumar, R. Abinaya, KG. Arunya, M. Gowsalya, and M. Malathy, *A new cement free binding material based on steatite powder activated through alkaline solution*, Master Builder, 2016.

[15] E. Vela, M. Peiteado, F. García, A. C. Caballero, and J. F. Fernández, "Sintering behaviour of steatite materials with barium carbonate flux," *Ceramics International*, vol. 33, no. 7, pp. 1325–1329, 2007.

[16] T. Zhang, L. J. Vandeperre, and C. R. Cheeseman, "Formation of magnesium silicate hydrate (M-S-H) cement pastes using sodium hexametaphosphate," *Cement and Concrete Research*, vol. 65, pp. 8–14, 2014.

[17] EFNARC, *Specification and Guidelines for Self-Compacting Concrete*, European Federation of Producers and Applicators of Specialist Products for Structures, 2002.

[18] H. Tran and A. Scott, "Strength and workability of magnesium silicate hydrate binder systems," *Construction and Building Materials*, vol. 131, pp. 526–535, 2017.

[19] W. R. EUBANK, "Calcination Studies of Magnesium Oxides," *Journal of the American Ceramic Society*, vol. 34, no. 8, pp. 225–229, 1951.

[20] T. Zhang, L. J. Vandeperre, and C. R. Cheeseman, "Magnesium-silicate-hydrate cements for encapsulating problematic aluminium containing wastes," *Journal of Sustainable Cement-Based Materials*, vol. 1, no. 1-2, pp. 34–45, 2012.

[21] C. Wu, W. Chen, H. Zhang et al., "The hydration mechanism and performance of Modified magnesium oxysulfate cement by tartaric acid," *Construction and Building Materials*, vol. 144, pp. 516–524, 2017.

[22] H. Ma, B. Xu, and Z. Li, "Magnesium potassium phosphate cement paste: degree of reaction, porosity and pore structure," *Cement and Concrete Research*, vol. 65, pp. 96–104, 2014.

[23] S. L. Sarkar, "X-ray mapping - A supplementary tool in clinker phase characterization," *Cement and Concrete Research*, vol. 14, no. 2, pp. 195–198, 1984.

[24] C. Roosz, S. Grangeon, P. Blanc et al., "Crystal structure of magnesium silicate hydrates (M-S-H): The relation with 2:1 Mg-Si phyllosilicates," *Cement and Concrete Research*, vol. 73, pp. 228–237, 2015.

18

Application of Ground Granulate Blast Furnace Slag-Steel Slag Composite Binder in a Massive Concrete Structure under Severe Sulphate Attack

Jin Liu and Dongmin Wang

School of Chemical and Environmental Engineering, China University of Mining and Technology, Beijing 100083, China

Correspondence should be addressed to Dongmin Wang; wangdongmin-2008@163.com

Academic Editor: Frederic Dumur

A composite mineral binder was prepared by mixing ground granulate blast furnace slag (GGBS) and steel slag (GGBS/steel slag ratios are 1 : 1 or 3 : 2 by mass). The application of a composite binder in a massive concrete structure under severe sulphate attack is discussed by determining the hydration heat, adiabatic temperature increase, compressive strength, elastic modulus, chloride ion permeability, and sulphate attack resistance. The results show that the hydration heat of the composite binder decreases greatly when the cement replacement ratio increases to 50% at 45°C. The adiabatic temperature rise of the concrete containing the composite mineral admixture decreases significantly. Concrete containing the composite mineral admixture has a lower early elastic modulus and satisfactory late-age compressive strength. The composite mineral admixture can improve the resistance to chloride ion permeability and sulphate attack resistance of concrete, especially during temperature match curing.

1. Introduction

According to the American Concrete Institute (ACI), massive concrete is cast in situ to a size capable of addressing the problem of hydration heat, which then causes volume deformation and has to reduce the influence of cracking. The hydration heat from the inner cement inside a mass concrete structure cannot dissipate immediately, causing a large temperature gradient and thermally induced stress cracking in the massive concrete structure [1, 2]. Therefore, to minimize the risk of cracking in massive concrete structures, high fractions of mineral admixtures are used to reduce the hydration heat of cement and the adiabatic temperature rise of the concrete [3]. Meanwhile, large numbers of massive concrete structures in soil with high saline concentrations suffer severe sulphate attacks, such as concrete foundations with a diameter larger than 1 m, or even 2 m for transmission line towers. Sulphates are abundant in saline soil and react with $Ca(OH)_2$ and C-A-H to form gypsum and ettringite, respectively, causing volume expansion in concrete [4]. This is especially true when concrete is subjected to dry-wet cycles. During this process, salt crystallization occurs, which also causes volume expansion and aggravates concrete corrosion [5]. To ensure the durability of massive concrete structures, the design strength of massive concrete is usually higher than 40 or 50 MPa.

Steel slag is a by-product from the process that transforms iron into steel [6]. The emission of steel slag makes up 13–20% of steel production [7]. Steel slag has cementitious properties, and the hydration process and hydration products of steel slag are similar to those of Portland cement (although its hydration activity is much lower than that of Portland cement) [8]. Used as a mineral admixture, steel slag can reduce the hydration heat of cement and significantly decrease the adiabatic temperature rise of concrete [9, 10]. However, the large particles in steel slag have few cementitious properties, and the interfaces between the large particles of steel slag and calcium silicate hydrate gels are weak points that may induce failures [11]. Additionally, large amounts of steel slag tend to result in a coarser pore structure for the hardened paste [12], yielding a negative effect on the strength and durability of the concrete [13]. In some studies, ground granulate blast furnace slag-steel slag composite binder was used as a composite mineral admixture and shown to prolong the setting time of the cement, improve the fluidity of fresh concrete, refine the pore structure of hardened paste, and result in a satisfactory compressive strength for the concrete [14, 15]. It is worth noting that the effect of the composite binder on decreasing the hydration heat of cement and the adiabatic temperature rise of concrete is more significant than that of fly ash [14].

The pozzolanic reaction of ground granulate blast furnace slag consumes $Ca(OH)_2$, decreasing the amount of gypsum formed by the reaction between sulphate and $Ca(OH)_2$ [4]. Additionally, the C-S-H gels formed by the pozzolanic reaction of the ground granulate blast furnace slag fills the pore structure of the hardened paste, causing the concrete to be denser and better resist sulphate attack [16, 17]. GGBS and steel slag both are industrial by-products and have reactivity. In this paper, the application of ground granulate blast furnace slag-steel slag composite binder in massive concrete structures under severe sulphate attack is discussed.

2. Raw Materials and Test Methods

2.1. Raw Materials. The cement used here was ordinary Portland cement with a specific surface area of $351\,m^2\,kg^{-1}$ and a strength grade of 42.5, which complied with the Chinese National Standard GB175-2007. Table 1 shows the major chemical components of the cement, GGBS, and steel slag. Figure 1 shows the particle size distributions of the cement, GGBS, and steel slag. The coarse aggregate used was limestone that was crushed to sizes between 5 mm and 25 mm. The fine aggregate used was natural river sand with a size smaller than 5 mm. To ensure the fluidity of the concrete, polycarboxylate superplasticizer was used. The steel slag was mixed with GGBS as a composite mineral admixture. Considering that both of GGBS and steel slag could be fully utilized and attempt to make concrete obtain a relatively higher performance, the mass ratios of the GGBS to steel slag are 1 : 1 and 3 : 2, which are defined as A and B, respectively.

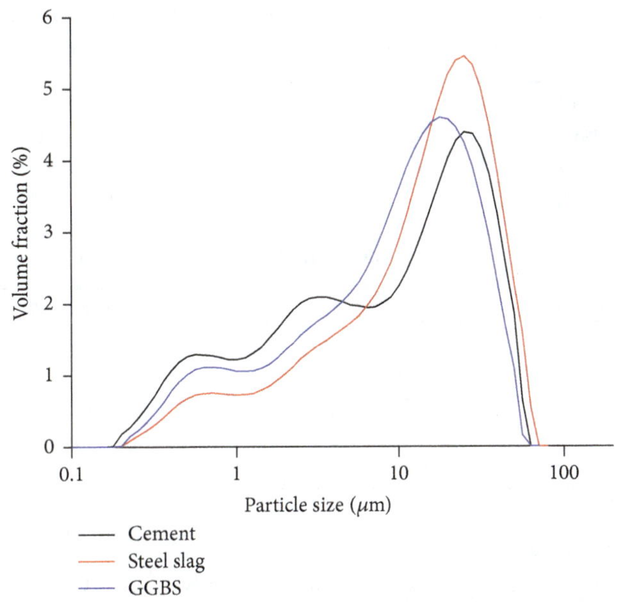

FIGURE 1: The particle size distributions of the cement, GGBS, and steel slag.

2.2. Test Methods. Table 2 lists the mixing proportions for the composite binders used for hydration heat evolution. An isothermal calorimeter was used to test the hydration heat of the composite binders within 72 h at constant temperatures of 20°C and 45°C, respectively. Table 3 lists the mixing proportions for the concrete. The adiabatic temperature increases for the concrete were tested using a specific adiabatic temperature measuring instrument with an accuracy of ±0.1°C.

Two curing methods were used for the concrete in this study: (1) standard curing and (2) temperature match curing. For standard curing, the concrete samples were cured in a room with a relative humidity higher than 95% and a temperature of 20 ± 1°C. For temperature match curing, the concrete samples were placed in curing boxes, whose temperatures were adjusted according to the adiabatic temperature rises for the concrete samples. After curing for 7 d, these samples were cooled naturally and placed in a room corresponding to the standard curing condition.

Concrete samples with sizes of 10 × 10 × 30 cm were cast for elastic moduli testing at ages of 7 d, 28 d, and 90 d. Concrete samples with sizes of 10 × 10 × 10 cm were cast for compressive strength testing at ages of 7 d, 28 d, and 90 d. According to ASTM C 1202 "Standard Test Method for Electrical Indication of Concretes Ability to Resist Chloride Ion Penetration," the chloride ion permeability of the concrete was evaluated by measuring the charge passed at ages of 28 d and 90 d.

The sulphate attack resistance of the concrete was tested using the dry-wet cycle method. The dry-wet cycle test was conducted according to "Chinese National Standard GBT50082-2009 Standard for test methods of long-term performance and durability of ordinary concrete." After standard curing and temperature match curing for 28 d, the concrete samples used for the sulphate attack group were placed into a 5% sodium sulphate solution and immersed for 15 h. Then, the concrete samples were dried in electric vacuum drying ovens at 80°C for 6 h. Finally, the concrete samples were cooled and each dry-wet cycle is 24 h. Concrete samples in the control group were cured using the standard curing conditions. After 90 and 120 cycles of this process, the compressive strengths of both groups were tested, and compressive strength loss rates were calculated using the compressive strengths of the sulphate attack group normalized by those of the control group.

3. Results and Discussion

3.1. Hydration Heat and Adiabatic Temperature Rise. Figures 2(a) and 2(b) show the exothermic rate curves and cumulative heat curves during the hydration of plain cement and the composite binders at 20°C, respectively. As shown in Figure 2(a), as the cement replacement ratio increased, the

TABLE 1: Chemical compositions of the cement, GGBS, and steel slag (%).

Sample	SiO_2	Al_2O_3	Fe_2O_3	CaO	MgO	MnO	Loss
Cement	22.36	7.73	3.66	57.21	3.10	—	2.31
GGBS	26.65	14.25	0.62	46.83	7.61	0.27	1.86
Steel slag	19.11	7.46	18.01	43.37	5.17	2.66	1.36

TABLE 2: Mixing proportions for the composite binders used for hydration heat evolution (g).

Samples	Cement	Composite mineral admixture A	Composite mineral admixture B	Water
P1	100	0	0	40
P2	60	40	0	40
P3	55	45	0	40
P4	50	50	0	40
P5	50	0	50	40

TABLE 3: Mixing proportions for the concrete (kg/m^3).

Samples	Cement	Composite mineral admixture A	Composite mineral admixture B	Water	Fine aggregates	Coarse aggregates
C1	380	0	0	152	785	1083
C2	190	190	0	152	785	1083
C3	190	0	190	152	785	1083

exothermic peak of the composite binder is delayed, and the exothermic peak value of the composite binder decreases. In addition, the exothermic rate of the composite binder is lower than that of the plain cement after approximately 2 d. As shown in Figure 2(b), the cumulative hydration heat of the composite binder is distinctly lower than that of the plain cement within 3 d. These results indicate that the hydration activities of the GGBS and steel slag are much lower than those of the cement during the early stages, which is consistent with previous findings [8, 12, 18].

Figures 3(a) and 3(b) show the exothermic rate curves and cumulative heat curves during the hydration of plain cement and the composite binders at 45°C, respectively. The exothermic rate of the composite binder at 45°C is clearly higher than that of the composite binder at 20°C. This high temperature has a significant activation effect on the hydration of the GGBS and steel slag. This is more pronounced for the GGBS, which is more sensitive to high temperatures than other mineral admixtures and cement. Additionally, the hydration degrees of the GGBS and steel slag increase dramatically at higher temperatures [18–20]. Additionally, the hydration heat of the GGBS is so large that the hydration heat of the binder containing 17–40% fine GGBS is greater than that of plain cement at room temperature, and the hydration heat of the binder containing 30–50% GGBS is close to or even higher than that of plain cement at 60°C [20, 21]. It is noteworthy that the exothermic rate of the composite binder is greater than that of plain cement after hydration for approximately 10 h. On the one hand, dense and thick C-S-H layers forms when the cement hydrates rapidly at high temperatures, hindering the hydration of the cement at later ages to some extent [22]. On the other hand, the hydration of the GGBS and steel slag promoted by the high temperature contributes significantly to the hydration heat of the composite binder. As shown in Figure 3(b), the cumulative heat of the binder containing 40% composite mineral admixture is greater than that of plain cement within 3 d, and the cumulative heat of the binder containing 45% composite mineral admixture is close to that of plain cement within 3 d. However, when the cement replacement ratio increases to 50%, the cumulative heat of the composite binder is clearly less than that of the plain cement within 3 d. This is because as cement replacement ratio increases, the hydration activity of the composite binder decreases with lower cement content and the activating effect on the hydration of GGBS is weakened with the decrease of $Ca(OH)_2$ in hydration products of cement [23].

The internal temperature of a massive concrete structure is usually higher than room temperature. Therefore, the hydration and exothermic properties of the binder at 45°C are more applicable to massive concrete than those of the binder at 20°C. According to the experimental results for the hydration heat of the binder at 45°C, the binder containing 50% composite mineral admixture is feasible for a massive concrete structure in order to minimize the temperature increase.

Figure 4 shows the adiabatic temperature rise curves of plain cement concrete and concrete containing the 50% composite mineral admixture. As shown in Figure 4, the adiabatic temperatures of the concrete containing 50% composite mineral admixture increase at a slower rate compared to plain cement concrete. Additionally, the adiabatic temperature increase of the concrete containing 50% composite mineral admixture is clearly less than that of plain cement concrete within 7 d. The GGBS content of composite mineral admixture A is less than that of composite mineral admixture B, so the effect of composite mineral admixture A on decreasing the hydration temperature increase is greater than that of composite mineral admixture B.

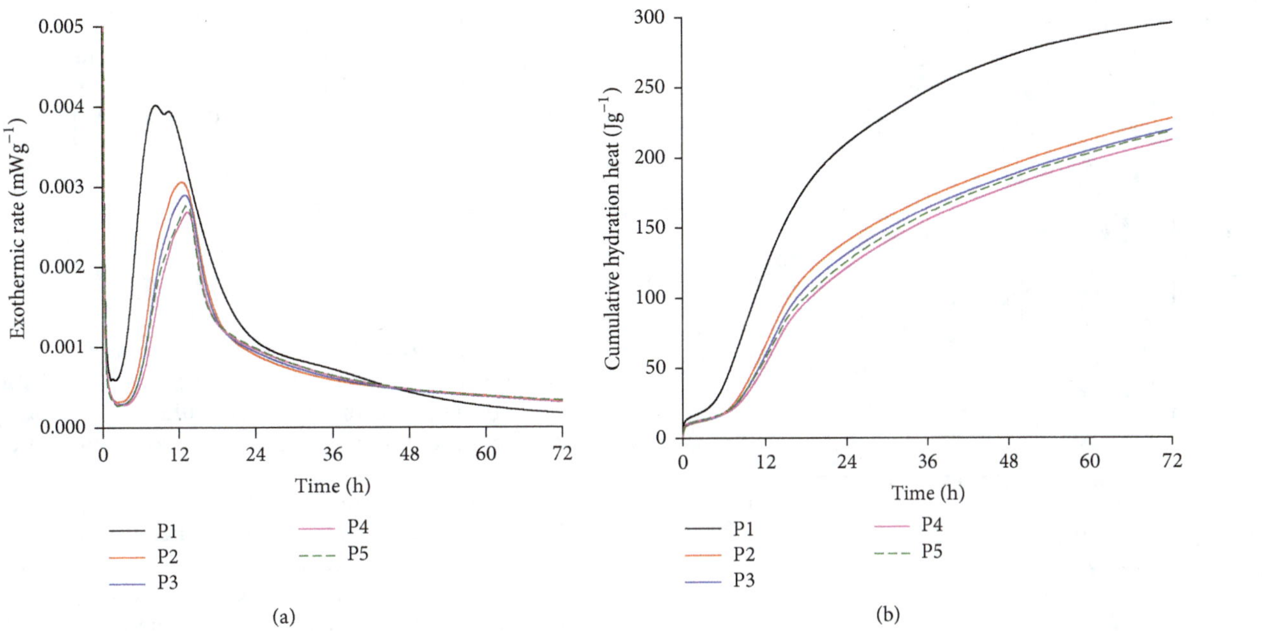

FIGURE 2: (a) Exothermic rate curves during the hydration of plain cement and composite binders at 20°C. (b) Cumulative heat curves during the hydration of plain cement and composite binders at 20°C.

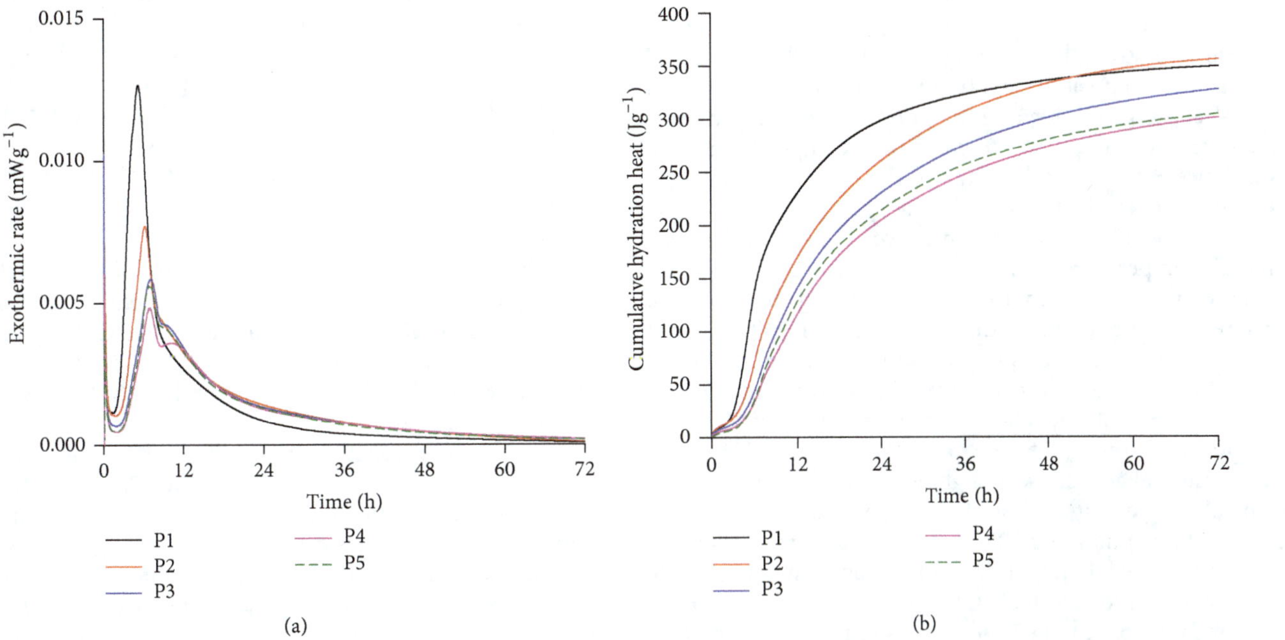

FIGURE 3: (a) Exothermic rate curves during the hydration of plain cement and composite binders at 45°C. (b) Cumulative heat curves during the hydration of plain cement and composite binders at 45°C.

3.2. *Mechanical Property*

3.2.1. Compressive Strength. Figure 5(a) shows the compressive strength of concrete for standard curing. The compressive strength of the concrete containing 50% composite mineral admixture is significantly less than that of plain cement concrete at 7 d. However, the compressive strength of the concrete containing 50% composite mineral admixture is close to that of plain cement concrete at 28 d, and the difference in the compressive strength between the plain cement concrete and concrete containing 50% composite mineral admixture is less than 3.6% at 90 d. These results indicate that although the early compressive strength of the concrete containing 50% composite mineral admixture is relatively low, the concrete containing 50% composite mineral admixture has a higher compressive strength gain rate and reaches a satisfactory

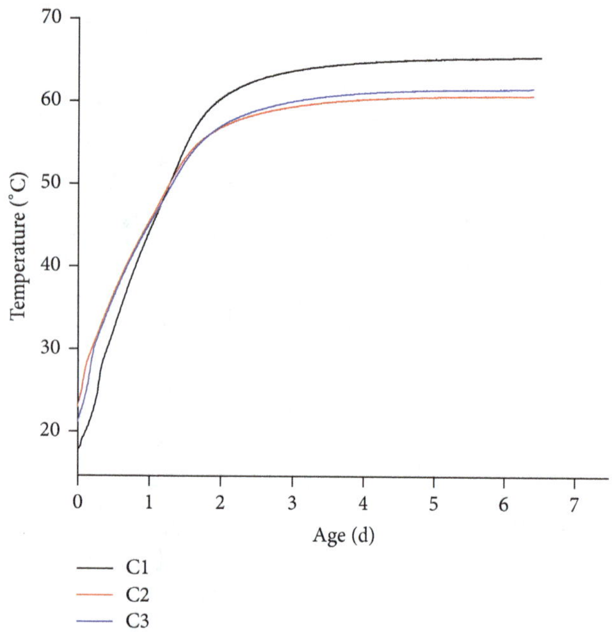

Figure 4: Adiabatic temperature rise curves of plain cement concrete and concretes containing 50% composite mineral admixture.

compressive strength at late ages. Additionally, the GGBS content of composite mineral admixture B is higher than that of composite mineral admixture A, and the compressive strength of the concrete containing composite mineral admixture B is higher than that of the concrete containing composite mineral admixture A at all ages.

Figure 5(b) shows the compressive strength of the concrete for temperature match curing. As shown in Figure 5(b), these samples benefited from the activating effect of the high temperature on the hydration of the composite mineral admixture and the compressive strength of the concrete containing 50% composite mineral admixture is only slightly less than that of the plain cement concrete at 7 d. The compressive strength of the concrete containing 50% composite mineral admixture is slightly greater than of the plain cement concrete at 28 d, and the compressive strength of the concrete containing 50% composite mineral admixture is clearly greater than that of plain cement concrete at 90 d. These results indicate that, for temperature match curing, the concrete containing 50% composite mineral admixture reaches a satisfactory early compressive strength, as well as higher late-age compressive strength, although the compressive strength gain rate of plain cement concrete is low at late ages. This is because concrete cured at a high temperature results in nonuniformly distributed hydration products and a coarser pore structure in hardened paste [24–26]. However, the hydration degree of the mineral admixture increases at high temperatures, and the C-S-H gels produced during the hydration of the mineral admixture fill the pore structure of the hardened paste, as a result of which the negative effect of high temperature on development of compressive strength is weakened and the concrete containing mineral admixture cured at high temperature at early ages has a relatively high compressive strength gain rate [26, 27]. In addition, the compressive strength of the concrete containing composite mineral admixture B is still greater than for the concrete containing composite mineral admixture A at all ages for temperature match curing.

3.2.2. Elastic Modulus. Stress is calculated as

$$\sigma = E \cdot \varepsilon, \qquad (1)$$

where σ is stress, E is elastic modulus, and ε is strain. Stress decreases with the decrease of elastic modulus for a constant strain. For massive concrete, the thermal stress will be relatively low if the elastic modulus is low at early ages, and the risk of thermally induced cracking tends to decrease under lower stress.

Figures 6(a) and 6(b) show the elastic modulus of the concrete for standard and temperature match curing, respectively. The elastic modulus of the concrete containing 50% composite mineral admixture is clearly less than that of plain cement concrete at 7 d, and the elastic modulus of the concrete containing 50% composite mineral admixture is still less than that of plain cement concrete at 28 d and 90 d. As shown in Figure 6(b), although the elastic modulus of the concrete containing 50% composite mineral admixture is slightly higher than that of plain cement concrete at 90 d, the elastic modulus of the concrete containing 50% composite mineral admixture is less than that of plain cement concrete at 7 d and 28 d. These results indicate that the early elastic modulus of the concrete decreases as the composite mineral admixture is added, which is helpful to decrease the thermally induced cracking.

3.3. Chloride Ion Permeability. According to ASTM C1202, the chloride ion permeability grade of concrete is "Moderate" if the charge passed is between 2000 and 4000 C. The chloride ion permeability grade of concrete is "Low" or "Very low" if the charge passed is between 1000 and 2000 C or less than 1000 C, respectively.

Figure 7(a) shows the chloride ion permeability for the concrete for standard curing. For either 28 d or 90 d, the charge passed for the concrete containing 50% composite mineral admixture is clearly less than that for plain cement concrete and the chloride ion permeability grade of the concrete containing 50% composite mineral admixture is less than that for plain cement concrete. These results indicate that the composite mineral admixture exhibits strong improvement effect on the resistance to chloride ion permeability for the concrete. In addition, the charge passed for the concrete containing composite mineral admixture B is slightly less than for the concrete containing composite mineral admixture B, indicating that composite mineral admixture B results in a greater resistance to chloride ion permeability of concrete than composite mineral admixture A. This could be explained that the pozzolanic reaction of GGBS consumes $Ca(OH)_2$, leading to the decrease of interconnected pores [28].

Figure 7(b) shows the chloride ion permeability of the concrete for temperature match curing. The charge passed for plain cement concrete for temperature match curing is higher

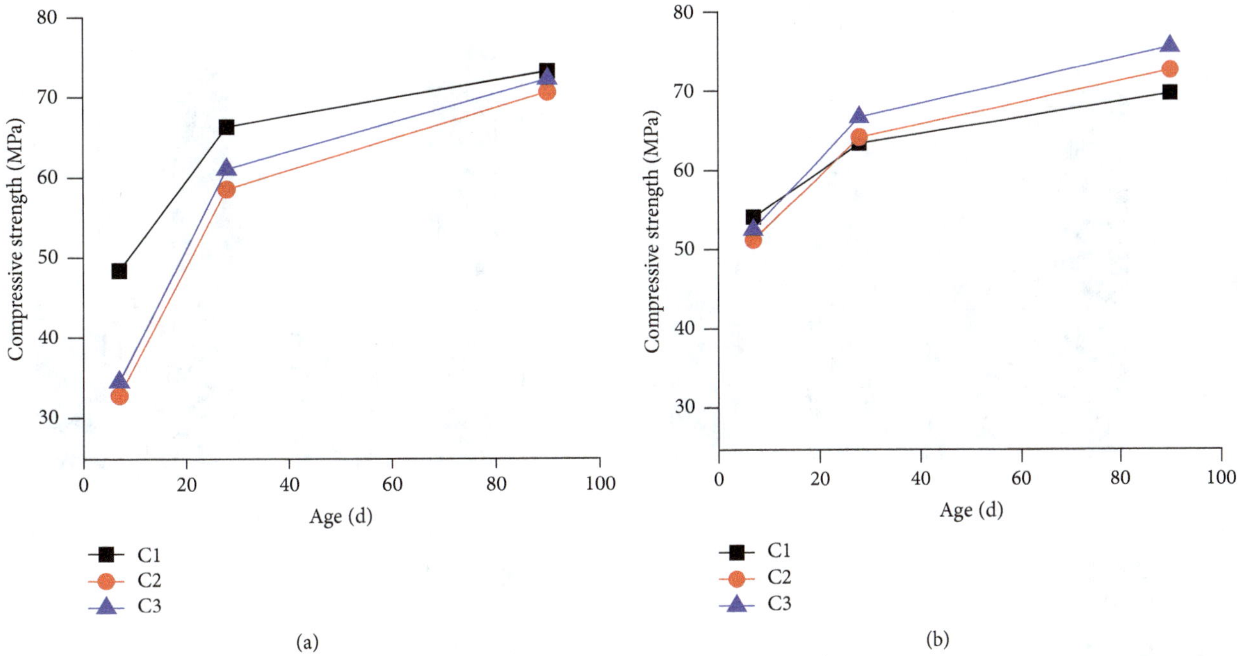

FIGURE 5: (a) Compressive strength of concrete for standard curing. (b) Compressive strength of concrete for temperature match curing.

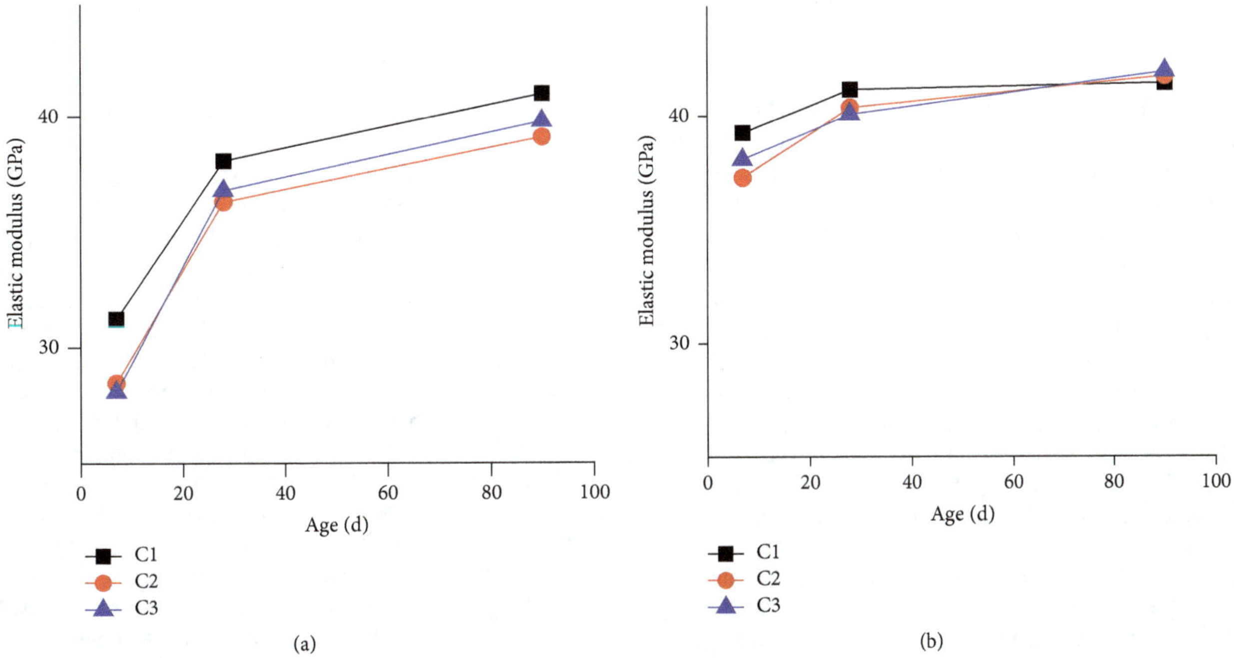

FIGURE 6: (a) Elastic modulus of concrete under standard curing. (b) Elastic modulus of concrete under temperature match curing.

than that for plain cement concrete for standard curing. However, the charge passed for concrete containing the composite mineral admixture for temperature match curing is less than that for concrete containing composite mineral admixture for standard curing. It is noteworthy that the chloride ion permeability grade of the concrete containing 50% composite mineral admixture B is "Very low" and the addition of 50% composite mineral admixture B improves the resistance to chloride ion permeability for the concrete by two levels. These results indicate that high temperatures have a negative effect on the resistance to chloride ion permeability in plain cement concrete, whereas the improvement effect of composite mineral admixture on the resistance to chloride ion permeability is enhanced at high temperatures. Therefore, the concrete containing the composite mineral admixture obtains satisfactory resistance to chloride ion permeability

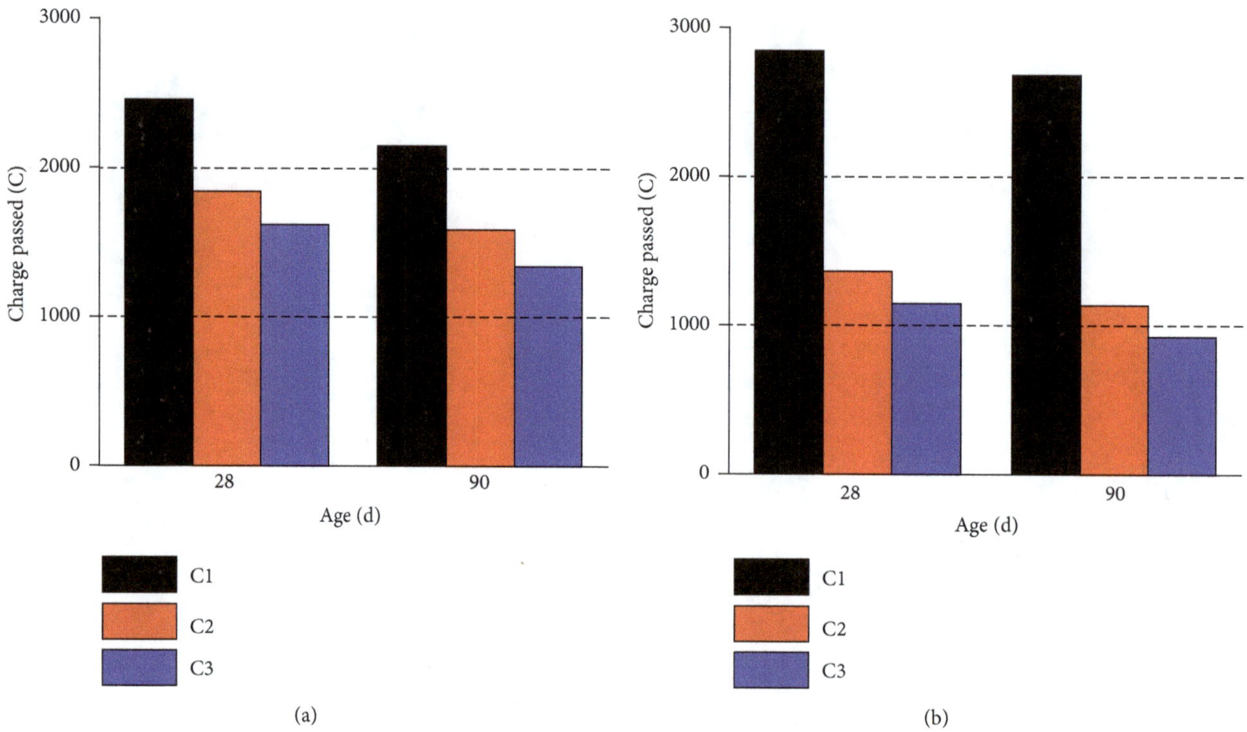

FIGURE 7: (a) Chloride ion permeability for the concrete for standard curing. (b) Chloride ion permeability for the concrete for temperature match curing.

when using temperature match curing. This is because the hydration degree of the mineral admixture increases at high temperature, as a result of which the pore structure of hardened paste is denser and the resistance to chloride ion permeability of concrete improves [26, 27, 29].

3.4. Sulphate Attack Resistance. Figure 8(a) shows the compressive strength of the concrete during a sulphate attack test when using standard curing. After 90 dry-wet cycles, the compressive strength of plain cement concrete exposed to sulphate attack decreases by 10.7% compared to concrete at 28 d. The strength loss rate of plain cement concrete is 21.0%. However, the compressive strengths of concrete containing 50% composite mineral admixtures A and B exposed to sulphate attack increase by 6.5% and 6.7%, respectively, compared to concrete at 28 d. Additionally, the strength loss rates for concrete containing 50% composite mineral admixtures A and B are only 15.0% and 13.0%, respectively. After 120 dry-wet cycles, the strength loss rate for the plain cement concrete is 37.9%, whereas the strength loss rates for concrete containing 50% composite mineral admixtures A and B are only 27.7% and 25.0%, respectively. These results indicate that the composite mineral admixture can improve concrete sulphate attack resistance significantly. In addition, the GGBS content of composite mineral admixture B is higher than that of composite mineral admixture A, and the improvement effect of composite mineral admixture B on the sulphate attack resistance for concrete is greater than that of composite mineral admixture A. This could be explained that the pozzolanic reaction of GGBS consumes $Ca(OH)_2$, leading to the reduction of the gypsum formation, and thus of ettringite [30].

Figure 8(b) shows the compressive strength of the concrete during a sulphate attack test when using temperature match curing. As shown in Figure 8(b), after 90 dry-wet cycles, the compressive strength of plain cement concrete exposed to sulphate attack is nearly the same as concrete at 28 d, and the strength loss rate for plain cement concrete is 22.5%. However, the compressive strengths of the concrete containing 50% composite mineral admixtures A and B exposed to sulphate attack increase by 26.0% and 29.5% compared to concrete at 28 d, and the strength loss rates of concrete containing 50% composite mineral admixtures A and B are only 12.9% and 11.4%, respectively. After 120 dry-wet cycles, the strength loss rate for the plain cement concrete is 40.4%, whereas the strength loss rates for the concrete containing 50% composite mineral admixtures A and B are only 25.0% and 23.0%, respectively. These results indicate that the composite mineral admixtures can still significantly improve the sulphate attack resistance of concrete when using temperature match curing and the improvement effect of composite mineral admixture B is stronger than that of composite mineral admixture A. It is worth noting that the strength loss rate for the plain cement concrete increases compared to concrete using standard curing, whereas the strength loss rate for the concrete containing composite mineral admixture decreases compared with standard cured concrete, indicating that high temperatures have a negative effect on the sulphate attack resistance of plain cement concrete. However, the improvement effect of composite mineral

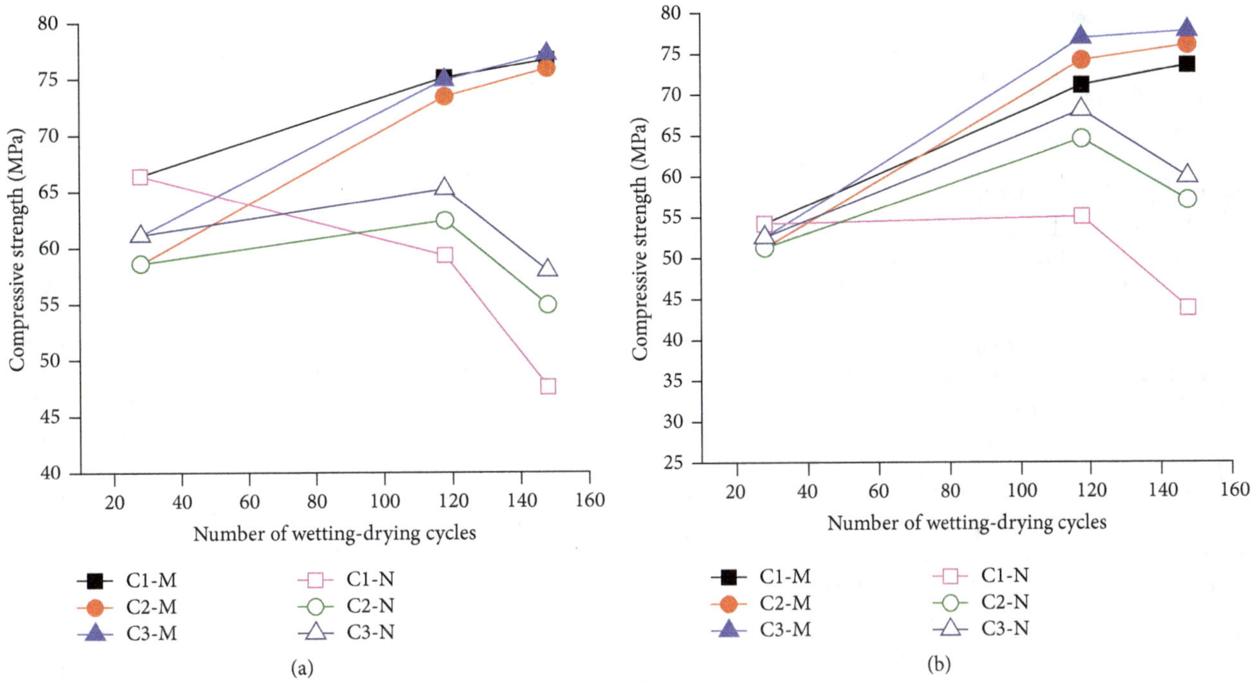

FIGURE 8: (a) Compressive strength of the concrete during a sulphate attack test when using standard curing. (b) Compressive strength of the concrete during a sulphate attack test when using temperature match curing.

admixture on the sulphate attack resistance is enhanced at high temperatures, and the concrete containing composite mineral admixture results in a satisfactory sulphate attack resistance for temperature match curing. These results are consistent with those from the chloride ion permeability test.

4. Conclusions

(1) The addition to granulate blast furnace slag-steel slag composite binder can reduce the hydration heat of the composite binder at 20°C. However, the hydration heat of the composite binder is decreased greatly only when the cement replacement ratio increases to 50% at 45°C. Therefore, the addition of the composite mineral admixture applied in massive concrete should be 50%.

(2) For either 28 d or 90 d, the addition of 50% composite mineral admixture can decrease the early elastic modulus of the concrete, which is helpful to decrease the risk of thermally induced cracking. In addition, the concrete containing 50% composite mineral admixture possesses satisfactory late-age compressive strength.

(3) The addition of 50% composite mineral admixture can improve the resistance to chloride ion permeability and the sulphate attack resistance of concrete, especially when using temperature match curing.

(4) Although the composite mineral admixture with a GGBS to steel slag mass ratio of 3 : 2 increases the adiabatic temperature rise by 2.7°C within 7 d compared to the composite mineral admixture with a GGBS to steel slag mass ratio of 1 : 1, the compressive strength, resistance to chloride ion permeability, and sulphate attack resistance of the concrete containing the composite mineral admixture with a GGBS to steel slag mass ratio of 3 : 2 are significantly higher.

Conflicts of Interest

The authors declare that there are no conflicts of interest regarding the publication of this paper.

Acknowledgments

The support of China National Natural Science Foundation Project (no. 51572293) is acknowledged.

References

[1] X. Liu, C. Zhang, X. Chang, W. Zhou, Y. Cheng, and Y. Duan, "Precise simulation analysis of the thermal field in mass concrete with a pipe water cooling system," *Applied Thermal Engineering*, vol. 78, no. 5, pp. 449–459, 2015.

[2] M. E. Shabab, K. Shahzada, B. Gencturk, M. Ashraf, and M. Fahad, "Synergistic effect of fly ash and bentonite as partial replacement of cement in mass concrete," *KSCE Journal of Civil Engineering*, vol. 20, no. 5, pp. 1987–1995, 2016.

[3] M. Batog and Z. Giergiczny, "Influence of mass concrete constituents on its properties," *Construction and Building Materials*, vol. 146, pp. 221–230, 2017.

[4] H. N. Atahan and D. Dikme, "Use of mineral admixtures for enhanced resistance against sulfate attack," *Construction and Building Materials*, vol. 25, no. 8, pp. 3450–3457, 2011.

[5] M. Najjar, M. Nehdi, A. Soliman, and T. Azabi, "Damage mechanisms of two-stage concrete exposed to chemical and physical sulfate attack," *Construction and Building Materials*, vol. 137, pp. 141–152, 2017.

[6] C. Shi, "Steel slag—its production, processing, characteristics, and cementitious properties," *Journal of Materials in Civil Engineering*, vol. 16, no. 3, pp. 230–236, 2004.

[7] E. Furlani, G. Tonello, and S. Maschio, "Recycling of steel slag and glass cullet from energy saving lamps by fast firing production of ceramics," *Waste Management*, vol. 30, no. 8-9, pp. 1714–1719, 2010.

[8] J. Li, Q. Yu, J. Wei, and T. Zhang, "Structural characteristics and hydration kinetics of modified steel slag," *Cement and Concrete Research*, vol. 41, no. 3, pp. 324–329, 2011.

[9] S. Liu and L. Li, "Influence of fineness on the cementitious properties of steel slag," *Journal of Thermal Analysis and Calorimetry*, vol. 117, no. 2, pp. 629–634, 2014.

[10] S. Mengxiao, W. Qiang, and Z. Zhikai, "Comparison of the properties between high-volume fly ash concrete and high-volume steel slag concrete under temperature matching curing condition," *Construction and Building Materials*, vol. 98, pp. 649–655, 2015.

[11] Q. Wang and P. Yan, "Hydration properties of basic oxygen furnace steel slag," *Construction and Building Materials*, vol. 24, no. 7, pp. 1134–1140, 2010.

[12] T. Zhang, Q. Yu, J. Wei, and J. Li, "Investigation on mechanical properties, durability and micro-structural development of steel slag blended cements," *Journal of Thermal Analysis and Calorimetry*, vol. 110, no. 2, pp. 633–639, 2012.

[13] J. Hu, "Comparison between the effects of superfine steel slag and superfine phosphorus slag on the long-term performances and durability of concrete," *Journal of Thermal Analysis and Calorimetry*, vol. 128, no. 3, pp. 1251–1263, 2017.

[14] Q. Wang, P. Yan, and G. Mi, "Effect of blended steel slag-GBFS mineral admixture on hydration and strength of cement," *Construction and Building Materials*, vol. 35, pp. 8–14, 2012.

[15] J. Zhao, D. Wang, and P. Yan, "Design and experimental study of a ternary blended cement containing high volume steel slag and blast-furnace slag based on Fuller distribution model," *Construction and Building Materials*, vol. 140, pp. 248–256, 2017.

[16] A. Attari, C. McNally, and M. G. Richardson, "A combined SEM - Calorimetric approach for assessing hydration and porosity development in GGBS concrete," *Cement and Concrete Composites*, vol. 68, pp. 46–56, 2016.

[17] E. Özbay, M. Erdemir, and H. I. Durmuş, "Utilization and efficiency of ground granulated blast furnace slag on concrete properties - A review," *Construction and Building Materials*, vol. 105, pp. 423–434, 2016.

[18] Y. Ballim and P. C. Graham, "The effects of supplementary cementing materials in modifying the heat of hydration of concrete," *Materials and Structures/Materiaux et Constructions*, vol. 42, no. 6, pp. 803–811, 2009.

[19] Q. Wang, M. Li, and G. Jiang, "The difference among the effects of high-temperature curing on the early hydration properties of different cementitious systems," *Journal of Thermal Analysis and Calorimetry*, vol. 118, no. 1, pp. 51–58, 2014.

[20] F. Han, R. Liu, D. Wang, and P. Yan, "Characteristics of the hydration heat evolution of composite binder at different hydrating temperature," *Thermochimica Acta*, vol. 586, pp. 52–57, 2014.

[21] P. Šiler, J. Krátký, I. Kolářová, J. Havlica, and J. Brandštetr, "Calorimetric determination of the effect of additives on cement hydration process," *Chemical Papers*, vol. 67, no. 2, pp. 213–220, 2013.

[22] M. S. Morsy, "Effect of temperature on electrical conductivity of blended cement pastes," *Cement and Concrete Research*, vol. 29, no. 4, pp. 603–606, 1999.

[23] S. C. Pal, A. Mukherjee, and S. R. Pathak, "Investigation of hydraulic activity of ground granulated blast furnace slag in concrete," *Cement and Concrete Research*, vol. 33, no. 9, pp. 1481–1486, 2003.

[24] E. Gallucci, X. Zhang, and K. L. Scrivener, "Effect of temperature on the microstructure of calcium silicate hydrate (C-S-H)," *Cement and Concrete Research*, vol. 53, pp. 185–195, 2013.

[25] K. O. Kjellsen, R. J. Detwiler, and O. E. Gjørv, "Development of microstructures in plain cement pastes hydrated at different temperatures," *Cement and Concrete Research*, vol. 21, no. 1, pp. 179–189, 1991.

[26] J. Liu and D. Wang, "The Role of Phosphorus Slag in Steam-Cured Concrete," *Advances in Materials Science and Engineering*, vol. 2017, Article ID 8392435, 14 pages, 2017.

[27] M. Li, Q. Wang, and J. Yang, "Influence of Steam Curing Method on the Performance of Concrete Containing a Large Portion of Mineral Admixtures," *Advances in Materials Science and Engineering*, vol. 2017, Article ID 9863219, 11 pages, 2017.

[28] A. Hadj-Sadok, S. Kenai, L. Courard, and A. Darimont, "Microstructure and durability of mortars modified with medium active blast furnace slag," *Construction and Building Materials*, vol. 25, no. 2, pp. 1018–1025, 2011.

[29] H. Jin, "Late-Age Properties of Concrete with Different Binders Cured under 45°C at Early Ages," *Advances in Materials Science and Engineering*, vol. 2017, pp. 1–13, 2017.

[30] A. Hadjsadok, S. Kenai, L. Courard, F. Michel, and J. Khatib, "Durability of mortar and concretes containing slag with low hydraulic activity," *Cement and Concrete Composites*, vol. 34, no. 5, pp. 671–677, 2012.

19

Influence of Chemical Admixtures on Fresh and Hardened Properties of Prolonged Mixed Concrete

Tarek Uddin Mohammed,[1] **Tanvir Ahmed,**[1] **Shibly Mostafiz Apurbo,**[1] **Tahir Absar Mallick,**[1] **Farhan Shahriar,**[2] **Abdul Munim,**[2] **and Mohammad Abdul Awal**[3]

[1]*Department of Civil and Environmental Engineering, Islamic University of Technology (IUT), Gazipur 1704, Bangladesh*
[2]*Housing and Building Research Institute (HBRI), Dhaka 1216, Bangladesh*
[3]*Structural Engineers Limited (SEL), West Panthapath, Dhaka 1209, Bangladesh*

Correspondence should be addressed to Tanvir Ahmed; tanvirbd@iut-dhaka.edu

Academic Editor: Jose M. Monzo

Effects of different chemical admixtures on fresh and hardened properties of prolonged mixed concrete and their cost-effectiveness were investigated. Influence of sand to aggregate volume ratio, cement content, and use of chilled mixing water on the properties of prolonged mixed concrete was studied as well. Different concrete mixtures were prepared using five different types of chemical admixture (one water reducer based on lignosulfonate and four superplasticizers based on sulfonated naphthalene polymer, polycarboxylic ether, second-generation polycarboxylic ether polymer, and organic polymer), varying s/a ratio (0.40 and 0.45) and cement content (340 kg/m^3 and 380 kg/m^3) and using chilled mixing water. Slump tests were performed at 15-minute intervals to assess the fresh performance of each prolonged mixed concrete mixture. 100 mm by 200 mm cylindrical concrete specimens were prepared and tested for compressive strength, Young's modulus, splitting tensile strength, and ultrasonic pulse velocity. Results indicate that concretes with sulfonated naphthalene polymer-based superplasticizer and second-generation polycarboxylic ether-based superplasticizer show best performances in both fresh and hardened states. Concrete with lignosulfonate-based water reducer exhibits poor performance in comparison with the concretes with superplasticizers. The cost per unit compressive strength of concrete with sulfonated naphthalene polymer-based superplasticizer is lower compared with the concretes with other types of chemical admixture.

1. Introduction

In recent years, the demand for ready-mix concrete (RMC) has increased rapidly in Dhaka city. The primary reasons behind this are convenience of using RMC in high-rise structures, shortage of space at construction site, saving of time related to the preparation of concrete on-site, and better quality of RMC compared to the conventional concrete. Nevertheless, because of heavy traffic congestions throughout Dhaka, especially during weekdays, the time required to haul RMC from plant to construction site is very high. Therefore, keeping concrete workable for such a long time without compromising the required strength has become one of the most challenging tasks in the RMC industry. Moreover, high ambient temperature in summer makes the situation worse, since high temperature adversely affects the workability of fresh concrete [1–5]. Selection of appropriate type and dosage of chemical admixture can play an important role in this regard.

A good number of researches were carried out to understand the effects of different types of chemical admixture on properties of concrete. For instance, Topçu and Ateşin [6] compared the slump flow results of concretes made with lignosulfonate-based admixture and naphthalenesulfonate-based admixture; their results indicate that naphthalenesulfonate-based admixture imparts better flowability to concrete compared to lignosulfonate-based admixture. Sugamata et al. [7] investigated the effect of chemical structure of polycarboxylic ether-based admixture on fresh performance of concrete and found that

TABLE 1: Properties and recommended ranges of dosage of chemical admixtures.

Chemical admixture	Composition	Appearance	Specific gravity at 25°C	Recommended dosage range (% weight of cement)
WR	Lignosulfonate based	Dark brown liquid	1.17	0.23–0.47
SP1	Sulfonated naphthalene polymer based	Dark brown liquid	1.24	0.87–2.23
SP2	Polycarboxylic ether based	Light brown liquid	1.05	0.42–1.26
SP3	Second-generation polycarboxylic ether polymer based	Light brown liquid	1.10	0.55–1.32
SP4	Organic polymer based	Dark brown liquid	1.19	0.71–1.31

polycarboxylic ether molecule with long side chains improved the fresh performance of concrete. Mardani-Aghabaglou et al. [8] studied different types of polycarboxylic ether-based admixture and observed that polycarboxylic ether-based admixture with low side chain density imparted high flowability to concrete. Papayianni et al. [9] studied the compressive strengths of concretes made with different admixtures and observed that the admixture based on modified polycarboxylic ether polymer imparted higher compressive strength to concrete compared to the admixtures based on sulfonated polymer and synthetic polymer at the end of 28 days. In another study, Mohammed and Hamada [10] examined the long-term behaviors of concrete made with chemical admixture having polycarboxyl group and concrete made with chemical admixture having naphthalene group in tidal environment. They observed, after 10 years, chemical admixture having naphthalene group imparted higher compressive strength as well as higher Young's modulus to concrete in comparison with admixture having polycarboxyl group. They also observed that the chloride ingress was higher in concrete made with admixture having polycarboxyl group compared to the concrete made with admixture having naphthalene group.

Although numerous studies were conducted with different types of chemical admixture, very few of them addressed the influence of chemical admixtures on properties of prolonged mixed concrete [11–16]. Nevertheless, most of the studies [11–15] did not address the influence of chemical admixtures on Young's modulus and tensile strength of concrete subjected to prolonged mixing. Moreover, most of the chemical admixtures studied in [11–13, 15] are old-generation plasticizers and are less commonly used in the RMC industry at present. Mohammed et al. [16] studied the effects of different types of chemical admixture on several mechanical properties of prolonged mixed concrete (e.g., compressive strength, Young's modulus, and splitting tensile strength). However, the effects of chemical admixtures on concrete properties at different dosages and for different mix design parameters (e.g., sand to aggregate volume (s/a) ratio and cement content (C)) are yet essential to be investigated.

With the viewpoint of the above discussion, a detailed experimental investigation has been carried out with some chemical admixtures which are used extensively in the ready-mix concrete industry of Bangladesh. The effects of these admixtures and their dosages on fresh properties (workability) and hardened properties (compressive strength, Young's modulus, splitting tensile strength, and ultrasonic pulse velocity (UPV)) of concrete have been studied so as to identify the best type of chemical admixture. The cost-effectiveness of using the chemical admixtures in concrete has also been analyzed. The effects of sand to aggregate volume (s/a) ratio, cement content (C), and use of chilled mixing water on fresh and hardened properties of prolonged mixed concrete have been investigated as well.

2. Experimental Method

2.1. Material Properties. In this study, five different types of chemical admixture were used. One of the admixtures was water reducer (WR) and the rest four were superplasticizers (SP1, SP2, SP3, and SP4). All the types of chemical admixture comply with ASTM C494. The chemical and physical properties of the chemical admixtures, and their dosage ranges as recommended by the manufacturers for the best performance, are mentioned in Table 1. Crushed stones were used as a coarse aggregate in concrete. The maximum size of the coarse aggregate was 19 mm. Natural river sand was used as a fine aggregate. Figure 1 shows the gradations of coarse and fine aggregates. Both the gradations of coarse and fine aggregates satisfy ASTM C33 requirements. The physical properties of coarse and fine aggregates are summarized in Tables 2 and 3, respectively. The physical properties of both coarse and fine aggregates were determined as per ASTM guidelines. CEM II/B-M 42.5 cement (as per BDS EN 197-1–2000 [17]) was used as the binding material. The cement contains 75% clinker and 25% mineral admixture including ground-granulated blast furnace slag, limestone, fly ash, and gypsum.

2.2. Mixture Proportion and Cases Studied. Seven different types of concrete mixture were prepared for this study, namely, M1, M2, M3, M4, M5, M6, and RM. The ambient temperature was 25°C ± 2°C when the mixtures were prepared. All the mixtures were prepared using a laboratory mixer having a capacity of 75 l. Each mixture was initially mixed for 5 minutes at a speed of 20 rpm to ensure its homogeneity. After that, the mixing was carried out at a speed of 6 rpm. This low rotational speed was chosen to simulate the mixing process of ready-mix concrete in truck mixer. The mixer was stopped briefly at 15-minute intervals in order to conduct slump tests. The mixing process of each concrete mixture was continued until the final slump became less than or equal to 2 cm.

Each M1 concrete mixture was prepared using the average of the maximum and minimum dosages of every

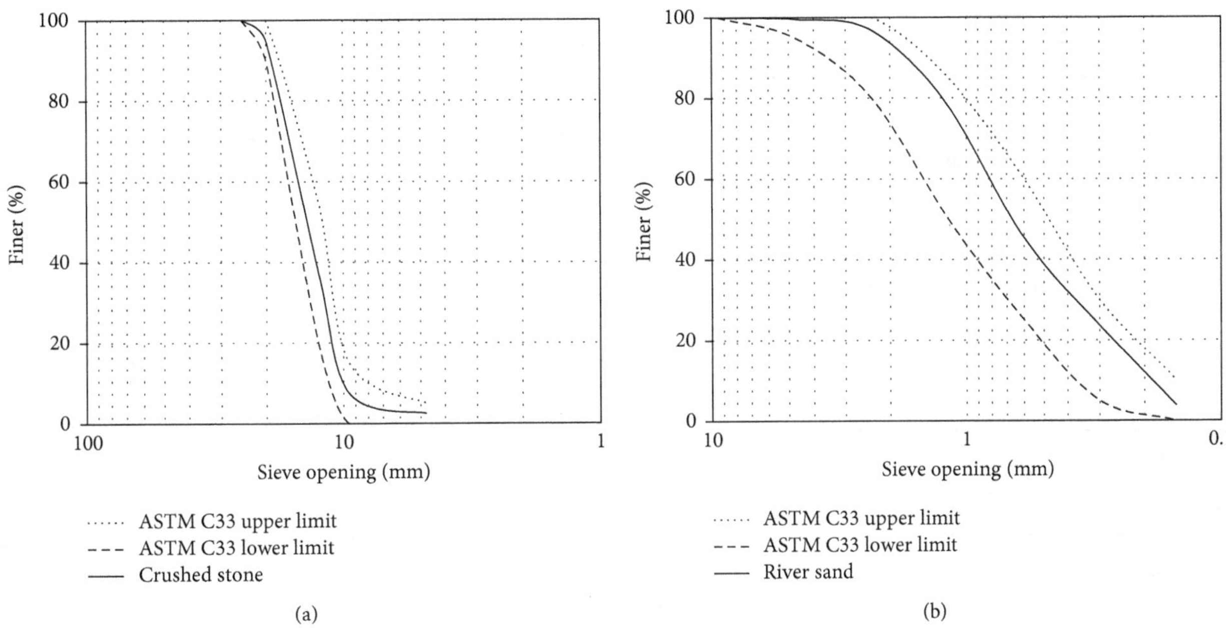

FIGURE 1: Gradations of (a) coarse aggregate (crushed stone) and (b) fine aggregate (river sand).

TABLE 2: Properties of coarse aggregate.

Aggregate type	Specific gravity	Absorption capacity (%)	Abrasion (%) (as per ASTM C131)	SSD unit weight (kg/m^3)	Fineness modulus
Crushed stone	2.56	2.39	38.30	1549	Controlled as per ASTM C33

TABLE 3: Properties of fine aggregate.

Aggregate type	Specific gravity	Absorption capacity (%)	SSD unit weight (kg/m^3)	Fineness modulus
River sand	2.45	3.30	1520	2.52

chemical admixture recommended by the respective manufacturer. On the other hand, M2, M3, M4, M5, and M6 mixtures were prepared using the maximum recommended dosages of admixtures. During the preparation of M1 and M2 mixtures, the entire dosage of admixture was applied with water at the beginning of mixing. However, in case of M3, M4, M5, and M6 mixtures, the total dosage of admixture was applied in two stages: at first, 2/3 of the dosage of admixture was applied with water, and the rest 1/3 was applied when the slump value of prolonged mixed concrete became less than or equal to 3 cm.

In M1, M2, and M3 mixtures, the water to cement (W/C) ratio, s/a ratio, and cement content (C) were, respectively, 0.40, 0.40, and 340 kg/m^3. In M4 mixture, the s/a ratio was increased to 0.45, but the W/C ratio and cement content were kept similar to M1, M2, and M3 mixtures. In M5 and M6 mixtures, the cement content was raised to 380 kg/m^3 of concrete, but W/C and s/a ratios were kept, respectively, to 0.40 and 0.40. In the RM (reference mix), no chemical admixture was used, and the W/C ratio, s/a ratio, and cement content were, respectively, 0.40, 0.40, and 340 kg/m^3. Prior to the preparation of each mixture, both the coarse and fine aggregates were brought to saturated surface dry (SSD) condition, so that the W/C ratio of the mix would remain unaffected.

In case of M6 mixture, instead of using plain water, chilled water was used (half of the total required water (by weight) as per mix design was plain water and the rest half was ice) to keep the temperature of the concrete mix low. The initial temperature of the chilled water was 0°C ± 1°C. It should be noted that, in M4, M5, and M6 mixtures, only the two types of admixture that gave best performances in M1, M2, and M3 mixtures were used. The details of different types of mixture are presented in Table 4.

Cylindrical concrete specimens of 100 mm diameter and 200 mm height were made for assessing the hardened properties of concrete. A total of 22 independent cases and 242 specimens were investigated. The mixture proportions of all 22 cases are summarized in Table 5.

2.3. Specimen Preparation and Test Setup. All the concrete mixtures with chemical admixtures (namely, M1, M2, M3, M4, M5, and M6) were subjected to prolonged mixing. Slump tests were done for concrete mixtures subjected to prolonged mixing at 15-minute intervals as per ASTM C143. When the slump of each mixture became less than or equal to 2 cm, the mixing process was stopped, and the mixture was poured into cylindrical molds to prepare 100 mm by

Table 4: Details of different types of concrete mixture.

Mix type	Details
M1	W/C = 0.40; s/a = 0.40; $C = 340\,kg/m^3$; average admixture dosage
M2	W/C = 0.40; s/a = 0.40; $C = 340\,kg/m^3$; maximum admixture dosage
M3	W/C = 0.40; s/a = 0.40; $C = 340\,kg/m^3$; maximum admixture dosage in two stages: 2/3 at the beginning and 1/3 when slump ≤ 3 cm
M4	W/C = 0.40; s/a = 0.45; $C = 340\,kg/m^3$; maximum admixture dosage in two stages: 2/3 at the beginning and 1/3 when slump ≤ 3 cm
M5	W/C = 0.40; s/a = 0.40; $C = 380\,kg/m^3$; maximum admixture dosage in two stages: 2/3 at the beginning and 1/3 when slump ≤ 3 cm
M6	W/C = 0.40; s/a = 0.40; $C = 380\,kg/m^3$; maximum admixture dosage in two stages: 2/3 at the beginning and 1/3 when slump ≤ 3 cm; chilled water used (1/2 of the total water (by weight) was ice)
RM (reference mix)	W/C = 0.40; s/a = 0.40; $C = 340\,kg/m^3$; no admixture

Table 5: Mixture proportion of concrete.

Concrete mix type	W/C	s/a	Admixture type	Unit content (kg/m^3)				Admixture dosage
				Cement	Sand	Aggregate	Water	(% weight of cement)
M1		0.40	WR	340	721	1130	136	0.35
			SP1	340	718	1125	136	1.55
			SP2	340	719	1127	136	0.84
			SP3	340	719	1127	136	0.94
			SP4	340	719	1127	136	1.01
M2		0.40	WR	340	721	1130	136	0.47
			SP1	340	716	1122	136	2.23
			SP2	340	718	1125	136	1.26
			SP3	340	718	1125	136	1.32
			SP4	340	718	1126	136	1.31
M3	0.40	0.40	WR	340	721	1130	136	0.47
			SP1	340	716	1122	136	2.23
			SP2	340	718	1125	136	1.26
			SP3	340	718	1125	136	1.32
			SP4	340	718	1126	136	1.31
M4		0.45	SP1	340	806	1029	136	2.23
			SP3	340	808	1032	136	1.32
M5		0.40	SP1	380	686	1075	152	2.23
			SP3	380	688	1079	152	1.32
M6		0.40	SP1	380	686	1075	152	2.23
			SP3	380	688	1079	152	1.32
RM (reference mix)		0.40	—	340	722	1132	136	—

Total number of cases = 22

Cylinders per case = 3 × 3 (compressive strengths at 7 days, 28 days, and 90 days) + 2 (splitting tensile strengths at 28 days) = 11

Total number of cylinders = 11 × 22 = 242

200 mm cylindrical concrete specimens. After casting of concrete specimens, they were cured initially for 24 hours by covering the cylindrical molds with wet clothes and polythene to prevent moisture loss. The specimens were demolded after 24 hours of casting and then cured under water till the age of testing as per ASTM C31.

The cylindrical specimens were tested for compressive strength and Young's modulus at 7 days, 28 days, and 90 days and for splitting tensile strength at 28 days according to ASTM specifications. Prior to compressive strength test, the UPV test was conducted for unloaded water-saturated concrete specimens according to ASTM C597.

3. Results and Discussions

3.1. Fresh Properties of Concrete. According to the JSCE 2007 guidelines for concrete [18], for normal weight ready-mix concrete, the minimum slump requirement at the place of unloading is 8 cm. Therefore, in this study, the workability of each concrete mix was judged based on the time up to which the slump value was greater than 8 cm. The more the time, the better the workability.

3.1.1. Influence of Chemical Admixture Type. The slump test results of RM, M1, M2, and M3 mixtures at 15-minute

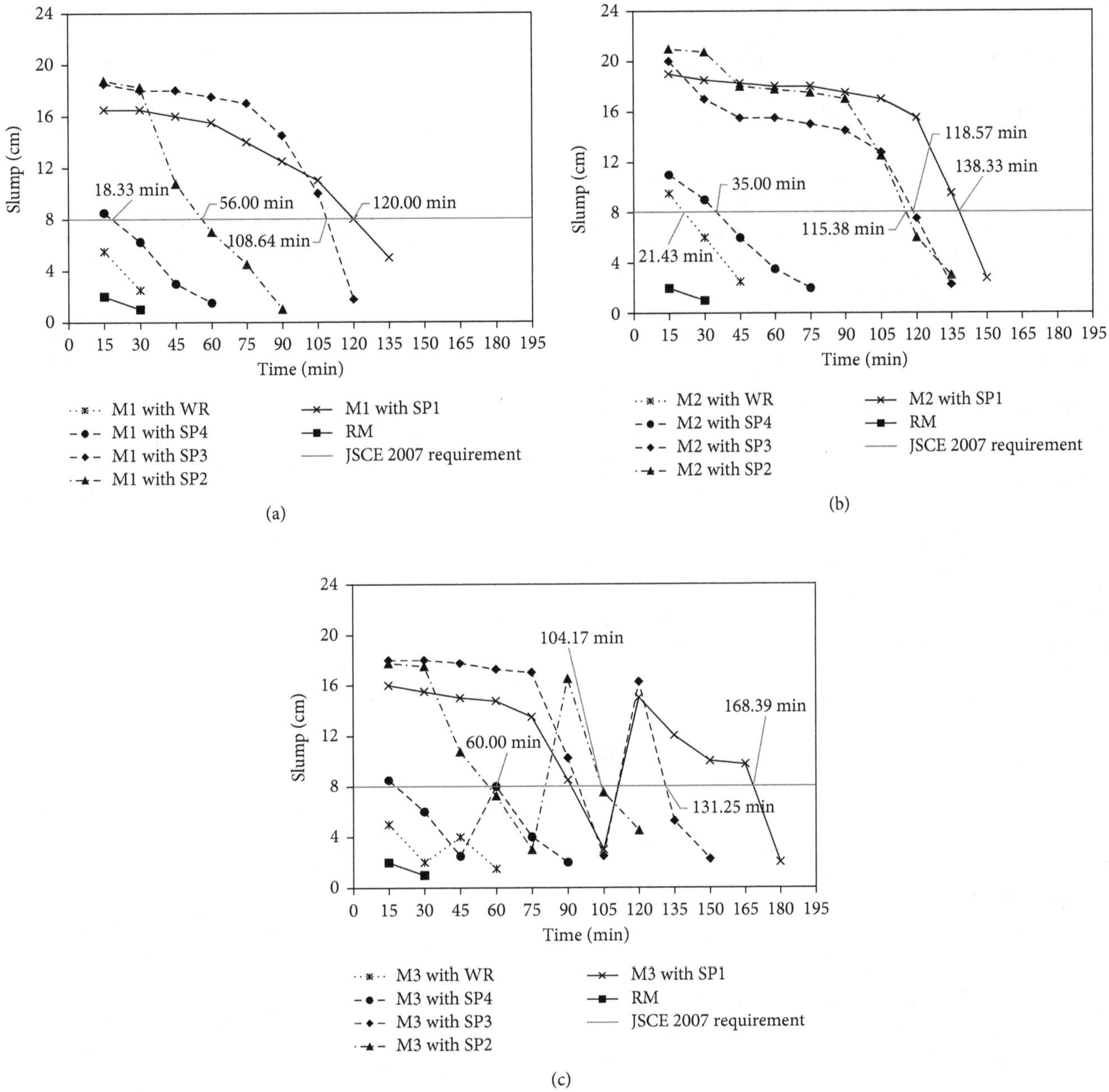

FIGURE 2: Slump test results of RM, M1, M2, and M3 mixtures (W/C = 0.40, s/a = 0.40, and C = 340 kg/m^3): (a) M1 = average dosage of admixture, RM = no admixture; (b) M2 = maximum dosage of admixture, RM = no admixture; (c) M3 = maximum dosage of admixture in two stages, RM = no admixture.

intervals are presented in Figure 2. Although M1, M2, and M3 concrete mixtures with sulfonated naphthalene polymer-based superplasticizer SP1 resulted in lower initial slumps compared to the mixtures with SP2 and SP3, concrete mixtures with SP1 remained workable for the longest durations. The second best performances were found for M1, M2, and M3 mixtures with second-generation polycarboxylic ether-based superplasticizer SP3. Polycarboxylic ether-based superplasticizer SP2 can be placed in the third position. However, M2 mixture with SP2 performed almost as good as M2 mixture with SP3. Concretes with organic polymer-based SP4 remained workable for shorter durations in comparison with the concretes with other types of superplasticizer. Fresh performances of concretes with superplasticizers were better in comparison with the performances of concrete mixtures with lignosulfonate-based water reducer WR. Results presented in Figure 2 also confirm that M1, M2, and M3 mixtures exhibited better performances in comparison with the reference mix RM (i.e., concrete without chemical admixture).

3.1.2. Influence of Increased Dosage of Chemical Admixture. Results presented in Figure 2 show that, for all the types of chemical admixture, M2 mixture remained workable for longer durations compared to M1 mixture (i.e., concrete remained workable for longer duration when the dosage of admixture was increased). For instance, in case of M2 mixture with SP3, slump became 8 cm at the end of 118.57 minutes, whereas in case of M1 concrete with SP3, slump became 8 cm at the end of 108.64 minutes.

3.1.3. Influence of Two-Stage Dosage of Chemical Admixture. Figure 2 shows that, irrespective of chemical admixture type, M3 mixture remained workable for longer time in comparison with M2 mixture, which implies that concrete would remain workable for longer duration when the dosage of admixture is applied in two stages instead of applying the entire dosage at the beginning of mixing.

3.1.4. Influence of Sand to Aggregate Volume Ratio. The fresh performances of M3, M4, M5, and M6 concrete mixtures with sulfonated naphthalene polymer-based SP1 and second-generation polycarboxylic ether-based SP3 are presented in Figure 3. For both SP1 and SP3, the slump values of M4 concrete mixture were greater than 8 cm for smaller time period compared to M3 mixture. So, the workability of fresh concrete decreased when the s/a ratio was increased from 0.40 to 0.45. The reason behind such results may be attributed to the increased volume of fine particles in concrete mix with the increase of s/a ratio. Increased volume of fine particles would increase the surface area; hence, more water would be required to wet this increased area, and consequently the workability of fresh concrete would reduce.

3.1.5. Influence of Cement Content. Figure 3 shows that, for both SP1 and SP3, M5 concrete mixture gave slump values more than 8 cm for longer period compared to M3 concrete mixture. The results indicate that, for given W/C and s/a ratios, the workability of concrete would increase with the increase of cement content. This is because, for a given W/C ratio, the paste content in concrete mix would increase with the increase of cement content which would eventually increase the workability of concrete.

3.1.6. Influence of Using Chilled Mixing Water. From Figure 3 it can be seen that M6 concrete mixture with SP1 and SP3 remained workable for longer durations in comparison with M5 mixture with SP1 and SP3. It can therefore be inferred from the results that the concrete would remain workable for longer time if chilled water is used instead of plain water. The slump of M6 mixture with SP1 became 8 cm at the end of 265.71 minutes, and the slump of M6 mixture with SP3 became 8 cm at the end of 222.5 minutes.

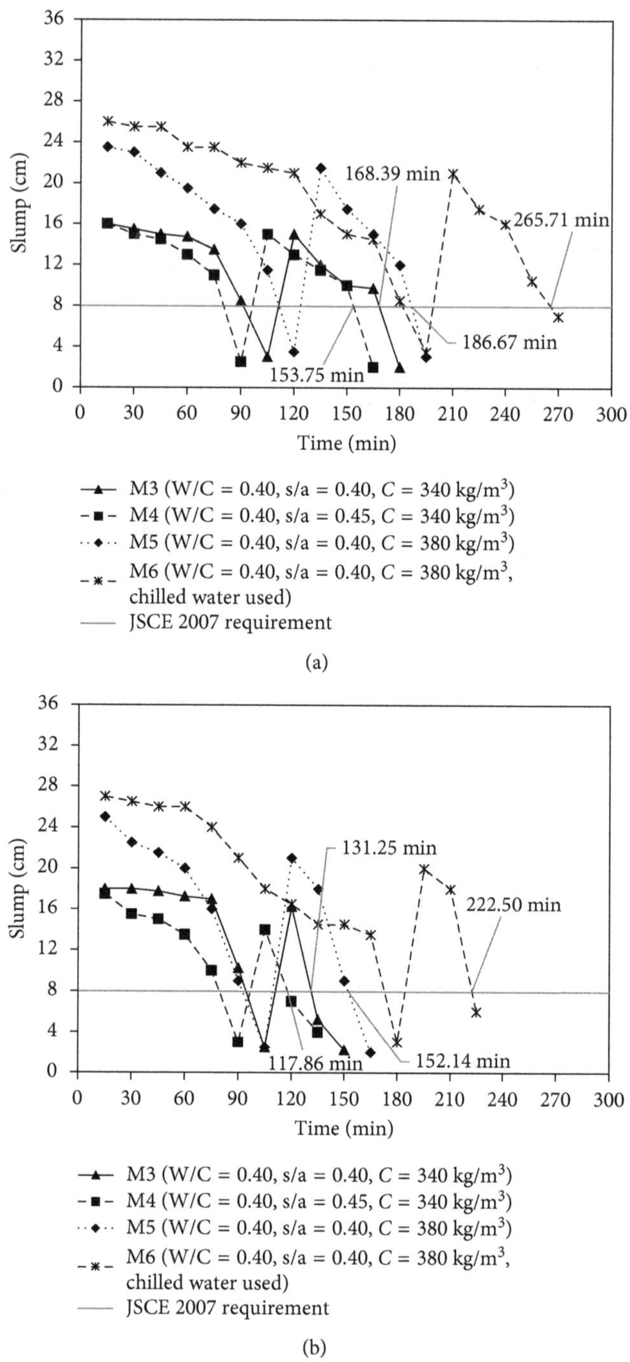

FIGURE 3: Slump test results of M3, M4, M5, and M6 mixtures: (a) sulfonated naphthalene polymer-based SP1 (maximum dosage of admixture in two stages) and (b) second-generation polycarboxylic ether-based SP3 (maximum dosage of admixture in two stages).

3.2. Hardened Properties of Concrete

3.2.1. Influence of Chemical Admixture Type. Figure 4 presents the hardened properties of RM, M1, M2, and M3 concretes at the age of 28 days. In all cases, concretes with chemical admixtures resulted in higher compressive strengths, Young's moduli, splitting tensile strengths, and UPVs compared to the concrete without chemical admixture. Concretes with sulfonated naphthalene polymer-based SP1 and second-generation polycarboxylic ether-based SP3 exhibited better hardened performances in comparison with the concretes with other types of chemical admixture. Concretes with lignosulfonate-based water reducer WR exhibited lower compressive strengths, Young's moduli, splitting tensile strengths, and UPVs

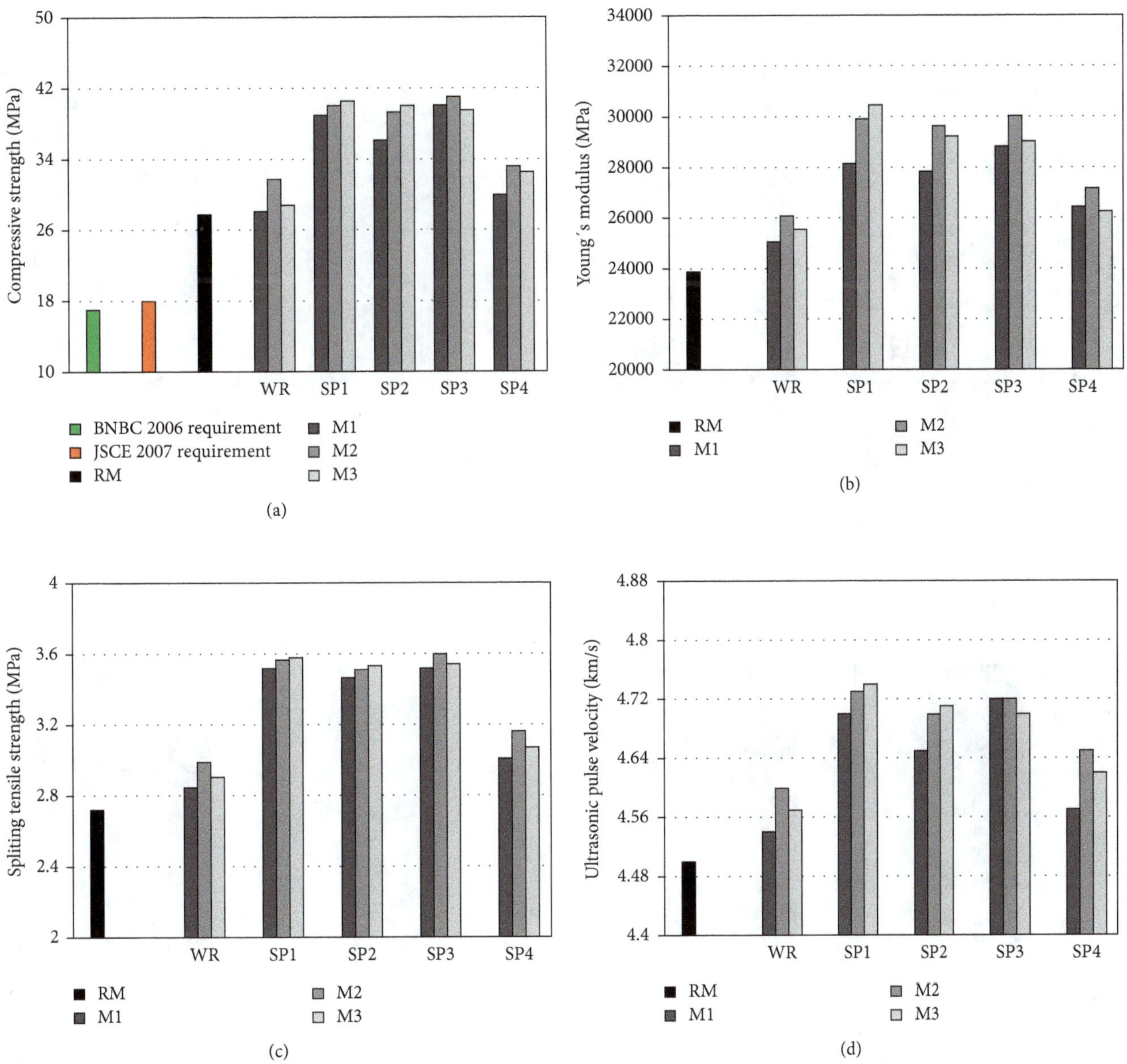

Figure 4: Hardened properties of RM, M1, M2, and M3 concretes (W/C = 0.40, s/a = 0.40, and $C = 340\ \text{kg/m}^3$) at the age of 28 days: (a) compressive strength, (b) Young's modulus, (c) splitting tensile strength, and (d) ultrasonic pulse velocity.

than the concretes with superplasticizers. The results comply with the conclusions drawn by Rixom and Mailvaganam [19]. Their study indicates that, for a given W/C, superplasticizers based on sulfonated naphthalene formaldehyde (SMF) and sulfonated melamine formaldehyde (SMF) impart substantially higher compressive strength to concrete in comparison with plasticizers based on lignosulfonate and hydroxycarboxylic acid.

The Bangladesh National Building Code (BNBC) 2006 [20] and JSCE 2007 guidelines for concrete [18] specify the lower limits of 28-day cylindrical compressive strength of normal weight concrete, respectively, to be 17 MPa and 18 MPa. Figure 4 shows that, in all cases, concretes with chemical admixtures satisfied the BNBC 2006 [20] and JSCE 2007 [18] requirements.

3.2.2. Influence of Increased Dosage of Chemical Admixture. It can be seen from Figure 4 that, for all types of chemical admixture, M2 concrete resulted in higher compressive strengths, Young's moduli, splitting tensile strengths, and UPVs than M3 concrete. This indicates that the hardened performance of concrete would improve when the dosage of admixture is increased. However, it should be noted that, in no concrete mixture investigated in this study, the dosage of admixture was higher than the maximum dosage recommended by the respective manufacturer. The results conform to the findings obtained by Topçu and Ateşin [6] and Gagné et al. [21]. Their studies imply that, for a given W/C, the compressive strength and flexural strength of concrete increase with the increase of dosage of chemical admixture up to an optimum limit. Topçu and Ateşin [6]

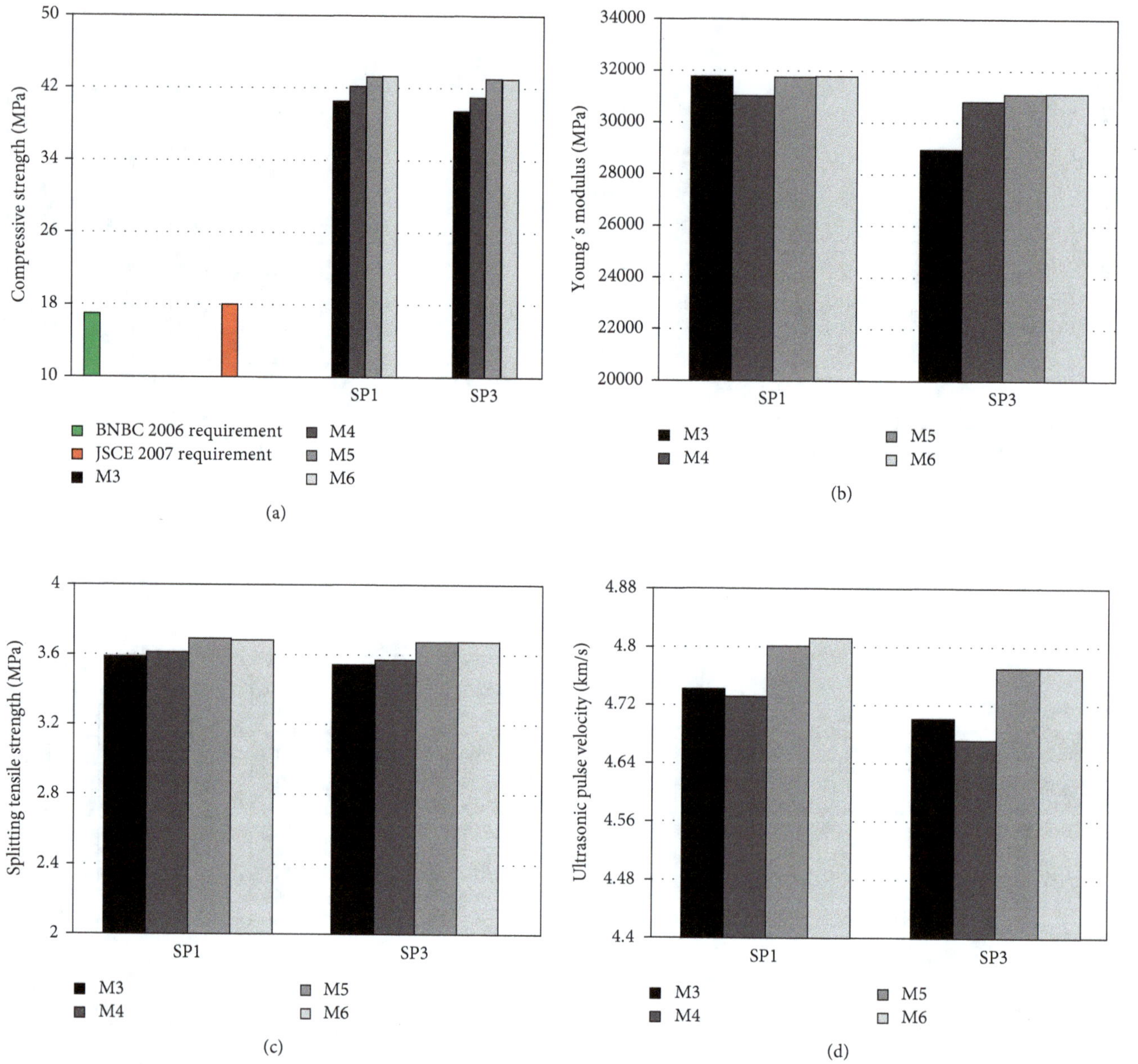

Figure 5: Hardened properties of M3, M4, M5, and M6 concretes at the age of 28 days: (a) compressive strength, (b) Young's modulus, (c) splitting tensile strength, and (d) ultrasonic pulse velocity.

examined the microstructures of concretes made with lignosulfonate and naphthalenesulfonate-based chemical admixtures, by means of a scanning electron microscope (SEM). They found that when the dosage of each chemical admixture was within the optimum limit, concrete with admixture was less porous than the concrete without admixture. The reason behind concrete with admixture exhibiting less porous formations and thus providing improved hardened properties may be attributed to the improved compactibility of concrete.

The present study focused on assessing the fresh and hardened performances of concrete for admixture dosages not exceeding the maximum limits suggested by respective manufacturers. Therefore, the study did not address the influence of overdosing of admixture on concrete properties. Nevertheless, the study conducted by Gagné et al. [21] suggests that dosing of admixture in concrete beyond an optimum limit may deteriorate the physical properties of concrete. This is because overdosing of chemical admixture may result in internal bleeding, which may reduce the quality of paste-aggregate interface.

3.2.3. Influence of Two-Stage Dosage of Chemical Admixture. Figure 4 shows that, although for some chemical admixtures, M3 concrete resulted in lower compressive strengths, Young's moduli, splitting tensile strengths, and UPVs compared to M1 and M2 concretes; in all cases, the compressive strengths, Young's moduli, splitting tensile strengths, and UPVs were higher than those resulted by RM concrete. Hence, it can be inferred that the hardened performance of concrete having dosage of chemical admixture applied in two stages would be better compared to the concrete without chemical admixture. However, the dosage of admixture should be within the range recommended by the manufacturer.

3.2.4. Influence of Sand to Aggregate Volume Ratio. Figure 5 presents the hardened properties of M3, M4, M5, and M6 concretes at the age of 28 days. For both SP1 and SP3, the compressive strengths, Young's moduli, and splitting tensile strengths of M4 concrete were greater than those of M3 concrete. However, the UPVs of M4 concrete were lower compared to M3 concrete. The results therefore indicate that, with the increase of s/a ratio from 0.40 to 0.45, the compressive strength, Young's modulus, and splitting tensile strength of concrete would increase, but the UPV through concrete would reduce. Similar conclusions were drawn by Lin et al. [22] and Mohammed et al. [23] based on experimental observations.

It should be noted that, with the increase of s/a ratio from 0.40 to 0.45, the workability of prolonged mixed concrete decreased, but the compressive strength of concrete increased. Therefore, when the workability of ready-mix concrete is of primary concern, it should be kept into consideration that the s/a ratio is not too high. On the other hand, if the strength of concrete is of primary concern, the s/a ratio should not be too low.

Results presented in Figure 5 also show that, for both SP1 and SP3, the 28-day compressive strengths of M4, M5, and M6 concretes satisfy the minimum requirements of the BNBC 2006 [20] and JSCE 2007 [18].

3.2.5. Influence of Cement Content. Figure 5 shows that compressive strengths, Young's moduli, splitting tensile strengths, and UPVs of M5 concrete were greater in comparison with M3 concrete, which implies that the hardened performance of concrete would improve if the cement content is increased.

3.2.6. Influence of Using Chilled Mixing Water. Results presented in Figure 5 show that the 28-day compressive strengths, Young's moduli, splitting tensile strengths, and UPVs of M6 concrete are almost similar to those of M5 concrete. So, the 28-day hardened properties of concrete did not change significantly when chilled water was used instead of plain water. Since use of chilled mixing water would improve the workability of fresh concrete significantly (Figure 3) without compromising its performance in hardened state, it is advisable to use ice partially with water in ready-mix concrete.

3.2.7. Influence of Age of Concrete. Figure 6 shows the compressive strength gains with age, for different types of concrete with sulfonated naphthalene polymer-based superplasticizer SP1. Results show that, for all types of concrete, the rates of strength gain were high before 28 days, beyond 28 days, the strength gain rates became lower. The 7-day strength of M6 concrete with SP1 was slightly lower compared to that of M5 concrete, this is because the reduced temperature of fresh M6 concrete due to the use of chilled water delayed the process of strength gain at the early age [24]. However, the 28-day compressive strength of M6 concrete was comparable to that of M5 concrete and the 90-day strength of M6 concrete was slightly higher compared to M5 concrete. The results comply with the conclusions drawn by Price [25] from experimental investigation.

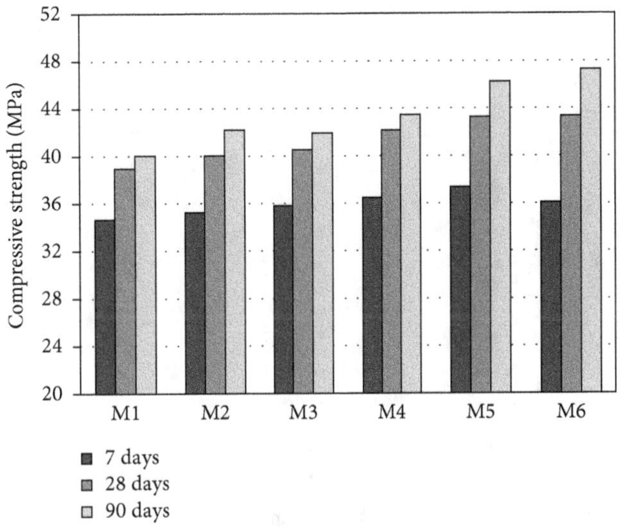

FIGURE 6: Compressive strength gains of different types of concrete made with sulfonated naphthalene polymer-based SP1 with respect to age.

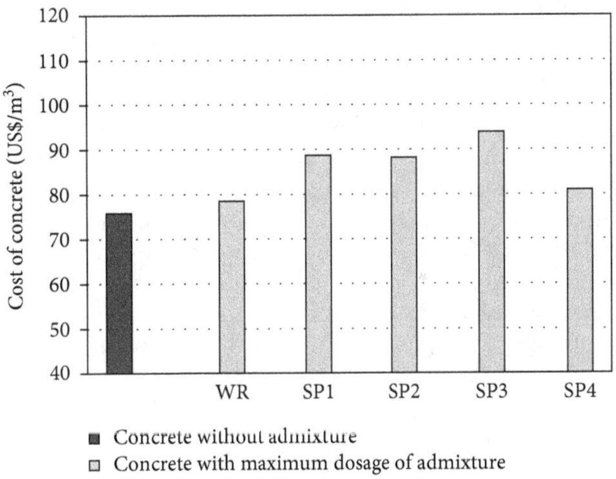

FIGURE 7: Costs of 1 m^3 of concretes (W/C = 0.40, s/a = 0.40, and C = 340 kg/m^3) prepared with maximum recommended dosages of different types of chemical admixture.

3.3. Cost Analysis. Costs of 1 m^3 of concretes made with maximum recommended dosages of different types of chemical admixture and cost of 1 m^3 of concrete made without admixture are shown in Figure 7. It should be noted that the costs presented in Figure 7 were calculated based on the local costs of materials used to prepare per cubic meter of concrete. From Figure 7, lignosulfonate-based water reducer WR seems to be more economical than other admixtures. On the other hand, second-generation

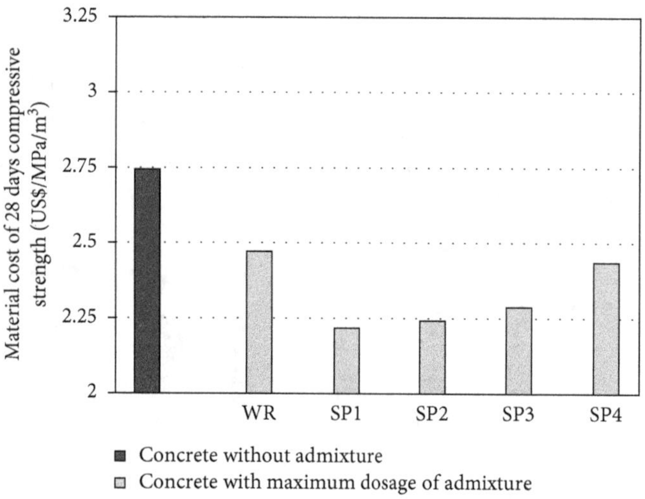

FIGURE 8: Unit strength costs of 1 m^3 of concretes (W/C = 0.40, s/a = 0.40, and C = 340 kg/m^3) prepared with maximum recommended dosages of different types of chemical admixture.

polycarboxylic ether-based superplasticizer SP3 seems to be the most expensive one.

The 28-day strengths of concretes prepared with different types of chemical admixture are different. Therefore, the costs per unit 28-day compressive strength (1 MPa) of 1 m^3 concretes prepared with different types of chemical admixture are presented in Figure 8. It can be seen that when the compressive strengths of concretes are taken into account, concrete mixtures with chemical admixtures are found to be more economical compared to the mixture without admixture, sulfonated naphthalene polymer-based superplasticizer SP1 is found to be the most economical admixture, and lignosulfonate-based water reducer WR is found to be the most expensive one.

4. Conclusions

From the scope of this investigation, the following conclusions can be drawn:

(1) Sulfonated naphthalene polymer-based superplasticizer shows best performance in improving workability of fresh concrete in comparison with other chemical admixtures. Second-generation polycarboxylic ether-based superplasticizer can be categorized as the second best chemical admixture in improving workability of fresh concrete.

(2) Concretes made with sulfonated naphthalene polymer-based superplasticizer and second-generation polycarboxylic ether-based superplasticizer exhibit higher compressive strengths, splitting tensile strengths, Young's moduli, and UPVs than concretes prepared with other admixtures.

(3) Superplasticizers show better performances in improving fresh and hardened properties of concrete compared to water reducers.

(4) The compressive strength of concrete increases with increase of admixture dosage, when the dosage of admixture is within the range recommended by the manufacturer.

(5) Applying dosage of chemical admixture in two stages helps to keep concrete workable for longer duration compared to applying the same dosage of admixture at the beginning of mixing process.

(6) Workability of fresh concrete decreases with the increase of s/a ratio from 0.40 to 0.45. However, the compressive strength of concrete increases when the s/a ratio is increased from 0.40 to 0.45.

(7) If chilled water is used in concrete mix, it remains workable for longer period in comparison with the concrete mix with plain water. Use of chilled water in fresh concrete causes reduction in early concrete strength but improves the long-term strength property of concrete.

(8) Unit strength cost of concrete made with sulfonated naphthalene polymer-based superplasticizer is lower compared to other chemical admixtures.

Conflicts of Interest

The authors declare that they have no conflicts of interest.

Acknowledgments

The authors express their sincere gratitude and appreciation to Islamic University of Technology (IUT), Gazipur, Bangladesh, and the Structural Engineers Limited (SEL), Bangladesh, for financing the research project. The cement used in this study was provided by Seven Circle Bangladesh Limited. The authors wish to express their gratitude to Seven Circle Bangladesh Limited as well.

References

[1] J. G. Munday, "Transportation effects on strength and consistency," in *Advanced Ready-Mix Concrete Technologies*, R. K. Dhir Ed., pp. 343–353, Pergamon Press, Oxford, UK, 1976.

[2] N. P. Mailvaganam, "Factors influencing slump loss in flowing concrete," *ACI Special Publication*, vol. 62, pp. 389–404, 1979.

[3] N. Ghafoori and H. Diawara, "Influence of temperature on fresh performance of self-consolidating concrete," *Construction and Building Materials*, vol. 24, no. 6, pp. 946–955, 2010.

[4] V. Sampebulu, "Influence of high temperatures on the workability of fresh ready-mixed concrete," *ITB Journal of Engineering Science*, vol. 44, no. 1, pp. 21–32, 2012.

[5] S. Wang, W. K. Guo, and Z. Y. Gao, "Influences of cement temperature on the performance of concrete mixed with superplasticizer," in *Advances in Engineering and Technology*, L. Xie and D. Huang Eds., pp. 243–250, Taylor & Francis Group, London, UK, 2014.

[6] I. B. Topçu and Ö. Ateşin, "Effect of High Dosage Lignosulphonate and Naphthalene Sulphonate Based Plasticizer Usage on Micro Concrete Properties," *Construction and Building Materials*, vol. 120, pp. 189–197, 2016.

[7] T. Sugamata, T. Sugiyama, and A. Ohta, "The effects of a new high-range water-reducing agent on the improvement of rheological properties," in *Proceedings of the Seventh*

CANMET/ACI International Conference on Superplasticizers and Other Chemical Admixtures in Concrete, pp. 343–359, Berlin, Germany, October 2003.

[8] A. Mardani-Aghabaglou, M. Tuyan, G. Yilmaz, Ö. Ariöz, and K. Ramyar, "Effect of different types of superplasticizer on fresh, rheological and strength properties of self-consolidating concrete," *Construction and Building Materials*, vol. 47, pp. 1020–1025, 2013.

[9] I. Papayianni, G. Tsohos, N. Oikonomou, and P. Mavria, "Influence of superplasticizer type and mix design parameters on the performance of them in concrete mixtures," *Cement and Concrete Composites*, vol. 27, no. 2, pp. 217–222, 2005.

[10] T. U. Mohammed and H. Hamada, "Durability of concrete made with different water-reducing chemical admixtures in tidal environment," *ACI Materials Journal*, vol. 100, no. 3, pp. 194–202, 2003.

[11] R. F. Adams, P. R. Stodola, and D. R. Mitchell, "Discussion of concrete retempering studies by Hawkins M. J.," *ACI Journal Proceedings*, vol. 59, pp. 1249–1250, 1962.

[12] D. Ravina, "Retempering of prolonged-mixed concrete with admixtures in hot weather," *ACI Journal Proceedings*, vol. 72, no. 6, pp. 291–295, 1975.

[13] R. W. Previte, "Concrete slump loss," *ACI Journal Proceedings*, vol. 74, no. 8, pp. 361–367, 1977.

[14] D. Ravina and I. Soroka, "Slump loss and compressive strength of concrete made with WRR and HRWR admixtures and subjected to prolonged mixing," *Cement and Concrete Research*, vol. 24, no. 8, pp. 1455–1462, 1994.

[15] A. Baskoca, M. H. Ozkul, and S. Artirma, "Effect of chemical admixtures on workability and strength properties of prolonged agitated concrete," *Cement and Concrete Research*, vol. 28, no. 5, pp. 737–747, 1998.

[16] T. U. Mohammed, T. Ahmed, T. A. Mallick, F. Shahriar, and A. Munim, "Influence of chemical admixtures on fresh and hardened properties of ready mix concrete," in *1st International Conference on Engineering Research Practice*, A. Rahman, F. Mashiri, M. M. Rahman, and R. Karim Eds., pp. 36–41, STAMCA, Dhaka, Bangladesh, 2017.

[17] BDS EN 197–1, *Cement-Part 1: Composition, Specifications and Conformity Criteria for Common Cements*, Bangladesh Standards and Testing Institution, Dhaka, Bangladesh, 2003.

[18] JSCE Guidelines for Concrete, *Standard Specifications for Concrete Structures*, Japan Society of Civil Engineers, Tokyo, Japan, 2007.

[19] M. R. Rixom and N. P. Mailvaganam, *Chemical Admixtures for Concrete*, E & FN Spon, London, UK, 3rd edition, 1999.

[20] *Bangladesh National Building Code*, Housing and Building Research Institute, and Bangladesh Standards and Testing Institution, Dhaka, Bangladesh, 2006.

[21] R. Gagné, A. Boisvert, and M. Pigeon, "Effect of superplasticizer dosage on mechanical properties, permeability, and freeze-thaw durability of high-strength concretes with and without silica fume," *ACI Materials Journal*, vol. 93, no. 2, pp. 111–120, 1996.

[22] Y. Lin, S.-F. Kuo, C. Hsiao, and C.-P. Lai, "Investigation of pulse velocity-strength relationship of hardened concrete," *ACI Materials Journal*, vol. 104, no. 4, pp. 344–350, 2007.

[23] T. U. Mohammed and M. N. Rahman, "Effect of types of aggregate and sand-to-aggregate volume ratio on UPV in concrete," *Construction and Building Materials*, vol. 125, pp. 832–841, 2016.

[24] A. M. Neville, *Properties of Concrete*, Pearson, London, UK, 5th edition, 2011.

[25] W. H. Price, "Factors influencing concrete strength," *ACI Journal Proceedings*, vol. 47, no. 2, pp. 417–432, 1951.

20

Comparative Study of Strength and Corrosion Resistant Properties of Plain and Blended Cement Concrete Types

Velu Saraswathy,[1,2] Subbiah Karthick,[1,3] Han Seung Lee,[1] Seung-Jun Kwon,[4] and Hyun-Min Yang[1]

[1] *Department of Architectural Engineering, Hanyang University, 1271 Sa 3-dong, Sangrok-gu, Ansan 426791, Republic of Korea*
[2] *Corrosion and Materials Protection Division, CSIR-Central Electrochemical Research Institute, Karaikudi, Tamil Nadu 630003, India*
[3] *PG and Research Department of Chemistry, Alagappa Government Arts College, Karaikudi, Tamil Nadu 630003, India*
[4] *Department of Civil Engineering, Hannam University, Daejeon 306-791, Republic of Korea*

Correspondence should be addressed to Seung-Jun Kwon; jjuni98@hannam.ac.kr

Academic Editor: Michael J. Schütze

The relative performances of mechanical, permeability, and corrosion resistance properties of different concrete types were compared. Concrete types were made from ordinary Portland cement (OPC), Portland pozzolana cement (PPC), and Portland slag cement (PSC). Compressive strength test, effective porosity test, coefficient of water absorption, short-term accelerated impressed voltage test, and rapid chloride permeability test (RCPT) were conducted on M30 and M40 grades of concrete designed with OPC, PPC, and PSC cements for 28- and 90-day cured concrete types. Long-term studies such as microcell and electrochemical evaluation were carried out to understand the corrosion behaviour of rebar embedded in different concrete types. Better corrosion resistant properties were observed for PSC concrete by showing a minimum current flow, lowest free chloride contents, and lesser porosity. Besides, PSC concrete has shown less coefficient of water absorption, chloride diffusion coefficient (CDC), and lower corrosion rate and thereby the time taken for initiation of crack extended.

1. Introduction

Reinforced concrete is protected from corrosion due to the high alkalinity existing in the concrete environment and the cover concrete which acts as a protective barrier to the access of chloride ions. Apart from this, reinforced concrete structures when exposed to severe marine environments lead to the premature failure of the structures [1]. Corrosion of embedded steel in concrete is one of the major causes of premature deterioration of concrete structures which affects the service life of the important structures in the chloride prone environment [2]. Therefore, the chloride permeability plays a significant role in determining the service life of the major infrastructures [3]. Due to the heterogeneous nature of the concrete, chloride ions enter the pores and attack the embedded steel in concrete causing cracking and spalling of the cover concrete portion and exposing the steel reinforcement when it is exposed to the marine environment [4]. Hence, to enhance the durability of the structures, waste materials like fly ash, rice husk ash, slag, and other pozzolanic materials are being used in the construction industry, not only from the durability point of view, but also to reduce the greenhouse gas emission and to save the environment [5]. The use of supplementary cementitious materials (SCM) and their effectiveness in making the blended cement concrete are well debated in the literature [6–12]. The composition of SCM may vary from source to source from which they are collected, because the nature of the source depends upon the raw materials used for making the manufacturing process and the geographical conditions [13].

Curing plays a major role in affecting the strength and permeability characteristics of the concrete structures. Mainly while using supplementary cementitious materials, care should be taken for continuous curing of the structures in the initial stages since hydration reactions are slow in using pozzolanic materials. The other factors which influence the strength characteristics are fineness modulus, water-cement ratio (w/c), chemical composition, and so on [14, 15]. Many researchers have discussed the different types of curing, pore structure refinement, and permeability characteristics of blended cements by using fly ash, slag, silica fume, and so on [16–19]. The results indicated that blended cement with longer duration has shown strength and other properties equal to those of ordinary Portland cement. It is also reported that blended cements enhanced the workability, durability, and mechanical properties of the concrete [20–24]. So far, the studies have focused on the particular aspect of corrosion like permeability, pore structure, chloride ingress, or physical and chemical characteristics by either short-term or long-term studies.

Hence, the aim of the present investigation is to evaluate the corrosion resistant properties of rich and lean mix concrete made from Portland pozzolana cement (PPC) and Portland slag cement (PSC) under accelerated and normal exposure conditions. The results were compared with ordinary Portland cement (OPC) concrete. Further, in this investigation, two curing periods, two grades of concrete, and two different exposure conditions, namely, normal and accelerated exposure conditions, were carried out. In addition to these, short-term and long-term tests, porosity and permeability studies, and mechanical and electrochemical studies were also carried out for plain and blended cements, and the results were presented in a detailed manner.

2. Experimental

2.1. Materials and Mix Proportions. Two mix proportions having characteristic compressive strength of 30 and 40 MPa concrete types were used for casting the concrete specimens. The details of concrete mix proportions are given in Table 1. Three cements, namely, 43 grade ordinary Portland cement conforming to IS:8112 [25] (equivalent to ASTM Type I cement), Portland pozzolana cement conforming to IS:1489–Part-I [26], and Portland slag cement conforming to IS:455 [27], were used. Chemical composition of cements used is given in Table 2. Graded river sand passing through 1.18 mm sieve with a fineness modulus of 2.85 and a specific gravity of 2.55 was used as fine aggregate. Locally available well graded aggregates of normal size greater than 4.75 mm and less than 20 mm having a fineness modulus of 2.72 were used as coarse aggregates. The specific gravity of the fine aggregate used was 2.78. Thermomechanically treated (TMT) rebar (Fe-415 grade) of 12 mm diameter conforming to IS:1786 [28] was used. Potable water was used for casting the concrete.

2.2. Methods

2.2.1. Compressive Strength. Compressive strength test was carried out in concrete cubes of size 150 × 150 × 150 mm using 1 : 1.8 : 3.69 mix with w/c ratio of 0.55 (M30 grade) and 1 : 1.18 : 3.36 with w/c ratio 0.45 (M40 grade) using OPC and blended cements (PPC and PSC). During casting, the moulds were mechanically vibrated. After 24 hours, the specimens were removed from the mould and subjected to water curing for 28 and 90 days. After a specified period of curing, the specimens were tested for compressive strength using AIMIL compression testing machine of 2000 kN capacity at a rate of loading of 210 kN/min. The tests were carried out on six specimens and the average compressive strength values were recorded for OPC, PPC, and PSC concrete types.

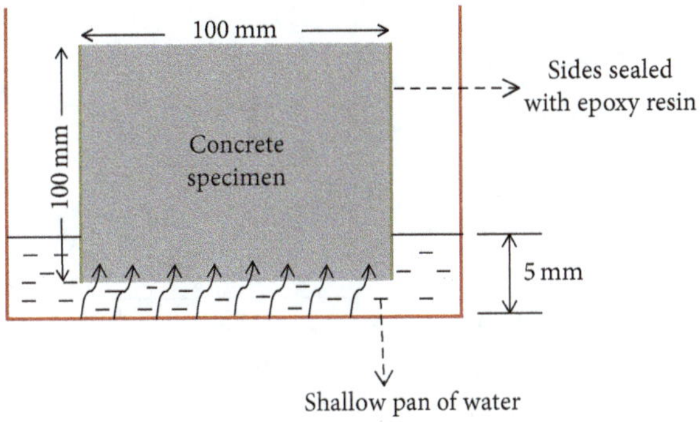

Figure 1: Schematic of coefficient of water absorption test.

2.2.2. Effective Porosity Test. Percentage of water absorption is a measure of the pore volume or porosity in hardened concrete, which is occupied by water in saturated condition. Water absorption test was carried out as per ASTM C642-97 [29]. The effective porosity for OPC, PPC, and PSC concrete types is calculated as follows:

$$\text{Effective porosity (\%)} = \frac{(B-A)}{V} \times 100, \qquad (1)$$

where A is mass of oven dried sample in air, B is saturated mass of the surface dry sample in air after immersion, and V is bulk volume of the sample.

2.2.3. Coefficient of Water Absorption. Coefficient of water absorption is suggested as a measure of permeability of water. This is measured by the rate of uptake of water by dry concrete over a period of 1 h. The concrete cube specimens of 100 × 100 × 100 mm were preconditioned by drying in an oven at 105°C for 48 hrs until constant weight was reached and then allowed to cool in an airtight sealed container for 3 days. The sides of the concrete samples were coated with transparent epoxy resin to allow the flow of water in unidirection. Then the samples were kept in a vertical position partially immersed up to a depth of 5 mm at one end while the rest of the portions were kept exposed to the atmosphere as shown in Figure 1. The quantity of water absorbed during the first 60 min was calculated. Coefficients of water absorption values for OPC, PPC, and PSC concrete specimens after 28 and 90 days of moisture curing were determined using the formula [29]

$$K_a = \left(\frac{Q}{A}\right)^2 \times \frac{1}{t}, \qquad (2)$$

where K_a is the coefficient of water absorption ($m^2 s^{-1}$), Q is the quantity of water absorbed (m^3) by the oven dry specimen

Table 1: Details of concrete mix proportions.

Grade	Type of cement	W/C ratio	Cement ($kg\,m^{-3}$)	Fine aggregate ($kg\,m^{-3}$)	Coarse aggregate ($kg\,m^{-3}$)	Water ($kg\,m^{-3}$)
M30	OPC	0.55	338	607	1247	186
	PPC	0.55	338	607	1247	186
	PSC	0.55	338	607	1247	186
M40	OPC	0.45	400	472	1247	180
	PPC	0.45	400	472	1247	180
	PSC	0.45	400	472	1247	180

Table 2: Chemical composition of cements.

Constituent	Weight/(%)		
	OPC	PPC	PSC
SiO_2	21.0	32.0	31.0
Al_2O_3	5.50	10.0	10.5
Fe_2O_3	4.60	5.50	2.80
CaO	63.0	44.0	47.0
MgO	0.60	1.50	3.90
SO_3	2.60	2.60	2.60
LOI	2.70	4.40	2.70

in time (t), t is 3600 s, and A is the surface area (m^2) of concrete specimen through which water penetrates.

2.2.4. Impressed Voltage Test. Cylindrical concrete specimens of size 50 mm diameter and 100 mm height were cast using M30 and M40 grades of concrete types. Rebar types of 12 mm diameter and 100 mm height were embedded centrally into the concrete specimens. During casting, the moulds were mechanically vibrated. After 24 h, the cylindrical concrete specimens were demoulded and cured for 28 and 90 days. Then the specimens were subjected to impressed voltage test. In this technique, the concrete specimens were immersed in 5% NaCl solution and embedded steel in concrete is made as anode with respect to an external stainless steel electrode serving as the cathode by applying a constant positive potential of 12 V to the system from a DC source. The variation of current is recorded with time. For each specimen, the time taken for initial crack and the corresponding maximum anodic current flow was recorded [30].

At the end of the 12 V impressed voltage test, the embedded steel specimens were broken and the sections were open, cleaned in pickling solution (concentrated HCl containing 20 g/L of stannous chloride and 15 g/L of antimony trioxide), then cleaned in distilled water, dried, and weighed as per ASTM G1-03 [31]. From the initial weight, the loss in weight of embedded steel due to accelerated corrosion in OPC, PPC, and PSC concrete types was calculated. From the weight loss data, corrosion rate in mmpy was calculated using the following equation [2]:

$$\text{Corrosion rate} = 87.6 \times \left(\frac{W}{D \times A \times T}\right), \quad (3)$$

where W is the weight loss in milligram, D is the density of the material used, g/cc, A is the area of the specimen (cm^2), and T is the duration of the test period in hours.

2.2.5. Rapid Chloride Ion Permeability Test (RCPT). The resistance to chloride ion penetration in terms of total charge that passed in coulombs through OPC and blended concrete specimens after 28 and 90 days of moisture curing was measured as per ASTM C1202-12 [32]. A concrete disc of size 100 mm diameter and 50 mm thickness was cast and allowed to cure for 28 and 90 days. Then the concrete specimens were subjected to the RCPT by impressing 60 V between two TSIA electrodes. Two halves of the specimens were sealed with PVC container of 95 mm diameter. One side of the container is filled with 3% NaCl solution (that side of the cell will be connected to the negative terminal of the power supply); the other side is filled with 0.3 N NaOH solution (which will be connected to the positive terminal of the power supply). Current flow is measured at every 30 min up to 6 h. From the results, using the current and time, chloride permeability is calculated in terms of coulombs at the end of 6 h.

2.2.6. Chloride Diffusion Coefficient. The amount of chloride ion migrating through the concrete specimens after 28 and 90 days of curing was monitored by removing small amount of solution and determining the chloride concentration of these samples after 120 h. Chloride diffusion coefficients were calculated using Nernst–Einstein equation [33].

$$D = \frac{JRTL}{ZFC_0E}, \quad (4)$$

where D is the chloride diffusion coefficient (cm^2/s), J is the flux of chloride ions ($mol/cm^2 s$), R is the gas constant (8.314 J/K mol), T is the absolute temperature (300 K), L is the thickness of the specimen (cm), Z is the valiancy of chloride ion ($Z = 1$), F is the Faradays constant (9.648×10^4 J/V mol), C_0 is the initial chloride ion concentration (mol/l), and E is the applied potential (60 V).

2.2.7. Potentiodynamic Polarization Method. Concrete cubes of size $100 \times 100 \times 100$ mm that were cast with TMT rebar of 12 mm diameter and 75 mm length (two rebar types) were embedded at a cover depth of 25 mm as shown in Figure 2. The cubes were cast for OPC, PPC, and PSC with

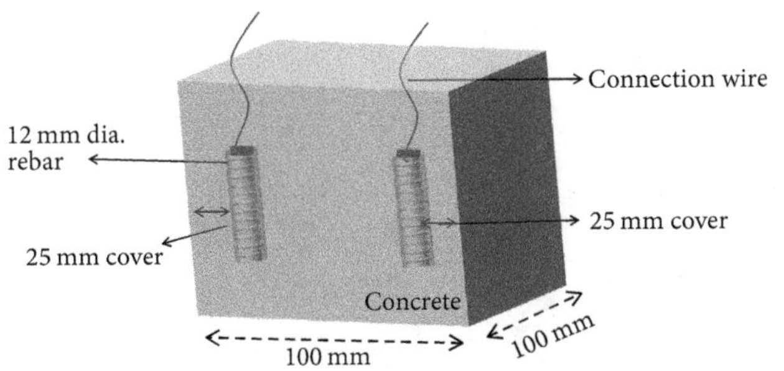

FIGURE 2: Schematic representation of concrete cube specimen used for potentiodynamic polarization and AC-impedance study.

M30 and M40 grade concrete and allowed to cure for 90 days. After 90-day curing the specimens were exposed to the open atmosphere under natural weathering conditions. After two years of exposure, the embedded rebar types in the concrete cube specimens were subjected to potentiodynamic polarization technique using the three-electrode assembly. The rebar embedded in concrete cube acts as a working electrode, stainless steel electrode acts as a counter electrode, and Saturated Calomel Electrode (SCE) acts as a reference electrode. From the polarization technique, the corrosion kinetic parameters such as corrosion current (I_{corr}), corrosion potential (E_{corr}), cathodic Tafel slope (b_c), and anodic Tafel slope (b_a) were obtained. Both cathodic and anodic polarization curves were recorded potentiodynamically using ACM Instruments, UK. The potentiodynamic conditions correspond to a potential sweep rate of $0.1\,mVmin^{-1}$ and potential ranges of +0.2 V to −0.2 V versus SCE from the OCP. All the experiments were carried out at the temperature of 32 ± 1°C.

2.2.8. AC-Impedance Measurement. The same cube specimens mentioned above were subjected to AC-impedance measurements. The real part (Z') and the imaginary part ($-Z''$) of the cell impedance were measured for various frequencies (30 kHz to 0.01 Hz). Plots of Z' versus $-Z''$ were made. The same experimental setup was used here also.

2.2.9. Macrocell Corrosion Studies. A rectangular concrete (M40 grade) specimen of size 279 mm × 152 mm × 114 mm was designed as per ASTM G 109-07 [34] for macrocell corrosion studies. TMT rebar of 12 mm diameter and 300 mm length was used as both anode and cathode in the same concrete. The top mat of rebar acts as the anode and the bottom mat of rebar acts as cathode. The anode to cathode area ratio was maintained as 1 : 2 to induce accelerated corrosion of the embedded anode. The configuration of macrocell specimen is given in Figure 3. In both the anode and cathode 250 mm length was embedded inside the concrete and the remaining length was used for taking electrical connections with proper insulations.

The specimens were mechanically vibrated. After 24 hours of setting, the specimens were demoulded and cured in distilled water for 90 days. Then the specimens were subjected to alternate wetting and drying cycles. One cycle consists of 15 days of wetting in 3% NaCl and 15 days of drying in open atmosphere. Measurements were carried out during the wetting cycle (15th day) as macrocell current showed maximum magnitude due to the low resistivity of the concrete. All the concrete specimens were subjected to 24 complete cycles of exposure period.

3. Results and Discussion

3.1. Compressive Strength. The compressive strength of OPC, PPC, and PSC concrete types after 28 and 90 days of curing for M30 and M40 grades of concrete are given in Figures 4 and 5, respectively.

It was found from Figure 4 that, in M30 grade concrete, the 28-day strength obtained was 37 MPa, 35.5 MPa, and 32.9 MPa for OPC, PPC, and PSC concrete types, respectively. There was a slight increase in the 90-day compressive strength values such as 40 MPa for OPC, 39 MPa for PPC, and 36.9 MPa for PSC, respectively.

On the other hand, in M40 grade concrete (Figure 5) the 28-day strength obtained was 43.7 MPa, 41.2 MPa, and 43.2 MPa for OPC, PPC, and PSC concrete types, respectively. There was a slight increase in the 90-day compressive strength values such as 47.7 MPa for OPC, 44.3 MPa for PPC, and 46.8 MPa for PSC, respectively.

The magnitude of increase in compressive strength from 28 days to 90 days was found to be similar in M30 grade and M40 grade concrete types. Comparable compressive strength values were obtained for PPC and PSC concrete types when compared with OPC concrete.

3.2. Effective Porosity and Coefficient of Water Absorption. Effective porosity and coefficient of water absorption values calculated for OPC, PPC, and PSC concrete types at 28- and 90-day cured specimens were reported in Table 3. It was observed that both 28- and 90-day cured M30 and M40 grade concrete types showed the effective porosity and coefficient of water absorption in the decreasing order as follows: OPC > PPC > PSC.

Among the three types of concrete used, PSC concrete showed least pores and lower coefficient of water absorption. The same trend is observed in both 28- and 90-day cured M30 and M40 grade concrete types. For example, in the case of PSC, $0.71 \times 10^{-7}\,m^2s^{-1}$ and $0.62 \times 10^{-7}\,m^2s^{-1}$ are the coefficient of water absorption values obtained for 28-day cured M30 and M40 grade concrete, respectively. Similarly, for M30 and M40 grade concrete types the coefficient of water absorption values obtained were $0.40 \times 10^{-7}\,m^2s^{-1}$ and $0.15 \times 10^{-7}\,m^2s^{-1}$ for 90-day cured concrete, respectively. The observed values of coefficient of water absorption was attributed to the impermeability nature of the pore structure refinement in PPC and PSC concrete types. The improvement in the permeability characteristics of PPC and PSC is due to the secondary hydration reaction.

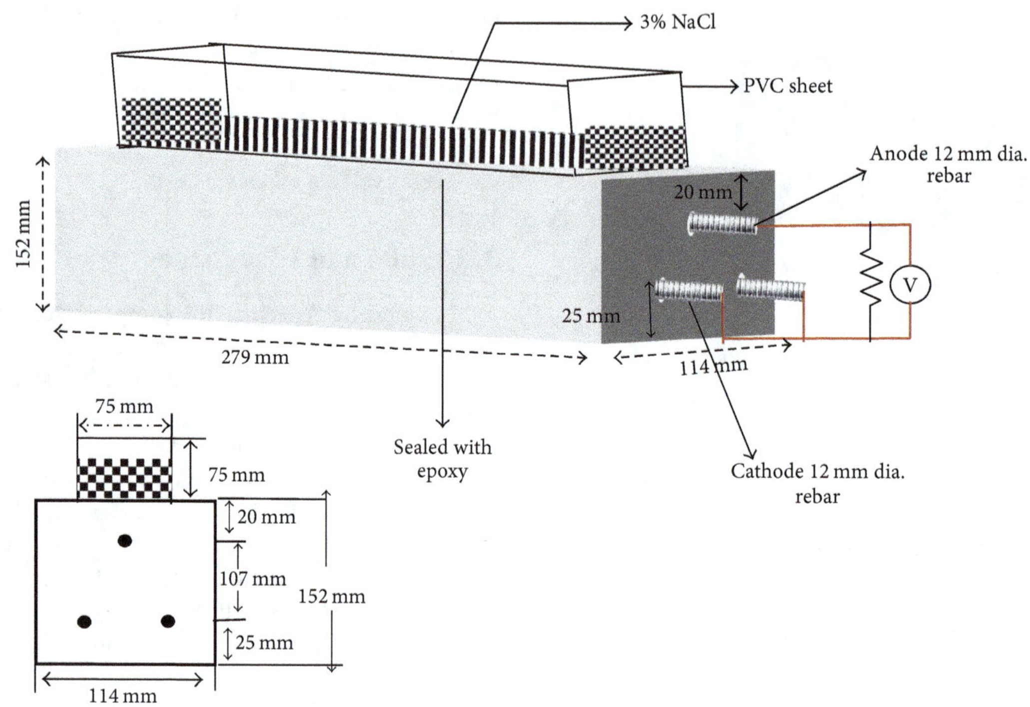

Figure 3: Schematic view of macrocell specimen.

Table 3: Permeability parameters for OPC, PPC, and PSC M30 and M40 grade concrete types.

Grade	Type of cement	Effective porosity/(%)		Coefficient of water absorption/(m^2s^{-1}) × 10^{-7}	
		28 days	90 days	28 days	90 days
M30	OPC	15.15	10.72	3.20	2.30
	PPC	14.28	9.62	2.50	2.00
	PSC	12.61	9.57	0.71	0.40
M40	OPC	13.28	9.43	3.00	1.90
	PPC	12.42	6.79	2.90	1.10
	PSC	10.61	6.32	0.62	0.15

In general, the primary hydration reaction in concrete is as follows:

$$2\left(3\mathrm{CaO}\cdot\mathrm{SiO_2}\right)+6\mathrm{H_2O} \longrightarrow 3\mathrm{CaO}\cdot2\mathrm{SiO_2}\cdot3\mathrm{H_2O}+3\mathrm{Ca}\left(\mathrm{OH}\right)_2 \quad (5)$$

But, in PPC and PSC concrete types, the additional secondary hydration reaction is

$$3\mathrm{Ca}\left(\mathrm{OH}\right)_2+2\mathrm{SiO_2}+\mathrm{H_2O}\longrightarrow 3\mathrm{CaO}\cdot2\mathrm{SiO_2}\cdot3\mathrm{H_2O} \quad (6)$$

In PPC and PSC concrete types, the $Ca(OH)_2$ content reduction is due to the secondary hydration reaction. During the hydration in PPC, lime is consumed but in OPC lime is produced. This is the main advantage of using blended cements to decrease the permeability of the concrete thereby increasing the corrosion resistance properties [35].

3.3. Impressed Voltage Test. The impressed voltage parameters of M30 and M40 grades of OPC, PPC, and PSC concrete types under 28 and 90 days of curing time were given in Table 4. In M30 grade concrete, the maximum anodic current flow measured was found to be 77, 65, and 43 mA for OPC, PPC, and PSC concrete types, respectively, at 28-day cured specimens. The corresponding reduction in current flow was observed as 15.6% for PPC and 44.2% for PSC when compared to OPC. These results clearly indicated that blended cements, namely, PPC and PSC, considerably decreased the anodic current flow. Similar trend was observed in M40 grade concrete. Among the three concrete types, the maximum reduction in current flow was observed for PSC concrete. The reason is that PSC cement consists of higher amount of Al_2O_3 (10.5%) when compared to OPC (5.5%) and PPC (10%), which interacted with $Ca(OH)_2$ and favoured more C_3A and C-S-H gel. The formation of such compounds reduced the micro- and macropores present in the concrete and this is due to the refinement of the microstructure, considerably reducing the current flow [36]. After 90 days of curing, OPC, PPC, and PSC concrete types showed only a slight decrease

Table 4: Impressed voltage parameters for rebar in OPC, PPC, and PSC M30 and M40 grade concrete types.

Grade and type of cement	Maximum anodic current (mA)	(%) reduction	Time to cracking (days)	Free chloride contents (ppm)	Weight loss (mg)	Reduction in corrosion rate (%)
			28-day cured concrete specimens			
OPC M30	77	—	07	4680	44.3210	—
PPC M30	65	15.6	08	3560	25.3120	42.9
PSC M30	43	44.2	10	3253	7.9140	82.1
OPC M40	72	—	10	3789	35.3020	—
PPC M40	56	22.2	12	3420	14.1410	59.7
PSC M40	39	45.8	14	3200	2.4650	93.0
			90-day cured concrete specimens			
OPC M30	74	—	08	4200	33.5780	—
PPC M30	58	21.62	09	3165	12.2290	63.6
PSC M30	43	41.89	11	2880	1.7030	94.9
OPC M40	60	—	14	3040	23.4530	—
PPC M40	43	28.30	18	2840	11.9420	49.0
PSC M40	26	56.66	20	2640	1.1470	95.1

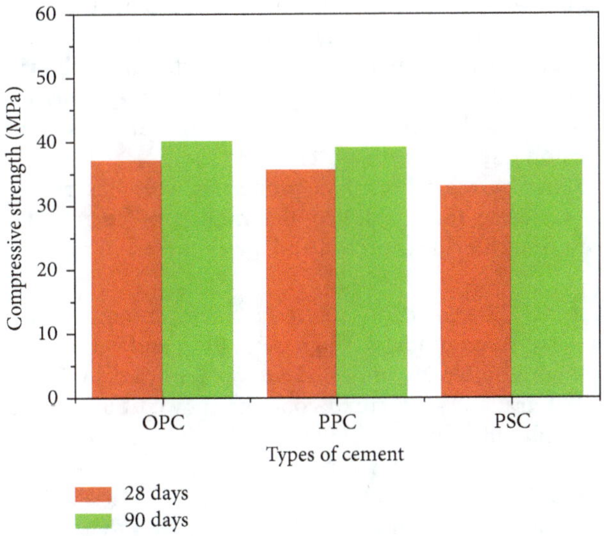

Figure 4: Comparison of compressive strength of M30 grade OPC, PPC, and PSC concrete types.

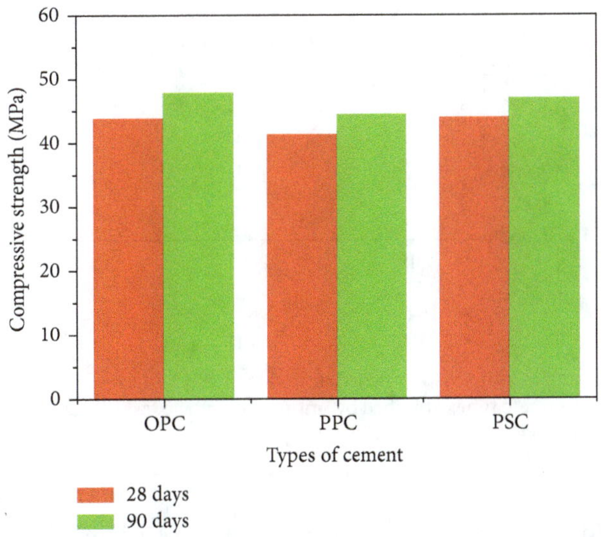

Figure 5: Comparison of compressive strength of M40 grade OPC, PPC, and PSC concrete types.

in the magnitude of current flow in both M30 and M40 grade concrete types. The percentage reduction in the current flow was 41.89% and 56.66% for PSC system at 28 and 90 days of curing period. These data confirmed the impermeability nature of blended cement concrete types.

3.3.1. Time Taken for Initial Crack. The time taken for initial crack measured for 28- and 90-day cured OPC, PPC, and PSC concrete types was given in Table 4. The results indicated that both 28- and 90-day cured M30 and M40 grade concrete types showed the following order with respect to time taken for initial crack.

$$\text{OPC} < \text{PPC} < \text{PSC}. \quad (7)$$

Among the three concrete types, PSC concrete in M30 and M40 grades, 28- as well as 90-day cured specimens showed the longest time taken for initial crack. For example, for 28-day cured M30 and M40 concrete, time taken for initial crack observed was at 10 and 14 days, respectively. Similar trend was observed for 90-day cured concrete.

The better performance of PSC is due to the following reason. Aluminate phase is the most responsible one for binding the chlorides into complex forms. Fixing of chloride in concrete by the formation of complex aluminate phases has been reported by Karthick et al. [37]. About 50% to 80% of chloride ions are bound by aluminate phases by the action of calcium

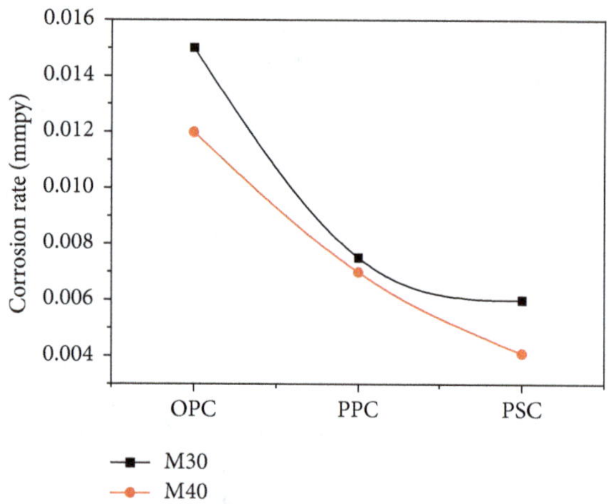

FIGURE 6: Comparison of corrosion rate of rebar in OPC, PPC, and PSC concrete types obtained by impressed voltage test (28-day cured specimens).

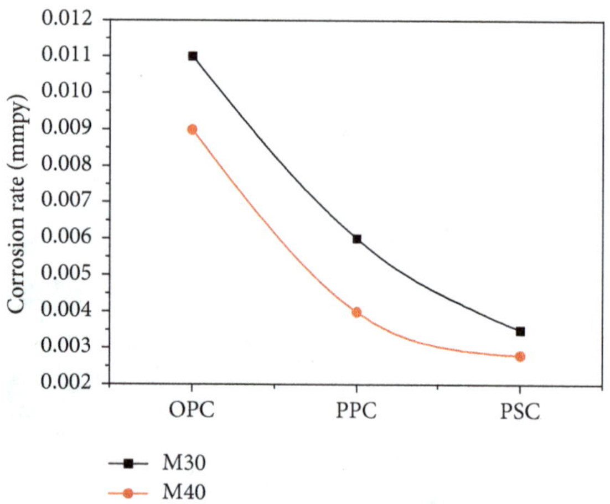

FIGURE 7: Comparison of corrosion rate of rebar in OPC, PPC, and PSC concrete types obtained by impressed voltage test (90-day cured specimens).

oxide and aluminate with chloride, forming calcium chloroaluminate complexes ($CaO{\cdot}Al_2O_3{\cdot}NaCl_2{\cdot}10H_2O$). With the result, the time taken for initial crack was increased.

The free chloride contents showed the lower values for PSC concrete. This can be attributed to the formation of free chlorides into complex form. The mechanism of the formation of complex, namely, Friedel salt, is by the precipitation method.

3.3.2. Corrosion Rate of Embedded Steel. The corrosion rates of embedded steel in OPC, PPC, and PSC were given in Figures 6 and 7, respectively.

The corrosion rates of steel calculated for 28 days cured M30 OPC, PPC, and PSC concrete types were found to be 0.0150, 0.0075, and 0.0060 mmpy, respectively. These results showed that blended cements, namely, PPC and PSC, considerably decreased the corrosion rate of embedded steel by 2 times and 2.5 times. Similar trend was also observed in M40 grade concrete. PPC and PSC concrete types showed 1.7 times and 3 times reduction in corrosion rate when compared to OPC concrete.

The corrosion rate values calculated for 90 days cured M30 OPC, PPC, and PSC concrete types were found to be 0.0110, 0.0060, and 0.0035 mmpy, respectively. These results showed that blended cements, namely, PPC and PSC, considerably decreased the corrosion rate of embedded steel by 2 and 3 times, respectively. A similar trend was also observed in M40 grade concrete. PPC and PSC concrete types showed 2.2 and 3.3 times reduction in corrosion rate when compared to OPC concrete.

For reinforcement corrosion to occur, a critical quantity of chloride is required at the steel/concrete interface. In OPC, the critical chloride is reached in a very short time, whereas it took more time for PPC and PSC concrete types as evidenced by the time taken for initial crack. The considerable reduction in corrosion rate of the blended cements was attributed to the secondary pozzolanic reaction in blended cements which favoured more C-S-H gel formation and interconnected the micropores present in the blended cements. Moreover, the PSC showed the least corrosion rate values because, in addition to the secondary pozzolanic reaction, the higher Al_2O_3 content adsorbed more free chloride and formed chloroaluminate complex [37]. Thus, the free chlorides mainly responsible for inducing the reinforcement corrosion were considerably reduced and hence a marked reduction in corrosion rate values was observed.

Better corrosion resistant properties were observed for PSC concrete that exhibited the minimum current flow, lowest free chloride contents, and lower corrosion rate.

3.4. Rapid Chloride Permeability Test. The amount of charge that passed (coulombs) through OPC, PPC, and PSC concrete types cured after 28 and 90 days was reported in Table 5. In M30 grade 28-day cured concrete showed 669, 281, and 226 coulombs for OPC, PPC, and PSC concrete types, respectively. The results obtained for M40 grade concrete showed 626, 261, and 204 coulombs for OPC, PPC, and PSC concrete types, respectively. This showed that, in M40 grade concrete, due to the higher cement content, the higher reactivity of $Ca(OH)_2$ with SiO_2 leads to the formation of C-S-H gel which reduced the micro- and macropores present in the concrete. Besides, PSC and PPC cement consists of higher amount of Al_2O_3 (10.0 to 10.5%) when compared to OPC (5.5%), which interacted with $Ca(OH)_2$ and favoured formation of calcium chloroaluminate complexes (chloride binding effect) which made the concrete more impermeable [38]. For this reason, the amount of charge that passed reduced considerably in M40 grade concrete when compared to M30 grade concrete. The results indicated that the amount of charge that passed considerably reduced in the case of both M30 and M40 grade PPC (2.38 and 2.39 times) and PSC (2.96 and 3.06 times) concrete types when compared to OPC concrete. The refinement of the pore structure occurring in blended cements considerably reduced the permeation of aggressive chloride ions. The same trend was observed by Saraswathy and Song, by replacing 15 with 30% RHA [30].

TABLE 5: Rapid chloride permeability test (RCPT) parameters for OPC, PPC, and PSC M30 and M40 grade concrete types.

Grade	28 days			90 days		
	Charge passed (coulomb)	Free chloride contents (ppm)	Chloride diffusion coefficient $(cm^2s^{-1}) \times 10^{-10}$	Charge passed (coulombs)	Free chloride contents (ppm)	Chloride diffusion coefficient $(cm^2s^{-1}) \times 10^{-10}$
OPC M30	669	749	2.4	630	579	1.9
PPC M30	281	681	1.8	220	528	1.5
PSC M30	226	481	1.0	200	380	1.4
OPC M40	626	540	1.8	600	520	1.4
PPC M40	261	532	1.5	180	483	0.8
PSC M40	204	596	0.5	200	393	0.6

3.4.1. Chloride Diffusion Coefficient. The chloride diffusion coefficient (CDC) calculated for 28- and 90-day cured OPC, PPC, and PSC concrete types was reported in Table 5. The results indicated that both 28- and 90-day cured M30 and M40 grade concrete types showed the decreasing trend as follows:

$$\text{OPC} > \text{PPC} > \text{PSC}. \quad (8)$$

In PSC M30 and M40 grade 28- as well as 90-day cured concrete showed the least chloride diffusion coefficient (CDC) values. For example, in M30 grade 28-day cured concrete the CDC values are found to be $1.8 \times 10^{-10}\ cm^2s^{-1}$ and $1.0 \times 10^{-10}\ cm^2s^{-1}$ for PPC and PSC concrete types, respectively. In M40 grade, 90-day cured concrete CDC values are $0.8 \times 10^{-10}\ cm^2s^{-1}$ and $0.6 \times 10^{-10}\ cm^2s^{-1}$ for PPC and PSC concrete types, respectively. For high performance concrete, the CDC values were found to be of the order of $10^{-12}\ cm^2\ s^{-1}$. The considerable lower CDC values for blended cements were attributed to their impermeability nature that is due to the refinement of their pore structure [39].

3.5. Potentiodynamic Polarization Technique. Potentiodynamic polarization curves obtained for the corrosion of rebar in various types of cement concrete were shown in Figure 8. The corrosion kinetic parameters for rebar embedded in OPC, PPC, and PSC for M30 and M40 grade and 90-day cured concrete specimens were given in Table 6.

The corrosion potential of rebar in M30 concrete types from OPC, PPC, and PSC are −397, −303, and −299 mV, respectively. PSC concrete improves the corrosion resistance of rebar when compared to OPC concrete. There is a difference of −97 mV which moves in the positive direction for PSC cement concrete. The corrosion rates of rebar types in OPC, PPC, and PSC are 6.9922×10^{-3}, 1.6902×10^{-3}, and 1.4448×10^{-3} mmpy, respectively. The corrosion rate of PPC and PSC was reduced 4.088 and 4.78 times when compared to OPC. The corrosion current density for rebar in three types of cement concrete at the end of exposure is as follows:

$$\underset{\text{PSC}}{1.4449 \times 10^{-4}\ \text{mA} \cdot \text{cm}^{-2}} < \underset{\text{PPC}}{1.6902 \times 10^{-4}\ \text{mA} \cdot \text{cm}^{-2}} < \underset{\text{OPC.}}{6.9922 \times 10^{-4}\ \text{mA} \cdot \text{cm}^{-2}} \quad (9)$$

In M40 grade concrete, the measured corrosion potential was −301, −263, and −258 mV versus SCE for OPC, PPC, and PSC, respectively. The variation in potential for M40 PPC and PSC concrete types when compared to OPC was found to be less than M30 grade concrete. The corrosion current density for M40 concrete follows the same order as M30 concrete:

$$\underset{\text{PSC}}{2.491 \times 10^{-4}\ \text{mA} \cdot \text{cm}^{-2}} < \underset{\text{PPC}}{1.9589 \times 10^{-4}\ \text{mA} \cdot \text{cm}^{-2}} < \underset{\text{OPC.}}{1.674 \times 10^{-4}\ \text{mA} \cdot \text{cm}^{-2}} \quad (10)$$

The PSC and PPC concrete types exhibit lower corrosion current density and lowest corrosion rate when compared to their OPC concrete counterpart.

3.6. AC-Impedance Measurements. Impedance diagrams obtained for the frequency range 0.01 Hz to 30 kHz at the OCP for the corrosion of rebar in different concrete types

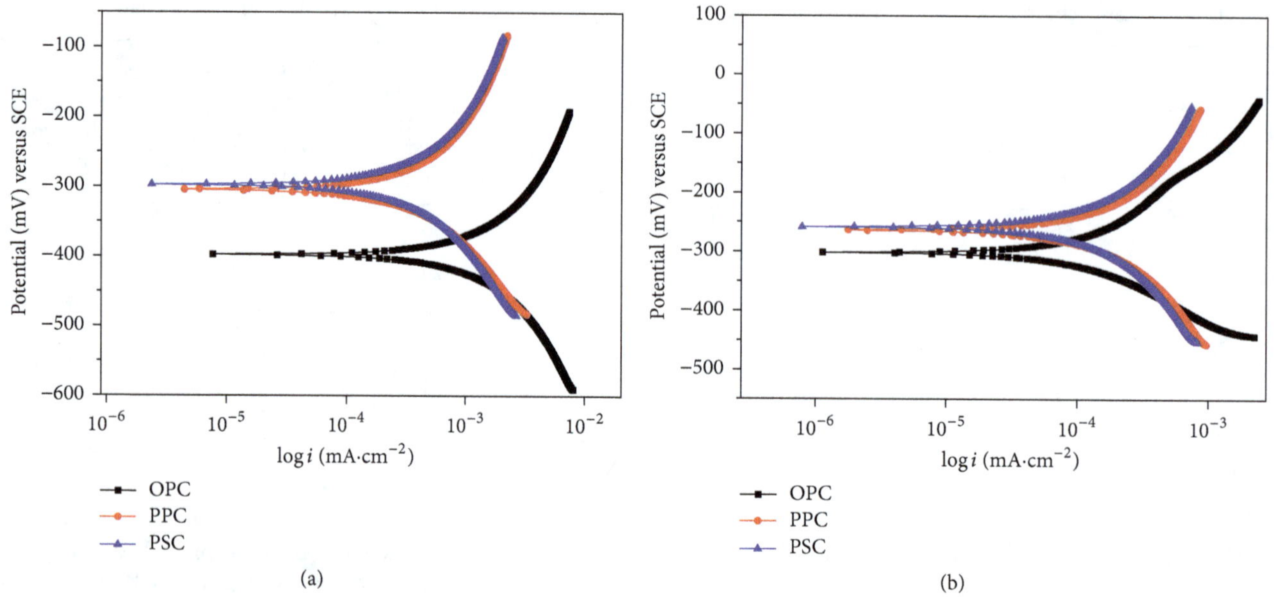

FIGURE 8: Potentiodynamic polarization curves for TMT rebar embedded in OPC, PPC, and PSC M30 (a) and M40 (b) grade concrete types (90-day cured specimens).

TABLE 6: Potentiodynamic polarization parameters for the corrosion of rebar embedded in OPC, PPC, and PSC M40 grade concrete (90-day cured specimens).

System	Various grades of concrete					
	M30			M40		
	E_{corr} (mV versus SCE)	I_{corr} $(mA{\cdot}cm^{-2}) \times 10^{-4}$	Corrosion rate (mmpy) $\times 10^{-3}$	E_{corr} (mV versus SCE)	I_{corr} $(mA{\cdot}cm^{-2}) \times 10^{-5}$	Corrosion rate (mmpy) $\times 10^{-4}$
OPC	−397	6.9922	8.0082	−301	2.1496	2.491
PPC	−303	1.6902	1.9589	−263	1.6902	1.9589
PSC	−299	1.4448	1.674	−258	1.4448	1.674

were shown in Figure 9. The corrosion parameters for rebar types in OPC, PPC, and PSC of M30 and M40 grade 90-day cured concrete types are given in Table 7. It is seen from the figure that the impedance diagram is not a perfect semicircle. This difference has been attributed to frequency dispersion.

From the table, it is found that R_{ct} (charge transfer resistance) values were found to be higher for both PPC and PSC when compared to OPC in both M30 and M40 grade concrete types, which indicates that PPC and PSC have better corrosion resistance than OPC. The corrosion rate values in M30 grade PPC and PSC concrete types were observed to be 1.12 and 1.27 times less when compared with OPC. In M40 grade concrete R_{ct} values observed for OPC, PPC, and PSC were 1.399×10^6, 1.594×10^6, and 1.714×10^6 ohm·cm^2, respectively. This indicated that M40 grade concrete has two orders of magnitude higher corrosion resistance than M30 grade concrete.

Nyquist plots clearly indicated that rebar types embedded in PSC and PPC concrete performed better than OPC. The corrosion rates of rebar in various concrete types at the end of 24 months' exposure in M30 and M40 grade concrete types are as follows:

$$\begin{gathered}\underset{\text{M30 PSC}}{8.728 \times 10^{-3}\,\text{mmpy}} < \underset{\text{PPC}}{7.782 \times 10^{-3}\,\text{mmpy}} < \underset{\text{OPC}}{6.873 \times 10^{-3}\,\text{mmpy}} \\ \underset{\text{M40 PSC}}{2.161 \times 10^{-4}\,\text{mmpy}} < \underset{\text{PPC}}{1.897 \times 10^{-4}\,\text{mmpy}} < \underset{\text{OPC.}}{1.764 \times 10^{-4}\,\text{mmpy}}\end{gathered} \tag{11}$$

TABLE 7: AC-impedance parameters for the corrosion of rebar embedded in OPC, PPC, and PSC M40 grade concrete (90-day cured specimens).

System	Various grades of concrete					
	M30			M40		
	R_{ct} (ohms·cm^2) × 10^4	I_{corr} (mA/cm^2) × 10^{-4}	Corrosion rate (mmpy) × 10^{-3}	R_{ct} (ohms·cm^2) × 10^6	I_{corr} (mA/cm^2) × 10^{-5}	Corrosion rate (mmpy) × 10^{-4}
OPC	3.464	7.531	8.728	1.399	1.865	2.161
PPC	3.885	6.715	7.782	1.594	1.637	1.897
PSC	4.399	5.930	6.873	1.714	1.522	1.764

From the results it is observed that PSC concrete showed the higher charge transfer resistance and lowest corrosion rate when compared to PPC and OPC concrete types.

3.7. Macrocell Corrosion Studies. The half-cell potential of steel measured periodically against SCE with time for OPC, PPC, and PSC for M30 and M40 grade 90-day cured concrete types was shown in Figures 10 and 11. It was observed from Figure 10 that steel embedded in M30 grade OPC, PPC, and PSC concrete types is showing active condition within 3, 4, and 5 cycles of exposure, respectively. After 5 cycles, all the three types of concrete types were showing the potential more negative than −275 mV versus SCE indicating the active condition of the rebar due to the ingress of chloride ions, whereas in M40 grade concrete the OPC, PPC, and PSC reached the active condition at 7, 9, and 11 cycles of exposure, respectively. In M30 grade concrete, the potential gets stabilized after 12 cycles of exposure for all the types of concrete types and it reached a maximum active potential of −610 mV, −530 mV, and −501 mV for OPC, PPC, and PSC, respectively, at the end of 24 cycles. In M40 grade concrete, after 11 cycles of exposure, both PPC and PSC have shown gradual increase in potential to the active state, reaching a potential of −420 mV and −400 mV at the end of 24th cycle. But OPC has shown a more negative potential of −520 mV at the end of 24th cycle. These results indicated that the PSC concrete performed better than PPC and OPC concrete types.

The macrocell current measured periodically with time for M30 and M40 grade OPC, PPC, and PSC concrete types is shown in Figures 12 and 13. From Figure 12, it is observed that M30 OPC concrete that crossed the critical current of 10 μA showed the higher magnitude of macrocell current flow at the end of 24 cycles of exposure indicating the severe corrosion of the embedded steel. The macrocell current measured at the end of the 24th cycle was 72 μA, 60 μA, and 54 μA for OPC, PPC, and PSC concrete types, respectively.

From Figure 13 it is observed that in M40 grade concrete OPC reached the threshold macrocell current of 10 μA within 2 cycles of exposure and PPC and PSC reached the microcell current at 4th and 6th cycles of exposure. The macrocell current measured at the end of 24th cycle was 60 μA, 45 μA, and 34 μA for OPC, PPC, and PSC, respectively.

The total integrated current calculated (ASTM G109) with time for M30 grade OPC, PPC, and PSC concrete types is illustrated in Figure 14. From the figure, it was observed that at the end of the exposure (24th cycle) OPC concrete showed the higher value of 1200 coulombs, whereas PPC and PSC exhibited 800 and 500 coulombs, respectively. Figure 15 shows the total integrated current versus number of cycles for M40 grade OPC, PPC, and PSC concrete types. From the figure, it is observed that the coulomb values observed for OPC, PPC, and PSC were 900, 700, and 600, respectively. In both M30 and M40 grades of concrete PPC and PSC showed a coulomb value of less than 1000 indicating the blended cements have very low chloride penetration. But OPC showed 1200 coulombs in M30 grade and 900 coulombs in M40 grade concrete types. In M30 grade concrete OPC is graded as a low permeable concrete. But in both the grades of concrete PPC and PSC were graded as a very low permeable concrete as per ASTM C1202-12 [32].

In the case of OPC system, the availability of chloride ions that occupied the defect position and reacted with ferrous ions initiated the corrosion process. But in the case of PPC and PSC concrete types, the availability of free chloride is significantly reduced due to the formation of chloroaluminate complexes [37]. As a result, there is a considerable reduction in the magnitude of current flow which was observed in PPC and PSC concrete types.

4. Conclusions

The following conclusions were drawn from the above investigation:

(i) Mechanical properties are not altered by using blended cements, but corrosion resistant properties are greatly improved.

(ii) The effective porosity and coefficient of water absorption values showed the decreasing order as follows: OPC > PPC > PSC. Among the three types of concrete used, PSC concrete showed the least pores and lower coefficient of water absorption.

(iii) Better corrosion resistant properties were observed for PSC concrete types by showing a minimum current flow, lowest free chloride contents, and lesser corrosion rate, and thereby the time taken for initiation of crack period was extended in accelerated impressed voltage technique.

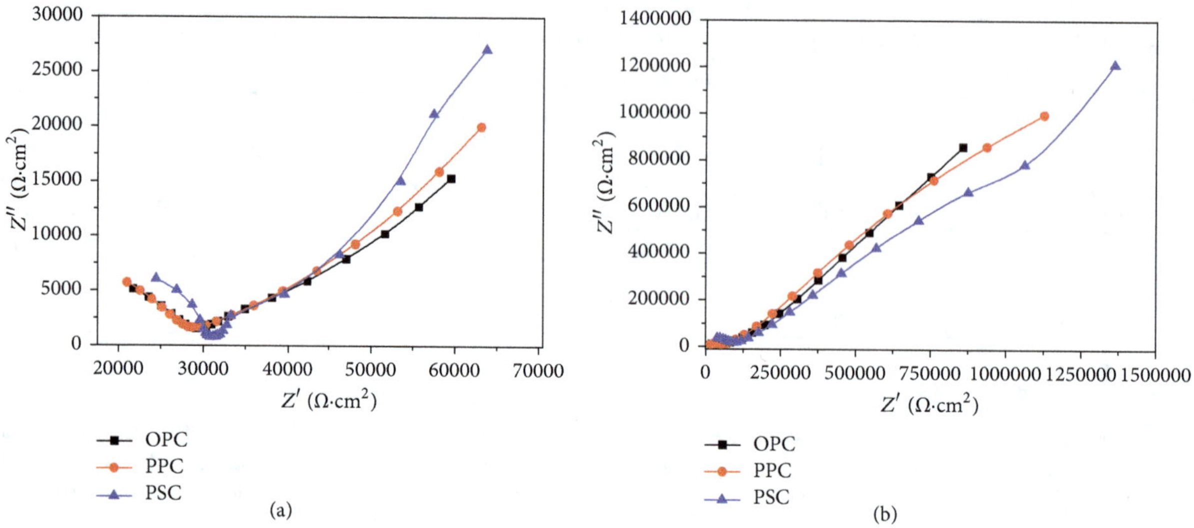

FIGURE 9: Nyquist plots for TMT rebar embedded in OPC, PPC, and PSC M30 (a) and M40 (b) grade concrete types (90-day cured specimens).

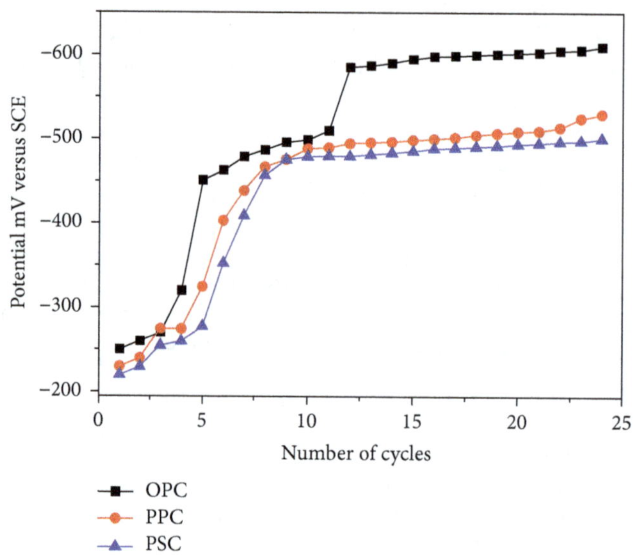

FIGURE 10: Potential versus number of cycles of exposure for rebar in OPC, PPC, and PSC M30 grade concrete types (90-day cured specimens) under macrocell condition.

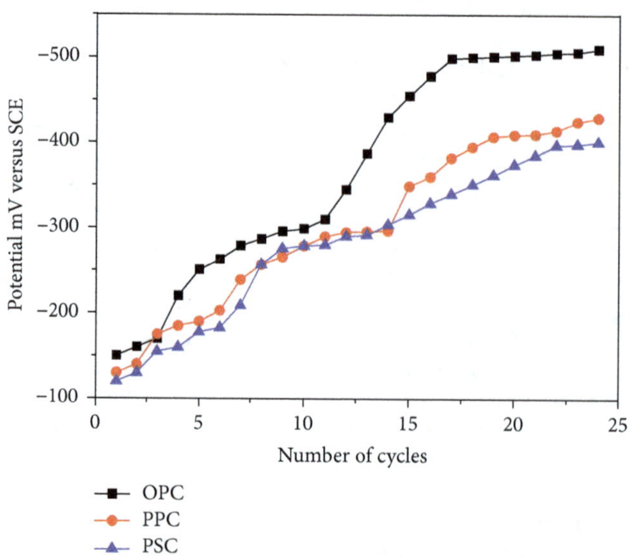

FIGURE 11: Potential versus number of cycles of exposure for rebar in OPC, PPC, and PSC M40 grade concrete types (90-day cured specimens) under macrocell condition.

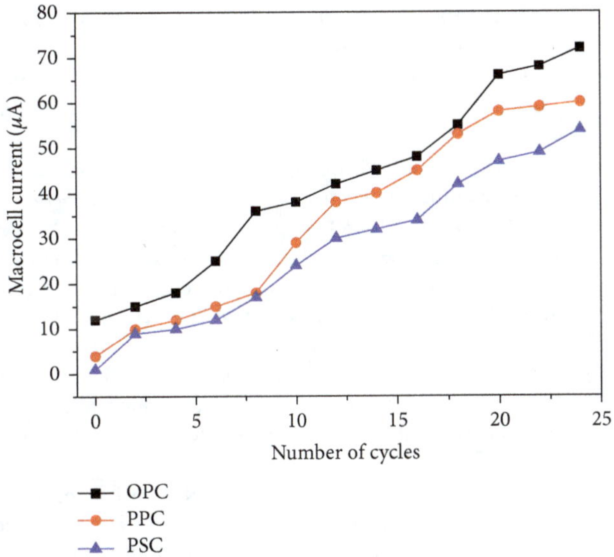

FIGURE 12: Macrocell current versus number of cycles of exposure for rebar in OPC, PPC, and PSC M30 grade concrete types (90-day cured specimens) under macrocell condition.

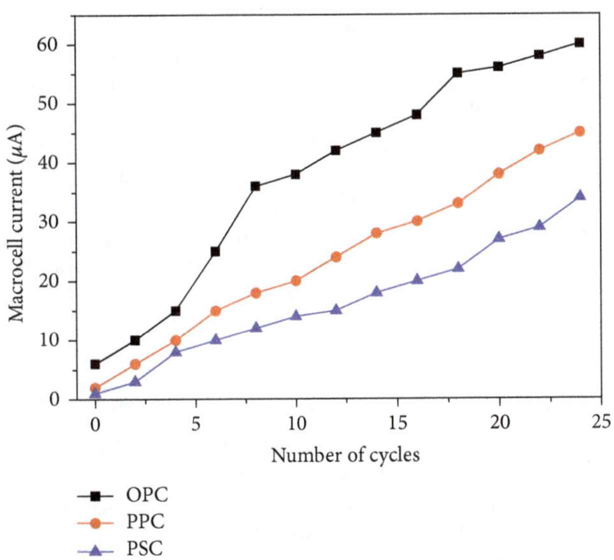

FIGURE 13: Macrocell current versus number of cycles of exposure for rebar in OPC, PPC and PSC M40 grade concrete types (90 days cured specimens) under macrocell condition.

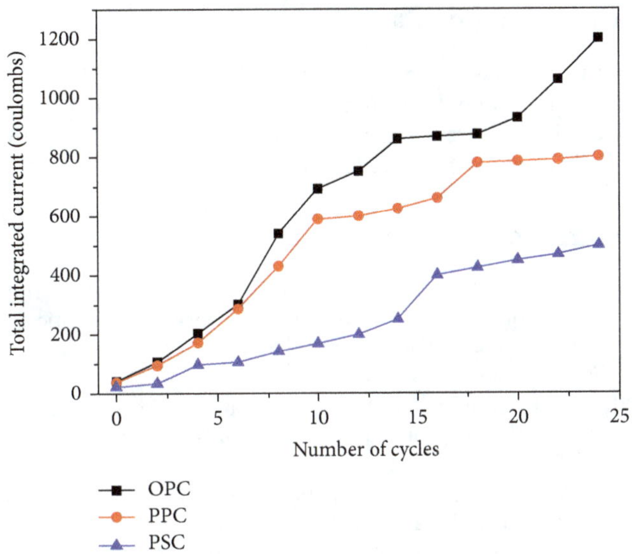

FIGURE 14: Total integrated current versus number of cycles of exposure for rebar in OPC, PPC, and PSC M30 grade concrete types (90-day cured specimens) under macrocell condition.

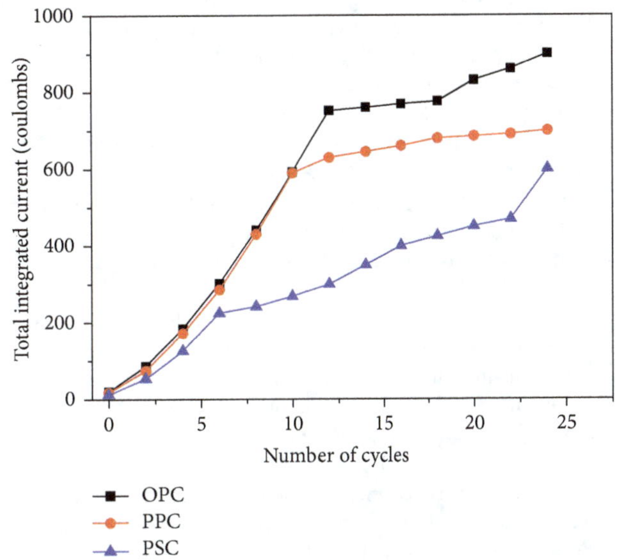

FIGURE 15: Total integrated current versus number of cycles of exposure for rebar in OPC, PPC, and PSC M40 grade concrete types (90-day cured specimens) under macrocell condition.

(iv) In the case of M30 grade concrete cured for 28 days, the corrosion rate was considerably decreased by 2 and 2.5 times for PPC and PSC, respectively.

(v) In the case of M40 grade concrete cured for 28 days, PPC and PSC concrete types showed 1.7 and 3 times reduction in corrosion rate when compared to the OPC concrete.

(vi) RCPT indicated that both 28- and 90-day cured M30 and M40 grade concrete types showed the decrease in CDC values as follows: OPC > PPC > PSC.

(vii) Macrocell corrosion studies revealed that PPC and PSC concrete types are graded as very low chloride permeability concrete in both M30 and M40 grade concrete types.

(viii) Electrochemical studies also proved beyond doubt that PSC concrete performed better than PPC and OPC concrete types.

Conflicts of Interest

The authors declare that there are no conflicts of interest regarding the publication of this paper.

Acknowledgments

This research was supported by the Basic Science Research Program through the National Research Foundation of Korea (NRF) funded by the Ministry of Science, ICT & Future Planning (no. 2015R1A5A1037548).

References

[1] E. Güneyisi, T. Özturan, and M. Gesoğlu, "Effect of initial curing on chloride ingress and corrosion resistance characteristics of concretes made with plain and blended cements," *Building and Environment*, vol. 42, no. 7, pp. 2676–2685, 2007.

[2] V. Saraswathy and H.-W. Song, "Improving the durability of concrete by using inhibitors," *Building and Environment*, vol. 42, no. 1, pp. 464–472, 2007.

[3] H.-W. Song and V. Saraswathy, "Corrosion monitoring of reinforced concrete structures: a review," *International Journal of Electrochemical Science*, vol. 2, no. 1, pp. 1–28, 2007.

[4] T. H. Wee, A. K. Suryavanshi, and S. S. Tin, "Evaluation of rapid chloride permeability test (RCPT) results for concrete containing mineral admixtures," *ACI Structural Journal*, vol. 97, no. 2, pp. 221–232, 2000.

[5] G. Chirag and J. Aakash, "Green concrete: efficient and eco-friendly construction materials," in *Proceedings of the nternational Journal of Research in Engineering and Technology*, vol. 2, pp. 259–264, 2014.

[6] V. M. Malhotra, *Role of supplementary cementing materials in reducing greenhouse gas emissions, Report MTL 98-03 (OP&) Natural resource Ottawa*, Ottawa, Canada, 1998.

[7] E. Guneyisi, T. Ozturan, and M. Gesoglu, "Laboratory investigation of chloride permeability for high performance concrete containing fly ash and silica fume," R. K. Dhir, Ed., pp. 295–305, Dundee, Scotland, 2002.

[8] V. Saraswathy, S. Muralidharan, R. M. Kalyanasundaram, K. Thangavel, and S. Srinivasan, "Evaluation of a composite corrosion-inhibiting admixture and its performance in concrete under macrocell corrosion conditions," *Cement and Concrete Research*, vol. 31, no. 5, pp. 789–794, 2001.

[9] V. Saraswathy, S. Muralidharan, K. Thangavel, and S. Srinivasan, "Activated fly ash cements: Tolerable limit of replacement for durable steel reinforced concrete," *Advances in Cement Research*, vol. 14, no. 1, pp. 9–16, 2002.

[10] V. Saraswathy, S. Muralidharan, K. Thangavel, and S. Srinivasan, "Influence of activated fly ash on corrosion-resistance and strength of concrete," *Cement and Concrete Composites*, vol. 25, no. 7, pp. 673–680, 2003.

[11] V. Saraswathy, S. Muralidharan, K. Thangavel, and S. Srinivasan, "Electrochemical studies on the corrosion performance of activated fly ash blended cements," *Materials Engineering*, vol. 14, pp. 261–283, 2003.

[12] S. Muralidharan, V. Saraswathy, S. P. M. Nima, and N. Palaniswamy, "Evaluation of a composite corrosion inhibiting admixtures and its performance in Portland pozzolana cement," *Materials Chemistry and Physics*, vol. 86, no. 2-3, pp. 298–306, 2004.

[13] G. A. Habeeb and H. B. Mahmud, "Study on properties of rice husk ash and its use as cement replacement material," *Materials Research*, vol. 13, no. 2, pp. 185–190, 2010.

[14] F. Sajedi and H. A. Razak, "Effects of curing regimes and cement fineness on the compressive strength of ordinary Portland cement mortars," *Construction and Building Materials*, vol. 25, no. 4, pp. 2036–2045, 2011.

[15] ACI 308-92 Standard Practice for curing concrete (ACI 308-92), *ACI Committee, American Concrete Institute*, Farmington Hills, Mich, USA, 1998.

[16] P. S. Mangat and M. C. Limbachiya, "Effect of initial curing on chloride diffusion in concrete repair materials," *Cement and Concrete Research*, vol. 29, no. 9, pp. 1475–1485, 1999.

[17] V. Bonavetti, H. Donza, V. Rahhal, and E. Irassar, "Influence of initial curing on the properties of concrete containing limestone blended cement," *Cement and Concrete Research*, vol. 30, no. 5, pp. 703–708, 2000.

[18] J. M. Khatib and P. S. Mangat, "Influence of high-temperature and low-humidity curing on chloride penetration in blended cement concrete," *Cement and Concrete Research*, vol. 32, no. 11, pp. 1743–1753, 2002.

[19] N. Shafiq and J. G. Cabrera, "Effects of initial curing condition on the fluid transport properties in OPC and fly ash blended cement concrete," *Cement and Concrete Composites*, vol. 26, no. 4, pp. 381–387, 2004.

[20] G. Erhan, O. Juran, and G. Mehmet, "A study of reinforcement corrosion and related properties of plain and blended cement concrete under different curing conditions," *Cement and Concrete Composites*, vol. 27, pp. 449–461, 2005.

[21] S. K. Antiohos, V. G. Papadakis, E. Chaniotakis, and S. Tsimas, "Improving the performance of ternary blended cements by mixing different types of fly ashes," *Cement and Concrete Research*, vol. 37, no. 6, pp. 877–885, 2007.

[22] T. R. Naik, S. S. Singh, and M. M. Hossain, "Enhancement in mechanical properties of concrete due to blended ash," *Cement and Concrete Research*, vol. 26, no. 1, pp. 49–54, 1996.

[23] T.-C. Lee, W.-J. Wang, P.-Y. Shih, and K.-L. Lin, "Enhancement in early strengths of slag-cement mortars by adjusting basicity of the slag prepared from fly-ash of MSWI," *Cement and Concrete Research*, vol. 39, no. 8, pp. 651–658, 2009.

[24] T. Nochaiya, W. Wongkeo, and A. Chaipanich, "Utilization of fly ash with silica fume and properties of Portland cement-fly ash-silica fume concrete," *Fuel*, vol. 89, no. 3, pp. 768–774, 2010.

[25] Specification for 43 grade Ordinary Portland Cement, IS 8112, BIS Standards, New Delhi, India, 2013.

[26] Specification for Portland Pozzolana Cement (fly ash based), IS 1489, BIS Standards, New Delhi, India, 1991.

[27] Specification for Portland slag cement, IS 455, BIS Standards, New Delhi, India, 2015.

[28] Specification for high strength deformed steel bars and wires for concrete reinforcement, IS 1786, BIS Standard, New Delhi, India, 2008.

[29] ASTM C642 – 97, "Standard Test Method for Density, Absorption, and Voids in Hardened Concrete," in *Proceedings of the ASTM International*, West Conshohocken, PA, USA, 1997.

[30] V. Saraswathy and H.-W. Song, "Corrosion performance of rice husk ash blended concrete," *Construction and Building Materials*, vol. 21, no. 8, pp. 1779–1784, 2007.

[31] ASTM G1-03, "Standard practice for preparing, cleaning and evaluating corrosion test specimens," in *Proceedings of the ASTM International*, West Conshohocken, PA, USA, 2011.

[32] ASTM C1202-12, "Standard test method for electrical indication of concretes ability to resist chloride ion penetration," in *Proceedings of the ASTM International*, West Conshohocken, PA, USA, 2012.

[33] H.-W. Song, V. Saraswathy, S. Muralidharan, C.-H. Lee, and K. Thangavel, "Corrosion performance of steel in composite concrete system admixed with chloride and various alkaline nitrites," *Corrosion Engineering Science and Technology*, vol. 44, no. 6, pp. 408–415, 2009.

[34] ASTM G109-07, "Standard Test Method for Determining Effects of Chemical Admixtures on Corrosion of Embedded Steel Reinforcement in Concrete Exposed to Chloride Environments," in *Proceedings of the ASTM International*, West Conshohocken, PA, USA, 2013.

[35] V. Saraswathy, S. Muralidharan, L. Balamurugan, P. Kathirvel, and A. S. S. Sekar, "Evaluation of composite cements using cyclic polarization techniques," *KSCE Journal of Civil Engineering*, vol. 15, no. 8, pp. 1415–1418, 2011.

[36] M. S. Meddah and A. Tagnit-Hamou, "Pore structure of concrete with mineral admixtures and its effect on self-desiccation shrinkage," *ACI Materials Journal*, vol. 106, no. 3, pp. 241–250, 2009.

[37] S. P. Karthick, S. Muralidharan, V. Saraswathy, and S. Kwon, "Effect of different alkali salt additions on concrete durability property," *Journal of Structural Integrity and Maintenance*, vol. 1, no. 1, pp. 35–42, 2016.

[38] K. Thangavel, S. Muralidharan, V. Saraswathy, K. Y. Ann, and L. Balamurugan, "Relationship between alumina and chloride content on their physical and corrosion resistance properties of concrete," *Arabian Journal for Science and Engineering*, vol. 35, no. 2 A, pp. 27–38, 2010.

[39] E. Berodier and K. Scrivener, "Evolution of pore structure in blended systems," *Cement and Concrete Research*, vol. 73, pp. 25–35, 2015.

21

3D Gradual Material Degradation Model for Progressive Damage Analyses of Unidirectional Composite Materials

Libin Zhao,[1] Tianliang Qin,[1] Jianyu Zhang,[2] and Yuli Chen[2]

[1]*School of Astronautics, Beihang University, Beijing 100191, China*
[2]*Institute of Solid Mechanics, Beihang University, Beijing 100191, China*

Correspondence should be addressed to Jianyu Zhang; jyzhang@buaa.edu.cn

Academic Editor: Chenfeng Li

A new 3D constitutive model for progressive damage analyses of unidirectional composite materials is presented, in which several important damage phenomena for the composite materials, such as the interfiber crack orientation, coupling of fiber failure and interfiber failure under longitudinal loads, closure effect for interfiber cracks, and longitudinal compressive behaviors under transversal constraints, have been considered comprehensively. A modified maximum stress failure criterion has been used for the damage onset prediction and a linear damage model has been adopted to establish the evolution rules of different damage. Numerical analyses with the model proposed have been implemented by using the subroutine UMAT in commercial software ABAQUS. Progressive damage analyses and static tensile experiments of a group of double-lap composite bolted joints have been carried out to validate the model proposed. Good agreements between the numerical and experimental results have been obtained.

1. Introduction

With the increasing application of composite materials in aircrafts, automobiles, ships, and so forth, there is an urgent requirement for composite structure design methodologies. During the past decades, the structural design has mainly relied on experimental data by using the building block approach [1], which results in extremely expensive cost due to a large number of tests at each structural level. Thus, initiatives bridging the tests and analyses to reduce the design cost have become a common view in the world [2].

Great efforts have been taken to develop reliable and accurate analysis methods for composite structures. Empirical methods [3–6] obtained from plenty of tests and experiences are usually only suitable for certain structure types or load cases, though they are convenient and beneficial to the structure design. Besides, series of tests are required to determine some parameters for the empirical methods when the materials, ply sequences, or structural configurations change, which also lead to expensive cost. Theoretical analysis methods can significantly reduce the composite structural design and analysis costs and always draw lots of attention in the engineering. Among these methods, some predict the structural strength only with failure criteria. These methods will result in too conservative predictions since they ignore the structural nonlinear damage propagation process and fail to reveal the structural failure mechanism conveniently [2, 7]. In the later of the 20th century, a progressive damage method, which builds nonlinear mechanics models for composite materials and has capacity of accurately simulating the structural failure process from initial damage to ultimate failure, attracts widespread attention in composite structure analyses [8–11]. A progressive damage model contains three parts: a stress analysis model to obtain structural stress distribution, a failure criterion to estimate the structural damage and failure, and a material degradation model to control the property changes of damaged materials. Furthermore, the last one has been a hotspot in the composite research during decades since it directly represents the nonlinear damage propagation.

The material degradation models can fall into two categories, sudden degradation models and gradual degradation models. In the sudden degradation models, the material properties degrade to zero or a certain proportion f of original values, which is called a degradation ratio. A total discounting model in which all the material properties degrade to zero when the damage is predicted is the simplest sudden degradation model. Based on the total discounting model, conservative predictions of structural strength are always obtained. Additionally, a usually used ply discounting model is one of extreme editions [12–15]. Another average practice is that the properties are degraded with targeted selection according to the failure modes. For example, Engelstad et al. [16] only reduced one material modulus at a time corresponding to a failure mode dominated by a stress component. More commonly, several material properties are reduced simultaneously for a failure mode, which is called interaction models [17–19]. Camanho and Matthews [10] distinguished the material degradation model by the tensile and compressive stress for the first time. Obviously, the sudden degradation models neglect the damage accumulation during the composite material failure process since they simply treat the material as undamaged and totally damaged.

To establish more accurate models, gradual degradation rules are good choices to describe the postfailure behavior of composite materials. The micromechanics theories are usually used for establishing gradual degradation models. Chang et al. [20–27] adopted the fiber bundle theory, while Lee et al. [28–30] used the shear lag theory to build the degradation model for the fiber failure. In contrast to the micromechanics methods, more researchers set the material property degradation rules with a predefined function as a matter of experience. In Lin and coworker's [31, 32] gradual degradation models, the material degradation models were chosen to make the stress reduce linearly after the damage initiation. Hwang et al. [33] adopted an exponential function to establish the composite material degradation model.

The continuum damage mechanics is another way to propose the gradual damage model for composite materials, in which the damage variable D is used to represent the damage scale in the materials. With the increasing damage variable D from zero to unit, the material properties degrade from the original value to zero. In a strict sense, the evolution laws for damage variables should obey the thermodynamics principles of irreversible processes. Since the continuum damage mechanics has a rigorous theory basis, establishing suitable 3D continuum damage model for composite materials has been the major goal of the Third World-Wide Failure Exercise [34]. In addition, the continuum damage mechanics based gradual degradation models are more appealing. However, due to the complexity and intricacy of composite material failure mechanisms, the developments of continuum damage model for composite materials are slow. Existing continuum damage mechanics based gradual degradation models commonly focused on partial characters of the failure. Maimí et al. [2] and Raimondo et al. [35] proposed 2D gradual degradation models for composite laminates separately, in which the crack direction and crack closure were considered. Pinho et al. [36] developed a 3D gradual degradation model for composite materials. The fiber failure was focused on, and the crack direction was taken into account. However, the crack closure was not mentioned.

For cases where the composite structures surfer from 3D complex stresses such as bolted composite joints, the more accurate 3D gradual degradation model that can consider more damage phenomena of composite materials is still required. In this paper, a 3D gradual degradation model that takes account of the matrix crack direction, matrix crack closure, interaction of the fiber damage, and matrix damage as well as mechanical behaviors of compressive fiber failure is proposed based on the physical damage phenomena of composite materials. Furthermore, the progressive damage analysis is implemented for composite structures by using subroutine UMAT in ABAQUS and validated by the experimental results of double-lap composite bolted joints.

2. 3D Progressive Damage Model Using Gradual Degradation Model

The degradation model of unidirectional composite materials is developed in this paper based on the characteristics of damage mechanisms of different failure modes. The modified maximum stress failure criteria [37] are used to predict the damage modes, initiation, and crack angles of composite materials. The linear damage evolution law is chosen to define the softening relationship for all failure modes. Finally, the model is embedded into the software ABAQUS with UMAT.

2.1. Failure Mechanisms of Composite Materials. The composite materials have complex damage mechanisms in the mesoscale such as fiber breakage, fiber buckling, matrix cracking, interface debonding, and fiber bridging [2, 38]. The propagation and connection of damage in the mesoscale lead to the damage in the macroscale. According to the loads inducing the damage, the composite damage in the macroscale include five modes: fiber tensile damage, fiber compressive damage, matrix tensile damage, matrix compressive damage, and fiber-matrix shearing damage. In virtue of the regular fiber distribution in the mesoscale, some characteristic of the damage in the macroscale can be tracked, which is helpful to establish the mechanics model of composite materials.

Under the longitudinal loading, fiber damage initiates by isolated fiber fractures in weak zones [38]. The localized fractures will induce the matrix cracking, interface debonding, and matrix shearing failure in the adjoining fibers on account of local stress concentrations caused by the localized fractures, as shown in Figure 1. With the increment of the longitudinal loading, more fibers fail and further lead to the ultimate failure of the material. For the longitudinal tensile case, a macroscopic crack perpendicular to the fiber will be formed [2], while for the longitudinal compressive case a damaged band as a result of fiber buckling, fiber kinking, matrix crack, and interface debonding is formed [39, 40]. The macroscopic damaged band is assumed to be perpendicular to the fiber for simplification.

The matrix tensile failure, matrix compressive failure, and fiber-matrix shearing failure will form a crack parallel to the fiber in the macroscale, called an interfiber crack [41, 42]. Under transverse tensile or fiber-matrix shearing loadings, the propagation and connection of internal flaws such as interface debonding, rich resin zones, and microvoids result in the interfiber crack perpendicular to the middle plane of

FIGURE 1: Damage of composite materials under longitudinal loading.

the ply [2, 41, 42], as shown in Figures 2(a) and 2(c). For the transverse compressive failure, the interfiber crack of most engineering composite materials usually has an angle of 53° inclined to the normal of the ply, as shown in Figure 2(b), rather than the maximum shear stress plane [41, 42]. Under the combination of a transverse compressive loading and a fiber-matrix shearing loading, the crack angle gradually reduces to 40° with the increase of the shearing loading but suddenly decreases to 0° when the shearing loads exceed a certain level [43], which is the same as that of a pure shearing crack. Generally, the interfiber crack for the matrix tensile failure, matrix compressive failure, and fiber-matrix shearing failure occurs in the matrix and fiber-matrix interface; as a result, it does not influence the properties in the fiber direction.

2.2. Constitutive Model of Composite Materials for 3D Progressive Damage Analysis. The progressive damage model proposed in the paper groups the unidirectional composite material damage into two classes: fiber cracks and interfiber cracks. The fiber cracks include the fiber tensile crack and fiber compressive crack, in which the latter is not an actual crack but a kink band. The interfiber cracks include the matrix tensile crack, matrix compressive crack, and fiber-matrix shearing crack.

The progressive damage model treats the fiber crack as a plane of symmetry because the fiber crack $P(D_1)$ shown in Figure 3 is generally perpendicular to fibers. The interfiber cracks $P(D_2)$ and $P(D_3)$ parallel to fibers will increase the anisotropy in the directions perpendicular to fibers. When an interfiber crack $P(D_2)$ occurs, the material can be assumed orthotropic with $P(D_2)$ and the plane perpendicular to $P(D_2)$ as a symmetric plane. In addition, the interfiber cracks $P(D_2)$ form a pair of free planes, on which only symmetric stresses exist and further result in symmetric damage. Generally, the symmetric damage leads to a new interfiber crack marked as $P(D_3)$, which is perpendicular to the interfiber crack $P(D_2)$ and fiber crack $P(D_1)$. As the damaged material is orthotropic and the normal of interfiber cracks may change when the stresses are different, it is very hard to establish the constitutive relation in the global coordinate directly for the damaged material with arbitrary interfiber cracks. Thus, in current work, a constitutive model under a local coordinate O-$1'2'3'$ fixed to the cracks, as shown in Figure 3, is provided and further deduced in the global coordinate.

Based on the continuum damage mechanics, three damage variables D_1, D_2, and D_3 are adopted to establish the model: D_1 is corresponding to the fiber crack, D_2 to the first interfiber crack, and D_3 to the second interfiber crack. The constitutive model in the local coordinate O-$1'2'3'$ is expressed with (1). $\boldsymbol{\varepsilon}'$ and $\boldsymbol{\sigma}'$ are the local strain and stress vectors, respectively, given by (2) and (3). $\mathbf{S}'$ is the flexibility matrix. $\widetilde{E}_{i'i'}$ and $\widetilde{G}_{i'j'}$ $(i, j = 1, 2, 3;\ i < j)$ are the effective elastic moduli. $\widetilde{\nu}_{i'j'}$ $(i, j = 1, 2, 3;\ i \neq j)$ are the effective Poisson's ratios. The model has considered the damage effects on the elastic moduli and Poisson's ratios, the interaction between the fiber crack and interfiber cracks, the crack closure effect, and the longitudinal compressive behaviors when the materials are laterally constrained:

$$\boldsymbol{\varepsilon}' = \mathbf{S}'\boldsymbol{\sigma}', \tag{1}$$

$$\boldsymbol{\varepsilon}' = \left[\varepsilon_{1'1'}\ \ \varepsilon_{2'2'}\ \ \varepsilon_{3'3'}\ \ \gamma_{1'2'}\ \ \gamma_{1'3'}\ \ \gamma_{2'3'}\right]^T, \tag{2}$$

$$\boldsymbol{\sigma}' = \left[\sigma_{1'1'}\ \ \sigma_{2'2'}\ \ \sigma_{3'3'}\ \ \sigma_{1'2'}\ \ \sigma_{1'3'}\ \ \sigma_{2'3'}\right]^T, \tag{3}$$

$$\mathbf{S}' = \begin{bmatrix} \dfrac{1}{\widetilde{E}_{1'1'}} & \dfrac{-\widetilde{\nu}_{2'1'}}{\widetilde{E}_{2'2'}} & \dfrac{-\widetilde{\nu}_{3'1'}}{\widetilde{E}_{3'3'}} & 0 & 0 & 0 \\ \dfrac{-\widetilde{\nu}_{1'2'}}{\widetilde{E}_{1'1'}} & \dfrac{1}{\widetilde{E}_{2'2'}} & \dfrac{-\widetilde{\nu}_{3'2'}}{\widetilde{E}_{3'3'}} & 0 & 0 & 0 \\ \dfrac{-\widetilde{\nu}_{1'3'}}{\widetilde{E}_{1'1'}} & \dfrac{-\widetilde{\nu}_{2'3'}}{\widetilde{E}_{2'2'}} & \dfrac{1}{\widetilde{E}_{3'3'}} & 0 & 0 & 0 \\ 0 & 0 & 0 & \dfrac{1}{\widetilde{G}_{1'2'}} & 0 & 0 \\ 0 & 0 & 0 & 0 & \dfrac{1}{\widetilde{G}_{1'3'}} & 0 \\ 0 & 0 & 0 & 0 & 0 & \dfrac{1}{\widetilde{G}_{2'3'}} \end{bmatrix}. \tag{4}$$

2.2.1. Damage Effect on Material Elastic Moduli and Poisson's Ratios. Longitudinal stresses will not produce a smooth crack but a damage band. There are not only fiber fracture (tension) and fiber buckling or kinking (compression) in the damage band, but also the matrix crack and fiber-matrix debonding. As a result, the fiber failure will decrease all the material properties. In contrast, interfiber damage seldom influences the load carrying capability in the fiber direction. To simplify the expression of the constitutive model, the effect of fiber failure on the transverse properties is considered by using the damage variables D_2 and D_3 in the interaction model of the fiber crack and interfiber crack. The effective elastic moduli and Poisson's ratios can be expressed as

$$\widetilde{E}_{1'1'} = E_{1'1'}\left(1 - D_1\right),$$

$$\widetilde{E}_{2'2'} = E_{2'2'}\left(1 - D_2\right),$$

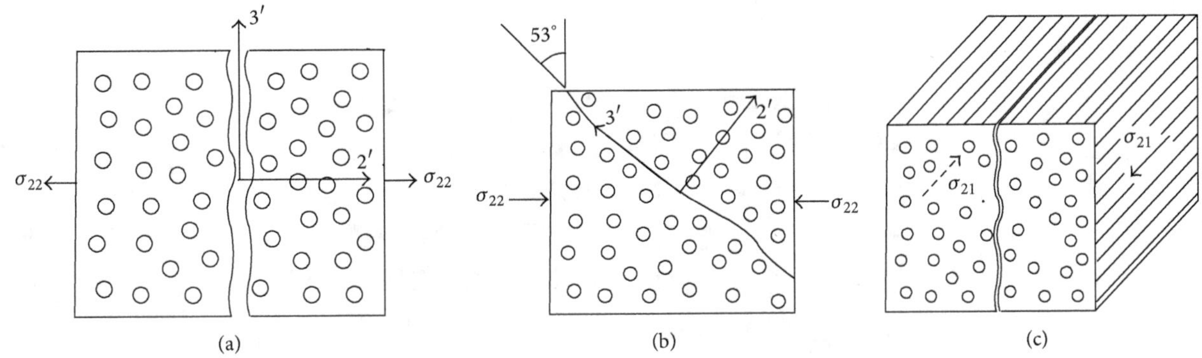

FIGURE 2: Interfiber failure of composite materials: (a) matrix crack in tension; (b) matrix crack in compression; (c) shear-out crack.

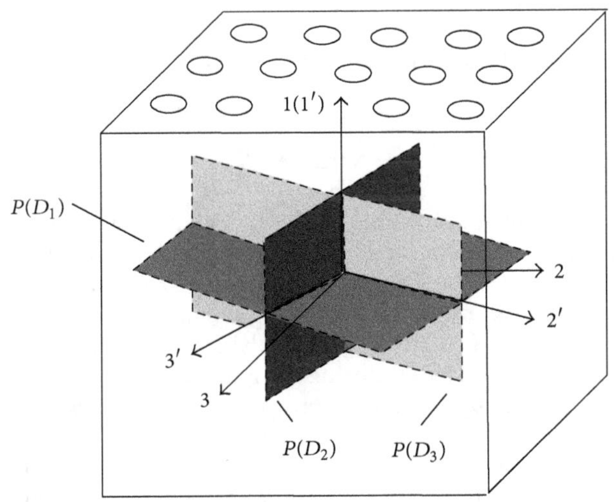

FIGURE 3: Definition of damage variables and local coordinate system.

$$
\begin{aligned}
\widetilde{E}_{3'3'} &= E_{3'3'}\left(1-D_3\right),\\
\widetilde{G}_{1'2'} &= G_{1'2'}\left(1-D_2\right),\\
\widetilde{G}_{1'3'} &= G_{1'3'}\left(1-D_3\right),\\
\widetilde{G}_{2'3'} &= G_{2'3'}\left(1-D_2\right)\left(1-D_3\right),\\
\widetilde{\nu}_{i'j'} &= \nu_{i'j'}\left(1-D_{i'}\right) \quad i',j'=1,2,3.
\end{aligned}
\tag{5}
$$

2.2.2. Interaction Model of the Fiber Failure and Interfiber Failure under Longitudinal Loads. Longitudinal loadings can induce not only the fiber crack but also the interfiber cracks, while transverse and shearing stresses can only result in the interfiber cracks. Thus, to simplify the constitutive model, this paper assumes that when D_1 of the fiber crack is produced, the interfiber cracks equaling to D_1 occur in the materials, which can be expressed as follows:

$$
\begin{aligned}
D_1 &= D_1; \qquad D_2 = \max\left(D_1, D_2\right);\\
D_3 &= \max\left(D_1, D_3\right).
\end{aligned}
\tag{6}
$$

2.2.3. Closure Effect for Interfiber Cracks. The normal stress reversal from the tension to compression will make the interfiber crack close as shown in Figure 4. This paper assumes the closed crack will make the material recover the normal load-bearing capacity, which means the effective damage variable of the interfiber crack for the normal stress becomes zero. The effective damage variables of the interfiber cracks D_2^n and D_3^n for the normal stresses can be provided by (7). $\langle x\rangle$ denotes the McCauley operator, defined by $\langle x\rangle = (x+|x|)/2$:

$$
D_2^n = D_2\frac{\langle\sigma_{22}\rangle}{\left|\sigma_{22}\right|}; \qquad D_3^n = D_3\frac{\langle\sigma_{33}\rangle}{\left|\sigma_{33}\right|}. \tag{7}
$$

On the other hand, this paper assumes the closed cracks cannot recover the shear load-bearing capacity, even though the frictions between the crack surfaces exist. Thus, the effective damage variables of the interfiber cracks D_2^t and D_3^t for the shear stresses are equal to D_2 and D_3, respectively, given by

$$
D_2^t = D_2; \qquad D_3^t = D_3. \tag{8}
$$

2.2.4. Consideration on the Longitudinal Stress Reversion. It is assumed that the effect of the fiber crack on the material properties under the longitudinal tensile stress is the same as that when it turns to compressive one. The effective damage variables D_1^+ and D_1^- of the fiber crack during the longitudinal stress reversion are given by

$$
D_1^+ = D_1^- = D_1. \tag{9}
$$

2.2.5. Model for the Longitudinal Compressive Behaviors considering Lateral Constraints. Under a longitudinal compressive loading, the composite material failure starts with the fiber kinking or buckling. With the load increases, the fiber kinking or buckling increases and induces the matrix cracks and fiber-matrix debonding. Consequently, the compressive load makes the fibers break into pieces [25].

The fully damaged composite materials have liquidity like sands [25, 26]. Thus, the behaviors of the composite materials under longitudinal compressive loads are related to the lateral constraint conditions. It is supposed that

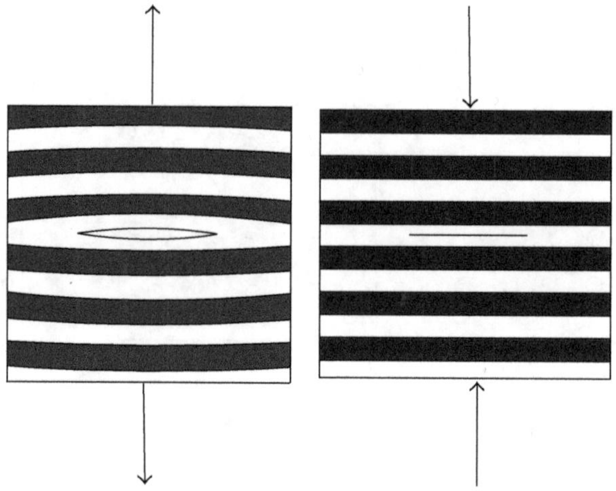

FIGURE 4: Interfiber crack status under reversal load conditions: (a) tension and (b) compression.

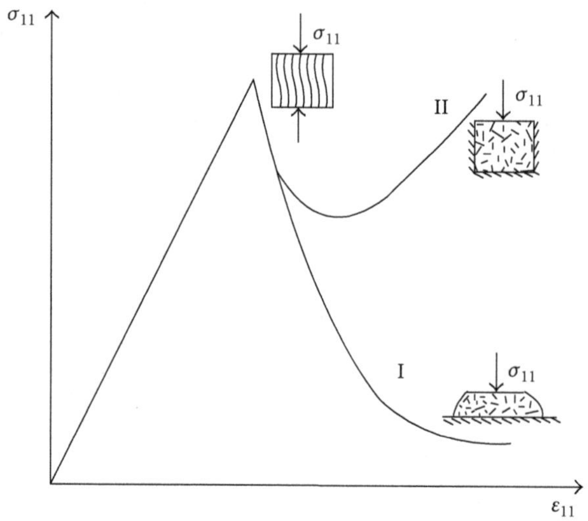

FIGURE 5: Compressive behaviors of composite materials.

the longitudinal load-carrying capability of the composite materials will continually decrease up to be zero approximately owing to the outflowing of damaged materials without lateral constraints, as the curve I shown in Figure 5. And a higher longitudinal load can be withstood for damage materials with lateral constraints, as the curve II.

Due to different mechanical behaviors of damaged composite materials under longitudinal tensile and compressive loads, it is hard to use the same model to depict longitudinal compressive behaviors as that to characterize longitudinal tensile behaviors. Inspired by Chang's model for the composite material longitudinal compressive behaviors [25], an approximately impressible constitutive model is utilized for the composite materials under longitudinal compressive loads in current work. As a result, the materials have a high load-carrying capability when lateral constraints are applied and a low load-carrying capability for case without lateral constraints by setting a relative low elastic modulus. Meanwhile, the damage variable D_1 is used to realize the transformation from the damage mechanics model to the approximately impressible model by controlling the elastic modulus and Poisson's ratios. The elastic modulus of impressible model is set as same as the transverse elastic modulus of composite materials, $E_{2'2'}$. The longitudinal elastic modulus changes from $E_{1'1'}$ to $E_{2'2'}$ and all Poisson's ratios change to 0.5 controlled by D_1 during the longitudinal compressive loading, given by (10). Meanwhile, a term of $1-D_2^n$ is added to last item of (10) to include the effect of the intercrack closure. In addition, a value of ν_{appr} near to 0.5 is set to Poisson's ratios in the impressible model since the impressible constitutive relation cannot be used in general finite elements. Here, ν_{appr} is set to be 0.49:

$$\begin{aligned}
\widetilde{E}_{1'1'} &= (E_{1'1'} - E_{2'2'})(1 - D_1) + E_{2'2'}, \\
\widetilde{\nu}_{1'2'} &= (\nu_{1'2'} - \nu_{\text{appr}})(1 - D_1) + \nu_{\text{appr}}, \\
\widetilde{\nu}_{1'3'} &= (\nu_{1'3'} - \nu_{\text{appr}})(1 - D_1) + \nu_{\text{appr}}, \\
\widetilde{\nu}_{2'3'} &= ((\nu_{2'3'} - \nu_{\text{appr}})(1 - D_1) + \nu_{\text{appr}})(1 - D_2^n).
\end{aligned} \tag{10}$$

Based on the assumptions above, the constitutive model of composite materials is established. The material parameters of the flexibility matrix $\mathbf{S}'$ in the model are listed in Table 1, from which the stiffness matrix $\mathbf{D}'$ in the local coordinate O-1′2′3′ can be further deduced. Then the stiffness matrix $\mathbf{D}$ in global coordinate could be obtained from $\mathbf{D}'$ and the interfiber crack angle.

2.3. Failure Criteria for Damage Onset. The modified maximum criterion proposed by Zhao et al. [37, 44] is adopted here to predict the onset of fiber crack and interfiber crack.

2.3.1. Fiber Crack Onset. If either of the fiber tensile failure criterion or the fiber compressive failure criterion is satisfied,

Table 1: Material parameters in 3D progressive damage model.

$D_1 = 0$ or $D_1 > 0, \sigma_{11} \geq 0$	$D_1 > 0, \sigma_{11} < 0$
	$\widetilde{E}_{1'1'} = (E_{1'1'} - E_{2'2'})(1 - D_1) + E_{2'2'}, \widetilde{E}_{2'2'} = E_{2'2'}(1 - D_2^n)$
$\widetilde{E}_{1'1'} = E_{1'1'}(1 - D_1)$	$\widetilde{E}_{3'3'} = E_{3'3'}(1 - D_3^n)$
$\widetilde{E}_{2'2'} = E_{2'2'}(1 - D_2^n)$, $\widetilde{E}_{3'3'} = E_{3'3'}(1 - D_3^n)$	$\widetilde{G}_{1'2'} = G_{1'2'}(1 - D_2^t), \widetilde{G}_{1'3'} = G_{1'3'}(1 - D_3^t)$
$\widetilde{G}_{1'2'} = G_{1'2'}(1 - D_2^t)$, $\widetilde{G}_{1'3'} = G_{1'3'}(1 - D_3^t)$	$\widetilde{G}_{2'3'} = G_{2'3'}(1 - D_2^t)(1 - D_3^t)$
$\widetilde{G}_{2'3'} = G_{2'3'}(1 - D_2^t)(1 - D_3^t)$	$\widetilde{\nu}_{1'2'} = (\nu_{1'2'} - \nu_{\text{appr}})(1 - D_1) + \nu_{\text{appr}}$
$\widetilde{\nu}_{1'2'} = \nu_{1'2'}(1 - D_1), \widetilde{\nu}_{1'3'} = \nu_{1'3'}(1 - D_1)$	$\widetilde{\nu}_{1'3'} = (\nu_{1'3'} - \nu_{\text{appr}})(1 - D_1) + \nu_{\text{appr}}$
$\widetilde{\nu}_{2'3'} = \nu_{2'3'}(1 - D_2^n)$	$\widetilde{\nu}_{2'3'} = ((\nu_{2'3'} - \nu_{\text{appr}})(1 - D_1) + \nu_{\text{appr}})(1 - D_2^n)$
$\widetilde{\nu}_{2'1'} = (\widetilde{\nu}_{1'2'}/\widetilde{E}_{1'1'})\widetilde{E}_{2'2'}, \widetilde{\nu}_{3'1'} = (\widetilde{\nu}_{1'3'}/\widetilde{E}_{1'1'})\widetilde{E}_{3'3'}$	$\widetilde{\nu}_{2'3'} = ((\nu_{2'3'} - \nu_{\text{appr}})(1 - D_1) + \nu_{\text{appr}})(1 - D_2^n)$,
$\widetilde{\nu}_{3'2'} = \left(\widetilde{\nu}_{2'3'}/\widetilde{E}_{2'2'}\right)\widetilde{E}_{3'3'}$	$\widetilde{\nu}_{2'1'} = (\widetilde{\nu}_{1'2'}/\widetilde{E}_{1'1'})\widetilde{E}_{2'2'}, \widetilde{\nu}_{3'1'} = (\widetilde{\nu}_{1'3'}/\widetilde{E}_{1'1'})\widetilde{E}_{3'3'}$
	$\widetilde{\nu}_{3'2'} = \left(\widetilde{\nu}_{2'3'}/\widetilde{E}_{2'2'}\right)\widetilde{E}_{3'3'}$

the fiber crack onset is detected. For fiber tensile ($\sigma_{11} > 0$) and compressive ($\sigma_{11} < 0$) failure, the criterion can be written as

$$\frac{\sigma_{11}}{X_t} = 1; \qquad \left|\frac{\sigma_{11}}{X_c}\right| = 1, \tag{11}$$

where X_t and X_c are the longitudinal tensile and compressive strength, respectively.

2.3.2. Interfiber Crack Onset. If any criterion among the matrix tensile failure criterion, matrix compressive failure criterion, and fiber-matrix shearing failure criterion is satisfied, the interfiber crack onset is detected. The criterion is expressed in (12) and (13), where Y_t, Y_c, and S_{12} are the transverse tensile strength, transverse compressive strength, and fiber-matrix shearing strength, respectively. θ_m^T, θ_m^C, and θ_S are the intercrack angles corresponding to the matrix tensile failure, matrix compressive failure, and fiber-matrix shearing failure, respectively.

Matrix tensile ($\sigma_m^{\max} \geq 0$) and compressive ($\sigma_m^{\min} \leq 0$) failure is

$$\begin{aligned} &\frac{\sigma_m^{\max}}{Y_t} = 1 \qquad \theta_m^T = \frac{1}{2}\text{arctg}\left(\frac{2\sigma_{23}}{\sigma_{22} - \sigma_{33}}\right), \\ &\left|\frac{\sigma_m^{\min}}{Y_c}\right| = 1 \qquad \theta_m^C = \frac{1}{2}\text{arctg}\left(\frac{2\sigma_{23}}{\sigma_2 - \sigma_3}\right) + 143^\circ, \end{aligned} \tag{12}$$

where $\sigma_m^{\max} = (\sigma_{22} + \sigma_{33})/2 + \sqrt{((\sigma_{22} - \sigma_{33})/2)^2 + \sigma_{23}^2}$, $\sigma_m^{\min} = (\sigma_{22} + \sigma_{33})/2 - \sqrt{((\sigma_{22} - \sigma_{33})/2)^2 + \sigma_{23}^2}$.

Fiber-matrix shearing failure is

$$\frac{\sigma_S^{\max}}{S_{12}} = 1 \qquad \theta_S = \text{arctg}\left(\frac{\sigma_{13}}{\sigma_{12}}\right), \tag{13}$$

where $\sigma_S^{\max} = \sqrt{\sigma_{12}^2 + \sigma_{13}^2}$.

2.4. Damage Evolution. After damage onset, the damage variable D increases from 0 to unit with the strains increase. A linear damage model is used to define the damage evolution as shown in Figure 6, in which the stress decreases linearly with the strain increase after the damage onset. σ^0 denotes the stress at damage onset. ε^f represents the final failure strain. The damage variable in the linear damage model is defined as

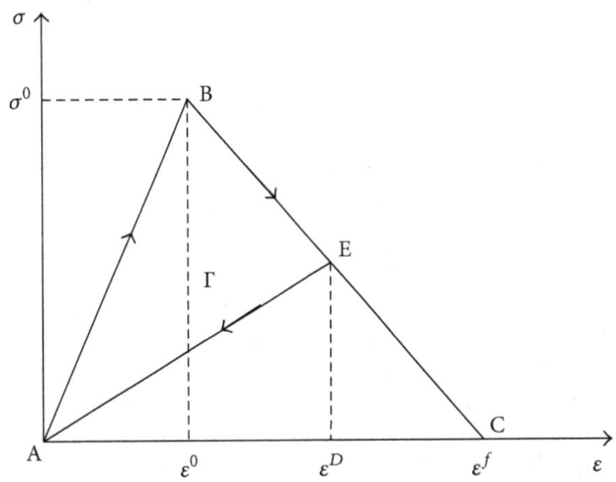

Figure 6: Stress-strain curve of linear damage model.

$$D = \max\left\{0, \min\left(1, \frac{\varepsilon^f\left(\varepsilon - \varepsilon^0\right)}{\varepsilon\left(\varepsilon^f - \varepsilon^0\right)}\right)\right\}. \tag{14}$$

The damage variable D_1 of the fiber crack could be provided as

$$\begin{aligned} D_1 &= \min\left(1, \frac{\varepsilon_{11}^{f+}\left(\varepsilon_{11} - \varepsilon_{11}^{0+}\right)}{\varepsilon_{11}\left(\varepsilon_{11}^{f+} - \varepsilon_{11}^{0+}\right)}\right), \quad \varepsilon_{11} > 0, \\ D_1 &= \min\left(1, \frac{\varepsilon_{11}^{f-}\left(\varepsilon_{11} - \varepsilon_{11}^{0-}\right)}{\varepsilon_{11}\left(\varepsilon_{11}^{f-} - \varepsilon_{11}^{0-}\right)}\right), \quad \varepsilon_{11} < 0, \end{aligned} \tag{15}$$

where ε_{11}^{0+} and ε_{11}^{f+} relate to the onset and final failure strains of the fiber tensile damage, respectively. ε_{11}^{0-} and ε_{11}^{f-} are the onset and final failure strains of the fiber compressive damage, respectively.

For interfiber cracks, the normal stress σ_N and two shear stresses τ_L and τ_T may lead to the damage propagation. Assume the most critical stress drives the interfiber crack

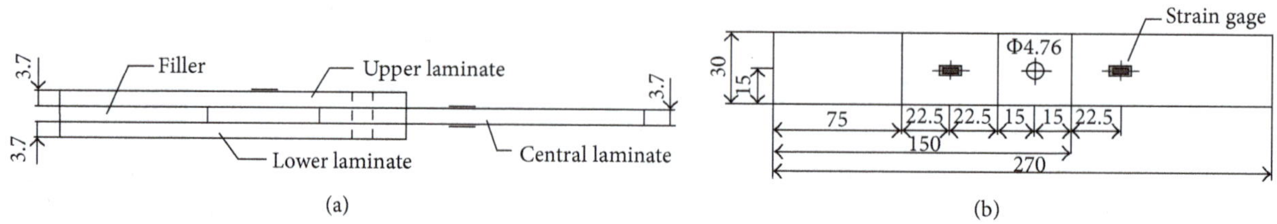

FIGURE 7: Configuration and dimensions of a composite bolted joint: (a) front view and (b) top view.

without regard to the stress interaction; then the damage variable is expressed as

$$D_i = \max\left(D_i^N, D_i^L, D_i^T\right) \quad i = 2, 3, \tag{16}$$

where $D_i^k = \min(1, (\varepsilon_i^{kf}(\varepsilon_i^k - \varepsilon_i^{k0})/\varepsilon_i^k(\varepsilon_i^{kf} - \varepsilon_i^{k0})))$ $k = N, L, T$.

For a complete continuum damage model, the damage onset strain and final failure strain are key parameters. The former is often obtained from the damage onset stress which is generally referred to the interface strength. Since the interface strength is difficult to be obtained from common tests, appropriate values are often used [45, 46]. This paper determines the damage onset strain by using the modified failure criteria and engineering material strengths. Moreover, the final failure strain could be provided with (17) according to the cohesive model proposed by Bažant and Oh [47], which applies the fracture toughness of crack to the damage mechanics model. In (17), G_C is the fracture toughness, and l^* is the characteristic length of the finite element. For the proposed damage mechanics model, the G_{1C}^+ for fiber crack under tensile stress, G_{1C}^- for fiber crack under compressive stress, G_{2C}^N for interfiber crack under tensile stress, G_{2C}^L for interfiber crack under longitudinal shear stress, and G_{2C}^T for interfiber crack under transverse shear stress are needed. G_{2C}^N can be examined by double cantilever beam tests. G_{2C}^L and G_{2C}^T are obtained by three-point bending tests or four-point bending tests. As measuring G_{2C}^T is difficult, let G_{2C}^T be equal to G_{2C}^L in this paper. There is a lack of standard tests for G_{1C}^+ and G_{1C}^- yet. Pinho et al. [48] used compact tension tests to examine G_{1C}^+ and compact compression tests to determine G_{1C}^-:

$$\varepsilon^f = \frac{2G_C}{\sigma^0 l^*}. \tag{17}$$

3. Validation of the Progressive Damage Model

To validate the capability of predicting damage propagation in the composite materials under complex loading conditions, the 3D progressive damage model was embedded into the software ABAQUS and applied to a double-lap composite bolted joint under a tensile load. Meanwhile, the tensile experiments of composite bolted joints were conducted to validate the numerical results. On considering the main object and the limited length of this paper, the implements of the 3D progressive damage model in ABAQUS will be described in a following independent paper.

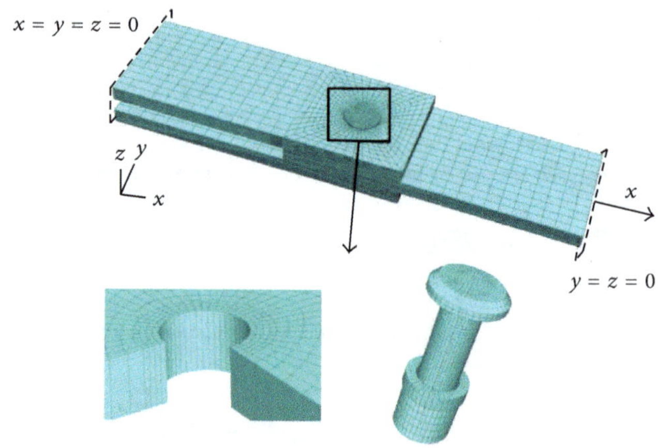

FIGURE 8: FEM of double-lap composite bolted joint.

3.1. Static Tensile Experiment of Composite Bolted Joints. A double-lap composite bolted joint was selected to verify the model proposed here. Its configuration and dimensions were designed as recommended in engineering and are shown in Figure 7. The laminate strips were made of T800 carbon/epoxy composites with lay-ups of $[45/0/-45/0/90/0/45/0/-45/0]_s$. Two outer strips were connected to the central one by using a titanium alloy protruding head bolt HST13-6-7 and a high locking collar HST1078 made of A280 high temperature alloy. Composite filler was added adhesively at the gripping section of the two outer strips.

The experiments were carried out by using a 250 kN Instron8803 testing machine. Five specimens were tested. The displacement load with rate of 1 mm/min was applied up to the catastrophic failure of the joint. The applied load and grip holder displacements were measured by sensors of the testing machine automatically, while the hole deformations were examined by an extensometer mounted on the specimens. More information about the experimental process can be found in [7].

3.2. Numerical Simulations. The FE model of the double-lap composite joint is shown in Figure 8, in which the composite laminates, fastener, and collar were established by using 3D element C3D8. Each layer of the laminates was established with one element through the thickness direction. Meshes around the hole of the composite laminates were refined to obtain accurate numerical predictions. Three meshes having 69260, 103100, and 150380 elements, named Model I, Model II, and Model III, respectively, were used to verify the mesh independence on the prediction results. Contact pairs were set in the potential contact regions which were between the outer strips and central strip, the hole and bolt, and the strips and bolt. The direct "hard" method was applied to calculate the normal behaviors. A model without friction for tangential behaviors was used to make the numerical simulation converge more easily. A preload of 4.7 kN calculated from the assembly tightening moment of 4.5 N-m by using empirical formulae [49, 50] was applied to the internal section of the bolt shank with a bolt load method in ABAQUS. Equivalent boundary constraints and displacement load shown in Figure 8 were established to decrease the element amounts. The ends of outer strips were fully constrained, while a displacement along the longitudinal direction was applied to the other end with transverse degrees constrained.

The elastic properties and strengths of the composite material are $E_1 = 195$ GPa, $E_2 = E_3 = 8.58$ GPa, $G_{12} = G_{13} = 4.57$ GPa. $\nu_{12} = \nu_{13} = 0.33$, $X_t = 3071$ MPa, $X_c = 1747$ MPa, $Y_t = Z_t = 88$ MPa, $Y_c = Z_c = 271$ MPa, and $S_{12} = S_{13} = 143$ MPa. The Poisson ratio ν_{23} is given by using Christensen's [51] equation: $\nu_{23} = \nu_{12}(1 - \nu_{12}E_2/E_1)/(1 - \nu_{12})$. S_{23} is calculated from the transverse compressive strength Y_c using the equation $S_{23} = Y_c/2\tan(\phi_0)$ [36]. ϕ_0 is the transverse compressive crack angle, usually 53°. G_{2C}^N and G_{2C}^L were tested as 0.368 and 1.480 N/mm, respectively. G_{2C}^T has the same value as G_{2C}^L. G_{1C}^+ and G_{1C}^- were set as 55.00 and 35.00 N/mm after tentative analysis within the reported range [35, 36, 48]. The elastic modulus and Poisson's ratio of the titanium alloy are 110 GPa and 0.29, while being 200 GPa and 0.30 with respect to the A280 high temperature alloy, respectively.

The gradual degradation model and failure criteria were applied to the FE model by developing a user defined subroutine UMAT for ABAQUS. At the beginning of each load increment, the UMAT will be called. The damage state will be checked using failure criteria. Once material damage is predicted, the gradual degradation model will be used to determine each damage variable according to the stress conditions and give the stiffness matrix of damaged material. This operation is repeated time and time again until the structure is fully destroyed.

3.3. Results and Discussion. Figure 9(a) gives the predicted load-hole deformation curves of double-lap composite joints by using the models having different meshes. It shows that the three curves are almost the same below 15 kN and slightly separate after that. With the element number increases, the load-hole deformation curve is higher before they reach each peak point. However, the differences between three curves are relatively small. The ultimate loads of three models are 19.78, 20.50, and 20.49 kN, respectively. The numerical predictions indicate that adopting model II will slightly affect the prediction of the damage process but have safely ignorable influence on the ultimate failure loads. Further, Figure 9(b) compares the numerical load-hole deformation of the double-lap composite joint by using Model II and experimental ones. The numerical curve has the similar

TABLE 2: Experimental and numerical ultimate loads of composite bolted joints.

Experimental ultimate load/kN	Avg. val./kN	Numerical ultimate load/kN
21.18	20.82	20.50
20.80	SD/kN	Error
20.78	0.35	–1.5%
21.06	C_v/%	/
20.26	1.69	/

trend as the experimental results. Both of them indicate that the double-lap composite joints exhibit large nonlinear behaviors after the elastic stage, although the predicted final failure deformation is smaller than the experimental ones. The predicted ultimate load from Model II is 1.5% lower than experimental average load 20.82 kN, as listed in Table 2. It follows that the proposed progressive damage model has successfully predicted the stiffness and strength of the composite bolted joint under tensile loads.

Figure 10(a) shows the experimental failure morphology of the central plate in the double-lap composite bolted joint. Large bearing damage occurred near the central plate hole. The materials were compressed into fragments and plumped up in the bearing damage zone. Close to the central plate end, shear-out damage mode can be found. The inner plies were squeezed out of the plate end in the zone. Figure 10(b) shows the predicted failure morphology of the central plate. Large hole deformation and bearing damage were predicted. The laminates bulge from the cut view indicates that the model has successfully simulated the bearing behavior of damaged composite materials. The model also shows similar damage morphology with the experimental one at the right end of central strip, where shear-out damage was formed with inner plies squeezed out.

Good agreements on the joint strength, stiffness, and failure morphology between the experimental outcomes and numerical predictions give evidence that the proposed PDM has great potential on simulating the mechanical behaviors of composite materials under complex loading and constraints.

4. Conclusion

A new 3D constitutive model for progressive damage analysis of unidirectional composite materials is proposed in this paper. The model has taken account of the possible fiber fracture planes caused by longitudinal tensile and compressive loads as well as interfiber fracture planes at any angles with respect to the ply thickness direction caused by transverse stresses and fiber-matrix shear stresses. Mathematical models are also presented to describe complex phenomena of damaged materials, such as crack closure due to load reversion, interaction of fiber damage and interfiber damage, effective elastic moduli, and Poisson's ratios of damaged materials. Besides, an approximate model has been created for the longitudinal compressive behaviors. The damage onset has been predicted by using the modified maximum stress failure

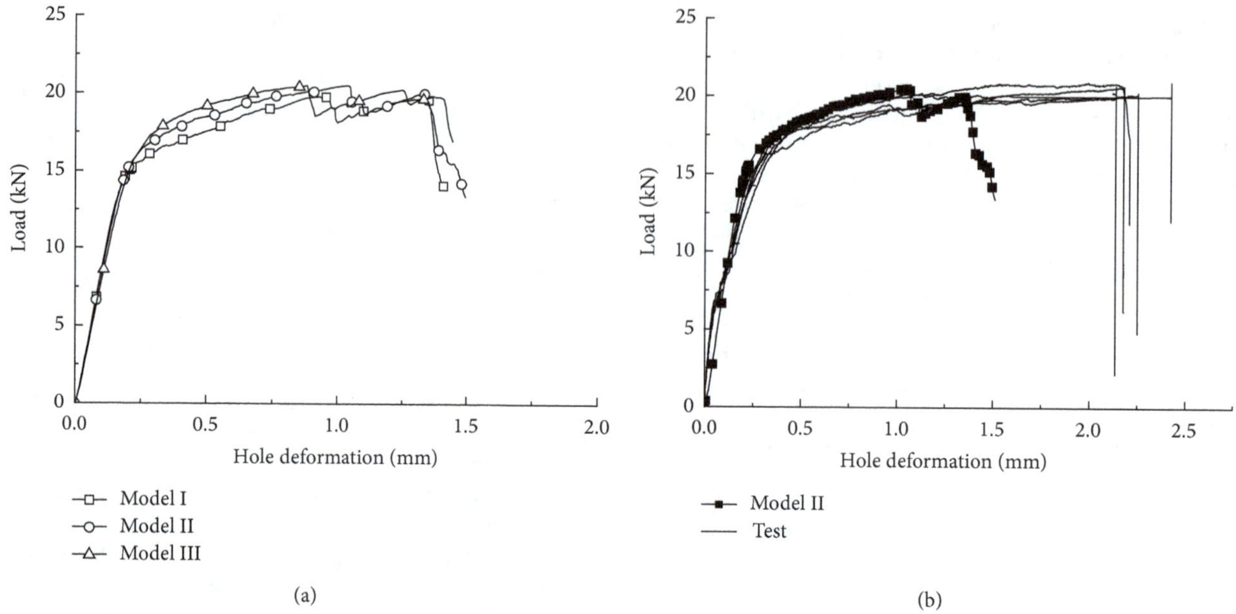

FIGURE 9: Load-hole deformation curves of specimens: (a) numerical predictions from different mesh schemes and (b) comparison between experimental and numerical results.

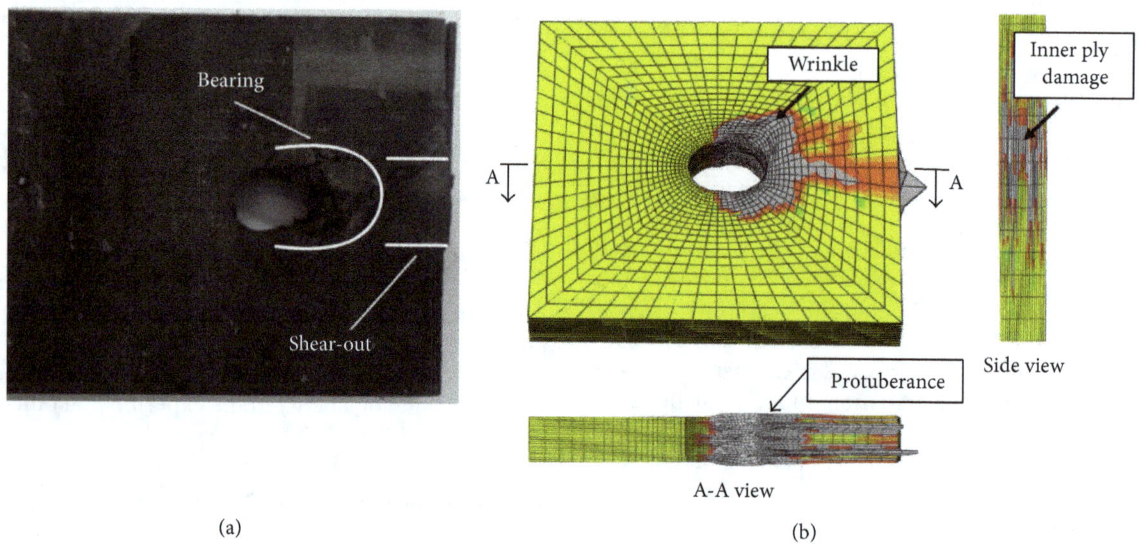

FIGURE 10: Failure morphologies of the central plate: (a) from experiments and (b) from numerical predictions.

criterion [37] and a linear damage model has been adopted to establish the evolution rules of different damage.

The proposed model has been embedded into ABAQUS using the subroutine UMAT and applied to predict the progressive damage accumulation and collapse failure of a double-lap composite bolted joint under tensile loads. The load-hole deformation curve, strength and stiffness information, progressive damage process, and final failure morphology can be obtained from the progressive damage simulation with the model proposed. The numerical prediction is in good agreement with the static tensile experimental outcomes of the composite bolted joint, which gives evidence of the effectiveness of the model proposed.

Conflict of Interests

The authors declare that there is no conflict of interests regarding the publication of this paper.

Acknowledgments

The research work is supported by the National High Technology Research and Development Program of China

(2012AA040209) and the National Science Foundation of China (11372020, 11202012, 11472027, and 10902004).

References

[1] MIL-HDBK-17-3F, *Military Handbook, Polymer Matrix Composites*, US Department of Defense, 2002.

[2] P. Maimí, P. P. Camanho, J. A. Mayugo, and C. G. Dávila, "A continuum damage model for composite laminates: part I—constitutive model," *Mechanics of Materials*, vol. 39, no. 10, pp. 897–908, 2007.

[3] L. J. Hart-Smith, *Bolted Joints in Graphite-Epoxy Composites*, Douglas Aircraft Company, Long Beach, Calif, USA, 1976.

[4] F. R. Liu, L. B. Zhao, S. Mehmood, J. Y. Zhang, and B. J. A. Fei, "A modified failure envelope method for failure prediction of multi-bolt composite joints," *Composites Science and Technology*, vol. 83, pp. 54–63, 2013.

[5] J. M. Whitney and R. J. Nuismer, "Stress fracture criteria for laminated composites containing stress concentrations," *Journal of Composite Materials*, vol. 8, pp. 253–265, 1974.

[6] J. Y. Zhang, F. R. Liu, L. B. Zhao, and B. J. Fei, "A novel characteristic curve for failure prediction of multi-bolt composite joints," *Composite Structures*, vol. 108, no. 1, pp. 129–136, 2014.

[7] T. L. Qin, L. B. Zhao, and J. Y. Zhang, "Fastener effects on mechanical behaviors of double-lap composite joints," *Composite Structures*, vol. 100, pp. 413–423, 2013.

[8] S. C. Tan, "Progressive failure model for composite laminates containing openings," *Journal of Composite Materials*, vol. 25, no. 5, pp. 556–577, 1991.

[9] C. Hühne, A.-K. Zerbst, G. Kuhlmann, C. Steenbock, and R. Rolfes, "Progressive damage analysis of composite bolted joints with liquid shim layers using constant and continuous degradation models," *Composite Structures*, vol. 92, no. 2, pp. 189–200, 2010.

[10] P. P. Camanho and F. L. Matthews, "A progressive damage model for mechanically fastened joints in composite laminates," *Journal of Composite Materials*, vol. 33, no. 24, pp. 2248–2280, 1999.

[11] M. R. Garnich and V. M. K. Akula, "Review of degradation models for progressive failure analysis of fiber reinforced polymer composites," *Applied Mechanics Reviews*, vol. 62, no. 1, pp. 1–33, 2009.

[12] J. P. Waszczak and T. A. Cruse, "Failure mode and strength predictions of anisotropic bolt bearing specimens," *Journal of Composite Materials*, vol. 5, no. 3, pp. 421–425, 1971.

[13] M. Di Sciuva, U. Icardi, and M. Villani, "Failure analysis of composite laminates under large deflection," *Composite Structures*, vol. 40, no. 3-4, pp. 239–255, 1997.

[14] P. Paul and C. Ray, "Progressive failure analysis of laminated composite plates by finite element method," *Journal of Reinforced Plastics and Composites*, vol. 21, no. 16, pp. 1505–1513, 2002.

[15] B. G. Prusty, "Progressive failure analysis of laminated unstiffened and stiffened composite panels," *Journal of Reinforced Plastics and Composites*, vol. 24, no. 6, pp. 633–642, 2005.

[16] S. P. Engelstad, J. N. Reddy, and N. F. J. Knight, "Postbuckling response and failure prediction of graphite-epoxy plates loaded in compression," *AIAA Journal*, vol. 30, no. 8, pp. 2016–2113, 1992.

[17] Y. S. Reddy and J. N. Reddy, "Three-dimensional finite element progressive failure analysis of composite laminates under axial extension," *Journal of Composites Technology and Research*, vol. 15, no. 2, pp. 73–87, 1993.

[18] W. C. Hwang and C. T. Sun, "Failure analysis of laminated composites by using iterative three-dimensional finite element method," *Computers and Structures*, vol. 33, no. 1, pp. 41–47, 1989.

[19] Y.-J. Lee and W.-H. Cheng, "Failure process and bolted joint strength of composite laminates," *Journal of the Chinese Society of Mechanical Engineers*, vol. 9, no. 3, pp. 169–182, 1988.

[20] F.-K. Chang and Z. Kutlu, "Strength and response of cylindrical composite shells subjected to out-of-plane loadings," *Journal of Composite Materials*, vol. 23, no. 1, pp. 11–31, 1989.

[21] F.-K. Chang and Z. Kutlu, "Study on the crushing response of cylindrical composite shells subjected to transverse loading," *Journal of Composites Technology and Research*, vol. 12, no. 4, pp. 239–244, 1990.

[22] I. Shahid and F. Chang, *Modeling of Failure and Response to Laminated Composites Subjected to in-Plane Loads*, Stanford University, Palo Alto, Calif, USA, 1994.

[23] I. Shahid and F.-K. Chang, "An accumulative damage model for tensile and shear failures of laminated composite plates," *Journal of Composite Materials*, vol. 29, no. 7, pp. 926–981, 1995.

[24] I. Shahid, H.-T. Sun, and F.-K. Chang, "Predicting scaling effect on the notched strength of prepreg and fiber tow-placed laminated composites," *Journal of Composite Materials*, vol. 29, no. 8, pp. 1063–1095, 1995.

[25] H.-T. Sun, F.-K. Chang, and X. Qing, "The response of composite joints with bolt-clamping loads, Part I: model development," *Journal of Composite Materials*, vol. 36, no. 1, pp. 47–67, 2002.

[26] H.-T. Sun, F.-K. Chang, and X. Qing, "The response of composite joints with bolt-clamping loads, Part II: model verification," *Journal of Composite Materials*, vol. 36, no. 1, pp. 69–92, 2002.

[27] X. Qing, H. T. Sun, L. Dagba et al., *Damage-Tolerance-Based Design of Bolted Composite Joints*, American Society for Testing and Materials, West Conshohocken, Pa, USA, 2000.

[28] J. W. Lee and I. M. Daniel, "Progressive transverse cracking of cross ply composite laminates," *Journal of Composite Materials*, vol. 24, no. 11, pp. 1225–1243, 1990.

[29] Y. W. Kim and C. S. Hong, "Progressive failure model for the analysis of laminated composites based on finite element approach," *Journal of Reinforced Plastics and Composites*, vol. 11, no. 10, pp. 1078–1092, 1992.

[30] S. G. Lim and C. S. Hong, "Prediction of transverse cracking and stiffness reduction in cross-ply laminated composites," *Journal of Composite Materials*, vol. 23, no. 7, pp. 695–713, 1989.

[31] W.-P. Lin and H.-T. Hu, "Nonlinear analysis of fiber-reinforced composite laminates subjected to uniaxial tensile load," *Journal of Composite Materials*, vol. 36, no. 12, pp. 1429–1450, 2002.

[32] W.-P. Lin and H.-T. Hu, "Parametric study on the failure of fiber-reinforced composite laminates under biaxial tensile load," *Journal of Composite Materials*, vol. 36, no. 12, pp. 1481–1503, 2002.

[33] T.-K. Hwang, C.-S. Hong, and C.-G. Kim, "Probabilistic deformation and strength prediction for a filament wound pressure vessel," *Composites Part B: Engineering*, vol. 34, no. 5, pp. 481–497, 2003.

[34] M. J. Hinton and A. S. Kaddour, "The background to the second world-wide failure exercise," *Journal of Composite Materials*, vol. 46, no. 19-20, pp. 2283–2294, 2012.

[35] L. Raimondo, L. Iannucci, P. Robinson, and P. T. Curtis, "A progressive failure model for mesh-size-independent FE analysis

of composite laminates subject to low-velocity impact damage," *Composites Science and Technology*, vol. 72, no. 5, pp. 624–632, 2012.

[36] S. T. Pinho, L. Iannucci, and P. Robinson, "Physically based failure models and criteria for laminated fibre-reinforced composites with emphasis on fibre kinking. Part II. FE implementation," *Composites Part A: Applied Science and Manufacturing*, vol. 37, no. 5, pp. 766–777, 2006.

[37] L. Zhao, T. Qin, J. Zhang, and R. A. Shenoi, "Modified maximum stress failure criterion for composite π joints," *Journal of Composite Materials*, vol. 47, no. 23, pp. 2995–3008, 2013.

[38] I. M. Daniel and O. Ishai, *Engineering Mechanics of Composite Materials*, Oxford University Press, Oxford, UK, 1994.

[39] N. A. Fleck and D. Liu, "Microbuckle initiation from a patch of large amplitude fibre waviness in a composite under compression and bending," *European Journal of Mechanics—A/Solids*, vol. 20, no. 1, pp. 23–37, 2001.

[40] C. R. Schultheisz and A. M. Waas, "Compressive failure of composites. Part I. Testing and micromechanical theories," *Progress in Aerospace Sciences*, vol. 32, no. 1, pp. 1–42, 1996.

[41] A. Puck and H. Schürmann, "Failure analysis of FRP laminates by means of physically based phenomenological models," *Composites Science and Technology*, vol. 58, no. 7, pp. 1045–1067, 1998.

[42] A. Puck and H. Schürmann, "Failure analysis of FRP laminates by means of physically based phenomenological models," *Composites Science and Technology*, vol. 62, no. 12-13, pp. 1633–1662, 2002.

[43] P. D. Soden, M. J. Hinton, and A. S. Kaddour, "A comparison of the predictive capabilities of current failure theories for composite laminates," *Composites Science and Technology*, vol. 58, no. 7, pp. 1225–1254, 1998.

[44] L. B. Zhao, T. L. Qin, Y. L. Chen, and J. Y. Zhang, "Three-dimensional progressive damage models for cohesively bonded composite π joint," *Journal of Composite Materials*, vol. 48, no. 6, pp. 707–721, 2014.

[45] L. B. Zhao, Y. Gong, J. Y. Zhang, Y. L. Chen, and B. J. Fei, "Simulation of delamination growth in multidirectional laminates under mode I and mixed mode I/II loadings using cohesive elements," *Composite Structures*, vol. 116, pp. 509–522, 2014.

[46] L. B. Zhao, Y. Gong, T. L. Qin, S. Mehmood, and J. Y. Zhang, "Failure prediction of out-of-plane woven composite joints using cohesive element," *Composite Structures*, vol. 106, pp. 407–416, 2013.

[47] Z. P. Bažant and B. H. Oh, "Crack band theory for fracture of concrete," *Materials and Structures*, vol. 16, no. 3, pp. 155–177, 1983.

[48] S. T. Pinho, P. Robinson, and L. Iannucci, "Fracture toughness of the tensile and compressive fibre failure modes in laminated composites," *Composites Science and Technology*, vol. 66, no. 13, pp. 2069–2079, 2006.

[49] H.-J. Park, "Effects of stacking sequence and clamping force on the bearing strengths of mechanically fastened joints in composite laminates," *Composite Structures*, vol. 53, no. 2, pp. 213–221, 2001.

[50] J. H. Oh, Y. G. Kim, and D. G. Lee, "Optimum bolted joints for hybrid composite materials," *Composite Structures*, vol. 38, no. 1–4, pp. 329–347, 1997.

[51] R. M. Christensen, "Tensor transformation and failure criteria for analysis of fiber composite materials," *Journal of Composite Materials*, vol. 22, no. 9, pp. 874–897, 1988.

22

Activated Carbon-Fly Ash-Nanometal Oxide Composite Materials: Preparation, Characterization, and Tributyltin Removal Efficiency

Olushola S. Ayanda,[1] Olalekan S. Fatoki,[1] Folahan A. Adekola,[2] and Bhekumusa J. Ximba[1]

[1] *Department of Chemistry, Faculty of Applied Sciences, Cape Peninsula University of Technology, P.O. Box 1906, Bellville 7535, South Africa*
[2] *Department of Chemistry, University of Ilorin, PMB 1515, Ilorin 240004, Nigeria*

Correspondence should be addressed to Olushola S. Ayanda; osayanda@gmail.com

Academic Editor: Alberto Ritieni

The physicochemical properties, nature, and morphology of composite materials involving activated carbon, fly ash, nFe_3O_4, $nSiO_2$, and nZnO were investigated and compared. Nature and morphology characterizations were carried out by means of scanning electron and transmission electron microscopy, X-ray diffraction, and Fourier transform infrared spectroscopy. Other physicochemical characterizations undertaken were CNH analysis, ash content, pH, point of zero charge, and surface area and porosity determination by BET. Experimental results obtained revealed that activated carbon, $nSiO_2$, activated carbon-fly ash, activated carbon-fly ash-nFe_3O_4, activated carbon-fly ash-$nSiO_2$, and activated carbon-fly ash-nZnO composite materials exhibited net negative charge on their surfaces while fly ash, nFe_3O_4, and nZnO possessed net positive charge on their surfaces. Relatively higher removal efficiency (>99%) of TBT was obtained for all the composite materials compared to their respective precursors except for activated carbon. These composite materials therefore offer great potential for the remediation of TBT in wastewaters.

1. Introduction

Fly ash, generated during the combustion of coal for energy production consists of fine, powdery particles predominantly spherical in shape, either solid or hollow, and mostly glassy (amorphous) in nature [1]. Fly ash has been proposed as a good adsorbent for NOx, SOx, and mercury removal from flue gases as well as adsorption of organic gas [1, 2]. Fly ash has a potential application in wastewater treatment because of its major chemical components, which are alumina, silica, ferric oxide, calcium oxide, magnesium oxide, and carbon, and its physical properties such as porosity, particle size distribution, and surface area. Hence, it has been used as a low-cost adsorbent for the removal of heavy metals [3–6], dyes [7], phenolic compounds [8], and humic acids [9] in wastewaters. Activated carbon, on the other hand, is also widely used in a variety of areas, namely, as an adsorbent in air and water pollution control, a catalyst in the chemical and petrochemical industries, and a purifier in the food and pharmaceutical industries [10]. In the water treatment field, activated carbon is often used as an adsorbent for the removal of various synthetic and naturally occurring organic chemicals in drinking water [11]. In wastewater treatment, activated carbon is a powerful adsorbent because of its large surface area and pore volume, which allows the removal of liquid-phase contaminants, including organic compounds, heavy metal ions, and colors. Adsorption on activated carbons has been investigated extensively due to their use in many applications including the removal of toxic volatile organic compounds (VOCs) and pollutants in water and air [12–15].

Within the last few years, intensive wide spread contamination of the atmosphere and surface water related to adverse industrial operations has been of great concern and call for the development of better adsorbents. Many researchers have therefore focused on the search for better adsorbents with very high adsorption capacities, use of nanometal oxides as adsorbents, and the surface modification of existing adsorbents. Limited work was thus reported on the use of composite materials for wastewater treatments. Some of the reported works in this area is by Zhang et al. [10] who reported the preparation of $CuFe_2O_4$/activated carbon magnetic adsorbents with mass ratio of 1 : 1, 1 : 1.5, and 1 : 2 for the adsorption of acid orange II (AO7) in water and subsequent separation of adsorbent from the medium by a magnetic technique. Their results suggest that the composite has much higher catalytic activity than that of activated carbon, and this is attributed to the presence of $CuFe_2O_4$. Shukla et al. [14] studied the synthesis of composites of carbon and natural zeolite with varying amounts of carbon as prospective adsorbents to adsorb organic contaminants from waste water such as phenol. They reported that the adsorption isotherm indicated an enhanced adsorption of phenol on the composites as compared with the natural zeolite, and that adsorption increased with increase in carbon content of the composite materials. The adsorption and degradation of trichloroethylene (TCE) through dechlorination using synthetic granular activated carbon and zerovalent iron (GAC-ZVI) composites was reported by Tseng et al. [16]. They reported that the usage of granular activated carbon-zerovalent iron composites liberated a greater amount of Cl than when zerovalent iron was used alone. Jha et al. [17] also investigated the preparation of composite materials of activated carbon, and zeolite by activating coal fly ash by fusion and reported that the composites of activated carbon and zeolite proved to be suitable for the uptake of toxic metal ions.

Researches have therefore focused on the enhancement of the effectiveness of activated carbon and fly ash by modifying their specific properties by chemical modification (treatment with acids or bases), thermal activation, impregnation, and/or surfactant modification [18–21] in order to enable the carbon to develop affinity for certain contaminants. No work has been reported on the preparation of composite materials involving activated carbon, fly ash, nFe_3O_4, $nSiO_2$, and nZnO as precursors, except for Fatoki et al. [22] who reported the preparation and characterization of activated carbon-nFe_3O_4, activated carbon-$nSiO_2$, and activated carbon-nZnO hybrid materials. Composite materials involving activated carbon, fly ash, and nanometal oxides are expected to have high adsorption capacity due to their nature, morphology, and properties, and due to the presence of nanooxides in the composite materials; it is also expected that the remediation mechanism by these materials will combine the synergistic effect of adsorption and oxidation during the adsorption processes and not adsorption alone.

The aim of this study is, therefore, to prepare activated carbon, fly ash, and nanometal oxide composite materials capable of enhancing the adsorption of pollutants from wastewaters and to carry out a detailed characterization of these materials in order to understand the properties that will be of great importance to environmental management.

2. Experimental

2.1. Materials. Fly ash from Matla power station, Mpumalanga, South Africa was used in this study. Matla power station was the first of the giant 3 600 MW coal-fired power stations in South Africa and was fully operational in 1983 [23]. Activated carbon (100–400 mesh), iron (II, III), oxide nanopowder (particle size < 50 nm), silica nanopowder (particle size 12 nm), zinc oxide NanoGard (particle size 40–100 nm APS powder), acetic acid, hexane, sodium acetate, $NaBEt_4$, and tributyltin chloride (TBT) were purchased from Sigma-Aldrich, USA. Sodium nitrate ($NaNO_3$) and potassium bromide (KBr) were supplied by Merck, Germany, while methanol was supplied by Industrial Analytical, South Africa. Milli-Q water was used for all analytical preparations.

2.2. Preparation of Composite Materials. Activated carbon, fly ash, and nanometal oxides in the ratio 1 : 1 : 1 were dispersed in 0.5 M HCl to form slurries. The slurries were stirred by means of a stirrer and evaporated to dryness in an oven. The composite materials obtained were washed with Milli-Q water, filtered, further dried in an oven at 100°C for 24 hours, and ground to fine powder using agate mortar and pestle [22, 24].

2.3. Instrumentation. FEI scanning electron microscope (Nova Nano SEM 230) and transmission electron microscope (TECNAI G^2 20) were used for the SEM and TEM analyses of the precursors and composite materials. Fourier transmission infrared (FTIR) absorption spectra of the precursors and composite materials were obtained by using the potassium bromide (KBr) pellet method of sample preparation and Perkin Elmer Spectrum 1000 instrument for analysis. Euro Ea elemental analyzer was use for carbon, nitrogen, and hydrogen (CNH) analyses. Phase identification of the precursors and activated carbon-fly ash-nanometal oxide composite materials were determined by X-ray diffractometry using a PANalytical PW 3830 diffractometer, while TriStar 3000 analyser (Micromeritics Instrument Corporation) was used for surface area and porosity determination.

2.4. pH and Point of Zero Charge (PZC) Determination. The pH was determined by gently boiling 50 mL of Milli-Q water in a flask containing 0.1 g of the samples for 5 mins. The pH was measured using a Mettler Toledo pH meter after the solution was cooled to room temperature. Mass titration technique was used to determine the PZC [22, 25]. Increasing amounts of sample from 0 to 2 g were added to 10 mL of 0.01 M $NaNO_3$ solution. The resulting pH of each suspension was measured after 24 hours. The pH plateau for the highest concentrations of solid in a successive series of mass titrations is taken as the PZC.

2.5. Ash Content Determination. Approximately ±0.1 g of the precursors and composite materials were measured into crucibles and heated in a muffle furnace at a temperature of about 500–600°C for 4 hours. The samples were withdrawn from the furnace after ashing, allowed to cool in a desiccator, and then reweighed. The ash content of the precursors and composite materials was calculated by difference, and this process was carried out in triplicate.

(a) (b)

FIGURE 1: (a) SEM of activated carbon. (b) TEM of activated carbon.

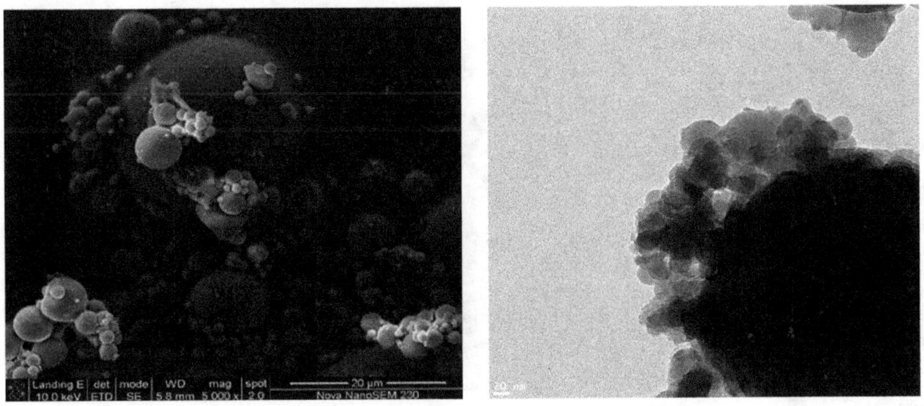

(a) (b)

FIGURE 2: (a) SEM of fly ash. (b) TEM of fly ash.

2.6. TBT Removal Efficiency and Analysis. The removal efficiency of TBT by these materials was tested by applying overall optimal conditions for the adsorption of TBT from TBT-contaminated artificial seawater. The removal efficiency (R) is defined as

$$R = \left(\frac{c_i - c_e}{c_i}\right) \times 100, \quad (1)$$

where c_i is the initial concentration of TBT (100 mg/L) in artificial seawater placed in a conical flask and shaken at 200 rpm for 60 min with 0.5 g of the adsorbents, and c_e is the equilibrium solution concentration.

The concentration of TBT was determined after derivatization by the addition of 2 mL of acetate buffer (pH = 4.5) and 1.0 mL of 1% $NaBEt_4$ and extraction into hexane by horizontal shaking in a separation funnel. The extracts were reduced to 1 mL and analyzed by the use of GC-FPD (Shimadzu GC-2010 Plus) with a capillary column HP 5 (5% phenyl methyl siloxane, 30 m × 0.25 mm, i.d., film thickness 0.25 μm), and the temperature was programmed as follows: initially at 60°C hold for 1 min, then heated to 280°C at 10°C/min, and hold for 4 min. The injection and detector temperatures were 270°C and 300°C, respectively, and the carrier gas was high purity helium.

A plot of the percent removal of TBT by the various adsorbents was obtained, and the results were compared.

3. Results and Discussion

3.1. SEM and TEM. The scanning electron micrograph (SEM) and transmission electron micrograph (TEM) of activated carbon (Figures 1(a) and 1(b)) showed that activated carbon exhibit aggregated irregular surfaces with a large number of micropores and crevices of various sizes at the surface. The SEM and TEM of activated carbon confirmed that the activated carbon is a better adsorbent for the adsorption of pollutants from wastewaters.

Figure 2(a) showed that the particles of Matla fly ash are spherical with smooth and regular surfaces. The TEM of fly ash (Figure 2(b)) presents agglomeration of different particle sizes. The fly ash particles showed different size distributions with spherical shapes.

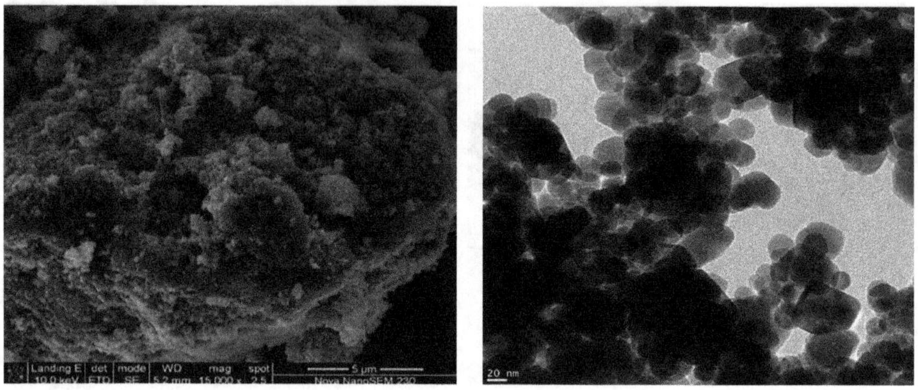

(a) (b)

Figure 3: (a) SEM of fly nFe_3O_4. (b) TEM of nFe_3O_4.

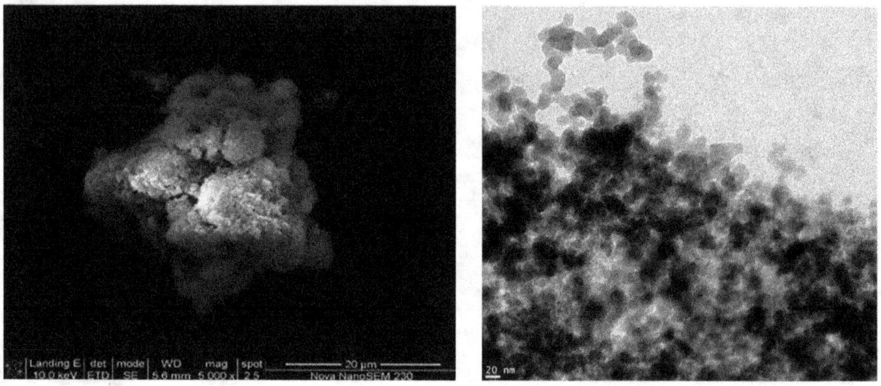

(a) (b)

Figure 4: (a) SEM of $nSiO_2$. (b) TEM of $nSiO_2$.

The SEM of nFe_3O_4 (Figure 3(a)) showed that nFe_3O_4 consists of agglomerated globules with irregular and rough surfaces. The TEM of nFe_3O_4 (Figure 3(b)) presents agglomeration of particles. The TEM thus showed that nFe_3O_4 is made up of different shapes including square, spherical, and hexagonal shapes.

The SEM of $nSiO_2$ (Figure 4(a)) showed that $nSiO_2$ exhibit agglomerated irregular surfaces with a large number of micropores and a few voids and crevices, while the TEM of $nSiO_2$ (Figure 4(b)) showed a bimodal distribution of particles size.

The SEM and TEM of nZnO (Figures 5(a) and 5(b)) showed that nZnO particles consist of nonuniform granules and more regular surfaces. The TEM of nZnO (Figure 5(b)) confirmed the various shapes and sizes of nZnO particles.

Figure 6(a) showed that activated carbon-fly ash composite material is made up of smooth surface materials (activated carbon) and spherical materials (fly ash) deposited at various position throughout the surfaces of the activated carbon. The TEM (Figure 6(b)) showed that the activated carbon (irregular surfaces) was aggregated with the spherical particle of fly ash.

The SEM and TEM (Figures 6(a) and 6(b)) showed that the fly ash particles maintained their spherical morphology after the preparation of activated carbon-fly ash composite material.

The SEM and TEM of activated carbon-fly ash-nFe_3O_4 composite material (Figures 7(a) and 7(b)) showed that the composite material exhibit aggregated irregular surfaces with large number of micropores and crevices at the surface. Fly ash and nFe_3O_4 were found at the surface of the activated carbon.

The SEM and TEM of activated carbon-fly ash-$nSiO_2$ composite material (Figures 8(a) and 8(b)) showed that the composite material also exhibited aggregated irregular surfaces with large number of micropores and crevices at the surface. The $nSiO_2$ and fly ash were distributed at the surface of the activated carbon.

The SEM of activated carbon-fly ash-nZnO composite material (Figure 9(a)) showed that the activated carbon, fly ash, and nZnO particles were fused together. Large intergranular voids and crevices were associated with the activated carbon-fly ash-nZnO composite material with fly ash still maintaining its spherical regular shape.

(a) (b)

FIGURE 5: (a) SEM of nZnO. (b) TEM of nZnO.

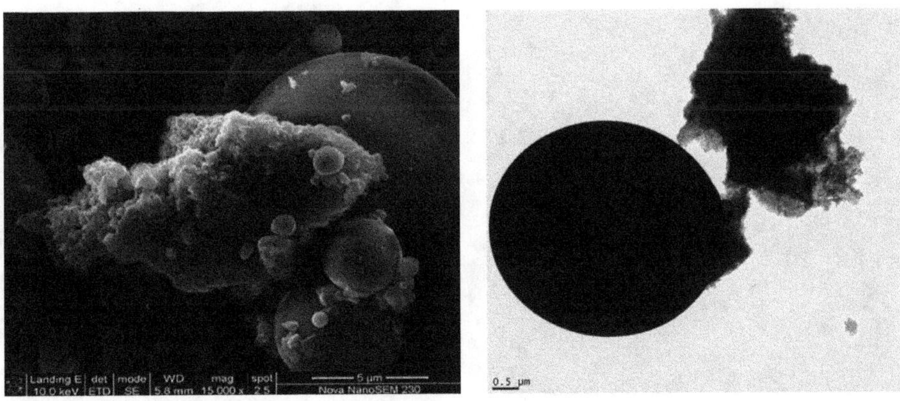

(a) (b)

FIGURE 6: (a) SEM of activated carbon-fly ash composite material. (b) TEM of activated carbon-fly ash composite material.

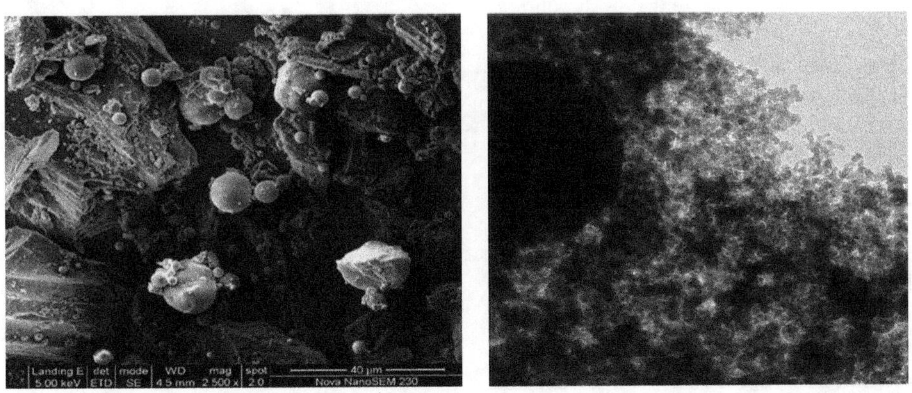

(a) (b)

FIGURE 7: (a) SEM of activated carbon-fly ash-nFe_3O_4 composite material. (b) TEM of activated carbon-fly ash-nFe_3O_4 composite material.

The TEM of activated carbon-fly ash-nZnO composite material (Figure 9(b)) thus showed a clustered activated carbon, fly ash, and nZnO composite material with large intergranular voids and crevices.

3.2. FTIR Absorption Spectra. In the FTIR spectrum of activated carbon, fly ash, and activated carbon-fly ash composite material (Figure 10), the absorption at 1616 cm^{-1} (curve (a)) is assigned to the C=C stretching of activated carbon

(a) (b)

FIGURE 8: (a) SEM of activated carbon-fly ash-nSiO_2 composite material. (b) TEM of activated carbon-fly ash-nSiO_2 composite material.

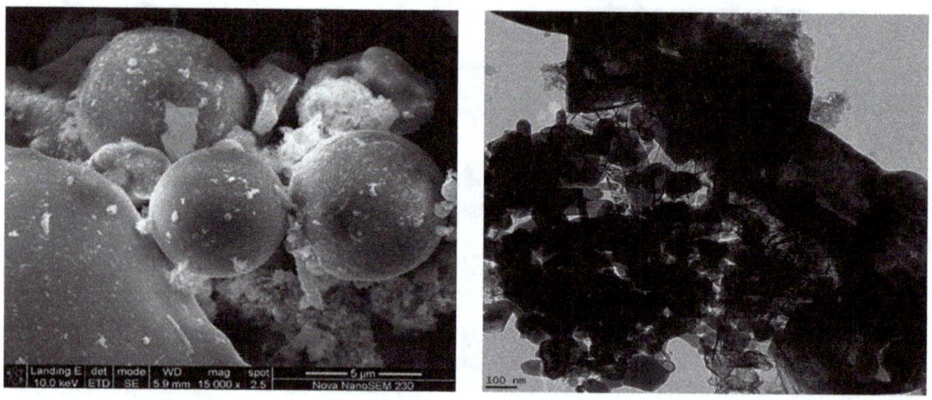

(a) (b)

FIGURE 9: (a) SEM of activated carbon-fly ash-nZnO composite material. (b) TEM of activated carbon-fly ash-nZnO composite material.

[26, 27], while the absorption at 1097 cm^{-1} (curve (b)) is assigned to the C–C stretching of fly ash. It was found that the wavenumber of C–C stretching of fly ash changed slightly from 1097 cm^{-1} of fly ash to 1093 cm^{-1} (curve (f)) of the activated carbon-fly ash composite material. The wavenumber of the absorption peak decreased by 4 cm^{-1}. The slight change in the wavenumber suggests that a new bond was formed during the preparation of the activated carbon-fly ash composite material. In the FTIR spectrum of activated carbon, fly ash, nFe_3O_4, and activated carbon-fly ash-nFe_3O_4 composite material (Figure 11), the absorption at 1616 cm^{-1} (curve (a)) is assigned to the C=C stretching of activated carbon, and the absorption at 1097 cm^{-1} (curve (b)) is assigned to the C–C stretching of fly ash, while the absorption at 586 cm^{-1} (curve (c)) is assigned to the Fe–O stretching of nFe_3O_4. It was found that the wavenumber of Fe–O stretching changed from 586 cm^{-1} of nFe_3O_4 to 560 cm^{-1} (curve (g)) of the activated carbon-fly ash-nFe_3O_4 composite material. The wavenumber of the absorption peak decreased by 26 cm^{-1}. Decrease in the wavenumber suggests

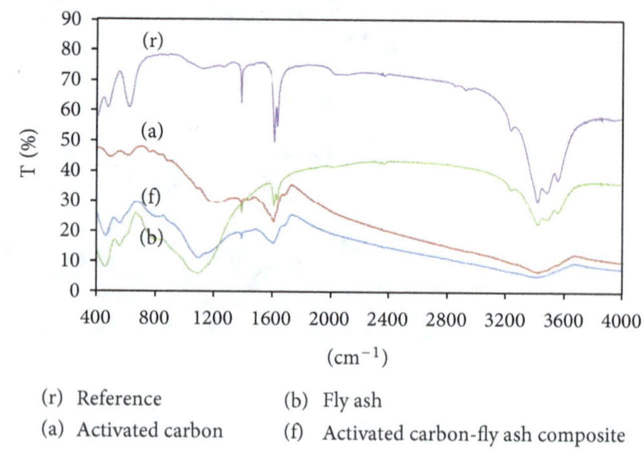

FIGURE 10: FTIR spectrum of precursors and activated carbon-fly ash composite material.

that a new bond was formed during the preparation of the activated carbon-fly ash-nFe_3O_4 composite material.

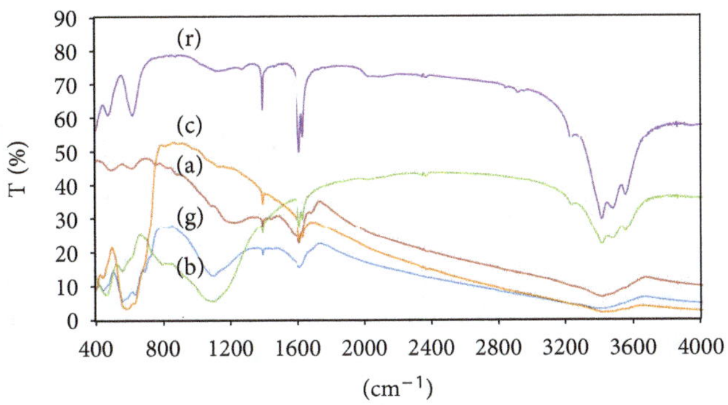

FIGURE 11: FTIR spectrum of precursors and activated carbon-fly ash-nFe_3O_4 composite material.

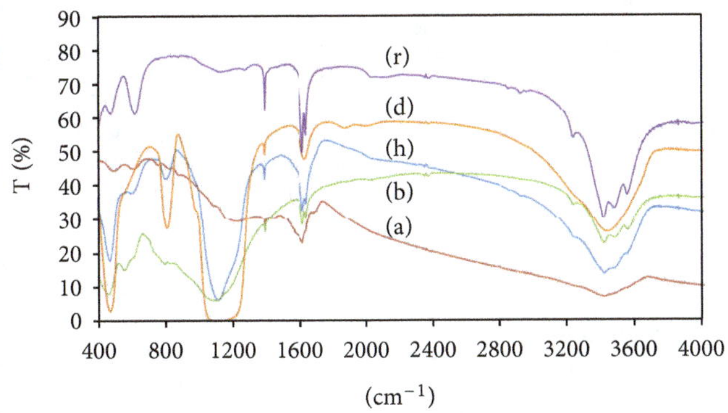

FIGURE 12: FTIR spectrum of precursors and activated carbon-fly ash-nSiO_2 composite material.

In the FTIR spectrum of activated carbon, fly ash, nSiO_2, and activated carbon-fly ash-nSiO_2 composite material (Figure 12), the absorption at 1616 cm^{-1} (curve (a)) is assigned to the C=C stretching of activated carbon, and the absorption at 1097 cm^{-1} (curve (b)) is assigned to the C–C stretching of fly ash, while the absorption at 1101 cm^{-1} (curve (d)) is assigned to the asymmetric vibration of Si–O. The absorption at 809 cm^{-1} (curve (d)) is assigned to the symmetric vibration of Si–O [28]. It was found that the wavenumber of the symmetric vibration of Si–O changed from 809 cm^{-1} of nSiO_2 to 805 cm^{-1} (curve (h)) of the activated carbon-fly ash-nSiO_2 composite material. The wavenumber of the absorption peak decreased by 4 cm^{-1}. A decrease in the wavenumber suggests that a new bond was formed during the preparation of the activated carbon-fly ash-nSiO_2 composite material.

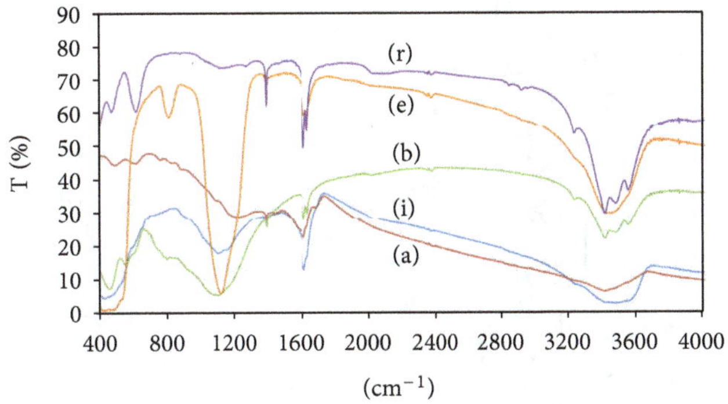

FIGURE 13: FTIR spectrum of precursors and activated carbon-fly ash-nZnO composite material.

In the FTIR spectrum of activated carbon, fly ash, nZnO, and activated carbon-fly ash-nZnO composite material (Figure 13), the absorption at 1616 cm^{-1} (curve (a)) is assigned to the C=C stretching of activated carbon, and the absorption at 1097 cm^{-1} (curve (b)) is assigned to the C–C stretching of fly ash, while the absorption at 1110 cm^{-1} (curve (e)) is assigned to the asymmetry vibration of Zn–O, and the absorption at 808 cm^{-1} (curve (e)) is assigned to the Zn–O stretching of nZnO. It was found that the wavenumber of Zn–O vibration changed from 1110 cm^{-1} of nZnO to 1094 cm^{-1} (curve (i)) of the activated carbon-fly ash-nZnO composite material. The wavenumber of the absorption peak decreased by 16 cm^{-1}. Decrease in the wavenumber suggests that a new bond was formed during the preparation of the activated carbon-fly ash-nZnO composite material.

The result obtained thus shows that the shift in the band is a function of the metal ions present in the composite materials. The FTIR data also confirm the absence of impurity in both the precursors and the prepared composite materials.

3.3. Carbon, Nitrogen, and Hydrogen Content. Figure 14 showed that the activated carbon–fly ash–nFe_3O_4, activated carbon–fly ash–nSiO_2, activated carbon–fly ash–nZnO, and activated carbon–fly ash composite materials contained 29.34%, 34.04%, 30.69%, and 36.83% carbon content, respectively. Values of 0.91%, 0.26%, and 0.20% were recorded for the nitrogen content of activated carbon-fly ash-nFe_3O_4, activated carbon-fly ash-nSiO_2, and activated carbon-fly ash-nZnO, respectively, while the nitrogen content of activated carbon-fly ash composite material was below the detection limit. The activated carbon-fly ash, activated carbon-fly ash-nFe_3O_4, activated carbon-fly ash-nSiO_2, and activated carbon-fly ash-ZnO thus contained 2.72%, 2.24%, 0.51%, and 0.89% hydrogen contents, respectively.

The result showed that the carbon content of activated carbon plays a dominant role in the carbon content of all the composite materials.

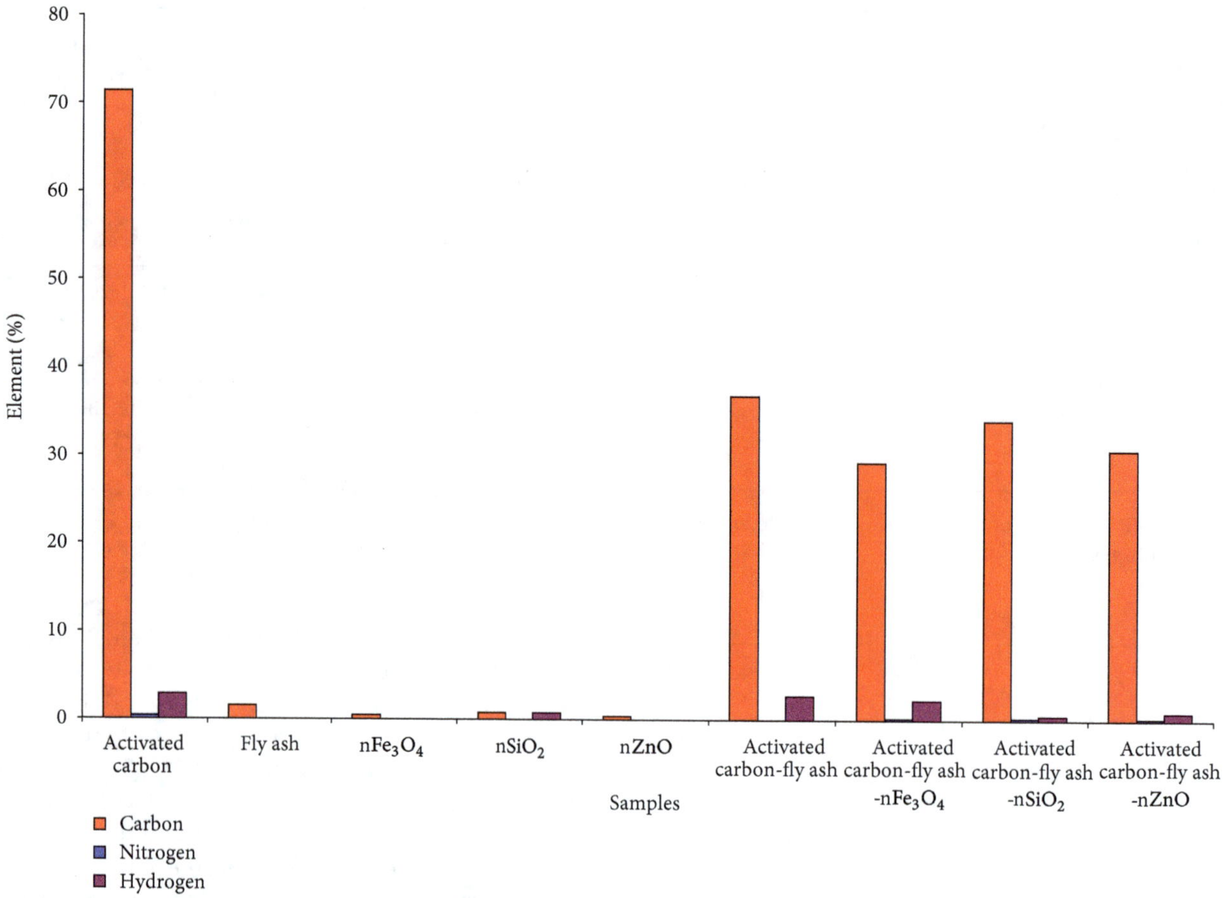

FIGURE 14: A plot of element (%) against the precursors and composite materials.

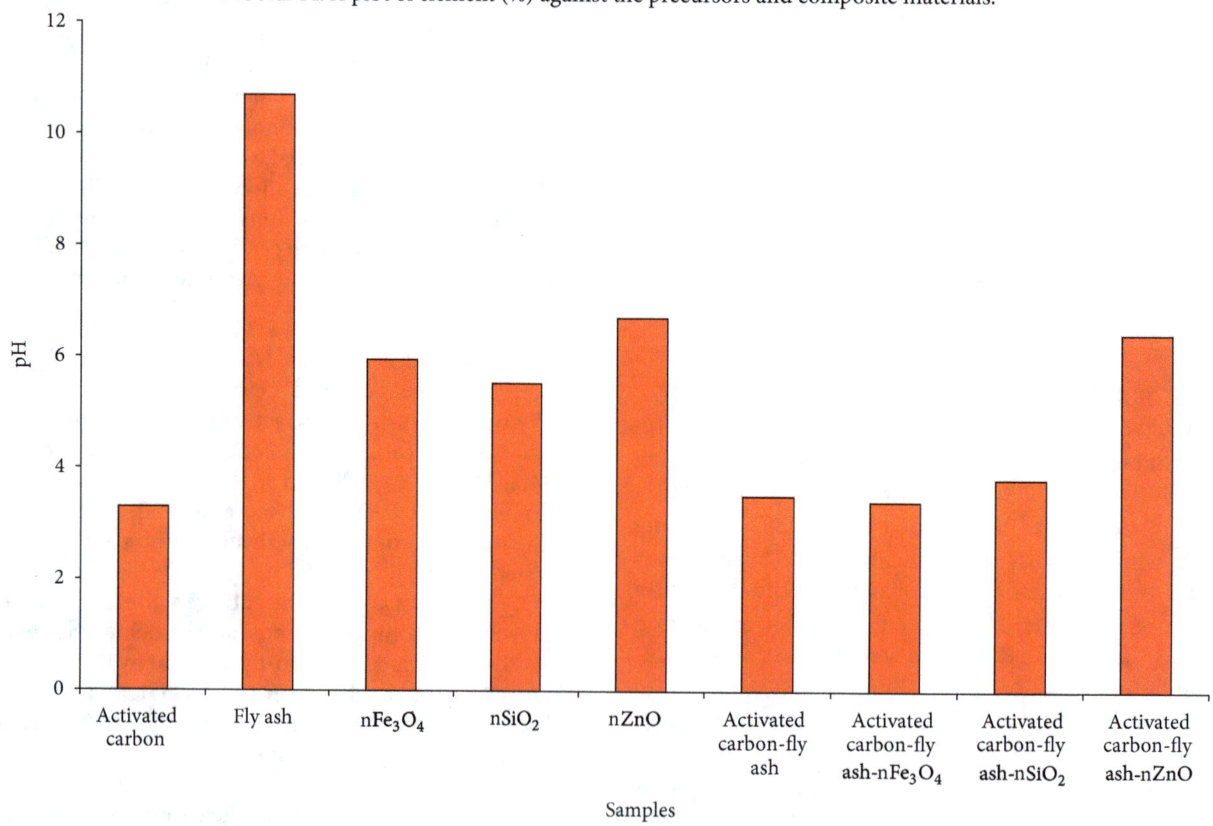

FIGURE 15: pH of activated carbon, fly ash, nano metal oxides, and composite materials.

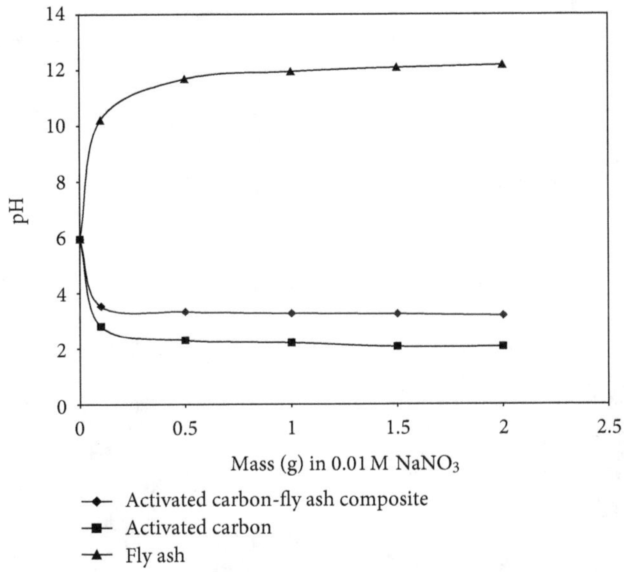

FIGURE 16: Result of mass titration experiments with activated carbon, fly ash, and activated carbon-fly ash composite material. Variation of pH versus mass of solid in 0.01 M $NaNO_3$.

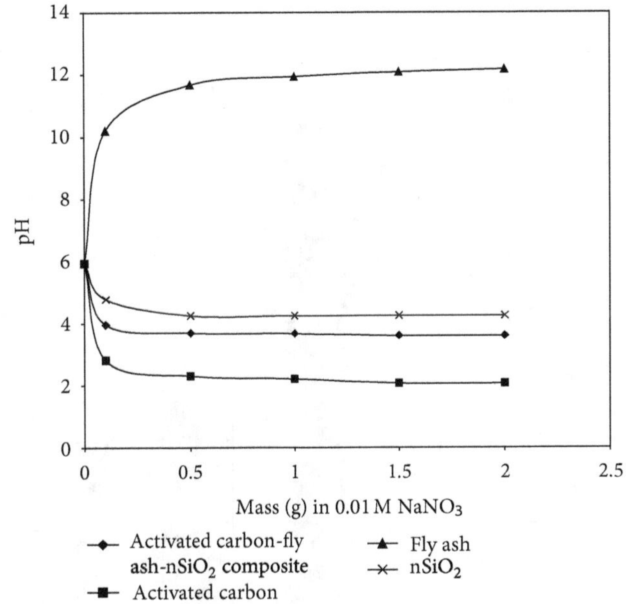

FIGURE 18: Result of mass titration experiments with activated carbon, fly ash, $nSiO_2$ and activated carbon-fly ash-$nSiO_2$ composite material. Variation of pH versus mass of solid in 0.01 M $NaNO_3$.

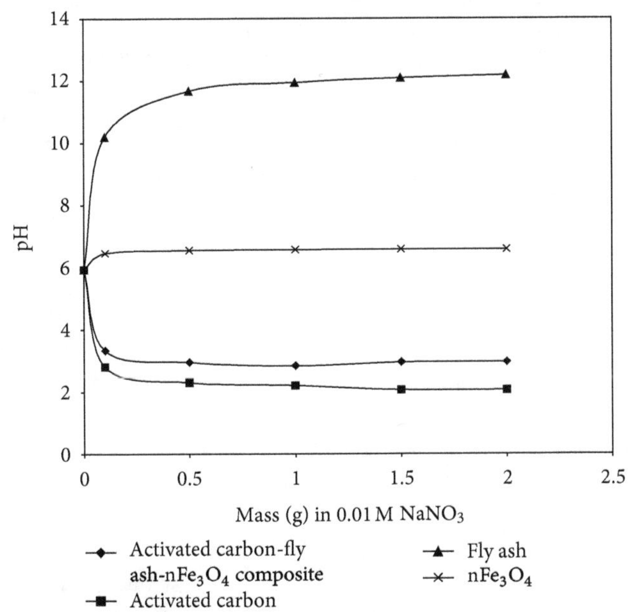

FIGURE 17: Result of mass titration experiments with activated carbon, fly ash, nFe_3O_4, and activated carbon-fly ash-nFe_3O_4 composite material. Variation of pH versus mass of solid in 0.01 M $NaNO_3$.

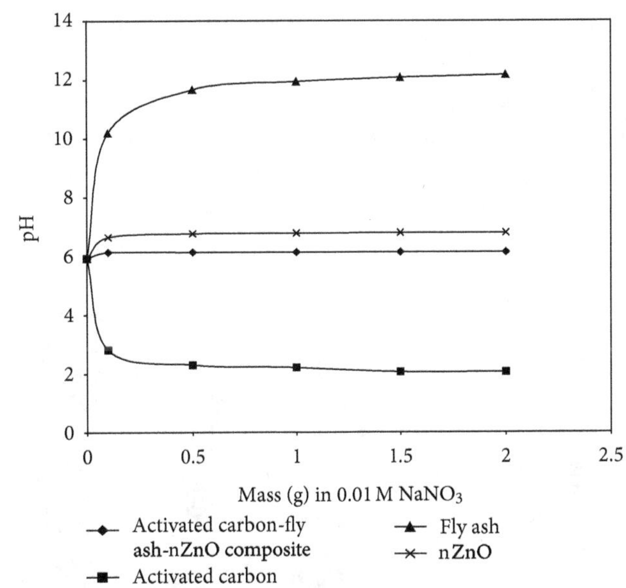

FIGURE 19: Result of mass titration experiments with activated carbon, fly ash, nZnO, and activated carbon-fly ash-nZnO composite material. Variation of pH versus mass of solid in 0.01 M $NaNO_3$.

3.4. pH and Point of Zero Charge (PZC) Measurement. From Figure 15, the preparation of activated carbon-fly ash composite material using activated carbon (pH 3.3) and fly ash (pH 10.70) as precursors resulted in activated carbon-fly ash composite material of pH 3.51. The pH was higher than the pH of activated carbon by 59.8% and lower than the pH of fly ash by 67.2%. The preparation of activated carbon-fly ash-nFe_3O_4 composite material using activated carbon (pH 3.3), fly ash (pH 10.70), and nFe_3O_4 (pH 5.95) as precursors resulted to activated carbon-fly ash-nFe_3O_4 composite material of pH 3.41. The pH was higher than pH of activated carbon by 32.3%, lower than pH of fly ash by 68.1%, and lower than pH of nFe_3O_4 by 42.7%.

The preparation of activated carbon-fly ash-$nSiO_2$ composite material using activated carbon (pH 3.3), fly ash (pH 10.70), and $nSiO_2$ (pH 5.53) as precursors resulted in

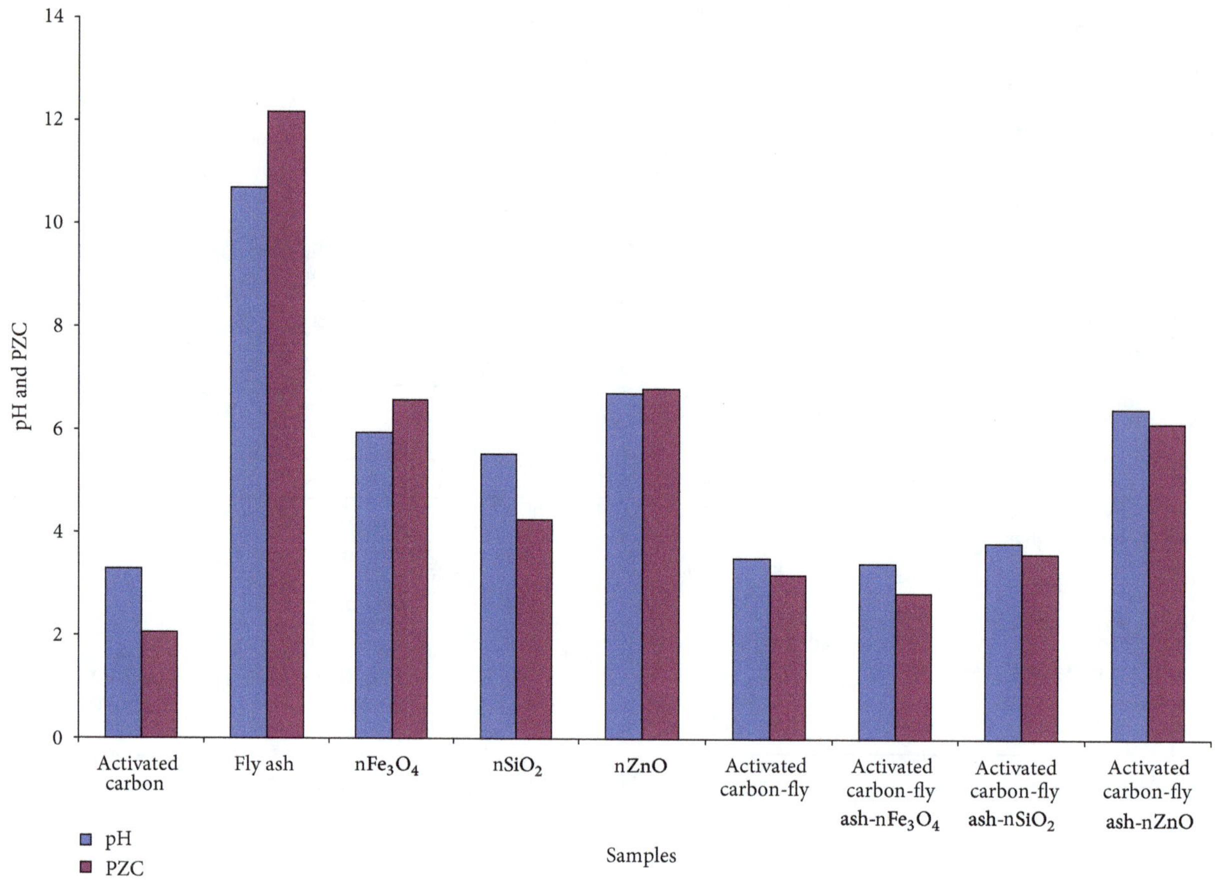

FIGURE 20: pH and PZC of activated carbon, fly ash, nanometal oxides, and composite materials.

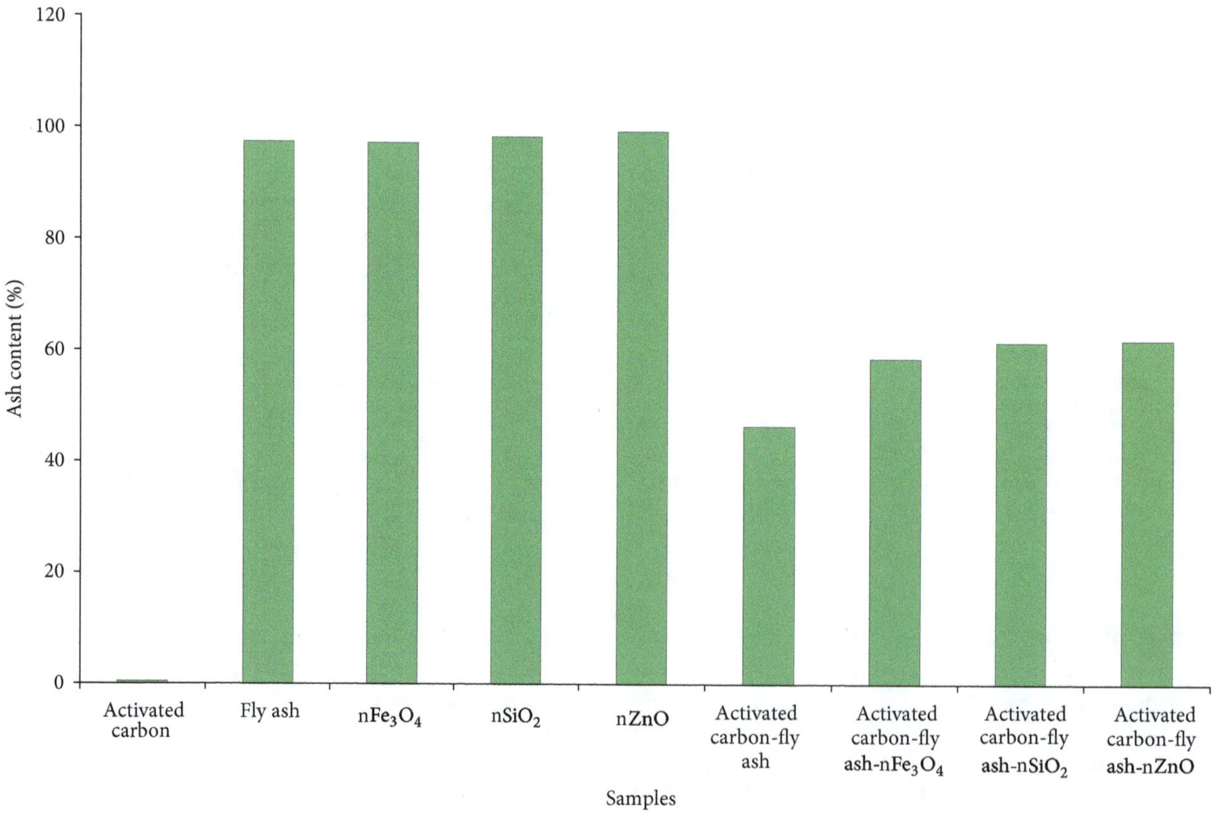

FIGURE 21: Ash content (%) versus activated carbon, fly ash, nanoparticles, and composite materials.

activated carbon-fly ash-$nSiO_2$ composite material of pH 3.34. The pH was higher than pH of activated carbon by 1.2%, lower than pH of fly ash by 68.8%, and lower than pH of $nSiO_2$ by 39.6%. The preparation of activated carbon-fly ash-nZnO composite material using activated carbon (pH 3.3), fly ash (pH 10.70), and nZnO (6.71) as precursors resulted to activated carbon-fly ash-nZnO composite material of pH 6.42. The pH was higher than pH of activated carbon by 48.6%, lower than pH of fly ash by 40.0%, and lower than pH of nZnO by 4.3%. The result obtained shows that the pH values of the composite materials were determined by the pH value of each of the precursors that made up the composite materials.

Figure 16 showed that the point of zero charge (PZC) of activated carbon, fly ash, and activated carbon-fly ash composite material are 2.06, 12.17, and 3.19, respectively. The PZC of activated carbon-fly ash composite material was higher than PZC of activated carbon by 35.42%, but lower than the PZC of fly ash by 73.79%. The graph showed that the presence of fly ash (high PZC value and basic) in the activated carbon (acidic) raised the PZC of activated carbon to form activated carbon-fly ash composite material of PZC of 3.19.

Figure 17 showed that the PZC of activated carbon, fly ash, nFe_3O_4, and activated carbon-fly ash-nFe_3O_4 composite material are 2.06, 12.17, 6.58, and 2.84, respectively. The PZC of activated carbon-fly ash-nFe_3O_4 composite material was higher than the PZC of activated carbon by 27.46%, lower than PZC of fly ash by 76.66%, and also lower than the PZC of nFe_3O_4 by 56.84%.

From Figure 18, the PZC of activated carbon, fly ash, $nSiO_2$, and activated carbon-fly ash-$nSiO_2$ composite material are 2.06, 12.17, 4.25, and 3.60, respectively. The PZC of activated carbon-fly ash-$nSiO_2$ composite material was therefore higher than the PZC of activated carbon by 42.78%, lower than PZC of fly ash by 70.42%, and also lower than the PZC of $nSiO_2$by 15.29%.

From Figure 19, the PZC of activated carbon, fly ash, nZnO, and activated carbon-fly ash-nZnO composite material are 2.06, 12.17, 6.80, and 6.14, respectively. The PZC of activated carbon-fly ash-nZnO composite material was therefore higher than the PZC of activated carbon by 66.45%, lower than PZC of fly ash by 49.55%, and also lower than the PZC of nZnO by 9.71%.

Comparing the PZC values of the precursors and the composite materials, it could be concluded that it is not the presence of the nanoparticles alone that determines the PZC changes, but the PZC of each of the component precursors that made up the composite materials.

Figure 20 thus showed that the pH values of activated carbon, $nSiO_2$, activated carbon-fly ash, activated carbon-fly ash-nFe_3O_4, activated carbon-fly ash-$nSiO_2$, and activated carbon-fly ash-nZnO were slightly higher than their corresponding PZC values. This suggests that the surface of these materials is negatively charged and will therefore attract cations. The pH values of fly ash, nFe_3O_4, and nZnO are slightly lower than their corresponding PZC values, hence, their surface is positively charged and will attract anions.

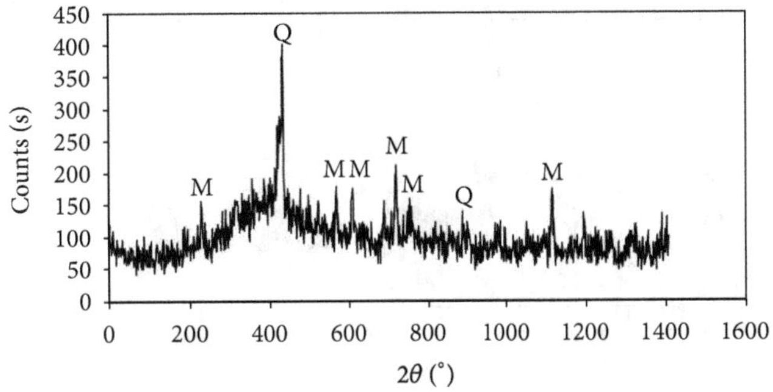

FIGURE 22: X-ray diffraction of activated carbon-fly ash composite material.

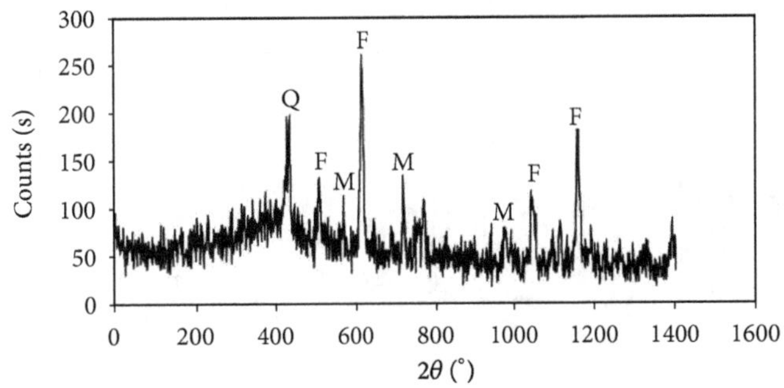

FIGURE 23: X-ray diffraction of activated carbon-fly ash-nFe_3O_4 composite material.

3.5. Ash Content. Figure 21 showed that the ash content of the activated carbon, fly ash, nFe_3O_4, $nSiO_2$, and nZnO is 0.45 ± 0.07%, 97.4 ± 0.14%, 97.2 ± 0.02%, 98.3 ± 0.07%, and 99.2 ± 0.14%, respectively, while 46.3 ± 0.14%, 58.5 ± 0.12%, 61.45 ± 0.07%, and 61.9 ± 0.14% were recorded as the ash contents of activated carbon-fly ash, activated carbon-fly ash-nFe_3O_4, activated carbon-fly ash-$nSiO_2$, and activated carbon-fly ash-nZnO composite materials, respectively.

The result showed that the percentage organic materials present in the activated carbon, fly ash, nFe_3O_4, $nSiO_2$, nZnO, activated carbon-fly ash, activated carbon-fly ash-nFe_3O_4, activated carbon-fly ash-$nSiO_2$ and activated carbon-fly ash-nZnO composite materials amount to 99.55%, 2.6%, 2.8%, 1.7%, 0.8%, 53.7%, 41.5%, 38.55%, and 38.1%, respectively. The result obtained in Figure 21 showed that the precursors have higher percentage of inorganic components as compared to the prepared composite materials except for activated carbon.

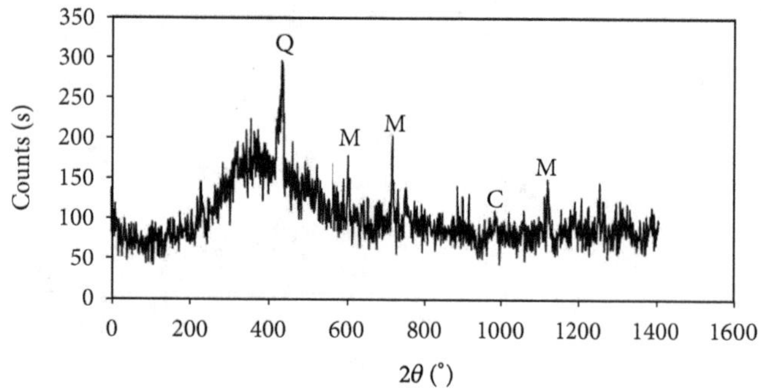

M: Mullite ($Al_6Si_2O_{13}$)
Q: Quartz (SiO_2)
C: Cristobalite (SiO_2)

FIGURE 24: X-ray diffraction of activated carbon-fly ash-nSiO_2 composite material.

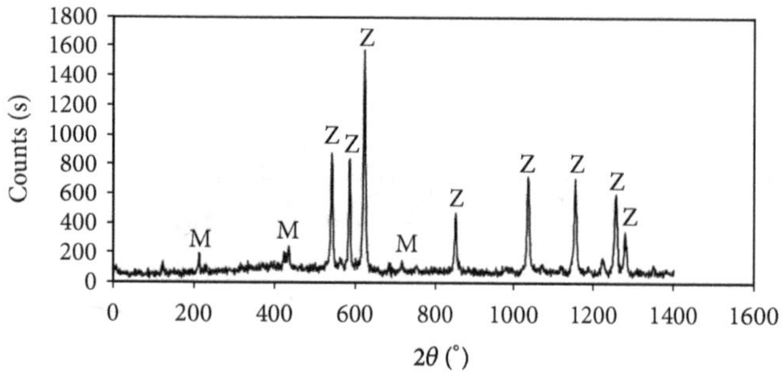

M: Mullite ($Al_6Si_2O_{13}$)
Q: Quartz (SiO_2)
Z: Zinc oxide (nZnO)

FIGURE 25: X-ray diffraction of activated carbon-fly ash-nZnO composite material.

3.6. X-Ray Diffraction. The diffractogram of activated carbon shows the absence of crystalline substances, while the fly ash is dominated mainly by crystalline minerals mullite and quartz with large characteristic peaks of quartz (SiO_2) as reported by Fatoki et al. [22] and Ayanda et al. [23], respectively. The x-ray diffractograms of nFe_3O_4, nSiO_2, and nZnO have also been reported by Fatoki et al. [22].

Figures 22 to 25 thus show the X-ray diffractograms of activated-fly ash, activated carbon-fly ash-nFe_3O_4, activated carbon-fly ash-nSiO_2, and activated carbon-fly ash-nZnO composite materials.

The diffractogram of activated carbon-fly ash (Figure 22) showed that the crystalline minerals mullite and quartz of fly ash are dominant. The X-ray diffractogram of activated carbon-fly ash-nFe_3O_4 composite material (Figure 23) consists of mullite ($Al_6Si_2O_{13}$), quartz (SiO_2), and magnetite (Fe_3O_4).

The x-ray diffractogram of activated carbon-fly ash-nSiO_2 composite material (Figure 24) consists of mullite ($Al_6Si_2O_{13}$), quartz (SiO_2), and cristobalite (SiO_2), while the X-ray diffractogram of activated carbon-fly ash-nZnO composite material (Figure 25) consists of mullite ($Al_6Si_2O_{13}$), quartz (SiO_2), and zinc oxide (nZnO).

All the diffractograms obtained showed defined characteristic peaks corresponding to the mineral constituents of the precursors and the composite materials. This showed that the precursors and all the prepared composite materials are pure.

3.7. Surface Area and Porosity Determination. Results obtained on the Brunauer, Emmett, and Teller (BET) surface area and porosity determinations of activated carbon-fly ash-nanometal oxide composite materials as well as their precursors are shown in Table 1 and Figure 26.

The surface areas of fly ash, activated carbon, nFe_3O_4, nSiO_2, and nZnO are 1.06 ± 0.003, 1156 ± 8.69, 37 ± 0.19, 217 ± 1.76, and 14 ± 0.039 m^2/g, respectively, while the surface areas of activated carbon-fly ash, activated carbon-fly ash-nFe_3O_4, activated carbon-fly ash-nSiO_2, and activated carbon-fly ash-nZnO are 5.3 ± 0.027, 299 ± 1.09, 352 ± 1.013 and 240 ± 1.15, respectively. The results showed that the use of activated carbon, fly ash, and nanometal oxides for the preparation of activated carbon-fly ash-nanometal oxide composite material greatly improve the surface area of fly ash and nanometal oxides. The surface area of fly ash was therefore improved by 99.65% for activated carbon-fly ash-nFe_3O_4, 99.70% for activated carbon-fly ash-nSiO_2, and 99.56% for activated carbon-fly ash-nZnO composites, while the surface area of nFe_3O_4, nSiO_2, and nZnO was increased by 87.60%, 38.28%, and 94.01% for the activated carbon-fly ash-nFe_3O_4, activated carbon-fly ash-nSiO_2 and activated carbon-fly ash-nZnO composites, respectively.

The micropore area of activated carbon-fly ash-nFe_3O_4 was 118.89 m^2/g, activated carbon-fly ash-nSiO_2 has a micropore area of 154.21 m^2/g, while activated carbon-fly ash-nZnO micropore area was 82.17 m^2/g. The micropore areas of fly ash, nFe_3O_4, nSiO_2 and nZnO which are 0.38, 3.98, 16.13, and 3.18 m^2/g, respectively, and were thus smaller than the micropore areas of the corresponding composite materials. It could therefore be concluded that the composition of activated carbon, nanometal oxide, and fly ash also improved the micropore area of fly ash and nano metal oxides.

3.8. Removal Efficiency of TBT by the Precursors and Composite Materials. The results obtained from TBT removal efficiency of these materials showed that the activated carbon, fly ash, nFe_3O_4, nSiO_2, nZnO were able to remove 99.3%, 94.5%, 81.9%, 79.9%, and 92.9% of the total TBT concentration in artificial seawater, respectively. However, activated carbon-fly ash, activated carbon-fly ash-nFe_3O_4, activated carbon-fly ash-nSiO_2, and activated carbon-fly ash-nZnO composite materials removed 99.78%, 99.98%, 99.97%, and 99.99% TBT, respectively. The results are illustrated in Figure 27.

It is therefore evident from the results presented in Figure 27 that apart from activated carbon which showed comparable result with the composite materials, all the composite materials exhibited higher (>99%) TBT removal efficiency than their respective precursors. These composite materials

TABLE 1: BET result of activated carbon-fly ash-nano metal oxide composite materials.

Samples	BET surface area m^2/g	Micropore volume cm^3/g	Micropore area m^2/g	External surface area m^2/g	Average pore diameter Å
Ac	1156 ± 8.69	0.182	442.75	713.89	48.89
Fly ash	1.06 ± 0.003	0.0001	0.38	0.68	89.43
nFe_3O_4	37 ± 0.19	0.002	3.98	33.19	217.42
$nSiO_2$	217 ± 1.76	0.006	16.13	201.49	88.08
nZnO	14 ± 0.039	0.001	3.18	11.23	98.50
Ac-fly ash	5.3 ± 0.027	0.00002	0.19	5.11	210.01
Ac-fly ash-nFe_3O_4	299 ± 1.09	0.048	118.89	180.86	63.55
Ac-fly ash-$nSiO_2$	352 ± 1.013	0.063	154.21	198.41	64.78
Ac-fly ash-nZnO	240 ± 1.15	0.033	82.17	158.64	51.84

Ac: Activated carbon.

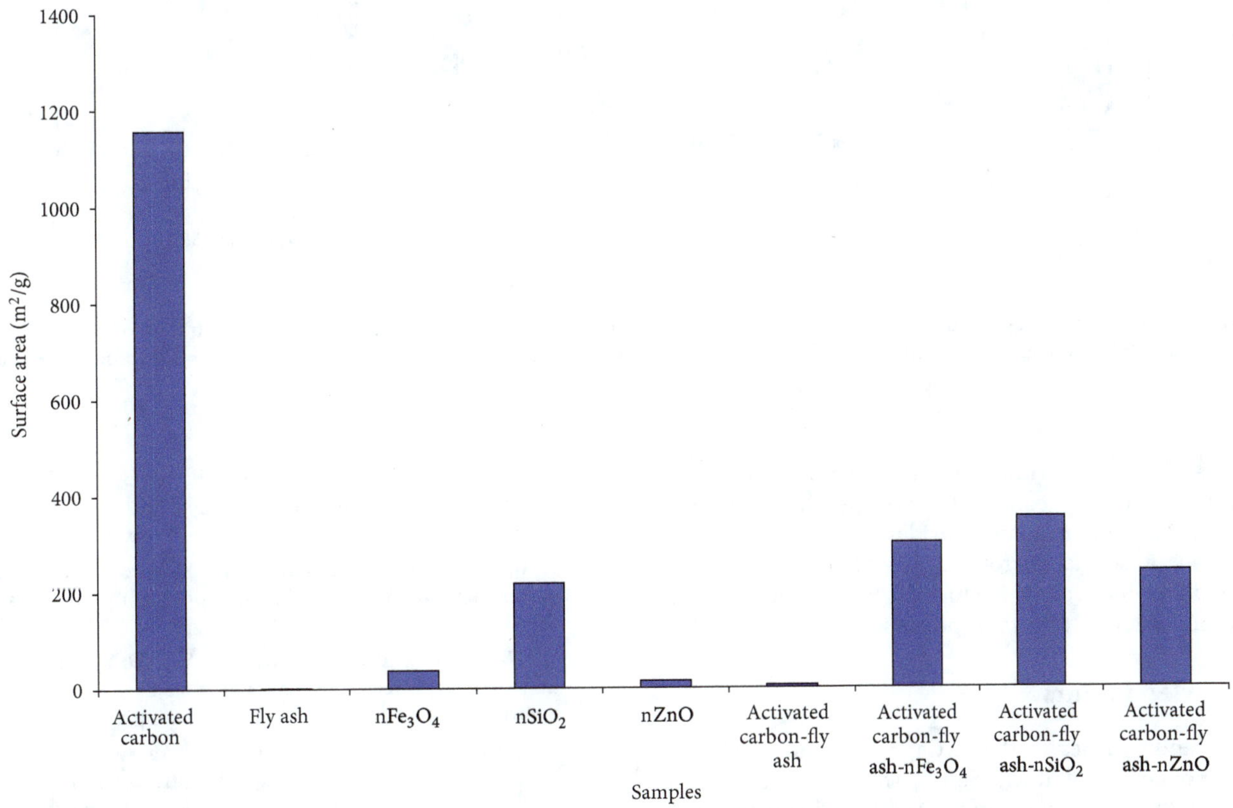

FIGURE 26: Surface area (m^2/g) versus precursors and activated carbon-fly ash-nanometal oxide composite materials.

are therefore potentially good materials for remediation application of TBT laden wastewater.

4. Conclusion

Experimental results showed that the pH values of activated carbon, $nSiO_2$, activated carbon-fly ash, activated carbon-fly ash-nFe_3O_4, activated carbon-fly ash-$nSiO_2$, and activated carbon-fly ash-nZnO are negatively charged and will therefore be suitable for the sorption of cationic complexes, while the pH values of fly ash, nFe_3O_4, and nZnO are slightly lower than their corresponding PZC values which suggest that their surfaces are positively charged and will therefore be favourable to the sorption of anionic complexes and heavy metals. The ash content determination also showed that the level of inorganic materials present in the adsorbent composite materials is a function of the precursors that make up the composite materials. The XRD and FTIR analyses confirmed the absence of impurity in the precursors and the prepared composite materials. The results of BET surface

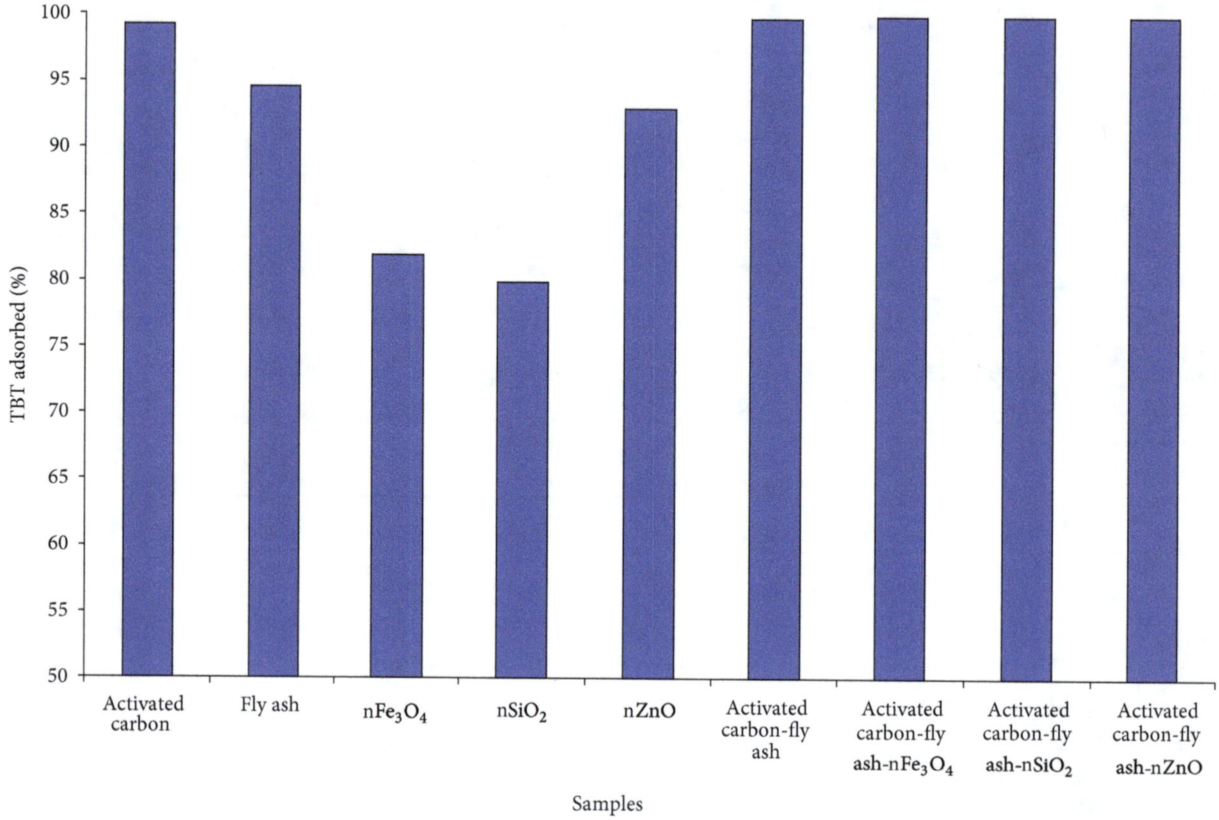

FIGURE 27: Removal efficiency of TBT by the precursors and composite materials. Experimental conditions: concentration of TBT = 100 ppm; pH = 8; volume of TBT solution = 25 mL; mass of activated carbon = 0.5 g; contact time = 60 min; stirring speed = 200 rpm; temperature = 22°C.

area and porosity determination also supported the higher sorption of TBT by the composite materials. The compositing of activated carbon, nanometal oxides, and fly ash increased the surface area and micropore area of fly ash and nano metal oxides which resulted in higher sorption capacity of the composite materials than their precursors.

Acknowledgments

O. S. Ayanda wishes to thank Cape Peninsula University of Technology, Cape Town, South Africa for the award of 2011 and 2012 bursary to study D. Tech. Chemistry in the institution. The author also thanks Professor L. Petrik and Dr. O. Fatoba (Department of Chemistry, University of Western Cape, South Africa) for providing the Matla fly ash.

References

[1] M. Ahmaruzzaman, "A review on the utilization of fly ash," *Progress in Energy and Combustion Science*, vol. 36, no. 3, pp. 327–363, 2010.

[2] G. Q. Lu and D. D. Do, "Adsorption properties of fly ash particles for NOx removal from flue gases," *Fuel Processing Technology*, vol. 27, no. 1, pp. 95–107, 1991.

[3] K. K. Panday, G. Prasad, and V. N. Singh, "Copper(II) removal from aqueous solutions by fly ash," *Water Research*, vol. 19, no. 7, pp. 869–873, 1985.

[4] P. Ricou, I. Lécuyer, and P. L. Cloirec, "Removal of Cu^{2+}, Zn^{2+} and Pb^{2+} adsorption onto fly ash and fly ash/lime mixing," *Water Science and Technology*, vol. 39, no. 10-11, pp. 239–247, 1999.

[5] P. Ricou-Hoeffer, I. Lecuyer, and P. L. Cloirec, "Experimental design methodology applied to adsorption of metallic ions onto fly ash," *Water Research*, vol. 35, no. 4, pp. 965–976, 2001.

[6] M. Nascimento, P. S. M. Soares, and V. P. D. Souza, "Adsorption of heavy metal cations using coal fly ash modified by hydrothermal method," *Fuel*, vol. 88, no. 9, pp. 1714–1719, 2009.

[7] I. D. Mall, V. C. Srivastava, and N. K. Agarwal, "Removal of Orange-G and Methyl Violet dyes by adsorption onto bagasse fly ash - Kinetic study and equilibrium isotherm analyses," *Dyes and Pigments*, vol. 69, no. 3, pp. 210–223, 2006.

[8] S. Wang and H. Wu, "Environmental-benign utilisation of fly ash as low-cost adsorbents," *Journal of Hazardous Materials*, vol. 136, no. 3, pp. 482–501, 2006.

[9] S. Wang, Q. Ma, and Z. H. Zhu, "Characteristics of coal fly ash and adsorption application," *Fuel*, vol. 87, no. 15-16, pp. 3469–3473, 2008.

[10] G. Zhang, J. Qu, H. Liu, A. T. Cooper, and R. Wu, "$CuFe_2O_4$/activated carbon composite: a novel magnetic adsorbent for the removal of acid orange II and catalytic regeneration," *Chemosphere*, vol. 68, no. 6, pp. 1058–1066, 2007.

[11] L. Li, P. A. Quinlivan, and D. R. U. Knappe, "Effects of activated carbon surface chemistry and pore structure on the adsorption of organic contaminants from aqueous solution," *Carbon*, vol. 40, no. 12, pp. 2085–2100, 2002.

[12] M. H. Stenzel, "Remove organics by activated carbon adsorption," *Chemical Engineering Progress*, vol. 89, no. 4, pp. 36–43, 1993.

[13] G. Newcombe, J. Morrison, C. Hepplewhite, and D. R. U. Knappe, "Simultaneous adsorption of MIB and NOM onto activated carbon II. Competitive effects," *Carbon*, vol. 40, no. 12, pp. 2147–2156, 2002.

[14] P. R. Shukla, S. Wang, H. M. Ang, and M. O. Tadé, "Synthesis, characterisation, and adsorption evaluation of carbon-natural-zeolite composites," *Advanced Powder Technology*, vol. 20, no. 3, pp. 245–250, 2009.

[15] C. T. Hsieh and H. Teng, "Langmuir and dubinin-radushkevich analyses on equilibrium adsorption of activated carbon fabrics in aqueous solutions," *Journal of Chemical Technology and Biotechnology*, vol. 75, no. 11, pp. 1066–1072, 2000.

[16] H. H. Tseng, J. G. Su, and C. Liang, "Synthesis of granular activated carbon/zero valent iron composites for simultaneous adsorption/dechlorination of trichloroethylene," *Journal of Hazardous Materials*, vol. 192, no. 2, pp. 500–506, 2011.

[17] V. K. Jha, M. Matsuda, and M. Miyake, "Sorption properties of the activated carbon-zeolite composite prepared from coal fly ash for Ni^{2+}, Cu^{2+}, Cd^{2+} and Pb^{2+}," *Journal of Hazardous Materials*, vol. 160, no. 1, pp. 148–153, 2008.

[18] Z. Sarbak and M. Kramer-Wachowiak, "Porous structure of waste fly ashes and their chemical modifications," *Powder Technology*, vol. 123, no. 1, pp. 53–58, 2002.

[19] C. Y. Yin, M. K. Aroua, and W. M. A. W. Daud, "Review of modifications of activated carbon for enhancing contaminant uptakes from aqueous solutions," *Separation and Purification Technology*, vol. 52, no. 3, pp. 403–415, 2007.

[20] G. G. Stavropoulos, P. Samaras, and G. P. Sakellaropoulos, "Effect of activated carbons modification on porosity, surface structure and phenol adsorption," *Journal of Hazardous Materials*, vol. 151, no. 2-3, pp. 414–421, 2008.

[21] P. Pengthamkeerati, T. Satapanajaru, and P. Chularuengoaksorn, "Chemical modification of coal fly ash for the removal of phosphate from aqueous solution," *Fuel*, vol. 87, no. 12, pp. 2469–2476, 2008.

[22] O. S. Fatoki, O. S. Ayanda, F. A. Adekola, B. J. Ximba, and B. O. Opeolu, "Preparation and Characterization of Activated carbon—nFe_3O_4, Activated carbon—$nSiO_2$ and Activated carbon—nZnO Hybrid Materials," *Particle & Particle Systems Characterization*, vol. 29, no. 3, pp. 178–191, 2012.

[23] O. S. Ayanda, O. S. Fatoki, F. A. Adekola, and B. J. Ximba, "Characterization of fly ash generated from matla power station in mpumalanga, South Africa," *E-Journal of Chemistry*, vol. 9, no. 4, pp. 1788–1795, 2012.

[24] P. Westerhoff, T. Karanfil, and J. Crittenden, *Aerogel and Iron-Oxide Impregnated Granular Activated Carbon Media For Arsenic Removal*, Awwa Research Foundation and Arsenic Water Technology Partnership, Denver, Colo, USA, 2006.

[25] F. Adekola, M. Fédoroff, H. Geckeis et al., "Characterization of acid-base properties of two gibbsite samples in the context of literature results," *Journal of Colloid and Interface Science*, vol. 354, no. 1, pp. 306–317, 2011.

[26] J. G. Collin, A. Bono, D. Krishnaiah, and K. O. Soon, "Sorption studies of methylene blue dye in aqueous solution by optimised carbon prepared from guava seeds (*Psidium guajava L.*)," *Materials Science*, vol. 13, no. 1, pp. 83–87, 2007.

[27] S. Mopoung and W. Nogklai, "Chemical and surface properties of longan seed activated charcoal," *International Journal of Physical Sciences*, vol. 3, no. 10, pp. 234–239, 2008.

[28] A. Beganskienė, V. Sirutkaitis, M. Kurtinaitienė, R. Juškėnas, and A. Kareiva, "FTIR, TEM and NMR investigations of stöber silica nanoparticles," *Journal of Materials Science*, vol. 10, pp. 287–290, 2004.

23

Multiscale Asymptotic Analysis and Parallel Algorithm of Parabolic Equation in Composite Materials

Xin Wang, Xi-liang Duan, and Yang Gao

Department of Mathematics, Shanghai University, Shanghai 200444, China

Correspondence should be addressed to Xin Wang; xinwang@shu.edu.cn

Academic Editor: Igor Andrianov

An efficient parallel multiscale numerical algorithm is proposed for a parabolic equation with rapidly oscillating coefficients representing heat conduction in composite material with periodic configuration. Instead of following the classical multiscale asymptotic expansion method, the Fourier transform in time is first applied to obtain a set of complex-valued elliptic problems in frequency domain. The multiscale asymptotic analysis is presented and multiscale asymptotic solutions are obtained in frequency domain which can be solved in parallel essentially without data communications. The inverse Fourier transform will then recover the approximate solution in time domain. Convergence result is established. Finally, a novel parallel multiscale FEM algorithm is proposed and some numerical examples are reported.

1. Introduction

Consider the parabolic equation with rapidly oscillating coefficients as follows:

$$\frac{\partial u^{\varepsilon}(x,t)}{\partial t}-\frac{\partial}{\partial x_i}\left(a_{ij}\left(\frac{x}{\varepsilon}\right)\frac{\partial u^{\varepsilon}(x,t)}{\partial x_j}\right)=f(x,t),\quad (x,t)\in\Omega\times(0,T),\quad (1)$$

$$u^{\varepsilon}(x,t)=0,\quad (x,t)\in\partial\Omega\times(0,T),$$

$$u^{\varepsilon}(x,0)=\overline{u}(x),\quad x\in\Omega,$$

where $\Omega\in\mathscr{R}^n$ is a bounded smooth domain or a bounded polygonal convex domain with a periodic microstructure and $(0,T)$ denotes the time domain. The unknown is the temperature increment $u^{\varepsilon}(x,t)$. $f(x,t)$ and $\overline{u}(x)$ are some given functions which denote heat source and initial temperature, respectively. $\varepsilon>0$ is a small parameter and it represents the relative size of a periodic cell.

Let $\xi = x/\varepsilon$ and the heat conductivity coefficient $(a_{ij}(\xi))_{1\le i,j\le n}$ is assumed to satisfy the following.

(A_1) $a_{ij}(x/\varepsilon) = a_{ij}(\xi)$ is Q-periodic in ξ, $\forall i,j=1,\dots,n$, where $Q=(0,1)^n$ is the reference cell.

(A_2) $(a_{ij}(\xi))_{1\le i,j\le n}$ is symmetric and satisfies the uniform elliptic condition; that is,

$$a_{ij}(\xi)=a_{ji}(\xi),\quad \forall i,j=1,\dots,n,$$
$$\lambda_1|\eta|^2\le a_{ij}(\xi)\eta_i\eta_j\le\lambda_2|\eta|^2,\quad \text{a.e. } \xi\in Q;\quad (2)$$

where is $\eta=(\eta_1,\dots,\eta_n)$ an arbitrary vector in $\mathscr{R}^n$, and $\lambda_1,\lambda_2>0$ are some constants independent of ε.

(A_3) Let $Q\subset\subset Q'$ and $Q'=(\bigcup_{m=1}^{L}\overline{D}_m)\backslash\partial Q'$. Suppose that the boundaries $\partial D_m\in C^{1,\gamma}$ for some $0<\gamma<1$ and $a_{ij}(\xi)\in C^{\mu}(\overline{D}_m)$, $\forall i,j=1,\dots,n$ for some $0<\mu<1$; see [1, 2].

In (1), the rapid spatial oscillations in the coefficients translate the periodic structure of the body which comes from assembling ε-scaled versions of the reference cell Q. This body has to be thought of as being made of a composite material.

An efficient analysis approach to such a problem induces a need for homogenization. The mathematical theory has been presented in a lot of works; see, for example, [3–6]. Several efficient numerical methods have been proposed and analyzed; for instance, see [7–9]. From these literatures , we found that the homogenized equation happens to be a system of the same type with constant coefficients. So the normal step-by-step time-marching algorithms such as the backward-Euler, Crank-Nicolson methods, or any higher order methods need to be applied to solve the homogenized equation step by step in time. Although these algorithms are effective in solving many practical problems, one of major drawbacks is that they are not easily parallelizable along the time axis, since they require the information of the solutions at previous time steps in order to advance to the next time step.

In this work, in order to propose a parallelizable multiscale numerical algorithm, we apply the Fourier transform and inverse Fourier transform to present the multiscale asymptotic analysis for (1). Numerical methods for time-dependent problems based on the use of Fourier transform have been considered in [10, 11], providing us with the starting point for using Fourier transform in multiscale asymptotic analysis. In [10, 11], Douglas et al. introduced an efficient parallel method for solving wave equations in the space-time domain after taking the Fourier transforms. The parallel method had been further developed and extended to viscoelasticity, parabolic problems, and Navier-Stokes equations; see [12, 13]. Later Sheen et al. also applied Laplace transform to solve parabolic problems; see [14, 15]. In addition, the early study of homogenization by employing means of integral transformation is [16], where Fourier and Laplace transforms were applied to prove the weak convergence theorems of the homogenization procedure for the time-dependent equations.

The main procedures in this paper are now briefly described. First, we take the Fourier transform with respect to the time variable in the parabolic equation and then present the multiscale asymptotic expansions of the solutions for the resulting equations in space-frequency domain at specific frequency points of interest, which are independent and may therefore be done in parallel. Finally, the approximate solution of parabolic equation in space-time domain is retrieved by the discrete inverse Fourier transform. It should be stated that our purpose of using Fourier transform is not just to prove the homogenization convergence theorems as done in [16], but to propose a parallelizable multiscale numerical algorithm. The new contributions presented in this study are the following. The strong convergence theorem of the second-order multiscale asymptotic solution in space-frequency domain is derived where we carefully estimate the bounds associated with frequency variable ω as it has a singularity for both zero and infinity in inverse Fourier transform. The convergence result for the approximate solution in time domain is obtained. A novel parallel multiscale FEM algorithm and numerical examples in 3D composite materials are given.

The remainder of this paper is organized as follows. In Section 2, the original parabolic equation is transformed to a set of complex-valued elliptic problems. The multiscale asymptotic expansions in space-frequency domain are then discussed. In Section 3, the multiscale truncation error estimates in space-frequency domain are derived under some assumptions. The approximate solution in space-time domain is inverted back by discrete inverse Fourier transform and the convergence result is proved in Section 4. Finally, a multiscale algorithm is proposed and numerical simulations are carried out to validate the presented method.

Throughout the paper, let C denote a positive constant independent of ε and ω. All complex-valued functions are assumed to have values in the complex field $\mathscr{C}$ and standard notations for function spaces and their associated norms will be used in this paper.

2. Fourier Transform and Multiscale Asymptotic Expansions

Denote the Fourier transform of a function $\rho(\cdot,t)$ by $\widehat{\rho}(\cdot,\omega)$, where

$$\widehat{\rho}(\cdot,\omega) = \int_{-\infty}^{+\infty} \rho(\cdot,t)\, e^{-i\omega t} dt. \tag{3}$$

In order to take the Fourier transform of (1), we first extend u^ε and f by zero when $t < 0$ and then we transfer (1) into the following complex-valued elliptic problem in space-frequency domain,

$$\begin{gathered} i\omega\widehat{u}^\varepsilon(x,\omega) - \frac{\partial}{\partial x_i}\left(a_{ij}(\xi)\frac{\partial \widehat{u}^\varepsilon(x,\omega)}{\partial x_j}\right) \\ = \widehat{f}(x,\omega) + \overline{u}(x), \quad x \in \Omega, \\ \widehat{u}^\varepsilon(x,\omega) = 0, \quad x \in \partial\Omega, \end{gathered} \tag{4}$$

where ω is some fixed parameter in frequency domain.

It is not difficult to check the existence and uniqueness of the problem (4) with an application of the Lax-Milgram lemma; see [17].

Proposition 1. *Let $\omega > 0$, under the assumptions (A_1)-(A_2); then, for any $\widehat{f}(\cdot,\omega)$, $\overline{u} \in H^{-1}(\Omega)$, (4) has a unique weak solution $\widehat{u}^\varepsilon(\cdot,\omega)$.*

The method requires solving a finite set of problem (4) at specific frequency point ω. It is natural to treat them simultaneously because they are independent. Moreover, observe that (4) is an equation with rapidly oscillating coefficients in spatial domain. The direct accurate numerical computation of the solution is difficult especially in 3D problems because it would require huge computational cost. We will apply the method of multiscale asymptotic expansion. Let the derivative with respect to x be

$$\frac{\partial}{\partial x_i} = \frac{\partial}{\partial z_i} + \varepsilon^{-1}\frac{\partial}{\partial \xi_i}, \qquad i = 1,\ldots,n, \tag{5}$$

where z denotes the macroscopic variable and ξ describes the microscopic one. Two scales describe the model: z gives the position of a point in domain Ω while ξ gives its position in the reference cell Q. On expanding $\widehat{u}^\varepsilon$ in asymptotic expansion of the form

$$\widehat{u}^\varepsilon(x,\omega) = \sum_{i=0}^{\infty}\varepsilon^i\widehat{u}_i(z,\xi,\omega), \tag{6}$$

where $\widehat{u}_i(z,\xi,\omega)$ is Q-periodic in ξ, the space-frequency (4) becomes

$$\begin{aligned}
&-\varepsilon^{-2}\left[\frac{\partial}{\partial\xi_i}\left(a_{ij}(\xi)\frac{\partial\widehat{u}_0(z,\xi,\omega)}{\partial\xi_j}\right)\right]\\
&-\varepsilon^{-1}\left[\frac{\partial}{\partial\xi_i}\left(a_{ij}(\xi)\left(\frac{\partial\widehat{u}_0(z,\xi,\omega)}{\partial z_j}+\frac{\partial\widehat{u}_1(z,\xi,\omega)}{\partial\xi_j}\right)\right)\right.\\
&\qquad\left.+\frac{\partial}{\partial z_i}\left(a_{ij}(\xi)\frac{\partial\widehat{u}_0(z,\xi,\omega)}{\partial\xi_j}\right)\right]+i\omega\widehat{u}_0(z,\xi,\omega)\\
&-\frac{\partial}{\partial z_i}\left(a_{ij}(\xi)\left(\frac{\partial\widehat{u}_0(z,\xi,\omega)}{\partial z_j}+\frac{\partial\widehat{u}_1(z,\xi,\omega)}{\partial\xi_j}\right)\right)\\
&-\frac{\partial}{\partial\xi_i}\left(a_{ij}(\xi)\left(\frac{\partial\widehat{u}_1(z,\xi,\omega)}{\partial z_j}+\frac{\partial\widehat{u}_2(z,\xi,\omega)}{\partial\xi_j}\right)\right)+O\left(\varepsilon^1\right)\\
&=\widehat{f}(z,\omega)+\overline{u}(z),\quad(z,\xi)\in\Omega\times R^n,\\
&\sum_{i=0}^{\infty}\varepsilon^i\widehat{u}_i(z,\xi,\omega)=0,\quad(z,\xi)\in\partial\Omega\times R^n,\\
&\qquad\widehat{u}_i(z,\xi,\omega)\quad\text{Q-periodic in }\xi.
\end{aligned}\tag{7}$$

On equating orders at $O(\varepsilon^{-2})$, with the well-known theorem of the existence and uniqueness of the solutions in quotient space, we deduce that

$$\widehat{u}_0(z,\xi,\omega)=\widehat{u}_0(z,\omega).\tag{8}$$

Equating terms at $O(\varepsilon^{-1})$, we have

$$\begin{aligned}
&-\frac{\partial}{\partial\xi_i}\left(a_{ij}(\xi)\frac{\partial\widehat{u}_1(z,\xi,\omega)}{\partial\xi_j}\right)\\
&\quad=\frac{\partial a_{ij}(\xi)}{\partial\xi_i}\frac{\partial\widehat{u}_0(z,\omega)}{\partial z_j},\quad(z,\xi)\in\Omega\times R^n,\\
&\qquad\widehat{u}_1(z,\xi,\omega)\quad\text{Q-periodic in }\xi,
\end{aligned}\tag{9}$$

with the method of separation of variables and we can obtain

$$\widehat{u}_1(z,\xi,\omega)=N_k(\xi)\frac{\partial\widehat{u}_0(z,\omega)}{\partial z_k},\tag{10}$$

where $N_k(\xi)$ $(k=1,\dots,n)$ are the solutions of the following problems:

$$\begin{aligned}
&-\frac{\partial}{\partial\xi_i}\left(a_{ij}(\xi)\frac{\partial N_k(\xi)}{\partial\xi_j}\right)=\frac{\partial a_{ik}(\xi)}{\partial\xi_i},\quad\xi\in R^n,\\
&\qquad N_k(\xi)\quad\text{Q-periodic.}
\end{aligned}\tag{11}$$

At (ε^0), we have

$$\begin{aligned}
&i\omega\widehat{u}_0(z,\omega)-a_{ij}(\xi)\frac{\partial^2\widehat{u}_0(z,\omega)}{\partial z_i\partial z_j}-a_{ij}(\xi)\frac{\partial N_k(\xi)}{\partial\xi_j}\frac{\partial^2\widehat{u}_0(z,\omega)}{\partial z_i\partial z_k}\\
&\quad-\frac{\partial\left(a_{ij}(\xi)N_k(\xi)\right)}{\partial\xi_i}\frac{\partial^2\widehat{u}_0(z,\omega)}{\partial z_j\partial z_k}\\
&\quad-\frac{\partial}{\partial\xi_i}\left(a_{ij}(\xi)\frac{\partial\widehat{u}_2(z,\xi,\omega)}{\partial\xi_j}\right)\\
&\quad=\widehat{f}(z,\omega)+\overline{u}(z),\quad(z,\xi)\in\Omega\times R^n,\\
&\qquad\widehat{u}_2(z,\xi,\omega)\quad\text{Q-periodic in }\xi.
\end{aligned}\tag{12}$$

On averaging (12) over Q, we get the homogenized equation

$$\begin{aligned}
&i\omega\widehat{u}_0(z,\omega)-\frac{\partial}{\partial z_i}\left(\widehat{a}_{ij}\frac{\partial\widehat{u}_0(z,\omega)}{\partial z_j}\right)\\
&\quad=\widehat{f}(z,\omega)+\overline{u}(z),\quad z\in\Omega,\\
&\widehat{u}_0(z,\omega)=0,\quad z\in\partial\Omega,
\end{aligned}\tag{13}$$

where the homogenized coefficients are defined by

$$\widehat{a}_{ij}=\frac{1}{|Q|}\int_Q a_{ij}(\xi)+a_{ik}(\xi)\frac{\partial N_j(\xi)}{\partial\xi_k}d\xi.\tag{14}$$

Remark 2. Under the assumptions (A_1)-(A_2) of the coefficients for the parabolic equation, one can prove that the homogenized coefficients $\widehat{a}_{ij}$ are also symmetric and uniform elliptic for some constants $\gamma_1,\gamma_2>0$; see, for example, [3, 5]. Therefore the existence and uniqueness of the solution $\widehat{u}_0$ for the homogenized (13) can be established.

Subtracting (13) from (12), we have

$$\begin{aligned}
&-\frac{\partial}{\partial\xi_i}\left(a_{ij}(\xi)\frac{\partial\widehat{u}_2(z,\xi,\omega)}{\partial\xi_j}\right)\\
&\quad=\left(a_{ij}(\xi)-\widehat{a}_{ij}\right)\frac{\partial^2\widehat{u}_0(z,\omega)}{\partial z_i\partial z_j}\\
&\qquad+a_{ij}(\xi)\frac{\partial N_k(\xi)}{\partial\xi_j}\frac{\partial^2\widehat{u}_0(z,\omega)}{\partial z_i\partial z_k}\\
&\qquad+\frac{\partial}{\partial\xi_i}\left(a_{ij}(\xi)N_k(\xi)\right)\frac{\partial^2\widehat{u}_0(z,\omega)}{\partial z_j\partial z_k},\\
&\qquad(z,\xi)\in\Omega\times R^n,\\
&\qquad\widehat{u}_2(z,\xi,\omega)\quad\text{Q-periodic in }\xi,
\end{aligned}\tag{15}$$

and using the method of separation of variables, we can obtain

$$\widehat{u}_2(z,\xi,\omega)=N_{kl}(\xi)\frac{\partial^2\widehat{u}_0(z,\omega)}{\partial z_k\partial z_l},\tag{16}$$

where $N_{kl}(\xi)$ $(k,l = 1,\dots,n)$ are the solutions of the following problems:

$$-\frac{\partial}{\partial\xi_i}\left(a_{ij}(\xi)\frac{\partial N_{kl}(\xi)}{\partial\xi_j}\right) = a_{kl}(\xi) - \hat{a}_{kl} + a_{kj}(\xi)\frac{\partial N_l(\xi)}{\partial\xi_j} + \frac{\partial}{\partial\xi_i}\left(a_{ik}(\xi)N_l(\xi)\right), \quad \xi \in R^n, \qquad N_{kl}(\xi) \quad Q\text{-periodic.} \tag{17}$$

Remark 3. Note that $N_k(\xi)$, $N_{kl}(\xi)$ $(k,l = 1,\dots,n)$ defined in (11) and (17) can be obtained by resolving the following cell problems in Q:

$$-\frac{\partial}{\partial\xi_i}\left(a_{ij}(\xi)\frac{\partial N_k(\xi)}{\partial\xi_j}\right) = \frac{\partial a_{ik}(\xi)}{\partial\xi_i}, \quad \xi \in Q, \qquad N_k(\xi) \quad Q\text{-periodic.} \tag{18}$$

$$-\frac{\partial}{\partial\xi_i}\left(a_{ij}(\xi)\frac{\partial N_{kl}(\xi)}{\partial\xi_j}\right) = a_{kl}(\xi) - \hat{a}_{kl} + a_{kj}(\xi)\frac{\partial N_l(\xi)}{\partial\xi_j} + \frac{\partial}{\partial\xi_i}\left(a_{ik}(\xi)N_l(\xi)\right), \quad \xi \in Q, \qquad N_{kl}(\xi) \quad Q\text{-periodic.} \tag{19}$$

and then extending their solutions by periodicity (see, e.g., [5]), so they are called cell functions as usual.

Therefore, we can obtain the first-order and second-order multiscale asymptotic solution for the space-frequency (4) as

$$\widehat{U}_1^{\varepsilon}(x,\omega) = \hat{u}_0(z,\omega) + \varepsilon N_k(\xi)\frac{\partial\hat{u}_0(z,\omega)}{\partial x_k},$$
$$\widehat{U}_2^{\varepsilon}(x,\omega) = \hat{u}_0(z,\omega) + \varepsilon N_k(\xi)\frac{\partial\hat{u}_0(z,\omega)}{\partial x_k} + \varepsilon^2 N_{kl}(\xi)\frac{\partial^2\hat{u}_0(z,\omega)}{\partial x_k\partial x_l}. \tag{20}$$

Remark 4. It is worthwhile to note that all the cell problems are real valued, while the homogenized equation is complex valued. Therefore, when we do numerical computation, we need to repeatedly solve the homogenized problem at a set of specific frequency points but solve all the cell problems only once.

3. Multiscale Truncation Error Estimate in Frequency Domain

In order to derive the truncation error estimate for multiscale asymptotic solutions, we first discuss the regularity for the homogenized solution $\hat{u}_0$. The key point is to derive precise estimates for the regularity bounds associated with ω since the behavior of these bounds as ω approaches zero and infinity is of critical importance in obtaining the multiscale truncation error estimate and when inverting the multiscale asymptotic solution back to time domain by the inverse Fourier transform.

Proposition 5. *Let $\hat{u}_0(\cdot,\omega)$ be the weak solution of the homogenized equation (13) whose coefficients are constants defined in (14). Then,*

(i) *for any $\hat{f}(\cdot,\omega), \bar{u} \in L^2(\Omega)$, it holds*

$$\|\hat{u}_0(\cdot,\omega)\|_{0,\Omega} \le \min\left\{C, \frac{1}{\omega}\right\}\|\hat{f}(\cdot,\omega) + \bar{u}\|_{0,\Omega},$$
$$\|\hat{u}_0(\cdot,\omega)\|_{1,\Omega} \le C\|\hat{f}(\cdot,\omega) + \bar{u}\|_{0,\Omega}, \tag{21}$$

where the constant C depending only on Ω and the coefficients;

(ii) *for any $\hat{f}(\cdot,\omega), \bar{u} \in H^m(\Omega)$ and any subdomain $\Omega^* \subset\subset \Omega$, it holds*

$$\|\hat{u}_0(\cdot,\omega)\|_{m+2,\Omega^*} \le C\|\hat{f}(\cdot,\omega) + \bar{u}\|_{m,\Omega}, \quad m \in \mathcal{Z}^+, \tag{22}$$

where the constant C is depending only on Ω^*, Ω and the coefficients. Moreover, if $\partial\Omega$ is C^{m+2}, then we have the global estimate (22) on $\Omega^* = \Omega$.

Proof. Define the conjugate bilinear form $\hat{a}(\cdot,\cdot) : H^1(\Omega) \times H^1(\Omega) \to \mathscr{C}$:

$$\hat{a}(u,v) = \int_\Omega \hat{a}_{ij}\frac{\partial u}{\partial z_j}\frac{\partial\bar{v}}{\partial z_i}dz, \tag{23}$$

where $\bar{v}$ denotes the complex conjugate of v. Then the variational formulation of the homogenized equation (13) is to find $\hat{u}_0 \in H_0^1(\Omega)$ such that

$$i\omega(\hat{u}_0, v) + \hat{a}(\hat{u}_0, v) = (\hat{f} + \bar{u}, v), \quad v \in H_0^1(\Omega). \tag{24}$$

Set $v = \hat{u}_0$ in (24) and take the imaginary part; one obtains

$$\omega\|\hat{u}_0\|_{0,\Omega}^2 = \mathrm{Im}\langle\hat{f} + \bar{u}, \hat{u}_0\rangle \le \|\hat{f} + \bar{u}\|_{0,\Omega}\|\hat{u}_0\|_{0,\Omega}, \tag{25}$$

which yields

$$\|\hat{u}_0\|_{0,\Omega} \le \frac{1}{\omega}\|\hat{f} + \bar{u}\|_{0,\Omega}. \tag{26}$$

Let $v = \hat{u}_0$ in (24) and consider the real part; with the coercivity of $\hat{a}_{ij}$ and the Poincare's inequality, we have

$$\gamma_1\|\nabla\hat{u}_0\|_{0,\Omega}^2 \le \hat{a}(\hat{u}_0,\hat{u}_0) = \mathrm{Re}(\hat{f} + \bar{u}, \hat{u}_0) \le \|\hat{f} + \bar{u}\|_{0,\Omega}\|\hat{u}_0\|_{0,\Omega} \le C\|\hat{f} + \bar{u}\|_{0,\Omega}\|\nabla\hat{u}_0\|_{0,\Omega}, \tag{27}$$

and thus

$$\|\widehat{u}_0\|_{1,\Omega} \le C\|\widehat{f}+\overline{u}\|_{0,\Omega}, \tag{28}$$

together with (26); we have

$$\|\widehat{u}_0\|_{0,\Omega} \le \min\left\{C, \frac{1}{\omega}\right\}\|\widehat{f}+\overline{u}\|_{0,\Omega}. \tag{29}$$

Next we investigate the interior regularity of the homogenized solution. Fix any open set $\Omega^* \subset\subset \Omega$, and choose an open set W such that $\Omega^* \subset\subset W \subset\subset \Omega$. Then select a cutoff function $\zeta \in C_0^\infty(R^n)$ satisfying

$$\begin{aligned} &\zeta = 1, \quad z \in \Omega^*; \\ &\zeta = 0, \quad z \in R^n - W; \\ &0 \le \zeta \le 1. \end{aligned} \tag{30}$$

Rewrite (24) as

$$\widehat{a}(\widehat{u}_0, v) = (\widehat{f}+\overline{u}, v) - i\omega(\widehat{u}_0, v) \equiv B, \tag{31}$$

and set $v = -D_k^{-h}(\zeta^2 D_k^h \widehat{u}_0)$, $k \in \{1,\dots,N\}$, where D_k^h denotes the difference quotient,

$$D_k^h u(z) = \frac{u(z+he_k)-u(z)}{h}, \quad h \in R,\ h \ne 0. \tag{32}$$

By a standard regularity technique (see, e.g., [18], pages 310–313), we have

$$\widehat{a}(\widehat{u}_0, v) \ge \frac{\gamma_1}{2}\int_\Omega \zeta^2 |D_k^h \nabla \widehat{u}_0|^2 dz - C\int_\Omega |\nabla \widehat{u}_0|^2 dz. \tag{33}$$

Cauchy's inequality with ϵ yields

$$\begin{aligned} |B| &\le C\int_\Omega \left(|\widehat{f}+\overline{u}| + \omega|\widehat{u}_0|\right)|v|\,dz \\ &\le \epsilon\int_\Omega |v|^2 dz + \frac{C}{\epsilon}\int_\Omega \left(|\widehat{f}+\overline{u}|^2 + \omega^2|\widehat{u}_0|^2\right)dz, \end{aligned} \tag{34}$$

and use the properties of the difference quotient for v,

$$\int_\Omega |v|^2 dz \le C\int_\Omega \left(|\nabla\widehat{u}_0|^2 + \zeta^2|D_k^h\nabla\widehat{u}_0|^2\right)dz, \tag{35}$$

and choose $\epsilon = \gamma_1/4$ that yields the bound

$$\begin{aligned} |B| &\le \frac{\gamma_1}{4}\int_\Omega \zeta^2|D_k^h\nabla\widehat{u}_0|^2 dz \\ &\quad + C\int_\Omega |\widehat{f}+\overline{u}|^2 + \omega^2|\widehat{u}_0|^2 + |\nabla\widehat{u}_0|^2 dz. \end{aligned} \tag{36}$$

Combining (33) and (36) implied

$$\begin{aligned} \int_{\Omega^*} |D_k^h\nabla\widehat{u}_0|^2 dz &\le C\int_\Omega \zeta^2|D_k^h\nabla\widehat{u}_0|^2 dz \\ &\le C\int_\Omega \left(|\widehat{f}+\overline{u}|^2 + \omega^2|\widehat{u}_0|^2 + |\nabla\widehat{u}_0|^2\right)dz. \end{aligned} \tag{37}$$

In view of the properties of the difference quotient, with (29) and (28), we deduce $\widehat{u}_0 \in H^2_{\text{loc}}$, with the estimate

$$\begin{aligned} \|\widehat{u}_0\|_{2,\Omega^*} &\le C\left(\|\widehat{f}+\overline{u}\|_{0,\Omega} + \omega\|\widehat{u}_0\|_{0,\Omega} + \|\nabla\widehat{u}_0\|_{0,\Omega}\right) \\ &\le C\|\widehat{f}+\overline{u}\|_{0,\Omega}. \end{aligned} \tag{38}$$

For higher regularity of $\widehat{u}_0$, insert $v = -\partial w/\partial z_k$ into (24), $k = 1,\dots,n$, where $w \in C^\infty(\Omega)$, and perform integration by parts; we have

$$i\omega\left(\frac{\partial\widehat{u}_0}{\partial z_k}, w\right) + \widehat{a}\left(\frac{\partial\widehat{u}_0}{\partial z_k}, w\right) = \left(\frac{\partial(\widehat{f}+\overline{u})}{\partial z_k}, w\right). \tag{39}$$

In view of (38), applying the interior H^2 regularity result to $\partial\widehat{u}_0/\partial z_k$, then

$$\left\|\frac{\partial\widehat{u}_0}{\partial z_k}\right\|_{2,\Omega^*} \le C\left\|\frac{\partial(\widehat{f}+\overline{u})}{\partial z_k}\right\|_{0,\Omega}, \quad k = 1,\dots,n, \tag{40}$$

which yields

$$\|\widehat{u}_0\|_{3,\Omega^*} \le C\|\widehat{f}+\overline{u}\|_{1,\Omega}. \tag{41}$$

Finally we can obtain (22) by induction on m. Given $\partial\Omega \in C^{m+2}$, following the standard boundary regularity technique we get the global estimate

$$\|\widehat{u}_0(\cdot,\omega)\|_{m+2,\Omega} \le C\|\widehat{f}(\cdot,\omega)+\overline{u}\|_{m,\Omega}. \tag{42}$$

□

Next we can derive the truncation error estimate for multiscale asymptotic solutions (20) in frequency domain.

Theorem 6. *Let $\widehat{u}^\varepsilon(x,\omega)$ be the weak solution of (4), and let $\widehat{U}_s^\varepsilon(x,\omega)$ $(s = 1,2)$ be the multiscale asymptotic solutions given in (20). Under the assumptions (A_1)–(A_3), with $\widehat{f}(\cdot,\omega) \in H^1(\Omega)$, $\overline{u} \in H^1(\Omega)$, $\widehat{u}_0(\cdot,\omega) \in H^3(\Omega)$, there exists a constant C independent of ε and ω such that*

$$\begin{aligned} &\|\widehat{u}^\varepsilon(\cdot,\omega) - \widehat{U}_2^\varepsilon(\cdot,\omega)\|_{1,\Omega} \\ &\quad \le C(1+\omega)\varepsilon^{1/2}\|\widehat{f}(\cdot,\omega)+\overline{u}\|_{1,\Omega}, \quad s = 1,2. \end{aligned} \tag{43}$$

Proof. For $s = 2$, set $Z_\varepsilon(x,\omega) = \widehat{u}^\varepsilon(x,\omega) - \widehat{U}_2^\varepsilon(x,\omega)$; one can check $Z_\varepsilon(x,\omega)$ that satisfy

$$\begin{aligned} &i\omega Z_\varepsilon(x,\omega) - \frac{\partial}{\partial x_i}\left(a_{ij}\left(\frac{x}{\varepsilon}\right)\frac{\partial Z_\varepsilon(x,\omega)}{\partial x_j}\right) \\ &\quad = \varepsilon F^\varepsilon(x,\omega), \quad x \in \Omega, \\ &Z_\varepsilon(x,\omega) = \varepsilon G^\varepsilon(x,\omega), \quad x \in \partial\Omega, \end{aligned} \tag{44}$$

where

$$F^\varepsilon \equiv F_0^\varepsilon + \varepsilon \frac{\partial F_i^\varepsilon}{\partial x_i},$$

$$F_0^\varepsilon = \left(a_{il}(\xi) N_k(\xi) + a_{ij}(\xi) \frac{\partial N_{kl}(\xi)}{\partial \xi_j} \right) \frac{\partial^3 \widehat{u}_0(z,\omega)}{\partial z_i \partial z_k \partial z_l} - i\omega N_k(\xi) \frac{\partial \widehat{u}_0(z,\omega)}{\partial z_k} - i\omega\varepsilon N_{kl}(\xi) \frac{\partial^2 \widehat{u}_0(z,\omega)}{\partial z_k \partial z_l}, \quad (45)$$

$$F_i^\varepsilon = a_{ij}(\xi) N_{kl}(\xi) \frac{\partial^3 \widehat{u}_0(z,\omega)}{\partial z_j \partial z_k \partial z_l},$$

$$G^\varepsilon = -N_k(\xi) \frac{\partial \widehat{u}_0(z,\omega)}{\partial z_k} - \varepsilon N_{kl}(\xi) \frac{\partial^2 \widehat{u}_0(z,\omega)}{\partial z_k \partial z_l}.$$

Under the assumptions (A_1)–(A_3), it follows from Theorem 1.1 of [2] (also see [1]) that

$$\|N_k\|_{1,\infty,Q} \le C, \qquad \|N_{kl}\|_{1,\infty,Q} \le C, \qquad k,l = 1,\dots,n. \quad (46)$$

Together with the regularity estimate of $\widehat{u}_0$, one can obtain

$$\|F_0^\varepsilon\|_{0,\Omega} \le C(1+\omega) \|\widehat{f} + \overline{u}\|_{1,\Omega}, \qquad \|F_i^\varepsilon\|_{0,\Omega} \le C \|\widehat{f} + \overline{u}\|_{1,\Omega}, \quad (47)$$

and therefore, we get $F^\varepsilon \in H^{-1}(\Omega)$ and

$$\|F^\varepsilon\|_{-1,\Omega} \le C(1+\omega) \|\widehat{f} + \overline{u}\|_{1,\Omega}. \quad (48)$$

In order to get the boundary estimate, introducing the following cutoff function $m_\varepsilon \in C_0^\infty(\Omega)$ such that

$$m_\varepsilon = 1, \quad \text{if } \operatorname{dist}(x,\partial\Omega) \le \varepsilon,$$
$$m_\varepsilon = 0, \quad \text{if } \operatorname{dist}(x,\partial\Omega) \ge 2\varepsilon, \quad (49)$$
$$\|\nabla m_\varepsilon\|_{L^\infty(\Omega)} \le C\varepsilon^{-1};$$

and setting $\Psi^\varepsilon = m_\varepsilon G^\varepsilon$, $\Omega_\varepsilon = \operatorname{supp} \Psi^\varepsilon$, with the regularity estimate of $\widehat{u}_0$, we can obtain

$$\|\Psi^\varepsilon\|_{0,\Omega_\varepsilon} \le C \|\widehat{f} + \overline{u}\|_{0,\Omega}, \qquad \left\| \frac{\partial \Psi^\varepsilon}{\partial x_i} \right\|_{0,\Omega_\varepsilon} \le C\varepsilon^{-1} \|\widehat{u}_0\|_{1,\Omega_\varepsilon} + C \|\widehat{u}_0\|_{3,\Omega_\varepsilon}. \quad (50)$$

Using the following estimate on boundary layer Ω_ε, (see [6])

$$\|\widehat{u}_0\|_{1,\Omega_\varepsilon} \le C\varepsilon^{1/2} \|\nabla \widehat{u}_0\|_{1,\Omega}, \quad (51)$$

we have

$$\|\Psi^\varepsilon\|_{1,\Omega_\varepsilon} \le C\varepsilon^{-1/2} \|\widehat{f} + \overline{u}\|_{1,\Omega}. \quad (52)$$

By the trace theorem, we know

$$\|G^\varepsilon\|_{1/2,\partial\Omega} = \|\Psi^\varepsilon\|_{1/2,\partial\Omega} \le C \|\Psi^\varepsilon\|_{1,\Omega_\varepsilon} \le C\varepsilon^{-1/2} \|\widehat{f} + \overline{u}\|_{1,\Omega}. \quad (53)$$

Similar to Proposition 5 for (44) with nonhomogeneous boundary condition, we can drive the following estimate:

$$\|Z_\varepsilon\|_{1,\Omega} \le C\varepsilon \|F^\varepsilon\|_{-1,\Omega} + C(1+\omega)\varepsilon \|G^\varepsilon\|_{1/2,\partial\Omega}, \quad (54)$$

together with (48) and (53) and we have

$$\left\| \widehat{u}^\varepsilon(\cdot,\omega) - \widehat{U}_2^\varepsilon(\cdot,\omega) \right\|_{1,\Omega} \le C(1+\omega)\varepsilon^{1/2} \|\widehat{f} + \overline{u}\|_{1,\Omega}, \quad (55)$$

where C is independent of ε and ω.

For $s = 1$, by means of Theorem 1.2 in [6], we can also derive the estimate (43). Therefore the proof is complete. □

Remark 7. It is worth noting that, for any $\Omega^* \subset\subset \Omega$, it is well known that homogenized solution $\widehat{u}_0 \in H^3(\Omega^*)$ if $\widehat{f}(\cdot,\omega)$, and $\overline{u} \in H^1(\Omega)$. If $\widehat{u}_0$ does not have the local H^3 regularity near the boundary $\partial\Omega$, follow the way in [19]; we can use the asymptotic expansion in the interior subdomain and construct the boundary layer near $\partial\Omega$.

4. Inverse Fourier Transform

In this section, we will construct the approximate solution of the parabolic equation (1) in space-time domain by the inverse Fourier transform. Recall the formula of the inverse Fourier transform as

$$v(\cdot,t) = \frac{1}{2\pi} \int_{-\infty}^{+\infty} \widehat{v}(\cdot,\omega) e^{i\omega t} d\omega = \frac{1}{\pi} \operatorname{Re} \int_0^\infty \widehat{v}(\cdot,\omega) e^{i\omega t} d\omega, \quad (56)$$

provided that $v(\cdot,t)$ is a real function.

Fix a sufficiently large $\omega^* > 0$ such that $u^\varepsilon(\cdot,\omega)$ is negligible for $|\omega| > \omega^*$. Set $\Delta\omega = \omega^*/M$, where $M \in \mathscr{Z}^+$ is a selected segment number for the interval $(0,\omega^*)$ and the space-time solution u^ε of (1) can be approximated by

$$U^\varepsilon_{s,\omega^*,\Delta\omega}(x,t) = \frac{1}{\pi} \operatorname{Re} \sum_{j=1}^{M} \widehat{U}_s^\varepsilon\left(x,\omega_{j-1/2}\right) e^{it\omega_{j-1/2}} \Delta\omega, \quad (57)$$

where $\omega_{j-1/2} = (j-1/2)\Delta\omega$, $j = 1,\dots,M$.

In order to get an estimate for $u^\varepsilon - U^\varepsilon_{s,\omega^*,\Delta\omega}$, let

$$u^\varepsilon_{\omega^*}(x,t) = \frac{1}{\pi} \operatorname{Re} \int_0^{\omega^*} \widehat{u}^\varepsilon(x,\omega) e^{i\omega t} d\omega, \qquad u^\varepsilon_{\omega^*,\Delta\omega}(x,t) = \frac{1}{\pi} \operatorname{Re} \sum_{j=1}^{M} \widehat{u}^\varepsilon\left(x,\omega_{j-1/2}\right) e^{it\omega_{j-1/2}} \Delta\omega, \quad (58)$$

and we get

$$\begin{aligned} u^\varepsilon(x,t) &- U^\varepsilon_{s,\omega^*,\Delta\omega}(x,t) \\ &= \left(u^\varepsilon(x,t) - u^\varepsilon_{\omega^*}(x,t)\right) \\ &\quad + \left(u^\varepsilon_{\omega^*}(x,t) - u^\varepsilon_{\omega^*,\Delta\omega}(x,t)\right) + \\ &\quad + \left(u^\varepsilon_{\omega^*,\Delta\omega}(x,t) - U^\varepsilon_{s,\omega^*,\Delta\omega}(x,t)\right) \\ &\equiv E_1(x,t) + E_2(x,t) + E_3(x,t). \end{aligned} \quad (59)$$

For E_1, from Proposition 5, we can check

$$\begin{aligned}\|E_1(\cdot,t)\|_{0,\Omega} &\le \frac{1}{\pi}\int_{\omega^*}^{\infty}\|\widehat{u}^{\varepsilon}(\cdot,\omega)\|_{0,\Omega}d\omega \\ &\le \frac{1}{\pi}\int_{\omega^*}^{\infty}\frac{1}{\omega}\left\|\widehat{f}(\cdot,\omega)+\overline{u}\right\|_{0,\Omega}d\omega.\end{aligned} \tag{60}$$

For the estimate of E_2, the following lemma was proved in [17].

Lemma 8. *Let*

$$\begin{aligned} F_k^0(t) &= \left(\int_0^t e^{t-\sigma}\left\|\sigma^k f(x,\sigma)\right\|^2 d\sigma\right)^{1/2}, \\ F_k^l(t) &\\ &= \left(\int_0^t\int_0^{\sigma_1}\cdots\int_0^{\sigma_l} e^{t-\sigma}\left\|\sigma^{k-l} f(x,\sigma)\right\|^2 d\sigma d\sigma_1\cdots d\sigma_l\right)^{1/2},\end{aligned} \tag{61}$$

and suppose that for $k = 0, 1, 2$, $F_k^l(t) \in L^2(J)$ for all $k \ge l \ge 0$. Then

$$\|E_2(\cdot,t)\|_{0,\Omega} \le C\Delta\omega^2\sum_{k=0}^{2}t^{2-k}\sum_{l=0}^{k}\left\|F_k^l(t)\right\|_{0,\Omega}. \tag{62}$$

For E_3, by Theorem 6, we have

$$\begin{aligned} &\|E_3(\cdot,t)\|_{0,\Omega} \\ &= \left\|\frac{1}{\pi}\operatorname{Re}\sum_{j=1}^{M}\left(\widehat{u}^{\varepsilon}\left(x,\omega_{j-1/2}\right)-\widehat{U}_s^{\varepsilon}\left(x,\omega_{j-1/2}\right)\right)e^{it\omega_{j-1/2}}\Delta\omega\right\|_{0,\Omega} \\ &\le C\Delta\omega\sum_{j=1}^{M}\left(1+\omega_{j-1/2}\right)\varepsilon^{1/2}\left\|\widehat{f}\left(\cdot,\omega_{j-1/2}\right)+\overline{u}\right\|_{1,\Omega} \\ &\le C\varepsilon^{1/2}\int_0^{\omega^*}(1+\omega)\left\|\widehat{f}(\cdot,\omega)+\overline{u}\right\|_{1,\Omega}d\omega.\end{aligned} \tag{63}$$

Combining (60), (62), and (63), we have the following theorem for the full error estimate.

Theorem 9. *Let $u^{\varepsilon}(\cdot,t)$ be the solution of (1), and let $U^{\varepsilon}_{s,\omega^*,\Delta\omega}(\cdot,t)$ $(s = 1,2)$ be the approximate solution given in (57). Under the assumptions (A_1)–(A_3), $\widehat{f}(\cdot,\omega)$, $\overline{u} \in H^1(\Omega)$, $\widehat{u}_0(\cdot,\omega) \in H^3(\Omega)$, and suppose that for $k = 0,1,2$, $F_k^l(t)$ defined in (61) belong to $L^2(J)$ for all $k \ge l \ge 0$; there exists a constant C independent of ε and ω such that*

$$\begin{aligned} &\left\|u^{\varepsilon}(\cdot,t)-U^{\varepsilon}_{s,\omega^*,\Delta\omega}(\cdot,t)\right\|_{0,\Omega} \\ &\le \frac{1}{\pi}\int_{\omega^*}^{\infty}\frac{1}{\omega}\left\|\widehat{f}(\cdot,\omega)+\overline{u}\right\|_{0,\Omega}d\omega \\ &+ C\Delta\omega^2\sum_{k=0}^{2}t^{2-k}\sum_{l=0}^{k}\left\|F_k^l(t)\right\|_{0,\Omega} \\ &+ C\varepsilon^{1/2}\int_0^{\omega^*}(1+\omega)\left\|\widehat{f}(\cdot,\omega)+\overline{u}\right\|_{1,\Omega}d\omega.\end{aligned} \tag{64}$$

5. Parallel Multiscale FEM Algorithm and Numerical Tests

In simulation, we will apply the following first-order and second-order difference quotients of some functions $f(x)$ given by

$$\begin{aligned} &\delta_{x_j}f^h\left(N_p\right) = \frac{1}{\tau\left(N_p\right)}\sum_{e\in\sigma(N_p)}\left[\frac{\partial f^h}{\partial x_j}\right]_e\left(N_p\right), \\ &\delta^2_{x_lx_m}f^h\left(N_p\right) \\ &= \frac{1}{\tau\left(N_p\right)}\sum_{e\in\sigma(N_p)}\left[\sum_{j=1}^{d}\delta_{x_l}f^h\left(P_j\right)\frac{\partial\chi_j}{\partial x_m}\right]_e\left(N_p\right), \\ &\qquad j,l,m = 1,\ldots,n,\end{aligned} \tag{65}$$

where $\sigma(N_p)$, the set of elements with the node N_p; $\tau(N_p)$, is the number of elements of $\sigma(N_p)$; d is the number of nodes on the element e; P_j are the nodes of the element e; $\chi_j(x)$ are the Lagrangian shape functions; f^h denotes the FEM solution of f; $[\partial f/\partial x_j]_e(N_p)$ is the value of $\partial f/\partial x_j$ at the node N_p associated with the element e.

The following parallel multiscale FEM algorithm are presented to solve (1).

Step 1. Solve the first-order cell problems of N_k defined in (18) for $k = 1,\ldots,n$ and the second-order cell problems of N_{kl} defined in (19) for $k,l = 1,\ldots,n$ in local fine grid of the periodic cell Q by FEM.

Step 2. Compute the homogenized coefficients $\widehat{a}_{ij}$ defined in (14) in the local fine grid of the periodic cell Q by numerical integration.

Step 3. Fix a sufficiently large $\omega^* > 0$ and set $\Delta\omega = \omega^*/M$, $M \in \mathscr{Z}^+$ and simultaneously solve the complex-valued homogenized equation (13) in the whole domain Ω in a coarse mesh at $\omega_{j-1/2} = (j-1/2)\Delta\omega$, $j = 1,\ldots,M$. Then compute the first-order and second-order partial derivatives of the homogenized solutions by high order difference quotients in the same coarse mesh.

Step 4. Extend all the cell functions by periodicity from the local fine grid of the periodic cell Q to the gird of the whole domain Ω. Then compute the multiscale asymptotic solutions given by (20) at frequency points $\omega_{j-1/2}$, $j = 1,\ldots,M$.

Step 5. Calculate the approximation solutions for any fixed time $t = t^*$ for the original equation (1) according to the formula (57).

Remark 10. Notice that cell problems are independent of the time variable t or the frequency variable ω, so they only need to be solved once.

Remark 11. It should be stated that Step 1 to Step 5 in the algorithm are serial. However, for the time-dependent problems, the main computation of the existing multiscale algorithms

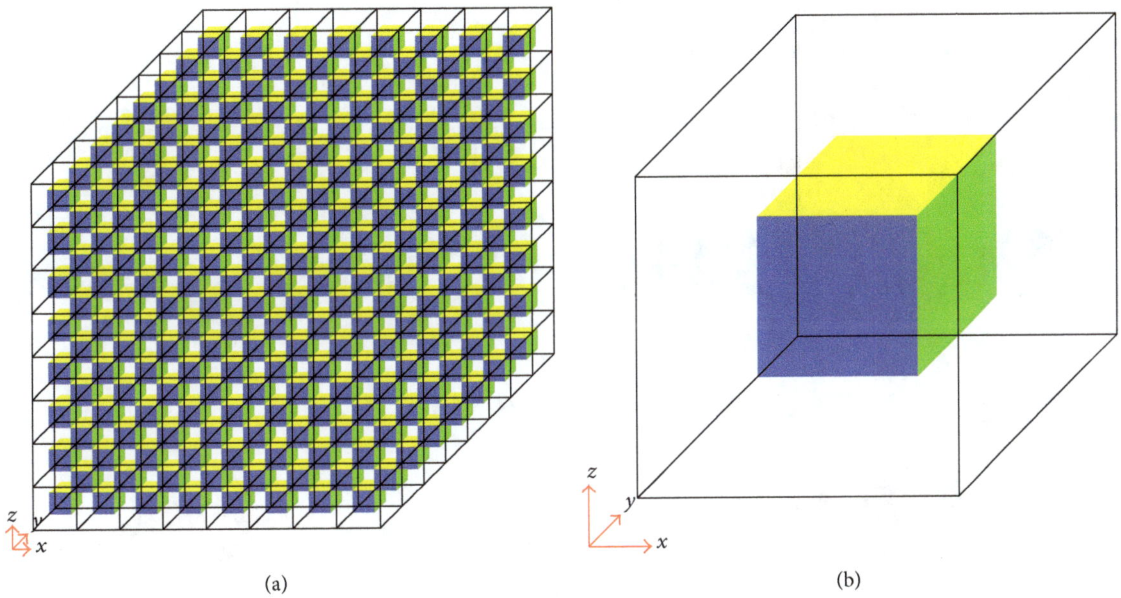

FIGURE 1: (a) A whole domain Ω, (b) the reference cell Q.

is in computing the time-dependent homogenized equation step by step. In our presented algorithm, we can simultaneously solve the homogenized equations at different frequent points, so the main computation of the algorithm is parallel. Also, notice that each step can be done simultaneously. In this respect, different from other multiscale algorithms for the time-dependent equation, we called it parallel algorithm.

Remark 12. If some conditions of the geometric symmetry are satisfied (see [19]), then we can employ the homogeneous Dirichlet boundary conditions on the boundary ∂Q for the cell problems (18) and (19). Following the way in [19], one can also obtain a similar convergence result to that of Theorem 6.

Now we discuss the finite element computation for the complex-valued homogenized Equation (13) in Ω. Let $\widehat{u}_{0,i} = \widehat{u}_{0,R} + \widehat{u}_{0,I}$ and $\widehat{f} = \widehat{f}_R + \widehat{f}_I$, where the subscript R denotes the real part of a function and the subscript I denotes the imaginary part.

The variational form of (13) is to find that $(\widehat{u}_{0,R}, \widehat{u}_{0,I}) \in H_0^1(\Omega) \times H_0^1(\Omega)$ satisfied

$$\begin{aligned} -\omega\left(\widehat{u}_{0,I}, v\right) + \widehat{a}\left(\widehat{u}_{0,R}, v\right) &= \left(\widehat{f}_R + \overline{u}, v\right), \\ \omega\left(\widehat{u}_{0,R}, v\right) + \widehat{a}\left(\widehat{u}_{0,I}, v\right) &= \left(\widehat{f}_I, v\right), \end{aligned} \tag{66}$$

for any $v \in H_0^1(\Omega)$, and $\widehat{a}(\cdot,\cdot)$ is given by (23). Let $S_h \subset H_0^1(\Omega)$ be the finite element space, the discrete variational form of (66) is to find that $(\widehat{u}_{0,R}^h, \widehat{u}_{0,I}^h) \in S_h \times S_h$ satisfied

$$\begin{aligned} -\omega\left(\widehat{u}_{0,I}^h, v^h\right) + \widehat{a}\left(\widehat{u}_{0,R}^h, v^h\right) &= \left(\widehat{f}_R + \overline{u}, v^h\right), \\ \omega\left(\widehat{u}_{0,R}^h, v^h\right) + \widehat{a}\left(\widehat{u}_{0,I}^h, v^h\right) &= \left(\widehat{f}_I, v^h\right), \end{aligned} \tag{67}$$

for any $v^h \in S_h$. Let $\{\phi_j(x)\}_{j=1}^{N_h}$ be a set of basis of S_h, and set

$$\widehat{u}_{0,R}^h(x,P) = \sum_{j=1}^{N_h} \Gamma_{R_j}\phi_j(x), \qquad \widehat{u}_{0,I}^h(x,P) = \sum_{j=1}^{N_h} \Gamma_{I_j}\phi_j(x), \tag{68}$$

and therefore, we solve the linear system in dimension $2 * N_h$ as follows:

$$\begin{pmatrix} A & -\omega B \\ \omega B & A \end{pmatrix} \begin{pmatrix} \mathbf{\Gamma}_R \\ \mathbf{\Gamma}_I \end{pmatrix} = \begin{pmatrix} \mathbf{F}_R \\ \mathbf{F}_I \end{pmatrix}, \tag{69}$$

where

$$\begin{gathered} A = \left\{\widehat{a}\left(\phi_j, \phi_k\right)\right\}_{1\le j,k\le N_h}, \\ B = \left\{\left(\phi_j, \phi_k\right)\right\}_{1\le j,k\le N_h}, \\ \mathbf{\Gamma}_R := \left(\Gamma_{R_1}, \ldots, \Gamma_{R_{N_h}}\right)^T, \qquad \mathbf{\Gamma}_I := \left(\Gamma_{I_1}, \ldots, \Gamma_{I_{N_h}}\right)^T, \\ \mathbf{F}_R := \left(\left(\widehat{f}_R + \overline{u}, \phi_1\right), \ldots, \left(\widehat{f}_R + \overline{u}, \phi_{N_h}\right)\right)^T, \\ \mathbf{F}_I := \left(\left(\widehat{f}_I, \phi_1\right), \ldots, \left(\widehat{f}_I, \phi_{N_h}\right)\right)^T. \end{gathered} \tag{70}$$

To validate the proposed method and to confirm the theoretical analysis reported in this paper, we present numerical simulations for the following cases.

Example. Consider the parabolic equation (1) in 3D composite materials, where whole domain Ω and the periodic cell Q are illustrated in Figures 1(a) and 1(b). In the configuration, we take $\varepsilon = 1/8$ and the blue cube $[0.25, 0.75]^3$ denotes

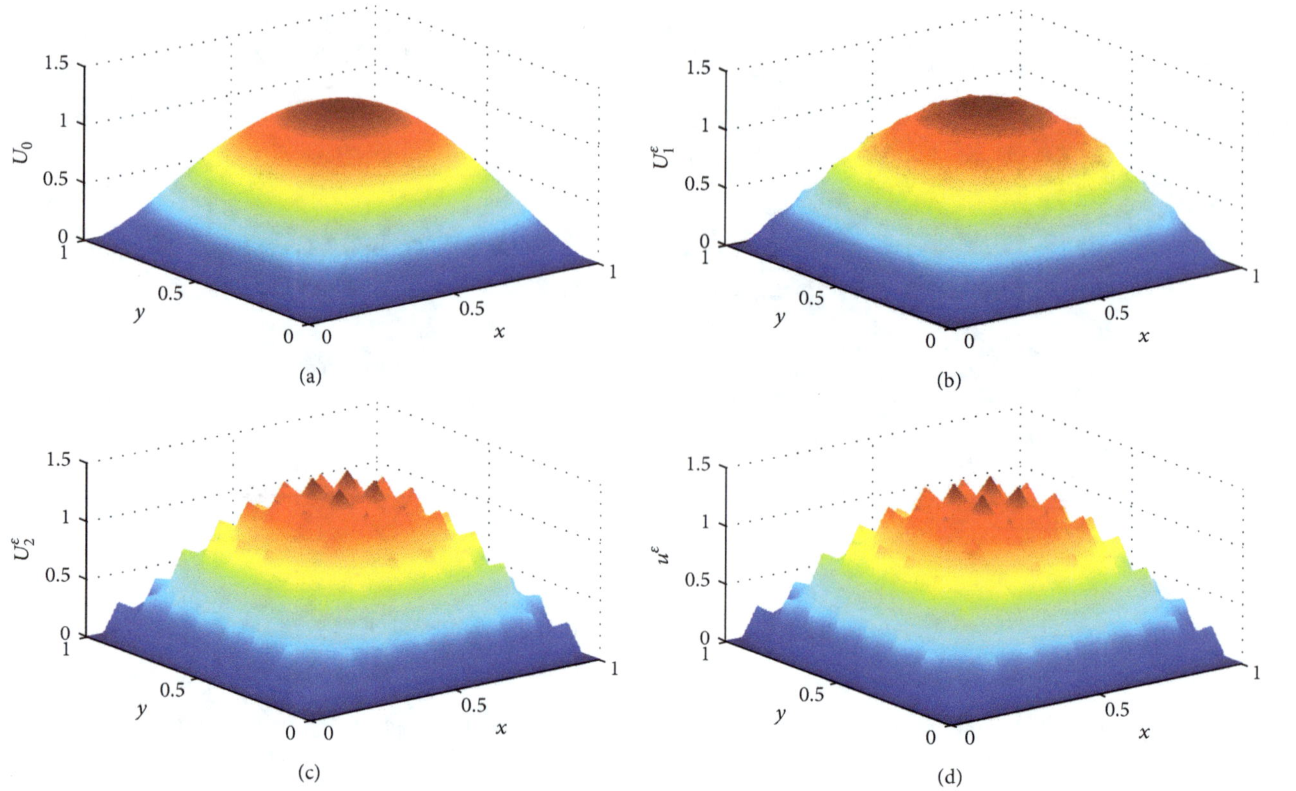

FIGURE 2: Case 1: The solutions in the cross-section $z = 0.5625$, $t = 0.1$, $\omega^* = 100$, and $M = 100$. (a) The homogenized solution U_0; (b) the first-order multiscale solution U_1^ε; (c) the second-order multiscale solution U_2^ε; (d) the reference solution u^ε.

the inclusion and other parts are the matrix. The parameters of the composite materials for two cases will be considered,

$$\text{Case 1: } a_{ij}(\xi) = \begin{cases} 10\delta_{ij} & \text{in inclusion} \\ 500\delta_{ij} & \text{in matrix,} \end{cases}$$
$$\text{Case 2: } a_{ij}(\xi) = \begin{cases} 500\delta_{ij} & \text{in inclusion} \\ 10\delta_{ij} & \text{in matrix,} \end{cases} \tag{71}$$

where $\xi = \varepsilon^{-1}x$. Let $f(x) = e^{-t}$ and $\overline{u}(x) = 0$.

To assess the validity of the presented method, the exact solution u^ε of the original parabolic equation (1) must be available. It is extremely difficult and even impossible to find the exact solution; we replace it by the numerical solution fully resolved by FEM in space and FDM in time using fine meshes (FD-FE method). It should be noted that more considerable computing time will be required to solve (1) using fully resolved FD-FE method when ε is small, so we only present numerical results for $\varepsilon = 1/8$. Here, the linear tetrahedral elements are employed for the semidiscrete problem of (1) using fine meshes, and the solution is computed by the Euler midpoint scheme.

Without confusion, let u^ε denote the numerical solution for the original problem (1) using the fully resolved FD-FE method. Let U_1^ε, U_2^ε denote the numerical solutions inverted back by the first-order and second-order multiscale asymptotic solutions in accordance with (57), respectively. Also, let U_0 denote the numerical solution inverted back by the homogenized solutions according to the following formula:

$$U_{0,\omega^*,\Delta\omega}(x,t) = \frac{1}{\pi}\,\mathrm{Re}\sum_{j=1}^{M}\widehat{u}_0\left(x,\omega_{j-1/2}\right)e^{it\omega_{j-1/2}}\Delta\omega. \tag{72}$$

All the codes are implemented on parallel hierarchical grid (PHG), which is a toolbox for developing parallel adaptive finite element programs. PHG is currently under active development at State Key Laboratory of Scientific and Engineering Computing of CAS. We use linear tetrahedron elements to solve the related problems numerically.

The numerical results in the cross-section $z = 0.5625$ and at time $t = 0.1$ for Case 1 and Case 2 are displayed in Figures 2 and 3, respectively. Although the first-order and second-order multiscale asymptotic solutions have the same convergence rate as stated in Theorem 9, numerical results displayed in Figure 2 confirm that the second-order corrector terms are crucial in some cases.

The relative errors vary with the segment number M in the numerical integration formula (57) for homogenized method (HM), first-order multiscale method (1st MsM), and second-order multiscale method (2nd MsM) corresponding to U_0, U_1^ε, and U_2^ε that are listed in Figure 4. We observe that the relative errors become stable after $M = 80$. Also, the relative errors evolution over time is listed in Figure 5. The results show that the accuracy of the proposed method especially the 2nd MsM is in good agreement with the fully

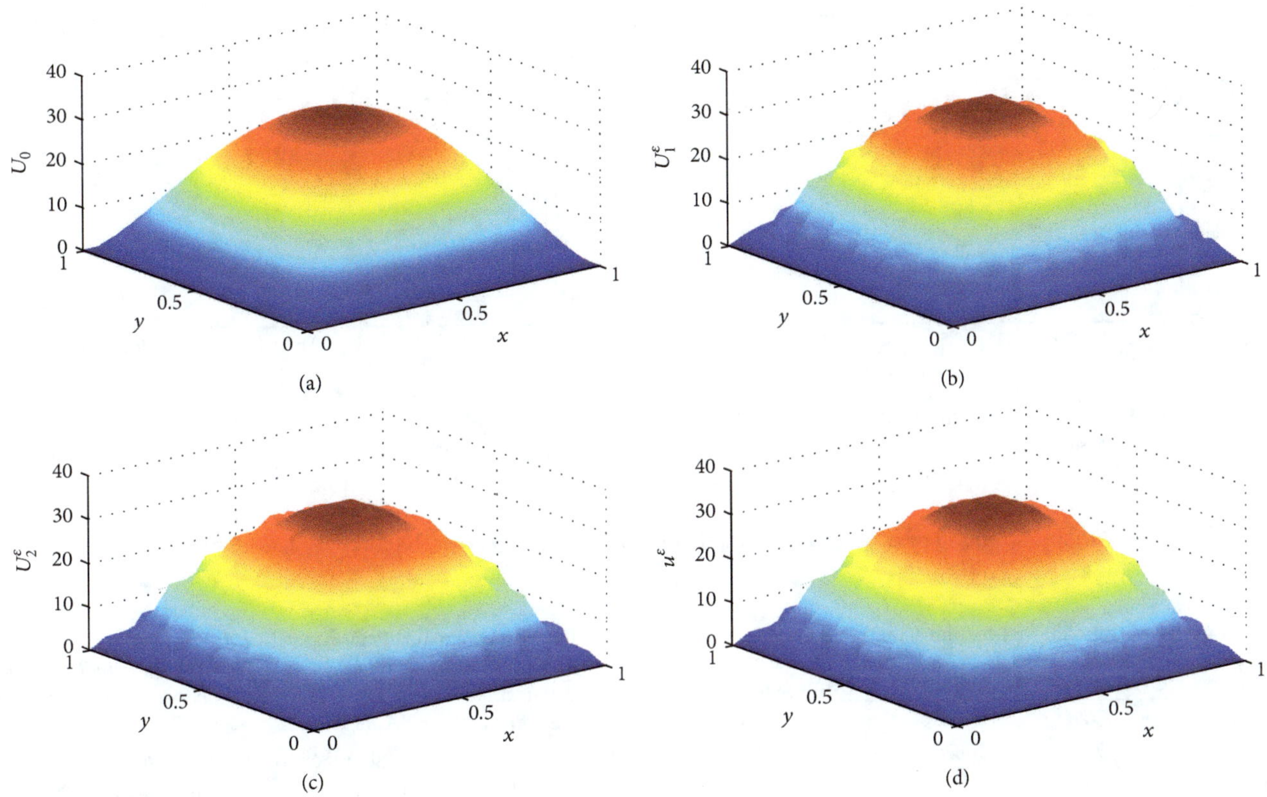

FIGURE 3: Case 2: The solutions in the cross-section $z = 0.5625$, $t = 0.1$, $\omega^* = 100$, and $M = 100$. (a) The homogenized solution U_0; (b) the first-order multiscale solution U_1^ε; (c) the second-order multiscale solution U_2^ε; (d) the reference solution u^ε.

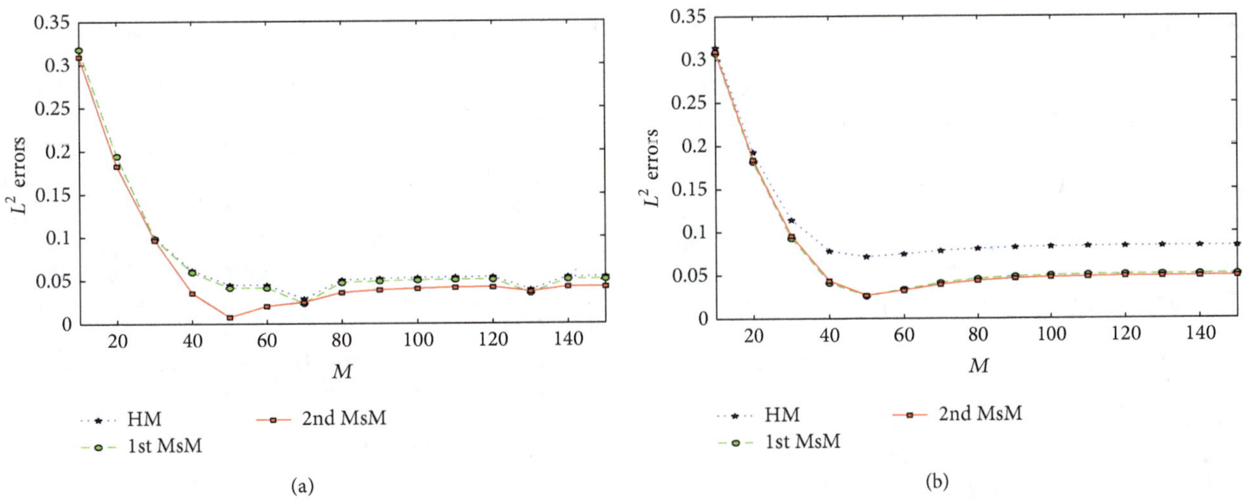

FIGURE 4: L^2-norm relative errors vary with the segment number M at $t = 0.1$, $\omega^* = 100$ (a) Case 1; (b) Case 2.

resolved case. It is worthwhile to note that the relative errors do not grow rapidly as time increases.

The costs required by the developed multiscale finite element method are not sensitive to the time variable but lie on the number of space-frequency equations which is determined by the segment number M, and each space-frequency equation is solved by multiscale finite element method with the computational cost listed in Table 1; for the fully resolved case, we need to compute $T/\Delta t$ steps and the cost for each time step is also listed in Table 1.

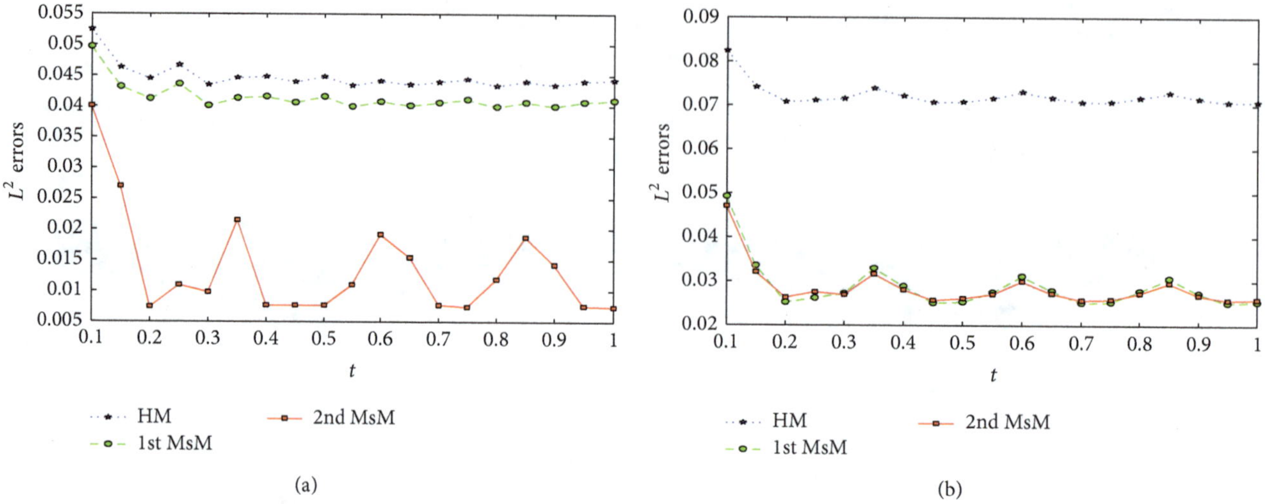

FIGURE 5: L^2-norm relative errors evolution over time at $\omega^* = 100$, $M = 100$ (a) Case 1; (b) Case 2.

TABLE 1: Computational cost for Case 1 and Case 2.

	Original Eqs.	Cell Eqs.	Homogenized Eqs.
Elements	3145728	393216	393216
Dof	536769	68705	137410
Time step	$\Delta t = 0.001$		

This clearly demonstrates that the present algorithm is very effective and tremendous saving in computing time is achieved.

Conflict of Interests

The authors declare that they have no conflict of interests regarding the publication of this paper.

Acknowledgments

This work is supported by National Natural Science Foundation of China (Grant no. 11301329) and the Grant of the First-class Discipline of Universities in Shanghai.

References

[1] Y. Li and L. Nirenberg, "Estimates for elliptic systems from composite material," *Communications on Pure and Applied Mathematics*, vol. 56, no. 7, pp. 892–925, 2003.

[2] Y. Y. Li and M. Vogelius, "Gradient estimates for solutions to divergence form elliptic equations with discontinuous coefficients," *Archive for Rational Mechanics and Analysis*, vol. 153, no. 2, pp. 91–151, 2000.

[3] A. Bensoussan, J.-L. Lions, and G. Papanicolaou, *Asymptotic Analysis for Periodic Structures*, vol. 5 of *Studies in Mathematics and Its Applications*, North-Holland, Amsterdam, The Netherlands, 1978.

[4] V. V. Jikov, S. M. Kozlov, and O. A. Oleĭnik, *Homogenization of Differential Operators and Integral Functionals*, Springer, Berlin, Germany, 1994.

[5] D. Cioranescu and P. Donato, *An Introduction to Homogenization*, vol. 17 of *Oxford Lecture Series in Mathematics and Its Applications*, The Clarendon Press, Oxford University Press, New York, NY, USA, 1999.

[6] O. A. Oleĭnik, A. S. Shamaev, and G. A. Yosifian, *Mathematical Problems in Elasticity and Homogenization*, vol. 26 of *Studies in Mathematics and Its Applications*, North-Holland, Amsterdam, The Netherlands, 1992.

[7] W. Allegretto, L. Cao, and Y. Lin, "Multiscale asymptotic expansion for second order parabolic equations with rapidly oscillating coefficients," *Discrete and Continuous Dynamical Systems. Series A*, vol. 20, no. 3, pp. 543–576, 2008.

[8] S. Chen, W. E, and C.-W. Shu, "The heterogeneous multiscale method based on the discontinuous Galerkin method for hyperbolic and parabolic problems," *Multiscale Modeling & Simulation*, vol. 3, no. 4, pp. 871–894, 2005.

[9] P. Ming and P. Zhang, "Analysis of the heterogeneous multiscale method for parabolic homogenization problems," *Mathematics of Computation*, vol. 76, no. 257, pp. 153–177, 2007.

[10] J. Douglas, Jr., D. Sheen, and J. E. Santos, "Approximation of scalar waves in the space-frequency domain," *Mathematical Models and Methods in Applied Sciences*, vol. 4, no. 4, pp. 509–531, 1994.

[11] J. Douglas, Jr., J. E. Santos, D. Sheen, and L. S. Bennethum, "Frequency domain treatment of one-dimensional scalar waves," *Mathematical Models and Methods in Applied Sciences*, vol. 3, no. 2, pp. 171–194, 1993.

[12] D. Kim, J. Kim, and D. Sheen, "Absorbing boundary conditions for wave propagation in viscoelastic media," *Journal of Computational and Applied Mathematics*, vol. 76, no. 1-2, pp. 301–314, 1996.

[13] C.-O. Lee, J. Lee, and D. Sheen, "Frequency domain formulation of linearized Navier-Stokes equations," *Computer Methods in Applied Mechanics and Engineering*, vol. 187, no. 1-2, pp. 351–362, 2000.

[14] D. Sheen, I. H. Sloan, and V. Thomée, "A parallel method for time-discretization of parabolic problems based on contour integral representation and quadrature," *Mathematics of Computation*, vol. 69, no. 229, pp. 177–195, 2000.

[15] D. Sheen, I. H. Sloan, and V. Thomée, "A parallel method for time discretization of parabolic equations based on Laplace transformation and quadrature," *IMA Journal of Numerical Analysis*, vol. 23, no. 2, pp. 269–299, 2003.

[16] E. Sánchez-Palencia, *Nonhomogeneous Media and Vibration Theory*, vol. 127 of *Lecture Notes in Physics*, Springer, Berlin, Germany, 1980.

[17] C.-O. Lee, J. Lee, D. Sheen, and Y. Yeom, "A frequency-domain parallel method for the numerical approximation of parabolic problems," *Computer Methods in Applied Mechanics and Engineering*, vol. 169, no. 1-2, pp. 19–29, 1999.

[18] L. C. Evans, *Partial Differential Equations*, vol. 19 of *Graduate Studies in Mathematics*, American Mathematical Society, Providence, RI, USA, 2nd edition, 2010.

[19] L.-Q. Cao and J.-Z. Cui, "Asymptotic expansions and numerical algorithms of eigenvalues and eigenfunctions of the Dirichlet problem for second order elliptic equations in perforated domains," *Numerische Mathematik*, vol. 96, no. 3, pp. 525–581, 2004.

Permissions

All chapters in this book were first published in AMSE, by Hindawi Publishing Corporation; hereby published with permission under the Creative Commons Attribution License or equivalent. Every chapter published in this book has been scrutinized by our experts. Their significance has been extensively debated. The topics covered herein carry significant findings which will fuel the growth of the discipline. They may even be implemented as practical applications or may be referred to as a beginning point for another development.

The contributors of this book come from diverse backgrounds, making this book a truly international effort. This book will bring forth new frontiers with its revolutionizing research information and detailed analysis of the nascent developments around the world.

We would like to thank all the contributing authors for lending their expertise to make the book truly unique. They have played a crucial role in the development of this book. Without their invaluable contributions this book wouldn't have been possible. They have made vital efforts to compile up to date information on the varied aspects of this subject to make this book a valuable addition to the collection of many professionals and students.

This book was conceptualized with the vision of imparting up-to-date information and advanced data in this field. To ensure the same, a matchless editorial board was set up. Every individual on the board went through rigorous rounds of assessment to prove their worth. After which they invested a large part of their time researching and compiling the most relevant data for our readers.

The editorial board has been involved in producing this book since its inception. They have spent rigorous hours researching and exploring the diverse topics which have resulted in the successful publishing of this book. They have passed on their knowledge of decades through this book. To expedite this challenging task, the publisher supported the team at every step. A small team of assistant editors was also appointed to further simplify the editing procedure and attain best results for the readers.

Apart from the editorial board, the designing team has also invested a significant amount of their time in understanding the subject and creating the most relevant covers. They scrutinized every image to scout for the most suitable representation of the subject and create an appropriate cover for the book.

The publishing team has been an ardent support to the editorial, designing and production team. Their endless efforts to recruit the best for this project, has resulted in the accomplishment of this book. They are a veteran in the field of academics and their pool of knowledge is as vast as their experience in printing. Their expertise and guidance has proved useful at every step. Their uncompromising quality standards have made this book an exceptional effort. Their encouragement from time to time has been an inspiration for everyone.

The publisher and the editorial board hope that this book will prove to be a valuable piece of knowledge for researchers, students, practitioners and scholars across the globe.

List of Contributors

Yun-Wang Choi, Sung-Rok Oh and Cheol-Gyu Kim
Department of Civil Engineering, Semyung University, Jecheon 27136, Republic of Korea

Jae-Heun Lee
Institute of Technology, Eugene Corporation Co. Ltd., Bucheon 14441, Republic of Korea

Wilson Webo and Leonard Masu
Department of Mechanical Engineering, Faculty of Engineering and Technology, Vaal University of Technology, Andries Potgieter Blvd., Vanderbijlpark 1911, South Africa

Maina Maringa
Department of Mechanical and Mechatronics Engineering, Faculty of Engineering and Technology, Central University of Technology, Bloemfontein 9300, South Africa

Regina Jeziorska, Agnieszka Szadkowska and Ewa Spasowka
Department of Polymer Technology and Processing, Industrial Chemistry Research Institute, Rydygiera 8 St., 01-793 Warsaw, Poland

Aneta Lukomska
Department of Biomedical Technology, Cosmetic Chemicals and Electrochemistry, Industrial Chemistry Research Institute, Rydygiera 8 St., 01-793 Warsaw, Poland

Michal Chmielarek
Faculty of Chemistry, Warsaw University of Technology, Koszykowa 75 St., 00-662 Warsaw, Poland

Qi Liu, Wenfeng Du and Zhiyong Zhou
College of Civil Engineering and Architecture, Henan University, Kaifeng, China

Nasim Uddin
Department of Civil, Construction and Environmental Engineering, University of Alabama at Birmingham, Birmingham, Alabama, USA

G. Pitchayyapillai and P. Seenikannan
Department of Mechanical Engineering, Sethu Institute of Technology, Kariapatti, Tamil Nadu, India

K. Raja
Department of Mechanical Engineering, University College of Engineering, Anna University, Dindigul, Tamil Nadu, India

K. Chandrasekaran
Department of Mechanical Engineering, MAM School of Engineering, Trichy, Tamil Nadu, India

Zhao Yang and Kun Wu
Institute of Urban Construction, Wuhan University of Science and Technology, Wuhan 430065, China

Weilai Yao, Wei Fei and Tao Cai
Civil Engineering Department, Logistical Engineering University, Chongqing 401311, China

Shiyong Jiang
Training Department, Logistical Engineering University, Chongqing 401311, China

Huashan Yang and Yujun Che
School of Materials and Architecture Engineering, Guizhou Normal University, Guiyang 550025, China

Dora Foti and Vitantonio Vacca
Department of Civil Engineering and Architecture, Polytechnic of Bari, Via Orabona 4, 70125 Bari, Italy

Michela Lerna
Department of Innovation Engineering, University of Salento, Via per Monteroni, 73100 Lecce, Italy

Yigang Tong and Zhibin Zhou
MOE Key Laboratory for Nonequilibrium Synthesis and Modulation of Condensed Matter, School of Science, Xi'an Jiaotong University, Xi'an 710049, China

Hui Cai
College of Materials Science and Engineering, Xi'an University of Science and Technology, Xi'an 710054, China

Xueliang Wang
MOE Key Laboratory of -ermo-Fluid Science and Engineering, School of Energy and Power Engineering, Xi'an Jiaotong University, Xi'an, Shaanxi 710049, China

Yaping Wang
MOE Key Laboratory for Nonequilibrium Synthesis and Modulation of Condensed Matter, School of Science, Xi'an Jiaotong University, Xi'an 710049, China
State Key Laboratory for Mechanical Behavior of Materials, Xi'an Jiaotong University, Xi'an 710049, China

Jing Xia, Shun Li, Shuxin Bai and Degan Xiong
Department of Materials Science and Engineering, National University of Defense Technology, Changsha 410073, China

Jiuchun Yan
State Key Laboratory of Advanced Welding and Joining, Harbin Institute of Technology, Harbin 150001, China

Hassan Ijaz, Waqas Saleem and Muhammad Zain-ul-Abdein
Mechanical Engineering Department, University of Jeddah, Jeddah, Saudi Arabia

Tarek Mabrouki
Mechanical Engineering Department, University of Tunis El Manar, ENIT, Tunis, Tunisia

Saeed Rubaiee
Industrial Engineering Department, University of Jeddah, Jeddah, Saudi Arabia

Abdullah Salmeen Bin Mahfouz
Chemical Engineering Department, University of Jeddah, Jeddah, Saudi Arabia

René Salgado-Delgado, Alfredo Olarte-Paredes, Areli Marlen Salgado-Delgado, Zully Vargas-Galarza and Israel Rico-Rodríguez
Division de Estudios de Posgrado e Investigación del Instituto Tecnológico de Zacatepec, Calzada Tecnológico No. 27, Col. Centro, 62780 Zacatepec, MOR, Mexico

Teresa Lopez-Lara and Juan Bosco Hernández-Zaragoza
Division de Estudios de Posgrado, Facultad de Ingeniería, Universidad Autónoma de Querétaro, Cerro de las Campanas S/N, Col. de lãs Campanas, 76010 Querétaro, QRO, Mexico

Gonzalo Martínez-Barrera
Laboratorio de Investigación y Desarrollo de Materiales Avanzados, Facultad de Química,
Universidad Autónoma del Estado de México, Km 12 Toluca-Atlacomulco, 50200 San Cayetano, MEX, Mexico

Sergei Alexandrov
Institute for Problems in Mechanics, Russian Academy of Sciences, 101-1 Prospect Vernadskogo, Moscow 119526, Russia

Yaroslav Erisov and Fedor Grechnikov
Samara State Aerospace University, 34 Moskovskoe Shosse, Samara 443086, Russia

Mahmoud Khalifeh, Arild Saasen and Helge Hodne
Department of Petroleum Engineering, University of Stavanger, 4036 Stavanger, Norway

Helge B. Larsen
Department of Mathematics and Natural Sciences, University of Stavanger, 4036 Stavanger, Norway

Kalyan Kumar Singh and Saurabh Singh
Department of Mechanical Engineering, Indian School of Mines, Dhanbad 826004, India

Anil Kumar Shrivastava
Department of Mechanical Engineering, Galgotias College of Engineering & Technology, Greater Noida 201306, India

P. Kumar
Department of Civil Engineering, P.S.R. Engineering College, Sivakasi, Tamil Nadu 626140, India

K. Sudalaimani
Department of Civil Engineering, Thiagarajar College of Engineering, Thiruparankundram, Madurai, Tamil Nadu 625015, India

M. Shanmugasundaram
Structural Engineering Division, School of Mechanical and Building Sciences, VIT University, Chennai Campus, Chennai, Tamil Nadu 600127, India

Jin Liu and Dongmin Wang
School of Chemical and Environmental Engineering, China University of Mining and Technology, Beijing 100083, China

Tarek UddinMohammed, Tanvir Ahmed, Shibly Mostafiz Apurbo and Tahir Absar Mallick
Department of Civil and Environmental Engineering, Islamic University of Technology (IUT), Gazipur 1704, Bangladesh

Farhan Shahriar and Abdul Munim
Housing and Building Research Institute (HBRI), Dhaka 1216, Bangladesh

Mohammad Abdul Awal
Structural Engineers Limited (SEL), West Panthapath, Dhaka 1209, Bangladesh

Velu Saraswathy
Department of Architectural Engineering, Hanyang University, 1271 Sa 3-dong, Sangrok-gu, Ansan 426791, Republic of Korea
Corrosion and Materials Protection Division, CSIR-Central Electrochemical Research Institute, Karaikudi, Tamil Nadu 630003, India

Subbiah Karthick
Department of Architectural Engineering, Hanyang University, 1271 Sa 3-dong, Sangrok-gu, Ansan 426791, Republic of Korea
PG and Research Department of Chemistry, Alagappa Government Arts College, Karaikudi, Tamil Nadu 630003, India

Han Seung Lee and Hyun-Min Yang
Department of Architectural Engineering, Hanyang University, 1271 Sa 3-dong, Sangrok-gu, Ansan 426791, Republic of Korea

Seung-Jun Kwo
Department of Civil Engineering, Hannam University, Daejeon 306-791, Republic of Korea

Libin Zhao and Tianliang Qin
School of Astronautics, Beihang University, Beijing 100191, China

Olushola S. Ayanda and Olalekan S. Fatoki
Department of Chemistry, Faculty of Applied Sciences, Cape Peninsula University of Technology, Bellville 7535, South Africa

Xin Wang, Xi-liang Duan and Yang Gao
Department of Mathematics, Shanghai University, Shanghai 200444, China

Xiaomei Hu, Lei Zhao and Wenlong Li
The Key Laboratory of Machinery Automation and Robotics, School of Mechatronic Engineering and Automation, Shanghai University, Yanchang Road, Shanghai 200072, China

Shi Yang Zhao and Pu Xue
School of Aeronautics, Northwestern Polytechnical University, Xi'an 710072, China

Xiaolin Chang, Xinping Yang, Wei Zhou and Shuhua Liu
State Key Laboratory of Water Resources and Hydropower Engineering Science, Wuhan University, Wuhan 430072, China

Guoshuai Xie
State Key Laboratory of Water Resources and Hydropower Engineering Science, Wuhan University, Wuhan 430072, China
Power China Huadong Engineering Corporation, Hangzhou 310014, China

Index

www.ingramcontent.com/pod-product-compliance
Lightning Source LLC
LaVergne TN
LVHW081027210926
848485LV00004B/967
* 9 7 8 1 6 3 9 8 9 2 5 5 6 *